Herrn Dr. W. Maysenhölder
mit besten Empfehlungen
C. Teodosiu

Cristian Teodosiu

# Elastic Models of Crystal Defects

With 58 Figures

Editura Academiei
București

Springer-Verlag
Berlin Heidelberg New York

1982

Cristian Teodosiu
Department of Solid Mechanics
Institute for the Physics and Technology of Materials
15, C. Mille str., Bucharest, R—70701, Romania

This is a revised and up-dated version of *Modele elastice ale defectelor cristaline*, Ed. Academiei, București, 1977.

Sole distribution rights for all non-socialist countries granted to Springer-Verlag Berlin Heidelberg New York

ISBN 3-540-11226-X Springer-Verlag Berlin Heidelberg New York
ISBN 0-387-11226-X Spinger-Verlag New York Heidelberg Berlin

# CONTENTS

# PREFACE TO THE ENGLISH EDITION

*This work deals with elastic models of crystal defects, a field situated at the boundary between continuum mechanics and solid state physics.*

*The understanding of the behaviour of crystal defects has become unavoidable for studying such processes as anelasticity, internal damping, plastic flow, rupture, fatigue, and radiation damage, which play a determining role in various fields of materials science and in top technological areas. On the other hand, the lattice distortion produced by a crystal defect can be calculated by means of elastic models, at least at sufficiently large distances from the defect. Furthermore, the interaction of a crystal defect with other defects and with applied loads is mainly due to the interaction of their elastic states. This explains the permanent endeavour to improve the elastic models of crystal defects, e.g. by taking into account anisotropic and non-linear elastic effects and by combining elastic with atomistic models in order to achieve a better description of the highly distorted regions near the defects.*

*This book has grown out of a two-semester course on "Continuum Mechanics with Applications to Solid State Physics" held by the author some ten years ago at the University of Stuttgart, which was an attempt to unify the topic with recent developments that have made continuum mechanics a highly deductive science. Since then, the extension of the application area and the development of new computing techniques have considerably enlarged the field and changed the plan of the work. However, the stress is still laid on theory and method: the problems solved are illustrative and intended to serve as background for approaching more complex or more specific applications. Moreover, their choice is inevitably influenced by the preference of the author for subjects to which personal contributions have been brought.*

*Chapter I concerns the basic concepts and laws of the kinematics, dynamics, and thermodynamics of deformable continuous media, the linear and non-linear elastic constitutive equations, as well as the formulation and solving of the boundary-value problems of linear elastostatics. Special attention is given to anisotropic elasticity, to the accurate formulation of boundary-value problems involving infinite domains and concentrated forces, and to the determination of Green's tensor function, in view of the importance of these topics for the simulation of crystal defects.*

*Chapter II contains a systematic study of the elastic states of single straight or curvilinear dislocations, of the elastic interactions between single dislocations, and of moving dislocations. The emphasis lies on the anisotropic elasticity theory of dislocations, especially on the powerful methods developed during the last ten years for the computation of the elastic states of dislocation loops by means of straight dislocation data.*

*Chapter III presents the main results obtained so far in describing non-linear effects in the elastic field of straight dislocations, as well as in the study of the core configuration of dislocations by using semidiscrete methods.*

*Chapter IV is devoted to the linear and non-linear theory of continuous distributions of dislocations and to its application to investigating the influence of dislocations on crystal density and on the low-temperature thermal conductivity of crystals.*

*Chapter V deals with the modelling of point defects as rigid or elastic inclusions in an elastic matrix, or as force multipoles. Finally, some of the results available on the interactions between point defects and other crystal defects are briefly reviewed.*

*Although the material in the text covers mainly the mathematical theory of crystal defects, the author has been constantly concerned with emphasizing the physical significance of the results and some of their possible applications. The reader can easily enlarge his information in these directions by reference to the standard books on crystal defects by Cottrell [84], Read [275], Friedel [124], Kröner [190], van Bueren [365], Indenbom [167], Nabarro [258], Hirth and Lothe [162], or to the review articles by Seeger [286], Eshelby [111], de Wit [385], and Bullough [50].*

*Printed jointly with Springer-Verlag, the English edition is a revised and up-dated version of the Romanian book "Modele elastice ale defectelor cristaline", published in 1977 by Editura Academiei. The present edition is supplemented by several subsections concerning the simulation of crystal dislocations by means of Volterra and Somigliana dislocations, the dislocation loops in anisotropic media, the interaction of crystal defects, and the flexible-boundary semidiscrete methods, as well as by a review of the main results published in the last four years.*

*The author expresses his deep gratitude to Prof. A. Seeger and Prof. E. Kröner for continuous encouragement to writing this book and for numerous discussions on the application of continuum mechanics to the simulation of crystal defects. The author is also greatly indebted to Dr. E. Soós for his valuable detailed criticism of the manuscript.*

CHAPTER I

# FUNDAMENTALS OF THE THEORY OF ELASTICITY

Before broaching the very subject of this chapter, we shall review briefly the basic elements of vector and tensor calculus that are necessary in the present work. This will also allow the reader to become familiar with the system of notation used in the following.

## 1. Vectors and tensors

### 1.1. Elements of vector and tensor algebra

We denote by $\mathscr{E}$ the three-dimensional Euclidean space; its elements $P, Q, \ldots$ are called *points*. The *translation vector space* associated with $\mathscr{E}$ is denoted by $\mathscr{V}$ and its elements $\mathbf{u}, \mathbf{v}, \cdots$ are called *vectors*.

The *scalar product* of the vectors $\mathbf{u}$ and $\mathbf{v}$ is denoted by $\mathbf{u}\cdot\mathbf{v}$. The magnitude of the vector $\mathbf{u}$ is the non-negative real number

$$\|\mathbf{u}\| = \sqrt{\mathbf{u}\cdot\mathbf{u}}. \tag{1.1}$$

Since $\mathscr{V}$ is also three-dimensional, any triplet of non-coplanar vectors is a *basis* of $\mathscr{V}$, and any vector of $\mathscr{V}$ can be written as a linear combination of the basis vectors. A *Cartesian co-ordinate frame* consists of an orthonormal basis $\{\mathbf{e}_k\} = \{\mathbf{e}_1, \mathbf{e}_2, \mathbf{e}_3\}$ and a point $O$ called the *origin*. Then

$$\mathbf{e}_k\cdot\mathbf{e}_m = \delta_{km}, \quad k, m = 1, 2, 3, \tag{1.2}$$

where

$$\delta_{km} = \begin{cases} 1 & \text{for } k = m \\ 0 & \text{for } k \neq m \end{cases} \tag{1.3}$$

is the *Kronecker delta*. The vector $\overrightarrow{OP} = \mathbf{x}$ is called the *position vector* of the point $P \in \mathscr{E}$. Clearly, the correspondence between points and their position vectors is one-to-one. Therefore, we shall sometimes label points by their position vectors, referring for conciseness to the point $P$ whose position vector is $\mathbf{x}$ as "the point $\mathbf{x}$".

The real numbers $u_1, u_2, u_3$, uniquely defined by the relation

$$\mathbf{u} = u_1\mathbf{e}_1 + u_2\mathbf{e}_2 + u_3\mathbf{e}_3 \tag{1.4}$$

are called the *Cartesian components* of the vector $\mathbf{u}$.

Both *direct notation*, using only vector and tensor symbols, and *indicial notation*, making use of vector and tensor components, will be employed throughout. Whenever indicial notation is used, the subscripts are assumed to range over the integers 1, 2, 3, and summation over twice repeated subscripts is implied, e.g.

$$\mathbf{u}\cdot\mathbf{v} = u_k v_k = u_1 v_1 + u_2 v_2 + u_3 v_3. \tag{1.5}$$

From (1.4) and (1.2) we see that the Cartesian components of $\mathbf{u}$ can be also defined by

$$u_k = \mathbf{u}\cdot\mathbf{e}_k. \tag{1.6}$$

The *vector product* of two vectors $\mathbf{u}$ and $\mathbf{v}$ is denoted by $\mathbf{u}\times\mathbf{v}$. In view of (1.4) we can write

$$\mathbf{e}_k \times \mathbf{e}_l = \epsilon_{klm}\mathbf{e}_m, \tag{1.7}$$

where $\epsilon_{klm}$ is the *alternator symbol*. A direct proof shows that

$$\epsilon_{klm} = \begin{cases} 1 \text{ for } klm = 123, 231, 312 \\ -1 \text{ for } klm = 132, 213, 321 \\ 0 \text{ for any other values of } klm. \end{cases} \tag{1.8}$$

From (1.4) and (1.7) it follows that

$$\mathbf{u} \times \mathbf{v} = \epsilon_{klm} u_l v_m \mathbf{e}_k. \tag{1.9}$$

We notice that the symbols $\epsilon_{klm}$ satisfy the identities

$$\epsilon_{ikl}\epsilon_{jmn} = \begin{vmatrix} \delta_{ij} & \delta_{im} & \delta_{in} \\ \delta_{kj} & \delta_{km} & \delta_{kn} \\ \delta_{lj} & \delta_{lm} & \delta_{ln} \end{vmatrix}, \tag{1.10}$$

$$\epsilon_{ikl}\epsilon_{imn} = \delta_{km}\delta_{ln} - \delta_{kn}\delta_{lm}. \tag{1.11}$$

A *second-order tensor* $\mathbf{A}$ is a linear mapping[1] that assigns to each vector $\mathbf{u}$ a vector

$$\mathbf{v} = \mathbf{A}\mathbf{u}. \tag{1.12}$$

We denote by $\mathscr{L}$ the set of all second-order tensors defined on $\mathscr{V}$. The *sum* $\mathbf{A} + \mathbf{B}$ of two tensors $\mathbf{A}, \mathbf{B} \in \mathscr{L}$ is defined by

$$(\mathbf{A} + \mathbf{B})\,\mathbf{u} = \mathbf{A}\mathbf{u} + \mathbf{B}\mathbf{u}, \tag{1.13}$$

and the *product of a tensor* $\mathbf{A} \in \mathscr{L}$ *and a real number* $\alpha$ by

$$(\alpha\mathbf{A})\mathbf{u} = \alpha(\mathbf{A}\mathbf{u}). \tag{1.14}$$

The space $\mathscr{L}$ endowed with the composition rules (1.13) and (1.14) is also a vector space.

The *unit tensor* $\mathbf{1}$ and the *zero tensor* $\mathbf{0}$ are defined by the relations

$$\mathbf{1}\mathbf{u} = \mathbf{u}, \quad \mathbf{0}\mathbf{u} = \mathbf{o} \quad \text{for every } \mathbf{u} \in \mathscr{V}, \tag{1.15}$$

where $\mathbf{o}$ is the *zero vector*.

The *tensor product* $\mathbf{u}\mathbf{v}$ of two vectors $\mathbf{u}$ and $\mathbf{v}$ is the second-order tensor defined by

$$(\mathbf{u}\mathbf{v})\mathbf{w} = \mathbf{u}(\mathbf{v}\cdot\mathbf{w}) \text{ for every } \mathbf{w} \in \mathscr{V}. \tag{1.16}$$

It can be shown that if $\mathbf{f}_k$ and $\mathbf{g}_m$ are two arbitrary bases of $\mathscr{V}$, then the tensor products $\mathbf{f}_k\mathbf{g}_m$, $k, m = 1, 2, 3$, are a basis of $\mathscr{L}$, which is thus a nine-dimensional vector space. In particular, the tensor products $\mathbf{e}_k\mathbf{e}_m$, $k, m = 1, 2, 3$, are a basis of $\mathscr{L}$, and we can write for every $\mathbf{A} \in \mathscr{L}$

$$\mathbf{A} = A_{km}\mathbf{e}_k\mathbf{e}_m. \tag{1.17}$$

The nine real numbers $A_{km}$, uniquely defined by (1.17), are called the *Cartesian components* of the tensor $\mathbf{A}$. From (1.17), (1.16), and (1.2), we deduce the relation

$$A_{km} = \mathbf{e}_k\cdot(\mathbf{A}\mathbf{e}_m), \tag{1.18}$$

which can be considered as an equivalent definition of the tensor components. In particular, by applying this definition to the unit tensor and taking into account $(1.15)_1$ and (1.2), we infer that $\delta_{km}$ are the Cartesian components of the unit tensor, i.e.

$$\mathbf{1} = \delta_{km}\mathbf{e}_k\mathbf{e}_m.$$

---

[1] This definition can still be applied when $\mathscr{V}$ is an arbitrary vector space.

If $\mathbf{v} = \mathbf{A}\mathbf{u}$, we also have by (1.17) and (1.16)

$$\mathbf{v} = (A_{km}\mathbf{e}_k\mathbf{e}_m)\,\mathbf{u} = A_{km}u_m\mathbf{e}_k,$$

and hence

$$v_k = A_{km}u_m. \tag{1.19}$$

The *product* $\mathbf{AB}$ *of two tensors* $\mathbf{A}$ *and* $\mathbf{B}$ is defined by the composition rule

$$(\mathbf{AB})\,\mathbf{u} = \mathbf{A}(\mathbf{Bu}) \text{ for every } \mathbf{u} \in \mathscr{V},$$

wherefrom it follows that

$$(\mathbf{AB})_{km} = A_{kp}B_{pm}. \tag{1.20}$$

The *transpose* of the tensor $\mathbf{A} = A_{km}\mathbf{e}_k\mathbf{e}_m$ is the tensor $\mathbf{A}^T = A_{mk}\mathbf{e}_k\mathbf{e}_m$. A second-order tensor $\mathbf{A}$ is called *symmetric* if $\mathbf{A}^T = \mathbf{A}$, and *skew* or *antisymmetric* if $\mathbf{A}^T = -\mathbf{A}$. By defining

$$\operatorname{sym}\mathbf{A} = \frac{1}{2}(\mathbf{A} + \mathbf{A}^T), \quad \operatorname{skw}\mathbf{A} = \frac{1}{2}(\mathbf{A} - \mathbf{A}^T)$$

as the *symmetric part* and the *skew part* of an arbitrary second-order tensor $\mathbf{A}$, we can always write

$$\mathbf{A} = \operatorname{sym}\mathbf{A} + \operatorname{skw}\mathbf{A}.$$

Given any skew tensor $\boldsymbol{\Omega}$, there exists a unique vector $\boldsymbol{\omega}$ such that

$$\boldsymbol{\Omega}\mathbf{u} = \boldsymbol{\omega} \times \mathbf{u} \text{ for every } \mathbf{u} \in \mathscr{V}. \tag{1.21}$$

Indeed, from (1.21), (1.9), and (1.11), it results that

$$\omega_i = -\frac{1}{2}\in_{ijk}\Omega_{jk}, \quad \Omega_{ij} = -\in_{ijk}\omega_k. \tag{1.22}$$

The vector $\boldsymbol{\omega}$, uniquely defined by $(1.22)_1$, is called the *axial vector* of the skew tensor $\boldsymbol{\Omega}$.

The trace of $\mathbf{A} \in \mathscr{L}$ is the real number

$$\operatorname{tr}\mathbf{A} = A_{mm}. \tag{1.23}$$

The passing from $\mathbf{A}$ to $\operatorname{tr}\mathbf{A}$ is called (tensor) *contraction*. It is easily seen that

$$\operatorname{tr}\mathbf{A}^T = \operatorname{tr}\mathbf{A}, \quad \operatorname{tr}(\mathbf{AB}) = \operatorname{tr}(\mathbf{BA}). \tag{1.24}$$

The *inner product* $\mathbf{A}\cdot\mathbf{B}$ of two second-order tensors $\mathbf{A}$ and $\mathbf{B}$ is the real number

$$\mathbf{A}\cdot\mathbf{B} = \operatorname{tr}(\mathbf{AB}^T) = A_{km}B_{km}, \tag{1.25}$$

while the *magnitude* of $\mathbf{A}$ is the real number

$$\|\mathbf{A}\| = \sqrt{\mathbf{A}\cdot\mathbf{A}} = \sqrt{A_{km}A_{km}}\,. \tag{1.26}$$

The *determinant* det $\mathbf{A}$ of the tensor $\mathbf{A}$ is defined by

$$\det \mathbf{A} = \det\,[A_{km}], \tag{1.27}$$

where $[A_{km}]$ denotes the matrix of the Cartesian components of $\mathbf{A}$. From this definition and some well-known rules of matrix algebra, we see that for every $\mathbf{A}, \mathbf{B} \in \mathscr{L}$:

$$\det \mathbf{A}^T = \det \mathbf{A}, \quad \det(\mathbf{AB}) = (\det \mathbf{A})(\det \mathbf{B}). \tag{1.28}$$

If $\det \mathbf{A} \neq 0$, there exists a unique inverse linear transformation $\mathbf{A}^{-1}$ of $\mathscr{V}$ on $\mathscr{V}$ such that if $\mathbf{v} = \mathbf{Au}$ then $\mathbf{u} = \mathbf{A}^{-1}\mathbf{v}$ for every $\mathbf{u}, \mathbf{v} \in \mathscr{V}$. From these two equations and $(1.15)_1$ it follows that

$$\mathbf{AA}^{-1} = \mathbf{A}^{-1}\mathbf{A} = \mathbf{1}. \tag{1.29}$$

The tensor $\mathbf{A}^{-1}$ is called the *inverse* tensor of $\mathbf{A}$.

A tensor $\mathbf{Q}$ is said to be *orthogonal* if

$$\mathbf{QQ}^T = 1, \quad Q_{kp}Q_{mp} = \delta_{km}. \tag{1.30}$$

By (1.30) and (1.28) we have $(\det \mathbf{Q})^2 = 1$, $\det \mathbf{Q} = \pm 1$. Hence, every orthogonal tensor admits an inverse and, by $(1.30)_1$, $\mathbf{Q}^{-1} = \mathbf{Q}^T$. The set of all orthogonal tensors forms a group, called the *orthogonal group;* the set of all orthogonal tensors with determinant equal to $+1$ forms a subgroup of the orthogonal group, called the *proper orthogonal group*.

A *tensor of n'th order* is a linear mapping that assigns to each vector $\mathbf{u} \in \mathscr{V}$ a tensor of $(n-1)$'st order, $n \geqslant 3$. Combining this definition with that of a second-order tensor given above allows the iterative introduction of tensors of an arbitrary order. We denote by $\mathscr{L}_n$ the space of all tensors of order $n$.

The tensor product $\mathbf{u}_1\mathbf{u}_2 \ldots \mathbf{u}_n$ is a tensor of $n$'th order defined as a linear mapping of $\mathscr{V}$ in $\mathscr{L}_{n-1}$ by the relation

$$(\mathbf{u}_1\mathbf{u}_2 \ldots \mathbf{u}_{n-1}\mathbf{u}_n)\,\mathbf{v} = \mathbf{u}_1\mathbf{u}_2 \ldots \mathbf{u}_{n-1}(\mathbf{u}_n\cdot\mathbf{v}) \quad \text{for every } \mathbf{v} \in \mathscr{V}.$$

It can be shown that the tens or products $\mathbf{e}_{k_1} \ldots \mathbf{e}_{k_n}$, $k_1, \ldots, k_n = 1, 2, 3$, form a basis of $\mathscr{L}_n$. Hence $\mathscr{L}_n$ is $3^n$-dimensional, and every tensor $\mathbf{\Phi} \in \mathscr{L}_n$ can be written uniquely in the form

$$\mathbf{\Phi} = \Phi_{k_1\ldots k_n}\mathbf{e}_{k_1} \ldots \mathbf{e}_{k_n}, \tag{1.31}$$

where $\Phi_{k_1\ldots k_n}$ are the Cartesian components of $\mathbf{\Phi}$. Moreover, if $\mathbf{\Psi} = \mathbf{\Phi u}$, the n

$$\Psi_{k_1\ldots k_{n-1}} = \Phi_{k_1\ldots k_{n-1}k_n}u_{k_n}.$$

Let us consider now the transformation rules of vector and tens or components when passing from the orthonormal basis $\{\mathbf{e}_k\}$ to another orthonormal basis $\{\mathbf{e}'_r\}$. Denote by

$$q_{kr} = \mathbf{e}_k \cdot \mathbf{e}'_r = \cos(\mathbf{e}_k, \mathbf{e}'_r), \qquad k, r = 1, 2, 3, \tag{1.32}$$

the direction cosines of the unit vectors $\mathbf{e}_k$ with respect to the unit vectors $\mathbf{e}'_r$. By (1.4) and (1.6), we obviously have

$$\mathbf{e}_k = q_{kr}\mathbf{e}'_r, \quad \mathbf{e}'_r = q_{kr}\mathbf{e}_k, \tag{1.33}$$

wherefrom

$$q_{kr}q_{ks} = \delta_{rs}, \quad q_{kr}q_{lr} = \delta_{kl}.$$

Substituting successively (1.33) into the relation

$$\mathbf{u} = u_k\mathbf{e}_k = u'_r\mathbf{e}'_r,$$

and taking into account the unicity of Cartesian components, we obtain the transformation rule of the vector components

$$u_k = q_{kr}u'_r, \qquad u'_r = q_{kr}u_k. \tag{1.34}$$

In a similar way, the transformation rule of the components of a second-order tensor $\mathbf{A}$ reads

$$A_{km} = q_{kr}q_{ms}A'_{rs}, \quad A'_{rs} = q_{kr}q_{ms}A_{km}, \tag{1.35}$$

the generalization for higher-order tensors being evident.

A real number $\lambda$ is said to be a *principal* or *characteristic value* of a second-order tensor $\mathbf{A}$ if there exists a unit vector $\mathbf{n}$ such that

$$\mathbf{A}\mathbf{n} = \lambda\mathbf{n}; \tag{1.36}$$

in this case $\mathbf{n}$ is called a *principal direction* corresponding to $\lambda$.

It can be shown (see, e.g. Halmos [151], Sect. 79) that if $\mathbf{A}$ is a symmetric second-order tensor, then there exists an orthonormal basis $\mathbf{n}_1$, $\mathbf{n}_2$, $\mathbf{n}_3$ and three (not necessarily distinct) principal values $\lambda_1$, $\lambda_2$, $\lambda_3$ of $\mathbf{A}$ such that

$$\mathbf{A} = \sum_{k=1}^{3} \lambda_k\mathbf{n}_k\mathbf{n}_k. \tag{1.37}$$

If $\lambda_1 = \lambda_2$, equation (1.37) reduces to

$$\mathbf{A} = \lambda_1\mathbf{n}_1\mathbf{n}_1 + \lambda_2(\mathbf{1} - \mathbf{n}_1\mathbf{n}_1). \tag{1.37a}$$

Finally, if $\lambda_1 = \lambda_2 = \lambda_3$, then

$$\mathbf{A} = \lambda_1\mathbf{1}. \tag{1.37b}$$

This theorem, called the *spectral theorem*, is of great importance for the elasticity theory. For instance, it implies the existence of the principal values of the strain tensor and of the Cauchy stress tensor for these are symmetric second-order tensors.

## 1.2. Elements of vector and tensor analysis

In this section we choose a *fixed* Cartesian co-ordinate frame in $\mathscr{E}$, with origin $O$ and orthonormal basis $\{\mathbf{e}_1, \mathbf{e}_2, \mathbf{e}_3\}$. Let $(x_1, x_2, x_3)$ denote the Cartesian co-ordinates of a point $P \in \mathscr{E}$ with respect to this frame. The position vector $\overrightarrow{OP} = \mathbf{x}$ can be written as

$$\mathbf{x} = x_k \mathbf{e}_k.$$

For the sake of simplicity we denote the partial derivative $\partial(.)/\partial x_k$ by $(.)_{,k}$.

Let $D$ be an open set in $\mathscr{E}$. A function $\mathbf{\Phi}$ that assigns to each point $P \in D$ a scalar, vector, or tensor $\mathbf{\Phi}(P)$ is called *scalar, vector,* or *tensor field* on $D$, respectively. A vector or tensor field is said to be of class $C^n$ on $D$ if its components with respect to the fixed co-ordinate frame are continuous on $D$ together with their partial derivatives up to the $n$'th order.

Let $\mathbf{\Phi}$ be a scalar, vector, or tensor field on $\mathscr{E}$. Denoting $\|\overrightarrow{OP}\| = r$, we shall write $\mathbf{\Phi}(P) = O(r^n)$ as $r \to \infty$, or $\mathbf{\Phi}(P) = o(r^n)$ as $r \to \infty$, according to whether the expression $\|r^{-n}\mathbf{\Phi}(P)\|$ is bounded or tends to zero as $r \to \infty$. The same system of notation will be used to describe analogous properties for $r \to 0$.

Consider a scalar field $F$ of class $C^1$. The *gradient* of $F$ is the vector field

$$\operatorname{grad} F = F_{,m}\, \mathbf{e}_m. \tag{1.38}$$

Let $\mathbf{u}$ be a vector field of class $C^1$ on $D$. The *gradient* of $\mathbf{u}$ is the second-order tensor field [1]

$$\operatorname{grad} \mathbf{u} = u_{k,m}\, \mathbf{e}_k \mathbf{e}_m, \tag{1.39}$$

the *curl* of $\mathbf{u}$ is the vector field

$$\operatorname{curl} \mathbf{u} = \in_{mrs} u_{r,s} \mathbf{e}_m, \tag{1.40}$$

and the *divergence* of $\mathbf{u}$ is the scalar field

$$\operatorname{div} \mathbf{u} = \operatorname{tr}(\operatorname{grad} \mathbf{u}) = u_{m,m}. \tag{1.41}$$

These operators, as well as those subsequently introduced in this section, can be also defined as linear mappings between scalar, vector, or tensor spaces (see, e.g. Gurtin [150], Sect. 4), and hence they are independent of the co-ordinate system.

---

[1] Note that we use throughout the so-called right-hand gradients, curls, and divergences of vector and tensor fields (cf. Malvern [227], Sect. 2.5, Jaunzemis [433], p. 88).

We shall also use the *symmetric gradient* of **u**, which is the symmetric part of grad **u**, i.e.

$$\operatorname{symgrad} \mathbf{u} = \tfrac{1}{2}(u_{k,m} + u_{m,k})\, \mathbf{e}_k \mathbf{e}_m. \tag{1.42}$$

Next, let **A** be a second-order tensor field of class $C^1$ on $D$. The *gradient* of **A** is the third-order tensor field

$$\operatorname{grad} \mathbf{A} = A_{km,r} \mathbf{e}_k \mathbf{e}_m \mathbf{e}_r, \tag{1.43}$$

the *curl* of **A** is the second-order tensor field

$$\operatorname{curl} \mathbf{A} = \in_{mrs} A_{kr,s} \mathbf{e}_k \mathbf{e}_m, \tag{1.44}$$

and the *divergence* of **A** is the vector field

$$\operatorname{div} \mathbf{A} = A_{km,m} \mathbf{e}_k. \tag{1.45}$$

For a tensor field **A** of class $C^2$ we shall also use the so-called *incompatibility* of **A**, which is the second-order tensor field [1]

$$\operatorname{inc} \mathbf{A} = -\in_{ikl} \in_{jmn} A_{ln,km} \mathbf{e}_i \mathbf{e}_j. \tag{1.46}$$

Finally, we define the *Laplacian* of a scalar field $F$ and that of a vector field **u**, both of class $C^2$ on $D$, by the relations

$$\Delta F = \operatorname{div}(\operatorname{grad} F) = F_{,mm}, \tag{1.47}$$

$$\Delta \mathbf{u} = \operatorname{div}(\operatorname{grad} \mathbf{u}) = u_{k,mm} \mathbf{e}_k. \tag{1.48}$$

It can be shown by a direct calculation that if $F$, **u**, and **A** are of class $C^2$ on $D$, then they satisfy the identities

$$\operatorname{curl}(\operatorname{grad} \mathbf{F}) = \mathbf{0}, \quad \operatorname{curl}(\operatorname{grad} \mathbf{u}) = \mathbf{0}, \tag{1.49}$$

$$\operatorname{div}(\operatorname{curl} \mathbf{u}) = 0, \quad \operatorname{div}(\operatorname{curl} \mathbf{A}) = \mathbf{0}, \tag{1.50}$$

and if **u** and **A** are of class $C^3$ on $D$, then

$$\operatorname{inc}(\operatorname{symgrad} \mathbf{u}) = \mathbf{0}, \quad \operatorname{div}(\operatorname{inc} \mathbf{A}) = \mathbf{0}. \tag{1.51}$$

---

[1] This operator has been introduced by Kröner [190]. Its name is justified by the fact that the compatibility equations in the linear theory of elasticity can be written as $\operatorname{inc} \mathbf{E} = \mathbf{0}$ where **E** is the infinitesimal strain tensor (cf. Sect. 2.7). In other words, infinitesimal strains are compatible only if the "incompatibility" of the infinitesimal strain tensor vanishes.

Let $V$ be a finite region in $\mathscr{E}$, bounded by a two-sided and piecewise smooth surface $S$, and designate by $\mathbf{n}$ the outward unit normal to $S$ (Fig. 1.1). It can be

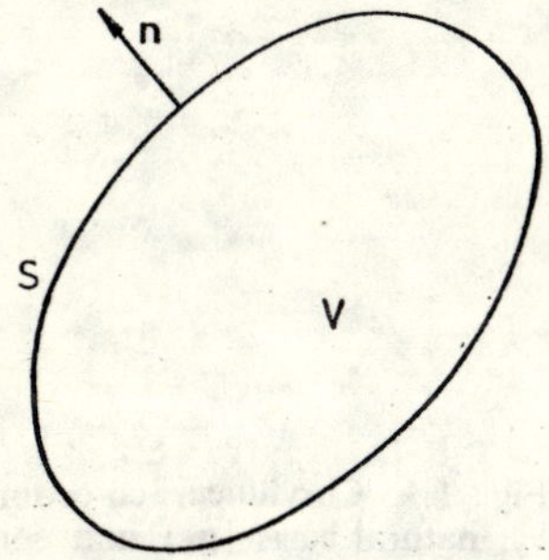


Fig. 1.1. On the application of Gauss' formulae.

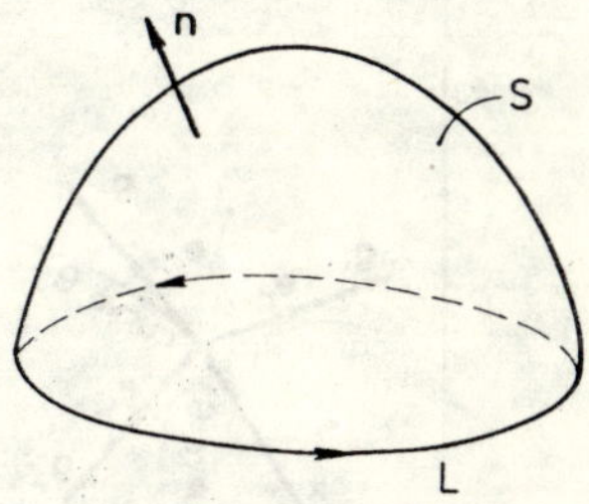


Fig. 1.2. On the application of Stokes' formulae.

shown that if $\mathbf{u}$ is a vector field and $\mathbf{A}$ is a second-order tensor field, both of which are of class $C^1$ on $\bar{V} = V \cup S$, then

$$\int_S \mathbf{u} \cdot \mathbf{n} \, \mathrm{d}s = \int_V \operatorname{div} \mathbf{u} \, \mathrm{d}v, \tag{1.52}$$

$$\int_S \mathbf{A}\mathbf{n} \, \mathrm{d}s = \int_V \operatorname{div} \mathbf{A} \, \mathrm{d}v. \tag{1.53}$$

These integral transformations, sometimes called Gauss' formulae, are still valid for unbounded regions, provided $\mathbf{u}$ and $\mathbf{A}$ are of bounded support.

Let $S$ be a closed two-sided and piecewise smooth surface, bounded by a closed simple and piecewise smooth curve (Fig. 1.2). Then, given any vector field $\mathbf{u}$ and any second-order tensor field $\mathbf{A}$, both assumed of class $C^1$ in some neighbourhood of $\bar{S} = S \cup L$, the following integral transformations, known as Stokes' formulae, hold

$$\oint_L \mathbf{u} \cdot \mathrm{d}\mathbf{x} = -\int_S (\operatorname{curl} \mathbf{u}) \cdot \mathbf{n} \, \mathrm{d}s, \tag{1.54}$$

$$\oint_L \mathbf{A} \, \mathrm{d}\mathbf{x} = -\int_S (\operatorname{curl} \mathbf{A}) \, \mathbf{n} \, \mathrm{d}s, \tag{1.55}$$

where $\mathbf{n}$ is one of the unit normals to $S$, and the integration sense on $L$ is chosen clockwise when looking down along $\mathbf{n}$.

## 1.3. Orthogonal curvilinear co-ordinates. Physical components of vectors and tensors

The formulation of the boundary-value problems in the elasticity theory can be sometimes significantly simplified by passing from the Cartesian co-ordinates $x_k$ to a suitably chosen system of curvilinear co-ordinates, say $\theta_\alpha$. In general, such a system is introduced by the transformation

$$x_k = x_k(\theta_\alpha), \qquad k, \alpha = 1,2,3, \tag{1.56}$$

which is supposed to be one-to-one and continuously differentiable, except some possible singular points or curves. Hence, excluding from $\mathscr{E}$ these possible singularities, we have

$$\det\left[\frac{\partial x_k}{\partial \theta_\alpha}\right] \neq 0, \tag{1.57}$$

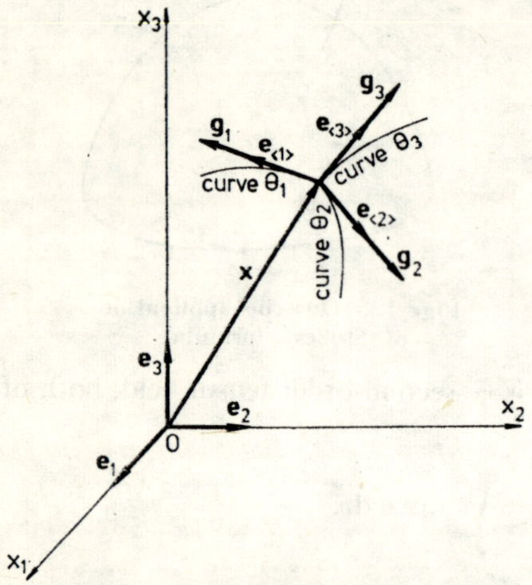


Fig. 1.3. Curvilinear co-ordinates $\theta_\alpha$, natural basis $\{\mathbf{g}_\alpha\}$, and corresponding orthonormal basis $\{\mathbf{e}_{<\alpha>}\}$ of physical components, $\alpha = 1, 2, 3$.

and there exists an inverse transformation

$$\theta_\alpha = \theta_\alpha(x_k), \tag{1.58}$$

which is also continuously differentiable.

If $\theta_1$ is held constant, the three equations (1.56) define parametrically a surface, giving its Cartesian co-ordinates as a function of the parameters $\theta_2$ and $\theta_3$. This surface is called a $\theta_1$-surface; $\theta_2$- and $\theta_3$-surfaces are defined in a similar way. The three co-ordinate surfaces intersect by pairs in three *co-ordinate curves*, on each of which varies only one curvilinear co-ordinate. Through any regular point of the space there passes one and only one co-ordinate curve of each family (Fig. 1.3).

To each point $P \in \mathscr{E}$ we can associate a so-called *natural basis* composed of the three vectors

$$\mathbf{g}_\alpha = \frac{\partial \mathbf{x}}{\partial \theta_\alpha} = \frac{\partial x_k}{\partial \theta_\alpha}\mathbf{e}_k, \tag{1.59}$$

which are tangent at $P$ to the co-ordinate curves, and are linearly independent by virtue of (1.57). However, when the curvilinear co-ordinates have different physical dimensions, the vector and tensor components with respect to the natural basis have also different physical dimensions, thus complicating the analysis. To avoid this difficulty, the vectors of the natural basis are usually replaced by the corresponding unit vectors

$$\mathbf{e}_{<\alpha>} = \frac{1}{h_\alpha}\mathbf{g}_\alpha \quad \text{(no sum)}, \tag{1.60}$$

where $h_\alpha = \|\mathbf{g}_\alpha\|$.

In the following we consider only *orthogonal* curvilinear co-ordinates. In this case the co-ordinate curves through any point $P$ are mutually orthogonal, and the basis $\{\mathbf{e}_\alpha\}$ is orthonormal (Fig. 1.3), while its orientation is generally point-dependent.

The vector and tensor components with respect to the basis $\{\mathbf{e}_\alpha\}$ are called *physical components;* they will be labelled by sharp brackets $\langle\ \rangle$. For example, the physical components $A_{\langle\alpha\beta\rangle}$ of a second-order tensor $\mathbf{A}$ will be defined by

$$\mathbf{A} = A_{\langle\alpha\beta\rangle}\mathbf{e}_{\langle\alpha\rangle}\mathbf{e}_{\langle\beta\rangle}. \tag{1.61}$$

Denoting as above by

$$q_{k\langle\alpha\rangle} = \mathbf{e}_k\cdot\mathbf{e}_{\langle\alpha\rangle}, \qquad k,\ \alpha = 1, 2, 3, \tag{1.62}$$

the direction cosines of the unit vectors $\mathbf{e}_k$ with respect to the unit vectors $\mathbf{e}_{\langle\alpha\rangle}$, we obtain from (1.33) the transformation formulae

$$\mathbf{e}_k = q_{k\langle\alpha\rangle}\mathbf{e}_{\langle\alpha\rangle}, \qquad \mathbf{e}_{\langle\alpha\rangle} = q_{k\langle\alpha\rangle}\mathbf{e}_k. \tag{1.63}$$

Next, from (1.34) and (1.35) it follows that the physical components of a vector $\mathbf{u}$ and of a second-order tensor $\mathbf{A}$ are connected with their respective Cartesian components by the relations

$$\begin{aligned} u_k &= q_{k\langle\alpha\rangle}u_{\langle\alpha\rangle}, \qquad & u_{\langle\alpha\rangle} &= q_{k\langle\alpha\rangle}u_k, \\ A_{km} &= q_{k\langle\alpha\rangle}q_{m\langle\beta\rangle}A_{\langle\alpha\beta\rangle}, \qquad & A_{\langle\alpha\beta\rangle} &= q_{k\langle\alpha\rangle}q_{m\langle\beta\rangle}A_{km}. \end{aligned} \tag{1.64}$$

By using these formulae and taking into account that now the $q_{k\langle\alpha\rangle}$'s are functions of $P\in\mathscr{E}$, it is possible to deduce the expressions in physical components of the operators defined in Sect. 1.2. In particular, by using (1.47), (1.39), and (1.45), it can be shown that

$$\Delta F = \partial_{\langle\alpha\rangle}\partial_{\langle\alpha\rangle}F + \langle\beta\alpha\beta\rangle\,\partial_{\langle\alpha\rangle}F, \tag{1.65}$$

$$\operatorname{grad}\mathbf{u} = (\partial_{\langle\beta\rangle}u_{\langle\alpha\rangle} + \langle\beta\gamma\alpha\rangle\,u_{\langle\gamma\rangle})\,\mathbf{e}_{\langle\alpha\rangle}\mathbf{e}_{\langle\beta\rangle}, \tag{1.66}$$

$$\operatorname{div}\mathbf{A} = (\partial_{\langle\beta\rangle}A_{\langle\alpha\beta\rangle} + \langle\beta\gamma\alpha\rangle\,A_{\langle\gamma\beta\rangle} + \langle\beta\gamma\beta\rangle\,A_{\langle\alpha\gamma\rangle})\,\mathbf{e}_{\langle\alpha\rangle}, \tag{1.67}$$

where $F(P)$ is a scalar field, $\mathbf{u}(P)$ is a vector field, and $\mathbf{A}(P)$ is a second-order tensor field, all of them of class $C^1$; the symbols $\langle\alpha\beta\gamma\rangle$ are defined by the relation

$$\partial_{\langle\alpha\rangle}\mathbf{e}_{\langle\beta\rangle} = \langle\alpha\beta\gamma\rangle\,\mathbf{e}_{\langle\gamma\rangle}, \tag{1.68}$$

where

$$\partial_{\langle\alpha\rangle} = \frac{1}{h_\alpha}\frac{\partial}{\partial\theta_\alpha} \text{ (no sum)}. \tag{1.69}$$

From (1.68) and (1.63) it follows that

$$\langle\alpha\beta\gamma\rangle = q_{k\langle\gamma\rangle}\partial_{\langle\alpha\rangle}q_{k\langle\beta\rangle}. \tag{1.70}$$

Since $q_{k\langle\beta\rangle}q_{k\langle\gamma\rangle} = \delta_{\beta\gamma}$, equation (1.70) can be rewritten as

$$\langle\alpha\beta\gamma\rangle = -\,q_{k\langle\beta\rangle}\partial_{\langle\alpha\rangle}q_{k\langle\gamma\rangle} = -\langle\alpha\gamma\beta\rangle,$$

and hence the symbol $\langle\alpha\beta\gamma\rangle$ is antisymmetric in the last two indexes. Moreover, it can be shown [227] that $\langle\alpha\beta\gamma\rangle = 0$, except when $\beta \neq \gamma$ and $\alpha = \beta$ or $\alpha = \gamma$, and that for $\alpha \neq \beta$

$$\langle \alpha\beta\alpha \rangle = - \langle \alpha\alpha\beta \rangle = \frac{1}{h_\alpha}\partial_{\langle\beta\rangle} h_\alpha \qquad \text{(no sum)}.$$

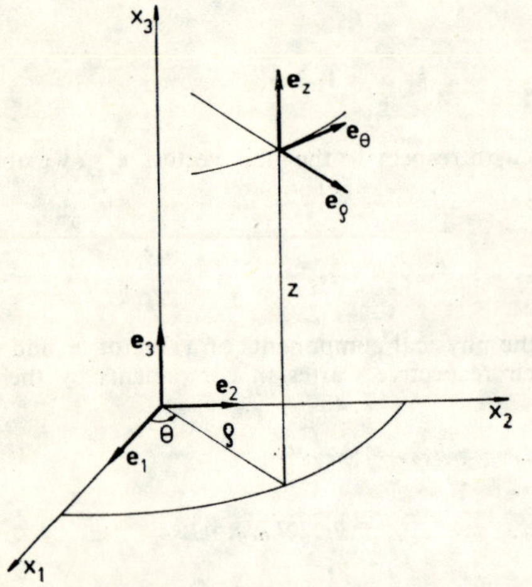


Fig. 1.4. Cylindrical co-ordinates $\rho$, $\theta$, $z$ and corresponding orthonormal basis $\{\mathbf{e}_\rho, \mathbf{e}_\theta, \mathbf{e}_z\}$ of physical components.

The last formula allows the calculation of all non-zero $\langle\alpha\beta\gamma\rangle$-symbols. However, the direct use of the definition (1.68) sometimes leads more easily to the same result.

We finally give the main results that are obtained when applying the above formalism to cylindrical and spherical co-ordinates. For the sake of simplicity we denote in this case the curvilinear co-ordinates $\theta_1$, $\theta_2$, $\theta_3$ by $(\rho, \theta, z)$, respectively $(\rho, \theta, \varphi)$, and thus the sharp brackets of the subscripts may be omitted without possible confusion.

We limit ourselves to indicate only the non-zero symbols $q_{k\langle\alpha\rangle}$ and $\langle\alpha\beta\gamma\rangle$, as well as the physical components of the fields $\mathbf{H} = \text{grad}\,\mathbf{u}$, $\mathbf{E} = \text{symgrad}\,\mathbf{u}$, $\text{div}\,\mathbf{A}$, and $\Delta F$, which occur systematically in the theory of elasticity.

*Cylindrical co-ordinates* (Fig. 1.4): $\theta_1 = \rho$, $\theta_2 = \theta$, $\theta_3 = z$.

$$x_1 = \rho\cos\theta, \qquad x_2 = \rho\sin\theta, \qquad x_3 = z. \tag{1.71}$$

$$h_\rho = 1, \qquad h_\theta = \rho, \qquad h_z = 1. \tag{1.72}$$

$$q_{3z} = 1, \qquad q_{1\rho} = q_{2\theta} = \cos\theta, \qquad q_{2\rho} = -q_{1\theta} = \sin\theta. \tag{1.73}$$

$$\langle\theta\rho\theta\rangle = -\langle\theta\theta\rho\rangle = \frac{1}{\rho}. \tag{1.74}$$

$$\left.\begin{aligned}
H_{\rho\rho} &= \frac{\partial u_\rho}{\partial\rho}, & H_{\rho\theta} &= \frac{1}{\rho}\frac{\partial u_\rho}{\partial\theta} - \frac{u_\theta}{\rho}, & H_{\rho z} &= \frac{\partial u_\rho}{\partial z},\\
H_{\theta\rho} &= \frac{\partial u_\theta}{\partial\rho}, & H_{\theta\theta} &= \frac{1}{\rho}\frac{\partial u_\theta}{\partial\theta} + \frac{u_\rho}{\rho}, & H_{\theta z} &= \frac{\partial u_\theta}{\partial z},\\
H_{z\rho} &= \frac{\partial u_z}{\partial\rho}, & H_{z\theta} &= \frac{1}{\rho}\frac{\partial u_z}{\partial\theta}, & H_{zz} &= \frac{\partial u_z}{\partial z}.
\end{aligned}\right\} \tag{1.75}$$

$$\left.\begin{aligned} E_{\rho\rho} &= \frac{\partial u_\rho}{\partial \rho}, & E_{\rho\theta} &= \frac{1}{2}\left(\frac{1}{\rho}\frac{\partial u_\rho}{\partial \theta} + \frac{\partial u_\theta}{\partial \rho} - \frac{u_\theta}{\rho}\right), \\ E_{\rho z} &= \frac{1}{2}\left(\frac{\partial u_\rho}{\partial z} + \frac{\partial u_z}{\partial \rho}\right), & E_{zz} &= \frac{\partial u_z}{\partial z}, \\ E_{\theta\theta} &= \frac{1}{\rho}\frac{\partial u_\theta}{\partial \theta} + \frac{u_\rho}{\rho}, & E_{\theta z} &= \frac{1}{2}\left(\frac{\partial u_\theta}{\partial z} + \frac{1}{\rho}\frac{\partial u_z}{\partial \theta}\right). \end{aligned}\right\} \tag{1.76}$$

$$\left.\begin{aligned} (\operatorname{div} \mathbf{A})_\rho &= \frac{\partial A_{\rho\rho}}{\partial \rho} + \frac{1}{\rho}\frac{\partial A_{\rho\theta}}{\partial \theta} + \frac{\partial A_{\rho z}}{\partial z} + \frac{1}{\rho}(A_{\rho\rho} - A_{\theta\theta}), \\ (\operatorname{div} \mathbf{A})_\theta &= \frac{\partial A_{\theta\rho}}{\partial \rho} + \frac{1}{\rho}\frac{\partial A_{\theta\theta}}{\partial \theta} + \frac{\partial A_{\theta z}}{\partial z} + \frac{1}{\rho}(A_{\rho\theta} + A_{\theta\rho}), \\ (\operatorname{div} \mathbf{A})_z &= \frac{\partial A_{z\rho}}{\partial \rho} + \frac{1}{\rho}\frac{\partial A_{z\theta}}{\partial \theta} + \frac{\partial A_{zz}}{\partial z} + \frac{1}{\rho}A_{z\rho}. \end{aligned}\right\} \tag{1.77}$$

$$\Delta F = \frac{1}{\rho}\frac{\partial}{\partial \rho}\left(\rho \frac{\partial F}{\partial \rho}\right) + \frac{1}{\rho^2}\frac{\partial^2 F}{\partial \theta^2} + \frac{\partial^2 F}{\partial z^2}. \tag{1.78}$$

*Spherical co-ordinates* (Fig. 1.5): $\theta_1 = r, \quad \theta_2 = \theta, \quad \theta_3 = \varphi.$

$$x_1 = r \sin\theta \cos\varphi, \qquad x_2 = r \sin\theta \sin\varphi, \qquad x_3 = r \cos\theta, \tag{1.79}$$

$$h_r = 1, \qquad h_\theta = r, \qquad h_\varphi = r \sin\theta. \tag{1.80}$$

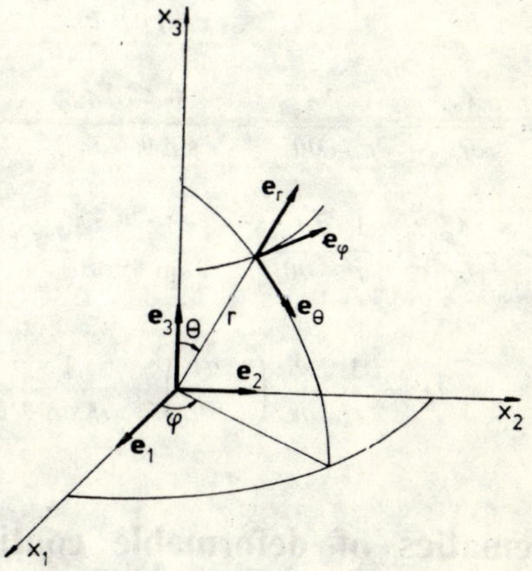


Fig. 1.5. Spherical co-ordinates $r$, $\theta$, $\varphi$ and corresponding orthonormal basis $\{\mathbf{e}_r, \mathbf{e}_\theta, \mathbf{e}_\varphi\}$ of physical components.

$$\left.\begin{aligned} q_{1r} &= \sin\theta \cos\varphi, & q_{2r} &= \sin\theta \sin\varphi, & q_{3r} &= \cos\theta, \\ q_{1\theta} &= \cos\theta \cos\varphi, & q_{2\theta} &= \cos\theta \sin\varphi, & q_{3\theta} &= -\sin\theta, \\ q_{1\varphi} &= -\sin\varphi, & q_{2\varphi} &= \cos\varphi, & q_{3\varphi} &= 0. \end{aligned}\right\} \tag{1.81}$$

$$\left.\begin{aligned}\langle\theta r\theta\rangle &= -\langle\theta\theta r\rangle = \langle\varphi r\varphi\rangle = -\langle\varphi\varphi r\rangle = \frac{1}{r},\\ \langle\varphi\theta\varphi\rangle &= -\langle\varphi\varphi\theta\rangle = \frac{1}{r}\operatorname{ctg}\theta.\end{aligned}\right\} \tag{1.82}$$

$$\left.\begin{aligned}
&H_{rr} = \frac{\partial u_r}{\partial r}, \quad H_{r\theta} = \frac{1}{r}\frac{\partial u_r}{\partial\theta} - \frac{u_\theta}{r}, \quad H_{r\varphi} = \frac{1}{r\sin\theta}\frac{\partial u_r}{\partial\varphi} - \frac{u_\varphi}{r},\\
&H_{\theta r} = \frac{\partial u_\theta}{\partial r}, \quad H_{\theta\theta} = \frac{1}{r}\frac{\partial u_\theta}{\partial\theta} + \frac{u_r}{r}, \quad H_{\theta\varphi} = \frac{1}{r\sin\theta}\frac{\partial u_\theta}{\partial\varphi} - \frac{u_\varphi}{r}\operatorname{ctg}\theta,\\
&H_{\varphi r} = \frac{\partial u_\varphi}{\partial r}, \quad H_{\varphi\theta} = \frac{1}{r}\frac{\partial u_\varphi}{\partial\theta}, \quad H_{\varphi\varphi} = \frac{1}{r\sin\theta}\frac{\partial u_\varphi}{\partial\varphi} + \frac{u_\theta}{r}\operatorname{ctg}\theta + \frac{u_r}{r}.
\end{aligned}\right\} \tag{1.83}$$

$$\left.\begin{aligned}
&E_{rr} = \frac{\partial u_r}{\partial r}, \quad E_{r\theta} = \frac{1}{2}\left(\frac{1}{r}\frac{\partial u_r}{\partial\theta} + \frac{\partial u_\theta}{\partial r} - \frac{u_\theta}{r}\right),\\
&E_{r\varphi} = \frac{1}{2}\left(\frac{1}{r\sin\theta}\frac{\partial u_r}{\partial\varphi} + \frac{\partial u_\varphi}{\partial r} - \frac{u_\varphi}{r}\right), \quad E_{\theta\theta} = \frac{1}{r}\frac{\partial u_\theta}{\partial\theta} + \frac{u_r}{r},\\
&E_{\theta\varphi} = \frac{1}{2}\left(\frac{1}{r\sin\theta}\frac{\partial u_\theta}{\partial\varphi} + \frac{1}{r}\frac{\partial u_\varphi}{\partial\theta} - \frac{u_\varphi}{r}\operatorname{ctg}\theta\right), \quad E_{\varphi\varphi} = \frac{1}{r\sin\theta}\frac{\partial u_\varphi}{\partial\varphi} + \frac{u_\theta}{r}\operatorname{ctg}\theta + \frac{u_r}{r}.
\end{aligned}\right\} \tag{1.84}$$

$$\left.\begin{aligned}
(\operatorname{div}\mathbf{A})_r &= \frac{\partial A_{rr}}{\partial r} + \frac{1}{r}\frac{\partial A_{r\theta}}{\partial\theta} + \frac{1}{r\sin\theta}\frac{\partial A_{r\varphi}}{\partial\varphi} + \frac{1}{r}(2A_{rr} - A_{\theta\theta} - A_{\varphi\varphi} + A_{r\theta}\operatorname{ctg}\theta),\\
(\operatorname{div}\mathbf{A})_\theta &= \frac{\partial A_{\theta r}}{\partial r} + \frac{1}{r}\frac{\partial A_{\theta\theta}}{\partial\theta} + \frac{1}{r\sin\theta}\frac{\partial A_{\theta\varphi}}{\partial\varphi} + \frac{1}{r}[2A_{\theta r} + A_{r\theta} + (A_{\theta\theta} - A_{\varphi\varphi})\operatorname{ctg}\theta],\\
(\operatorname{div}\mathbf{A})_\varphi &= \frac{\partial A_{\varphi r}}{\partial r} + \frac{1}{r}\frac{\partial A_{\varphi\theta}}{\partial\theta} + \frac{1}{r\sin\theta}\frac{\partial A_{\varphi\varphi}}{\partial\varphi} + \frac{1}{r}[2A_{\varphi r} + A_{r\varphi} + (A_{\theta\varphi} + A_{\varphi\theta})\operatorname{ctg}\theta].
\end{aligned}\right\} \tag{1.85}$$

$$\Delta F = \frac{1}{r^2}\frac{\partial}{\partial r}\left(r^2\frac{\partial F}{\partial r}\right) + \frac{1}{r^2\sin\theta}\frac{\partial}{\partial\theta}\left(\sin\theta\frac{\partial F}{\partial\theta}\right) + \frac{1}{r^2\sin^2\theta}\frac{\partial^2 F}{\partial\varphi^2}. \tag{1.86}$$

## 2. Kinematics of deformable continuous media

### 2.1. Configuration, motion

Continuum mechanics assumes that any body "fills" the spatial region it occupies at a given time. Therefore, each *material point* or *particle* $X$ of a body $\mathscr{B}$ may be identified with its *place* $P \in \mathscr{E}$ in an arbitrary but fixed configuration $(K)$ of the body, called *reference configuration* [1]. Let $\mathbf{X}$ denote the position vector of $P$.

The motion of a body $\mathscr{B}$ may be described by the mapping

$$\mathbf{x} = \boldsymbol{\chi}(\mathbf{X}, t), \tag{2.1}$$

where $\mathbf{x}$ is the position vector of the place occupied by the particle $X$ in the *current configuration* $(k)$ of the body at time $t$. Thus, the motion is a one-parameter family of configurations, with the time $t$ as real parameter. We assume that the function $\boldsymbol{\chi}$ is one-to-one and of class $C^3$. The Cartesian components $X_k$ of $\mathbf{X}$ are called *material* or *Lagrangian co-ordinates*, whereas the components $x_k$ of $\mathbf{x}$ are called *spatial* or *Eulerian co-ordinates*.

Any time-dependent scalar, vector, or tensor field defined on $\mathscr{B}$ may be considered either as a function of the particle $\mathbf{X}$ and time $t$, or as a function of the current position vector $\mathbf{x}$ and time $t$, provided that a definite motion (2.1) is given. These two possible descriptions are called *material description* and *spatial description*, respectively [2]. *Material derivatives* and *spatial derivatives* are defined accordingly. We shall use the symbols Grad, Div, Curl, and grad, div, curl for the gradient, the divergence, and the curl, calculated with respect to the co-ordinates $X_k$ and $x_k$, respectively.

*Material time derivatives* are denoted by $\mathrm{d}/\mathrm{d}t$ or by superposing dots; they are partial time derivatives with the material co-ordinates $X_k$ held constant. In particular, the *velocity* $\mathbf{v}$ and the *acceleration* $\mathbf{a}$ of the particle $X$ are, respectively, the first and second material time derivatives of the motion $\boldsymbol{\chi}(\mathbf{X}, t)$, i.e.

$$\mathbf{v} = \dot{\mathbf{x}} = \frac{\mathrm{d}}{\mathrm{d}t}\boldsymbol{\chi}(\mathbf{X}, t), \tag{2.2}$$

$$\mathbf{a} = \dot{\mathbf{v}} = \frac{\mathrm{d}^2}{\mathrm{d}t^2}\boldsymbol{\chi}(\mathbf{X}, t). \tag{2.3}$$

*Spatial time derivatives* are denoted by $\partial/\partial t$; they are partial time derivatives with the spatial co-ordinates $x_k$ held constant.

If $\boldsymbol{\Phi}$ denotes a vector or tensor field depending on $\mathbf{x}$ and $t$, then its material time derivative can be expressed, by using the chain rule, as

$$\frac{\mathrm{d}}{\mathrm{d}t}\boldsymbol{\Phi}(\mathbf{x}, t) = \frac{\partial \boldsymbol{\Phi}}{\partial t} + (\mathrm{grad}\, \boldsymbol{\Phi})\, \mathbf{v}. \tag{2.4}$$

---

[1] Through this identification, the topology of material manifolds reduces to that of the spatial differentiable manifolds of the three-dimensional Euclidean space $\mathscr{E}$. In particular, we understand by *material neighbourhood* of a particle $X$ the set of all particles that occupied a spatial neighbourhood of $X$ in the reference configuration $(K)$.

[2] The spatial description is especially useful in fluid mechanics where we may observe a flow in a fixed region of the space. In the elasticity theory, however, the material description is generally preferred, since the reference configuration can be chosen as the initial unstressed state, to which the body will return when it is unloaded.

Hence, the rate of change of the field $\boldsymbol{\Phi}$ can be decomposed into a *local rate of change*, $\partial\boldsymbol{\Phi}/\partial t$, which would be measured by an observer located at the fixed place $\mathbf{x}$, and a *convective rate of change*, $(\text{grad}\,\boldsymbol{\Phi})\mathbf{v}$, which is generated by the motion of the particle $X$ to places where the field $\boldsymbol{\Phi}$ has different values. Replacing $\boldsymbol{\Phi}$ by $\mathbf{v}$ in (2.4) yields [1]

$$\mathbf{a} = \frac{\partial \mathbf{v}}{\partial t} + (\text{grad}\,\mathbf{v})\,\mathbf{v}, \qquad a_k = \frac{\partial v_k}{\partial t} + v_m \frac{\partial x_k}{\partial v_m}, \tag{2.5}$$

where now $\mathbf{v}$ is considered as function of $\mathbf{x}$ and $t$.

## 2.2. Deformation tensors

The mapping defined by a motion at any fixed time is called a *deformation*. Differentiating (2.1) for $t = \text{const}$ gives

$$\mathrm{d}\mathbf{x} = \mathbf{F}\,\mathrm{d}\mathbf{X}, \qquad \mathrm{d}x_k = F_{km}\mathrm{d}X_m, \tag{2.6}$$

where

$$\mathbf{F} = \text{Grad}\,\boldsymbol{\chi}(\mathbf{X}, t), \qquad F_{km} = \frac{\partial \chi_k(\mathbf{X}, t)}{\partial X_m}. \tag{2.7}$$

The second-order tensor field $\mathbf{F}$ is called the *deformation gradient*. Since $\boldsymbol{\chi}$ is one-to-one, we have [2]

$$J = \det \mathbf{F} > 0. \tag{2.8}$$

By inverting (2.6) it follows that

$$\mathrm{d}\mathbf{X} = \mathbf{F}^{-1}\,\mathrm{d}\mathbf{x}, \qquad \mathrm{d}X_k = F^{-1}_{km}\mathrm{d}x_m, \tag{2.9}$$

where

$$\mathbf{F}^{-1} = \text{grad}\,\boldsymbol{\chi}^{-1}(\mathbf{x}, t), \qquad F^{-1}_{km} = \frac{\partial \chi^{-1}_k(\mathbf{x}, t)}{\partial x_m}. \tag{2.10}$$

---

[1] We shall generally write all major formulae in both direct notation and component form. In the last case we shall always use rectangular Cartesian components and denote by $X_k$ and $x_k$ material and spatial co-ordinates, respectively, with respect to a common Cartesian frame.

[2] Indeed, since $\det \mathbf{F}$ is by hypothesis a non-vanishing and continuous function throughout the motion for any fixed $\mathbf{X}$, it must have a constant sign for any $\mathbf{X}$ and $t$. On the other hand, we have in the reference configuration $\mathbf{F} = \mathbf{1}$, and hence $J = 1$. Consequently, (2.8) must hold for any $\mathbf{X}$ and $t$.

The keystone of the theory of finite deformation is the following theorem[1], which we state without proof.

**Polar decomposition theorem.** *Any invertible second-order tensor* **F** *has two unique multiplicative decompositions*

$$\mathbf{F} = \mathbf{R}\mathbf{U}, \qquad \mathbf{F} = \mathbf{V}\mathbf{R}, \tag{2.11}$$

*in which* **R** *is orthogonal and* **U** *and* **V** *are symmetric and positive-definite, i.e.*

$$\mathbf{U} = \mathbf{U}^T, \quad \mathbf{V} = \mathbf{V}^T, \quad \mathbf{R}\mathbf{R}^T = \mathbf{R}^T\mathbf{R} = \mathbf{1}. \tag{2.12}$$

When **F** is the deformation gradient, **R** is called the *rotation tensor*, **U** the *right stretch tensor*, and **V** the *left stretch tensor* of the deformation. The tensors **C** and **B**, defined by the relations

$$\mathbf{C} = \mathbf{U}^2 = \mathbf{F}^T\mathbf{F}, \quad \mathbf{B} = \mathbf{V}^2 = \mathbf{F}\mathbf{F}^T, \tag{2.13}$$

are called the *right* and the *left Cauchy-Green tensors* of the deformation. The motivation of this terminology will result from the discussion in the next subsection.

## 2.3. Length and angle changes

Let $\mathrm{d}L$ and $\mathrm{d}l$ denote the distance between the particles **X** and **X** + d**X** in the configurations $(K)$ and $(k)$, respectively. From (2.6) and $(2.13)_1$ it follows that

$$\mathrm{d}l^2 = \mathrm{d}\mathbf{x}\cdot\mathrm{d}\mathbf{x} = C_{km}\mathrm{d}X_k\mathrm{d}X_m. \tag{2.14}$$

The variation of the squared length of the infinitesimal material vector d**X** may now be written in Cartesian components as

$$\mathrm{d}l^2 - \mathrm{d}L^2 = \mathrm{d}\mathbf{x}\cdot\mathrm{d}\mathbf{x} - \mathrm{d}\mathbf{X}\cdot\mathrm{d}\mathbf{X} = (C_{km} - \delta_{km})\,\mathrm{d}X_k\mathrm{d}X_m.$$

By introducing the second-order tensor

$$\mathbf{D} = \tfrac{1}{2}(\mathbf{C} - \mathbf{1}) = \tfrac{1}{2}(\mathbf{F}^T\mathbf{F} - \mathbf{1}), \tag{2.15}$$

the last relation becomes

$$\mathrm{d}l^2 - \mathrm{d}L^2 = 2D_{km}\mathrm{d}X_k\mathrm{d}X_m. \tag{2.16}$$

---

[1] See, e.g. Truesdell and Toupin [357], Sect. 43 of the Appendix.

The tensor $\mathbf{D}$, which gives the change in the squared length of material vectors around a given particle $X$, is called the *finite strain tensor*.

A motion of a given body is said to be *rigid* if it leaves unchanged the distances between the particles of that body. In the case of a rigid motion, the general form of (2.1) is

$$\mathbf{x} = \mathbf{c} + \mathbf{Q}(\mathbf{X} - \mathbf{a}), \qquad x_k = c_k + Q_{km}(X_m - a_m), \tag{2.17}$$

where $\mathbf{c}$ is a time-dependent vector, $\mathbf{Q}$ is a time-dependent orthogonal tensor, and $\mathbf{a}$ is a fixed position vector. It can be shown that a motion of a body is rigid if and only if $\mathbf{D} = \mathbf{0}$ for any particle of the body. When this is the case, it results from (2.15), (2.13), (2.17), and (2.11) that $\mathbf{C} = \mathbf{B} = \mathbf{U} = \mathbf{V} = \mathbf{1}$ and $\mathbf{R} = \mathbf{Q}$.

Now let $\mathbf{N}$ be the unit vector of the infinitesimal material vector $\mathrm{d}\mathbf{X}$ in the configuration $(K)$, i.e. $\mathbf{N} = \mathrm{d}\mathbf{X}/\mathrm{d}L$. The ratio $\Lambda_{\mathbf{N}} = \mathrm{d}l/\mathrm{d}L$ is called the *stretch* of $\mathrm{d}\mathbf{X}$. Dividing (2.14) by $\mathrm{d}L^2$, we obtain

$$\Lambda_{\mathbf{N}}^2 = C_{km}N_kN_m. \tag{2.18}$$

The ratio

$$\epsilon_{\mathbf{N}} = \frac{\mathrm{d}l - \mathrm{d}L}{\mathrm{d}L} = \Lambda_{\mathbf{N}} - 1 \tag{2.19}$$

is called the *unit extension* of the material vector $\mathrm{d}\mathbf{X}$. In particular, if the material vector was parallel in the reference configuration to the unit vector $\mathbf{e}_1(1, 0, 0)$, then its stretch $\Lambda_{(1)}$ and its unit extension $\epsilon_{(1)}$ are given, respectively, by the relations

$$\Lambda_{(1)}^2 = C_{11} = 1 + 2D_{11}, \qquad \epsilon_{(1)} = \sqrt{C_{11}} - 1 = \sqrt{1 + 2D_{11}} - 1. \tag{2.20}$$

We will consider now the change in angle produced by deformation. Let $\mathrm{d}\mathbf{X}'$ and $\mathrm{d}\mathbf{X}''$ be two infinitesimal material vectors, which had in the reference configuration $(K)$ the unit vectors $\mathbf{N}'$, $\mathbf{N}''$ and the lengths $\mathrm{d}L'$, $\mathrm{d}L''$, respectively. Assume that these vectors become in the current configuration $(k)$ the vectors $\mathrm{d}\mathbf{x}'$, $\mathrm{d}\mathbf{x}''$, with the unit vectors $\mathbf{n}'$, $\mathbf{n}''$, and the lengths $\mathrm{d}l'$, $\mathrm{d}l''$, respectively. By (2.6) and $(2.13)_1$, we have

$$\cos(\mathbf{n}', \mathbf{n}'') = \mathbf{n}' \cdot \mathbf{n}'' = \frac{\mathrm{d}\mathbf{x}' \cdot \mathrm{d}\mathbf{x}''}{\mathrm{d}l'\mathrm{d}l''} = \frac{C_{km}\mathrm{d}X_k'\mathrm{d}X_m''}{\mathrm{d}l'\mathrm{d}l''},$$

wherefrom, by taking into account that

$$\mathrm{d}X_k' = N_k'\mathrm{d}L', \quad \mathrm{d}X_m'' = N_m''\mathrm{d}L'', \quad \mathrm{d}l' = \Lambda_{\mathbf{N}'}\mathrm{d}L', \quad \mathrm{d}l'' = \Lambda_{\mathbf{N}''}\mathrm{d}L''$$

and removing the common term $\mathrm{d}L'\mathrm{d}L''$ from both terms of the last fraction, we obtain

$$\cos(\mathbf{n}', \mathbf{n}'') = \frac{C_{km}N_k'N_m''}{\Lambda_{\mathbf{N}'}\Lambda_{\mathbf{N}''}}. \tag{2.2}$$

Since cos (N′, N″) can be easily calculated when the directions of the infinitesimal material vectors in the configuration $(K)$ are known, the two angles determined from the two cosines can be subtracted to give the change in angle produced by deformation. For example, if the material vectors dX′ and dX″ were parallel in the reference

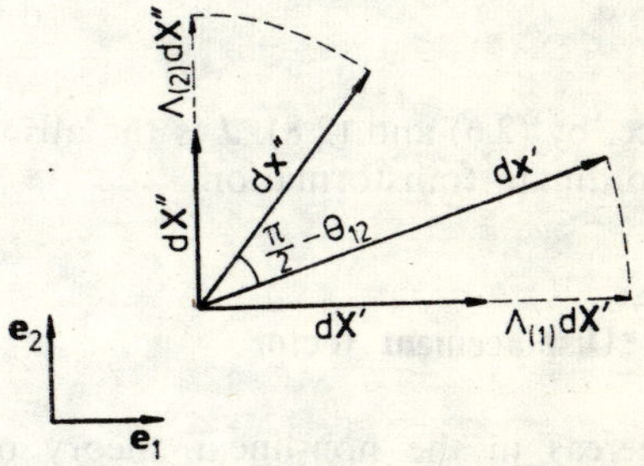


Fig. 2.1. Stretches and change in angle of two infinitesimal material vectors dX′ and dX″ that were parallel in the reference configuration to the unit vectors $\mathbf{e}_1$ and $\mathbf{e}_2$, respectively.

configuration to the unit vectors $\mathbf{e}_1(1, 0, 0)$ and $\mathbf{e}_2(0, 1, 0)$, respectively, then the angle change produced by deformation (Fig. 2.1) is $\pi/2 - \theta_{12}$, where

$$\cos\theta_{12} = \frac{C_{12}}{\Lambda_{(1)}\Lambda_{(2)}} = \frac{C_{12}}{\sqrt{C_{11}C_{22}}} = \frac{2D_{12}}{\sqrt{(1+2D_{11})(1+2D_{22})}}. \tag{2.22}$$

## 2.4. Material curves, surfaces, and volumes

*Material curves, surfaces, and volumes* are sets of particles that occupy in the reference configuration spatial curves, surfaces, and volumes, respectively. Thus, a material curve may be defined in the reference configuration by a relation of the form

$$\mathbf{X} = \mathbf{f}(u), \quad u \in [a, b], \tag{2.23}$$

where $u$ is a real parameter, and the vector-valued function $\mathbf{f}$ has to satisfy the same regularity conditions as in the case of a spatial curve. At the current time $t$, the material curve coincides with the spatial curve given by the equation

$$\mathbf{x} = \boldsymbol{\chi}(\mathbf{f}(u), t) \equiv \mathbf{g}(u, t). \tag{2.24}$$

Material surfaces and volumes can be defined in a similar way, by letting the function $\mathbf{f}$ depend on two or three parameters, respectively.

Let us consider now in the reference configuration an infinitesimal material vector $\mathrm{d}\mathbf{X}$, an infinitesimal oriented area $\mathbf{N}\,\mathrm{d}S$ with unit normal $\mathbf{N}$, and an infinitesimal volume $\mathrm{d}V$, and denote by $\mathrm{d}\mathbf{x}$, $\mathbf{n}\,\mathrm{d}s$, and $\mathrm{d}v$ the corresponding elements in the current configuration of the body at time $t$.

The relation between dX and dx is given by (2.6). The change of the oriented element of area is given by Nanson's formula ([357], p. 249)

$$J\mathbf{N}\,\mathrm{d}S = \mathbf{F}^T\mathbf{n}\,\mathrm{d}s, \qquad JN_k\mathrm{d}S = F_{mk}n_m\mathrm{d}s. \tag{2.25}$$

Finally, by considering the deformation $x_k = \chi_k(X_m, t)$ for a fixed value of $t$ as a change of co-ordinates from the material co-ordinates to the spatial co-ordinates of the particle $X$, and taking into account the change-of-variable rule of the integral calculus, we obtain

$$\mathrm{d}v = J\,\mathrm{d}V, \tag{2.26}$$

since, by (2.6) and (2.8), $J$ is the absolute value of the Jacobian determinant of the co-ordinate transformation.

## 2.5. Displacement vector

Whereas in the non-linear theory of elasticity the most important kinematical quantity is the motion $\chi(\mathbf{X}, t)$, in the linearized theory it proves advantageous to use the *displacement vector* $\mathbf{u}(\mathbf{X}, t)$, which is defined as the translation carrying the particle $X$ from its place $\mathbf{X}$ in the reference configuration to its place $\mathbf{x} = \chi(\mathbf{X}, t)$ in the current configuration (Fig. 2.2). In Cartesian co-ordinates we have then

$$x_k = \chi_k(\mathbf{X}, t) = X_k + u_k(\mathbf{X}, t). \tag{2.27}$$

Differentiating with respect to $X_m$ and taking into account (2.7) gives

$$\mathbf{F} = \mathbf{1} + \mathbf{H}, \qquad F_{km} = \delta_{km} + H_{km}, \tag{2.28}$$

where

$$\mathbf{H} = \operatorname{Grad} \mathbf{u}(\mathbf{X}, t), \qquad H_{km} = \frac{\partial u_k(\mathbf{X}, t)}{\partial X_m}. \tag{2.29}$$

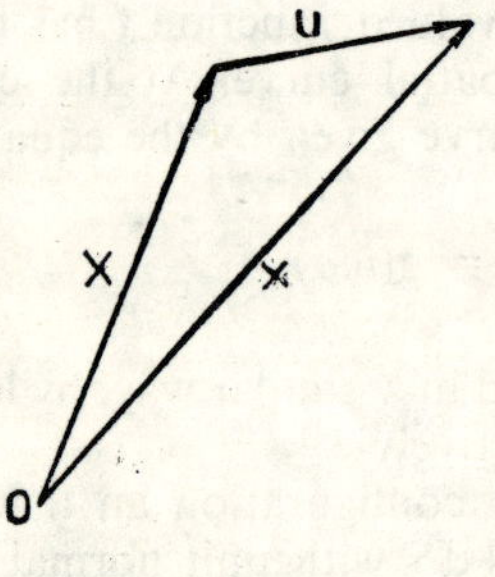


Fig. 2.2. On the definition of the displacement vector **u**.

Finally, from $(2.15)_2$ and (2.28), we deduce the expression of the finite strain tensor $\mathbf{D}$ in terms of the displacement gradient $\mathbf{H}$

$$\mathbf{D} = \tfrac{1}{2}(\mathbf{H} + \mathbf{H}^T + \mathbf{H}^T\mathbf{H}), \quad D_{km} = \frac{1}{2}\left(\frac{\partial u_k}{\partial X_m} + \frac{\partial u_m}{\partial X_k} + \frac{\partial u_p}{\partial X_k}\frac{\partial u_p}{\partial X_m}\right). \tag{2.30}$$

In particular, it follows from $(2.30)_2$ that

$$D_{11} = \frac{\partial u_1}{\partial X_1} + \frac{1}{2}\left\{\left(\frac{\partial u_1}{\partial X_1}\right)^2 + \left(\frac{\partial u_2}{\partial X_1}\right)^2 + \left(\frac{\partial u_3}{\partial X_1}\right)^2\right\},$$

$$D_{12} = \frac{1}{2}\left(\frac{\partial u_1}{\partial X_2} + \frac{\partial u_2}{\partial X_1}\right) + \frac{1}{2}\left(\frac{\partial u_1}{\partial X_1}\frac{\partial u_1}{\partial X_2} + \frac{\partial u_2}{\partial X_1}\frac{\partial u_2}{\partial X_2} + \frac{\partial u_3}{\partial X_1}\frac{\partial u_3}{\partial X_2}\right).$$

## 2.6. Linearization of the kinematic equations

If the magnitude of the displacement gradient is small compared to unity we may neglect in (2.30) the squares and the products of its components in comparison to the linear terms. More precisely, in order to linearize the kinematic equations we assume that $\varepsilon = \|\mathbf{H}\| \ll 1$ and neglect all terms of order $O(\varepsilon^2)$ or higher as $\varepsilon \to 0$

From (2.28), (2.30), and (2.11–13) we conclude that *to within an error of order* $O(\varepsilon^2)$ *as* $\varepsilon \to 0$, the following relations hold [1]

$$\mathbf{C} \approx \mathbf{B} \approx \mathbf{1} + 2\mathbf{E}, \qquad \mathbf{U} \approx \mathbf{V} \approx \mathbf{1} + \mathbf{E}, \tag{2.31}$$

$$\mathbf{D} \approx \mathbf{E}, \qquad \mathbf{R} \approx \mathbf{1} + \boldsymbol{\Omega}, \tag{2.32}$$

where

$$\mathbf{E} = \tfrac{1}{2}(\mathbf{H} + \mathbf{H}^T), \qquad E_{km} = \frac{1}{2}\left(\frac{\partial u_k}{\partial X_m} + \frac{\partial u_m}{\partial X_k}\right) \tag{2.33}$$

is the *infinitesimal strain tensor*, and

$$\boldsymbol{\Omega} = \tfrac{1}{2}(\mathbf{H} - \mathbf{H}^T), \qquad \Omega_{km} = \frac{1}{2}\left(\frac{\partial u_k}{\partial X_m} - \frac{\partial u_m}{\partial X_k}\right) \tag{2.34}$$

is the *infinitesimal rotation tensor*.

Next, from $(2.20)_2$ and (2.22), it follows that

$$\epsilon_{(1)} \approx E_{11},$$

$$\frac{\pi}{2} - \theta_{12} \approx \sin\left(\frac{\pi}{2} - \theta_{12}\right) = \cos\theta_{12} \approx 2E_{12}.$$

---

[1] From this point onward in this subsection, the sign $\approx$ is to be interpreted as equality to within an error of order $O(\varepsilon^2)$ or greater in $\varepsilon$.

Hence, to within an error of $O(\varepsilon^2)$, the diagonal components of the infinitesimal strain tensor coincide with the stretches of material vectors initially parallel to the co-ordinate axes, and the off-diagonal components of the same tensor coincide with half the changes of the angles between material vectors initially parallel to the co-ordinate axes.

It is worth noting that, when the gradient of an arbitrary vector or tensor field $\mathbf{\Phi}$ is of the order $O(\varepsilon)$, we may write

$$\frac{\partial \mathbf{\Phi}}{\partial X_k} = \frac{\partial \mathbf{\Phi}}{\partial x_m} F_{mk} = \frac{\partial \mathbf{\Phi}}{\partial x_m} (\delta_{mk} + H_{mk}) \approx \frac{\partial \mathbf{\Phi}}{\partial x_k} = \mathbf{\Phi}_{,k}, \tag{2.35}$$

and hence we may identify Grad $\mathbf{\Phi}$ with grad $\mathbf{\Phi}$. In particular, equations $(2.33)_2$ and $(2.34)_2$ may be written in the linearized theory as

$$E_{km} = \frac{1}{2}(u_{k,m} + u_{m,k}), \tag{2.36}$$

$$\Omega_{km} = \frac{1}{2}(u_{k,m} - u_{m,k}). \tag{2.37}$$

To the same approximation, it results from (2.37), $(1.22)_1$, and (1.40), that the axial vector $\boldsymbol{\omega}$ of $\Omega$, called the *infinitesimal rotation vector*, which is defined by

$$\omega_r = -\frac{1}{2}\in_{rkm}\Omega_{km}, \qquad \Omega_{km} = -\in_{kmr}\omega_r, \tag{2.38}$$

may be expressed in terms of the displacement vector as

$$\boldsymbol{\omega} = -\frac{1}{2}\operatorname{curl}\mathbf{u}, \qquad \omega_r = -\frac{1}{2}\in_{rkm}u_{k,m}. \tag{2.39}$$

It should be remembered that the above linearization holds for small displacement gradients. Starting from a different hypothesis, for example assuming that either the stretches or the rotations are small, may lead to different results.

The following identities relating the fields $\mathbf{u}$, $\boldsymbol{\omega}$, $\mathbf{E}$, $\Omega$ may be easily proved by making use of (2.36—38)

$$u_{k,m} = E_{km} + \Omega_{km} = E_{km} - \in_{kmr}\omega_r, \tag{2.40}$$

$$\Omega_{rs,k} = E_{kr,s} - E_{ks,r},$$

$$\omega_{k,m} = -\in_{krs}E_{mr,s}. \tag{2.41}$$

Finally, it can be shown (see, e.g. Gurtin [150], p. 31) that the infinitesimal strain tensor $\mathbf{E}$ vanishes at a given time if and only if the displacement field has the form

$$\mathbf{u}(\mathbf{x}) = \mathbf{u}^0 + \boldsymbol{\omega}^0 \times (\mathbf{x} - \mathbf{x}^0), \qquad u_k(\mathbf{x}) = u_k^0 + \in_{klm}\omega_l^0(x_m - x_m^0), \tag{2.42}$$

where $\mathbf{u}^0$ and $\boldsymbol{\omega}^0$ are constant vectors, and $\mathbf{x}^0$ is the position vector of a fixed point. A vector field of the form (2.42) defined on a body $\mathscr{B}$ is called a rigid displacement field. It obviously consists of a translation of vector $\mathbf{u}^0$ and a rotation of vector $\boldsymbol{\omega}^0$ around the point $\mathbf{x}^0$.

## 2.7. Compatibility conditions

When all kinematic quantities are calculated in terms of the functions $\chi(\mathbf{X}, t)$ or $\mathbf{u}(\mathbf{X}, t)$, the continuity of the deformed body is assured by the assumption that $\chi(\mathbf{X}, t)$ is one-to-one and continuously differentiable. Alternatively, if the strain tensor is used as fundamental kinematic quantity in the formulation of the boundary-value problems, then (2.30) are six independent partial differential equations for the three unknown components of the displacement vector. This overdeterminate system will not admit a solution unless the functions $D_{km}$ satisfy some integrability conditions, known as *compatibility equations* [1].

In the non-linear theory of elasticity the easiest way to infer the compatibility equations is by using the geometry of Riemann spaces (see, e.g. Eringen [105], Sect. 13, Malvern [227], pp. 193—195). However, since we shall not make use in the following of these rather sophisticated equations, we confine ourselves to indicating here the reduced form assumed by the compatibility equations in the linear case.

First, we note that, in view of the definition (1.42), we may rewrite (2.36) as

$$\mathbf{E} = \operatorname{sym} \operatorname{grad} \mathbf{u}.$$

If we apply the operator inc to this equation and take into account the identity $(1.51)_1$, we see that a *necessary* condition for the existence of a displacement field $\mathbf{u}$ is that $\mathbf{E}$ satisfy the following *equations of compatibility* [2]

$$\operatorname{inc} \mathbf{E} = \mathbf{0}, \qquad -\epsilon_{ikl}\,\epsilon_{jmn} E_{ln,km} = 0, \qquad i, j = 1, 2, 3. \tag{2.43}$$

Since $\mathbf{E}$ is symmetric, inc $\mathbf{E}$ is also symmetric, and hence there are only six distinct equations $(2.43)_2$. Moreover, since inc$\mathbf{E}$ must satisfy the tensor identity $(1.51)_2$, which is equivalent to three scalar equations, it follows that only three of the six compatibility equations $(2.43)_2$ are independent [3].

By virtue of (1.44) and of the symmetry of the strain tensor, equation $(2.43)_1$ may be also written in the equivalent form

$$-\operatorname{curl}\,(\operatorname{curl} \mathbf{E})^T = \mathbf{0}. \tag{2.44}$$

---

[1] The physical meaning of the compatibility conditions may be seen by imagining that the body is cut up into small volume elements, and then each element is given a certain strain. In general, the strained volume elements cannot be fitted back together to form a continuous body, unless the strain of each element is related to the strain of its neighbours according to the compatibility equations.

[2] The equations of compatibility were first derived by Saint-Venant in 1864.

[3] For a detailed analysis of this problem, see Washizu [374] and Malvern [227], p. 187.

Next, by (1.10), equations $(2.43)_2$ become after some calculation and rearranging terms

$$\Delta E_{kl} - E_{lm,km} - E_{km,lm} + (E_{mp,mp} - \Delta E_{mm})\,\delta_{kl} + E_{mm,kl} = 0.$$

By using this relation and considering also its trace

$$E_{mp,mp} - \Delta E_{mm} = 0,$$

we obtain

$$\Delta E_{kl} - E_{lm,km} - E_{km,lm} + E_{mm,kl} = 0. \tag{2.45}$$

As first shown by Beltrami in 1886, if the body is simply-connected [1], conditions (2.43) are also *sufficient* for the integrability of system (2.36), i.e. for the existence of a displacement field. More precisely, we have the following theorem.

**Theorem.** *Assume that the strain field* E *is single-valued, of class* $C^2$, *and satisfies equations* (2.43) *in a simply-connected region* $\mathscr{V}$. *Then there exist single-valued vector fields* $\mathbf{u}$ *of class* $C^3$ *and* $\boldsymbol{\omega}$ *of class* $C^2$ *that satisfy* (2.36), *and respectively* (2.41), *in* $\mathscr{V}$.

*Proof.* The reasoning proceeds along the lines of Cesaro [58] [2]. Let $P_0$ be a fixed point and $P$ a current point in $\mathscr{V}$ with position vectors $\mathbf{x}_0 = x_k^0\mathbf{e}_k$ and $\mathbf{x} = x_k\mathbf{e}_k$, respectively. From (2.41) it follows that

$$\omega_k(\mathbf{x}) = \omega_k^0 - \int_{P_0P} \epsilon_{krs} E_{mr,s}(\mathbf{y})\, \mathrm{d}y_m, \tag{2.46}$$

or, by (1.44),

$$\boldsymbol{\omega}(\mathbf{x}) = \boldsymbol{\omega}^0 - \int_{\mathbf{x}_0}^{\mathbf{x}} [\mathrm{curl}\ \mathbf{E}(\mathbf{y})]^T\, \mathrm{d}\mathbf{y}, \tag{2.47}$$

where $\boldsymbol{\omega}^0$ is a constant vector and $\mathbf{y} = y_m\mathbf{e}_m$ is the position vector of a current point on the integration path joining $P_0$ and $P$. The vector field $\boldsymbol{\omega}(\mathbf{x})$ is single-valued in $\mathscr{V}$ if the line integral in (2.47) is independent of the path in $\mathscr{V}$ from $\mathbf{x}_0$ to $\mathbf{x}$, or, equivalently, if it vanishes for every closed curve $L$ in the body, a condition which is always fulfilled in our case. Indeed, since $\mathscr{V}$ is simply-connected, there exists a surface $\Sigma$ bounded by $L$ and lying entirely in $\mathscr{V}$. Then, by applying Stokes, formula (1.55) and considering (2.44), we conclude that

$$\oint_L [\mathrm{curl}\ \mathbf{E}(\mathbf{y})]^T\, \mathrm{d}\mathbf{y} = -\int_{\Sigma} \{\mathrm{curl}\ [\mathrm{curl}\ \mathbf{E}(\mathbf{y})]^T\}\, \mathbf{n}\, \mathrm{d}s = 0, \tag{2.48}$$

where $\mathbf{n}$ is one of the unit normals to $\Sigma$, and the integration sense on $L$ is chosen clockwise when looking down along $\mathbf{n}$. Hence, the infinitesimal rotation vector $\boldsymbol{\omega}(\mathbf{x})$ is uniquely defined in $\mathscr{V}$; in particular, it follows from (2.47) that $\boldsymbol{\omega}^0 = \boldsymbol{\omega}(\mathbf{x}_0)$.

---

[1] An open region is said to be *simply-connected* if every closed curve in the region can be continuously deformed to a point without leaving the region; such curves will be called *reducible circuits*. A reducible circuit has the important property that there exists at least a surface bounded by the circuit and lying entirely in the region. Simply-connected regions are e.g. a solid sphere or a cube.

[2] In this connection, see also Volterra [373], Love [222], Sect. 156 A, Nabarro [258], Sect. 1.2, de Wit [486], Lurie [447], § 2, and Gurtin [150], Sect. 14.

In order to show that $\boldsymbol{\omega}(\mathbf{x})$ given by (2.47) satisfies (2.41) we will first prove a preliminary formula for the differentiation of a line integral. Let $\mathbf{U}(\mathbf{y}, \mathbf{x})$ be a second-order tensor field of class $C^1$ on $\mathscr{V}\times\mathscr{V}$. Then, for each fixed $m$, we have

$$\frac{\partial}{\partial x_m}\int_{\mathbf{x}_0}^{\mathbf{x}} \mathbf{U}(\mathbf{y}, \mathbf{x})\, d\mathbf{y} = \lim_{\alpha\to 0}\frac{1}{\alpha}\left\{\int_{\mathbf{x}_0}^{\mathbf{x}+\alpha\mathbf{e}_m} \mathbf{U}(\mathbf{y}, \mathbf{x}+\alpha\mathbf{e}_m)\, d\mathbf{y} - \int_{\mathbf{x}_0}^{\mathbf{x}} \mathbf{U}(\mathbf{y}, \mathbf{x})\, d\mathbf{y}\right\} =$$

$$= \frac{d}{d\alpha}\int_{\mathbf{x}_0}^{\mathbf{x}+\alpha\mathbf{e}_m} \mathbf{U}(\mathbf{y}, \mathbf{x}+\alpha\mathbf{e}_m)\, d\mathbf{y}\,\bigg|_{\alpha=0},$$

wherefrom it follows (Goursat [420], Sect. 94) that

$$\frac{\partial}{\partial x_m}\int_{\mathbf{x}_0}^{\mathbf{x}} \mathbf{U}(\mathbf{y}, \mathbf{x})\, d\mathbf{y} = \int_{\mathbf{x}_0}^{\mathbf{x}} \frac{\partial \mathbf{U}(\mathbf{y}, \mathbf{x})}{\partial x_m}\, d\mathbf{y} + \mathbf{U}(\mathbf{x}, \mathbf{x})\,\mathbf{e}_m. \tag{2.49}$$

Taking into account this formula, we obtain from (2.47)

$$\boldsymbol{\omega}_{,m}(\mathbf{x}) = -\,[\operatorname{curl}\ \mathbf{E}(\mathbf{x})]^T\mathbf{e}_m, \tag{2.50}$$

which coincides with (2.41) by virtue of (1.44). Clearly, $\boldsymbol{\omega}$ is of class $C^2$ since $\mathbf{E}$ is also of class $C^2$ in $\mathscr{V}$.

Next, $(2.40)_2$ yields

$$u_k(\mathbf{x}) = u_k^0 + \int_{P_0P} E_{km}(\mathbf{y})\, dy_m + \int_{P_0P} \epsilon_{krm}\omega_r(\mathbf{y})\, dy_m, \tag{2.51}$$

or

$$\mathbf{u}(\mathbf{x}) = \mathbf{u}^0 + \int_{\mathbf{x}_0}^{\mathbf{x}} \mathbf{E}(\mathbf{y})\, d\mathbf{y} + \int_{\mathbf{x}_0}^{\mathbf{x}} \boldsymbol{\omega}(\mathbf{y})\times d\mathbf{y}, \tag{2.52}$$

where $\mathbf{u}^0$ is a constant vector. Integrating by parts and taking into account (2.50) the last integral in (2.52) can be transformed as follows

$$\int_{\mathbf{x}_0}^{\mathbf{x}} \boldsymbol{\omega}(\mathbf{y})\times d\mathbf{y} = \int_{\mathbf{x}_0}^{\mathbf{x}} \boldsymbol{\omega}(\mathbf{y})\times d(\mathbf{y}-\mathbf{x}) = \int_{\mathbf{x}_0}^{\mathbf{x}} d[\boldsymbol{\omega}(\mathbf{y})\times(\mathbf{y}-\mathbf{x})] +$$

$$+\int_{\mathbf{x}_0}^{\mathbf{x}} (\mathbf{y}-\mathbf{x})\times d\boldsymbol{\omega}(\mathbf{y}) = \boldsymbol{\omega}^0\times(\mathbf{x}-\mathbf{x}_0) + \int_{\mathbf{x}_0}^{\mathbf{x}} (\mathbf{x}-\mathbf{y})\times\{[\operatorname{curl}\ \mathbf{E}(y)]^T\, d\mathbf{y}\}.$$

Substituting this result into (2.52) gives

$$\mathbf{u}(\mathbf{x}) = \mathbf{u}^0 + \boldsymbol{\omega}^0\times(\mathbf{x}-\mathbf{x}_0) + \int_{\mathbf{x}_0}^{\mathbf{x}} \mathbf{U}(\mathbf{y}, \mathbf{x})\, d\mathbf{y}, \tag{2.53}$$

where

$$\mathbf{U}(\mathbf{y}, \mathbf{x}) = \mathbf{E}(\mathbf{y}) + (\mathbf{x}-\mathbf{y})\times[\operatorname{curl}\ \mathbf{E}(\mathbf{y})]^T, \tag{2.54}$$

or, in component form,

$$U_{kr}(\mathbf{y}, \mathbf{x}) = E_{kr}(\mathbf{y}) + \epsilon_{kpq}(x_p - y_p)\epsilon_{qij}E_{ri,j}(\mathbf{y}) =$$

$$= E_{kr}(\mathbf{y}) + (x_p - y_p)\,[E_{rk,p}(\mathbf{y}) - E_{rp,k}(\mathbf{y})]\,. \tag{2.55}$$

In order to prove that (2.53) defines a single-valued vector field $\mathbf{u}(\mathbf{x})$, it is necessary and sufficient to show that the line integral in the right-hand side vanishes for every closed line in the region. Denote as above by $L$ a closed curve in $\mathscr{V}$ and let $\Sigma$ be a surface bounded by $L$ and lying entirely in $\mathscr{V}$. Then, by (1.55), (1.44), and (2.55), and taking into account (1.46), we successively find

$$\oint_L U_{kr}(\mathbf{y}, \mathbf{x})\, \mathrm{d}y_r = -\int_\Sigma \epsilon_{mrs} \frac{\partial U_{kr}(\mathbf{y}, \mathbf{x})}{\partial y_s} n_m \mathrm{d}s =$$

$$= \int_\Sigma \{-\epsilon_{mrs}E_{kr,s}(\mathbf{y}) + \epsilon_{mrs}\delta_{sp}[E_{rk,p}(\mathbf{y}) - E_{rp,k}(\mathbf{y})] -$$

$$-\epsilon_{kpq}(x_p - y_p)\epsilon_{qij}\epsilon_{mrs}E_{ri,js}(\mathbf{y})\}\, n_m \mathrm{d}s = \int_\Sigma \epsilon_{kpq}(x_p - y_p)\,(\mathrm{inc}\,\mathbf{E})_{qm} n_m\, \mathrm{d}s,$$

and hence, by $(2.43)_1$,

$$\oint_L \mathbf{U}(\mathbf{y}, \mathbf{x})\, \mathrm{d}\mathbf{y} = \int_\Sigma (\mathbf{x} - \mathbf{y}) \times \{[\mathrm{inc}\,\mathbf{E}(\mathbf{y})]\,\mathbf{n}\}\, \mathrm{d}s = \mathbf{0}. \tag{2.56}$$

Thus, $\mathbf{u}(\mathbf{x})$ is uniquely defined in $\mathscr{V}$; in particular, (2.53) yields $\mathbf{u}^0 = \mathbf{u}(\mathbf{x}_0)$. To show that $\mathbf{u}(\mathbf{x})$ given by (2.53) satisfies (2.36) we again apply (2.49) and obtain, in view of $(2.55)_1$ and (2.46),

$$u_{k,m}(\mathbf{x}) = E_{km}(\mathbf{x}) + \epsilon_{kmq}\left[\int_{\mathbf{x}_0}^{\mathbf{x}} \epsilon_{qij}E_{r,ij}(\mathbf{y})\, \mathrm{d}y_r - \omega_q^0\right] = E_{km}(\mathbf{x}) - \epsilon_{kmq}\omega_q(\mathbf{x}). \tag{2.57}$$

The symmetric part of this equation yields (2.36). Moreover, $\mathbf{u}$ is of class $C^3$ since both $\mathbf{E}$ and $\boldsymbol{\omega}$ are of class $C^2$ in $\mathscr{V}$ and this completes the proof.

Equations (2.47) and (2.53), which allow computation of an infinitesimal rotation field $\boldsymbol{\omega}(\mathbf{x})$ and of an infinitesimal displacement field $\mathbf{u}(\mathbf{x})$ corresponding to a given infinitesimal strain field $\mathbf{E}(\mathbf{x})$ in a simply-connected region, are called *Cesàro's formulae*.

Finally, we notice that the compatibility equations ensure only the *existence* of the fields $\mathbf{u}$ and $\boldsymbol{\omega}$, but not their *uniqueness*. Indeed, the displacement field is not unique, since we can always superimpose a rigid displacement field, which does not change the strains. Specifically, Cesàro's formulae show that if $\mathbf{u}^*(\mathbf{x})$ and $\boldsymbol{\omega}^*(\mathbf{x})$ are particular solutions of equations (2.36) and (2.41) then the general solutions of these equations are

$$\mathbf{u}(\mathbf{x}) = \mathbf{u}^*(\mathbf{x}) + \mathbf{u}^0 + \boldsymbol{\omega}^0 \times (\mathbf{x} - \mathbf{x}_0), \tag{2.58}$$

$$\boldsymbol{\omega}(\mathbf{x}) = \boldsymbol{\omega}^*(\mathbf{x}) + \boldsymbol{\omega}^0, \tag{2.59}$$

where $\mathbf{u}^0$ and $\boldsymbol{\omega}^0$ are arbitrary constant vectors, and $\mathbf{x}_0$ is the position vector of an arbitrary fixed point.

In the next subsection we shall consider the significance of the compatibility equations for multiply-connected bodies.

## 2.8. Volterra dislocations

Consider a multiply-connected body[1] occupying a region $\mathscr{V}$ and a single-valued infinitesimal strain field $\mathbf{E}$ of class $C^2$ in $\mathscr{V}$ that satisfies the compatibility conditions (2.43) in $\mathscr{V}$. We can still construct solutions of equations (2.36) and (2.41) by using Cesàro's formulae (2.47) and (2.53). However, when $L$ is an irreducible circuit we can no longer apply Stokes' formula to prove (2.50)

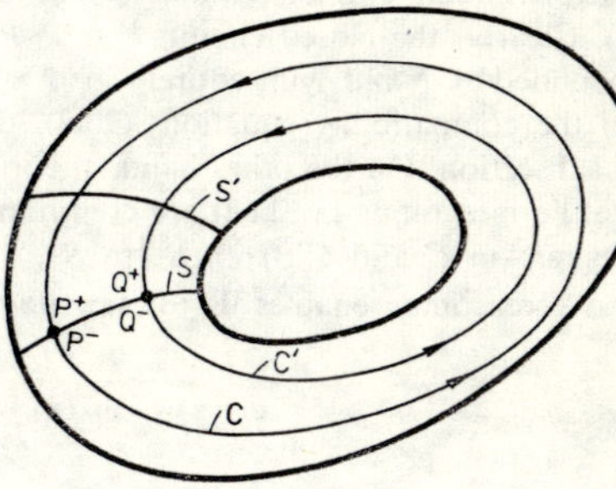


Fig. 2.3. Toroidal doubly-connected region. $S$ and $S'$ are cuts rendering the region simply-connected. $C$ and $C'$ denote irreducible circuits.

and (2.56). Consequently, the line integrals in (2.47) and (2.53) are not necessarily independent of path, and the fields $\boldsymbol{\omega}(\mathbf{x})$ and $\mathbf{u}(\mathbf{x})$ defined by these equations may be multiple-valued. For example, assume that $\mathscr{V}$ is the toroidal doubly-connected region shown in Fig. 2.3. Let $S$ be a two-sided barrier transforming $\mathscr{V}$ into a simply-connected region $\mathscr{V} \setminus S$. Arbitrarily choose a positive side $S^+$ and a negative side $S^-$ of the barrier and denote by $P^+$ and $P^-$ the points where an irreducible circuit $C$ intersects $S^+$ and $S^-$, respectively.

In the simply-connected region $\mathscr{V} \setminus S$ we can still apply the theorem proved in the preceding subsection to obtain *single-valued* vector fields $\boldsymbol{\omega}(\mathbf{x})$ and $\mathbf{u}(\mathbf{x})$. However, these fields may now be *discontinuous* across the barrier $S$. Indeed, by (2.47) and (2.53), we have

$$\boldsymbol{\omega}(P^+) - \boldsymbol{\omega}(P^-) = -\oint_C [\text{curl}\,\mathbf{E}(\mathbf{y})]^T\, d\mathbf{y}, \qquad \mathbf{u}(P^+) - \mathbf{u}(P^-) = \oint_C \mathbf{U}(\mathbf{y}, \mathbf{x})\, d\mathbf{y}, \tag{2.60}$$

where $\mathbf{x}$ denotes the common position vector of the points $P^+$ and $P^-$, and the integration sense on $C$ is taken from $P^-$ to $P^+$. In view of (2.54), equations (2.60) may be rewritten as

$$\boldsymbol{\omega}(P^+) - \boldsymbol{\omega}(P^-) = \mathbf{d}, \qquad \mathbf{u}(P^+) - \mathbf{u}(P^-) = \mathbf{b} + \mathbf{d} \times \mathbf{x}, \tag{2.61}$$

[1] An open region is said to be *multiply-connected* if it contains at least an *irreducible circuit*, i.e. a closed curve that cannot be contracted to a point without passing out of the region. Multiply-connected regions are for instance a torus or a hollow cylinder. A multiply-connected region can be reduced to a simply-connected one by means of a system of cuts or barriers. For example, the region between the bounding cylindrical surfaces of a hollow cylinder can be rendered simply-connected by a plane barrier passing through the axis of the cylinder and having that axis for an edge. If $n - 1$ simple non-intersecting cuts are necessary to transform a multiply-connected region into a simply-connected one, we say that $\mathscr{V}$ is *n-tuply connected.* Accordingly, the torus and the hollow cylinder are doubly-connected regions.

where **d** and **b** are constant vectors defined by

$$\mathbf{d} = -\oint_C [\text{curl}\ \mathbf{E}(\mathbf{y})]^T\, d\mathbf{y}, \qquad \mathbf{b} = \oint_C \{\mathbf{E}(\mathbf{y}) - \mathbf{y}\times[\text{curl}\ \mathbf{E}(\mathbf{y})]^T\}\, d\mathbf{y}. \tag{2.62}$$

It can be proved that **d** and **b** do not depend on the choice of $C$. Indeed, let $C'$ denote another irreducible circuit and let $Q^+$ and $Q^-$ be the points where $C'$ cuts $S^+$ and $S^-$, respectively (Fig. 2.3). Clearly, the closed circuit $\Gamma = P^-P^+Q^+Q^-P^-$ is reducible, and hence there exists a surface $\Sigma$ bounded by $\Gamma$ and lying entirely in $\mathscr{V}\setminus S$. Then, applying Stokes' formula (1.55) and making use of the compatibility equations (2.43), the line integrals in (2.62) vanish on $\Gamma$, like in the preceding subsection. On the other hand, the joint contribution of the paths $P^+Q^+$ and $Q^-P^-$ vanishes, since the integrands in (2.62) are continuous across $S$, and this implies the equality of the integrals taken on $C$ and $C'$ from $S^-$ to $S^+$.

The above reasoning enables us to rewrite (2.61) as

$$\boldsymbol{\omega}^+(\mathbf{x}) - \boldsymbol{\omega}^-(\mathbf{x}) = \mathbf{d} \quad \text{on } S, \tag{2.63}$$

$$\mathbf{u}^+(\mathbf{x}) - \mathbf{u}^-(\mathbf{x}) = \mathbf{b} + \mathbf{d}\times\mathbf{x} \quad \text{on } S, \tag{2.64}$$

where the superscripts + and − denote the limiting values taken by the corresponding fields on the positive and negative sides of the cut at an *arbitrary* point, whereas **d** and **b** are given by (2.62), where $C$ is an *arbitrary* irreducible closed circuit connecting the negative side of the cut with the positive one. Clearly, (2.64) implies that the jump of the displacement vector across $S$ is an infinitesimal rigid displacement consisting of a small translation of vector **b** and of a small rotation of vector **d** around the origin. Finally, it is worth noting that the jumps of $\boldsymbol{\omega}(\mathbf{x})$ and $\mathbf{u}(\mathbf{x})$ across the barrier $S$ do not depend on the choice of the barrier, for the integrands in (2.62) are continuous across the cuts that render $\mathscr{V}$ simply-connected. From this point of view the cuts $S$ and $S'$ shown in Fig. 2.3 are, therefore, equivalent.

The extension of the above results to regions with arbitrary connectivity is straightforward, leading to the following theorem, due to Weingarten [380].

**Weingarten's theorem.** *Let $\mathscr{V}$ be an n-tuply connected region and let $S_1, S_2, \ldots, S_{n-1}$ be $n-1$ non-intersecting barriers rendering $\mathscr{V}$ simply-connected. Assume that the infinitesimal strain field* E *is single-valued and of class $C^2$ in $\mathscr{V}$. Then the jump of the displacement vector across any barrier $S_k$, $k = 1, \ldots, n-1$, is a rigid displacement given by*

$$\mathbf{u}^+(\mathbf{x}) - \mathbf{u}^-(\mathbf{x}) = \mathbf{b}_k + \mathbf{d}_k \times \mathbf{x} \quad \text{on } S_k, \tag{2.65}$$

$$\mathbf{d}_k = -\oint_{C_k} [\text{curl}\ \mathbf{E}(\mathbf{y})]^T\, d\mathbf{y}, \qquad \mathbf{b}_k = \oint_{C_k} \{\mathbf{E}(\mathbf{y}) - \mathbf{y}\times[\text{curl}\ \mathbf{E}(\mathbf{y})]^T\}\, d\mathbf{y}, \tag{2.66}$$

*where* **x** *is the position vector of a current point on $S_k$, and $C_k$ is an arbitrary irreducible circuit intersecting only the barrier $S_k$ and oriented from the negative side to the positive side of $S_k$.*

Volterra [373] was the first[1] to consider displacement fields that are discontinuous across some surfaces although the corresponding infinitesimal strain fields are continuous together with their partial derivatives of first and second orders across these surfaces. From the theorem given in the preceding section it is apparent that such deformations, which are presently called Volterra dislocations, are impossible in a simply-connected region. Specifically, a deformation of an $n$-tuply connected body is said to be a *Volterra dislocation* when it has the following properties:

(i) if $S_k$, $k = 1, \ldots, n-1$, are $n-1$ regular and non-intersecting surfaces rendering simply-connected the region $\mathscr{V}$ occupied by the body, then $\mathbf{u}$ is discontinuous across these barriers;

(ii) the infinitesimal strain field $\mathbf{E}$ corresponding to $\mathbf{u}$ is continuous across the surfaces $S_k$ and the extension (by continuity) of $\mathbf{E}$ to $\mathscr{V}$ is of class $C^2$.

Clearly, by Weingarten's theorem, (ii) implies that the jump of $\mathbf{u}$ across any surface $S_k$ is a rigid displacement given by (2.65) and (2.66) in terms of $\mathbf{E}$. As already mentioned, these jumps are independent of the system of barriers chosen. Consequently, we can regard the displacement as either single-valued and discontinuous at the barriers in the simply-connected region $\mathscr{V} \setminus \bigcup_{k=1}^{n-1} S_k$, or as multiple-valued and of class $C^3$ in the multiply-connected region $\mathscr{V}$, supposed without barriers. In the latter case, the displacement vector may be again represented by (2.53). However, the line integral in (2.53) generally depends on the path in $\mathscr{V}$, and the multivaluedness of $\mathbf{u}$ is determined by the vectors $\mathbf{d}_k$ and $\mathbf{b}_k$, which now play the role of (vector) cycling constants[2]. From the discussion above it follows also that the compatibility equations are no longer sufficient for the existence of a *single-valued* displacement corresponding to an infinitesimal strain field $\mathbf{E}$ of class $C^2$ in a multiply-connected region $\mathscr{V}$. To assure the single-valuedness of $\mathbf{u}$ it is in fact necessary and sufficient to require the vanishing of the line integrals in (2.66) for $k = 1, \ldots, n-1$, together with the fulfilment of the compatibility conditions (2.43) in $\mathscr{V}$[3].

Volterra dislocations can describe real states of self-strain in multiply-connected bodies[4]. Indeed, assume that a body $\mathscr{B}$ occupying a multiply-connected region is rendered simply-connected by a system of non-intersecting cuts. If the two faces of each cut are given a small rigid relative displacement and then the continuity of the body is re-established by eventually adding or removing material and joining the faces of the cuts, the body will be again multiply-connected, but in a state of self-strain. As already pointed out, the position of the cuts is, to a great extent, immaterial. Thus, in a multiply-connected body which has suffered a Volterra dislocation, there is, in general, nothing to show the seat of the cuts.

# 3. Dynamics of deformable continuous media

## 3.1. Mass. Continuity equations

In non-relativistic mechanics one associates with each body $\mathscr{B}$ a positive scalar quantity $m(\mathscr{B})$ called the *mass* of $\mathscr{B}$, which is assumed as being constant throughout the motion. In continuum mechanics, however, we need supplementary concepts that are applicable to arbitrary small parts of a body. In particular, we assume that

---

[1] Weingarten [380] required only the continuity of the strain tensor across the barriers; dislocations of Weingarten's type are possible also in simply-connected regions, but under rather artificial restrictions on the admissible form of the cut faces (cf. also Pastori [458]). More general dislocations, which are possible in simply-connected bodies and correspond to more realistic mechanical conditions, will be discussed in Sects. 7.5. and 10.6.

[2] In this connection, see also Muskhelishvili [254], Sect. 15 and App. II, and Lurie [447], Sect. 2.4.

[3] For plane multiply-connected regions such conditions have been derived as early as 1900 by Michell [246] (cf. also Gurtin [150], Sect. 47).

[4] For the application of Volterra dislocations to the modelling of single crystal dislocations see Sect. 7.3.

the mass distribution $m$ in any configuration of the body is a positive scalar measure and an absolutely continuous and additive function of volume. Consequently, there exists a configuration-dependent *mass density* $\rho$, which is the ultimate ratio of mass to volume. More precisely, let $\mathscr{P}_n$ be a sequence of measurable parts of $\mathscr{B}$ having only one particle $X$ in common and such that $\lim_{n\to\infty} v(\mathscr{P}_n) = 0$, where $v(\mathscr{P}_n)$ denotes the volume of $\mathscr{P}_n$ in a given configuration $(k)$ of the body. Then, the mass density at $X$ in the configuration $(k)$ is defined by the relation

$$\rho(\mathbf{x}) \equiv \rho_{(k)}(X) = \lim_{n\to\infty} \frac{m(\mathscr{P}_n)}{v(\mathscr{P}_n)},$$

where $\mathbf{x}$ is the position vector of the place occupied by $X$ in the configuration $(k)$, the dependence of the mass density on the configuration being pointed out by the subscript $(k)$.

By hypothesis, the total mass of a system of bodies equals the sum of the masses of those bodies. Thus we can calculate the mass of a material volume $\mathscr{P}$ that occupies, respectively, the regions $\mathscr{V}_0$ and $\mathscr{V}$ in the reference configuration $(K)$ and in the current configuration $(k)$, by either of the formulae

$$m(\mathscr{P}) = \int_{\mathscr{V}_0} \rho_0 \mathrm{d}V = \int_{\mathscr{V}} \rho\, \mathrm{d}v, \tag{3.1}$$

where $\rho_0(\mathbf{X})$ and $\rho(\mathbf{x})$ are the corresponding mass densities. In (3.1), $\mathrm{d}V$ and $\mathrm{d}v$ designate the infinitesimal volume elements occupied by the same material neighbourhood of the particle $X$ in the configurations $(K)$ and $(k)$, respectively. Hence, by virtue of the *conservation of mass*, we have the relation

$$\rho_0 \mathrm{d}V = \rho \mathrm{d}v, \tag{3.2}$$

which may be also written, by (2.26) and (2.8), in the alternative forms

$$\rho_0 = \rho J = \rho \det \mathbf{F}. \tag{3.3}$$

By differentiating (3.3) with respect to $t$ and using the differentiation rule of a determinant, we obtain

$$\dot{\rho} + \rho \dot{F}_{km} F^{-1}_{mk} = 0. \tag{3.4}$$

On the other hand, differentiating (2.6) with respect to $t$ and considering (2.2) and (2.9) gives

$$\mathrm{d}v_k = \dot{F}_{km} \mathrm{d}X_m = \dot{F}_{km} F^{-1}_{mp} \mathrm{d}x_p,$$

wherefrom it follows that

$$\operatorname{grad} \mathbf{v} = \dot{\mathbf{F}}\mathbf{F}^{-1},$$

and hence

$$\dot{F}_{km} F^{-1}_{mk} = \operatorname{tr}(\dot{\mathbf{F}}\mathbf{F}^{-1}) = \operatorname{tr}(\operatorname{grad} \mathbf{v}) = \operatorname{div} \mathbf{v}.$$

Substituting this result into (3.4), we obtain

$$\dot{\rho} + \rho \operatorname{div} \mathbf{v} = 0. \tag{3.5}$$

Equations (3.2), (3.3), and (3.5), which express mathematically the law of mass conservation of a continuous body, are called *continuity equations.*

Finally, we notice for further use that if $\mathscr{V}_0$ and $\mathscr{V}$ denote as above the regions occupied by a *material* volume $\mathscr{P}$ in the reference configuration $(K)$ and the current configuration $(k)$, respectively, then, by virtue of (3.2), we may write

$$\frac{\mathrm{d}}{\mathrm{d}t}\int_{\mathscr{V}} \rho \boldsymbol{\Phi}\,\mathrm{d}v = \frac{\mathrm{d}}{\mathrm{d}t}\int_{\mathscr{V}_0} \rho_0 \boldsymbol{\Phi}\,\mathrm{d}V = \int_{\mathscr{V}_0} \rho_0 \dot{\boldsymbol{\Phi}}\,\mathrm{d}V = \int_{\mathscr{V}} \rho \dot{\boldsymbol{\Phi}}\,\mathrm{d}v, \tag{3.6}$$

where $\boldsymbol{\Phi}$ designates any continuously differentiable scalar, vector, or tensor field. Clearly, in (3.6), $\boldsymbol{\Phi}$ is considered alternatively as a function of $\mathbf{X}$ or $\mathbf{x}$ according as the integral is taken over $\mathscr{V}_0$ or $\mathscr{V}$.

## 3.2. Forces and stresses. Principles of continuum mechanics

In continuum mechanics the concept of force describes the interaction between different bodies or between different parts of the same body. We assume that the force $\mathbf{f}(\mathscr{B})$ exerted by the outside world on a body $\mathscr{B}$ in the current configuration $(k)$ consists of *body forces*, which act on the elements of volume or mass inside the body, and *surface forces*, which are contact forces acting on the boundary of the body. More precisely, we suppose that $\mathbf{f}(\mathscr{B})$ may be written in the form

$$\mathbf{f}(\mathscr{B}) = \int_{\mathscr{V}} \rho \mathbf{b}\,\mathrm{d}v + \int_{\mathscr{S}} \mathbf{t}\,\mathrm{d}s, \tag{3.7}$$

where $\mathscr{V}$ is the region occupied by $\mathscr{B}$ in the configuration $(k)$, $\mathscr{S}$ is the boundary of $\mathscr{V}$, $\mathbf{b}$ is the *body force per unit mass*, and $\mathbf{t}$ is the *surface force per unit area*, or *surface traction.*

Furthermore, we assume that there are neither body couples nor surface couples acting on the body. Consequently, the resultant moment of the forces exerted on $\mathscr{B}$ with respect to the origin is

$$\mathbf{m}_0(\mathscr{B}) = \int_{\mathscr{V}} \rho \mathbf{x} \times \mathbf{b}\,\mathrm{d}v + \int_{\mathscr{S}} \mathbf{x} \times \mathbf{t}\,\mathrm{d}s, \tag{3.8}$$

where $\mathbf{x}$ denotes as before the position vector of the particle $X$ in the configuration $(k)$.

The principles of continuum mechanics have been formulated by Euler as early as 1775 in the form of two integral balance equations, which generalize the corresponding principles of the dynamics of particle systems, namely the *balance equation of the momentum*

$$\frac{\mathrm{d}}{\mathrm{d}t}\int_{\mathscr{V}} \rho \mathbf{v}\,\mathrm{d}v = \mathbf{f}(\mathscr{B}) \tag{3.9}$$

and the *balance equation of the moment of momentum*

$$\frac{\mathrm{d}}{\mathrm{d}t}\int_{\mathscr{V}} \rho \mathbf{x} \times \mathbf{v}\,\mathrm{d}v = \mathbf{m}_0(\mathscr{B}). \tag{3.10}$$

Obviously, the left-hand sides of (3.9) and (3.10) are the rates of the total momentum and moment of momentum, respectively. Substituting (3.7) into (3.9), (3.8) into (3.10), and making use of (3.6), we obtain

$$\int_{\mathscr{V}} \rho \dot{\mathbf{v}}\,\mathrm{d}v = \int_{\mathscr{V}} \rho \mathbf{b}\,\mathrm{d}v + \int_{\mathscr{S}} \mathbf{t}\,\mathrm{d}s, \tag{3.11}$$

$$\int_{\mathscr{V}} \rho \mathbf{x} \times \dot{\mathbf{v}}\,\mathrm{d}v = \int_{\mathscr{V}} \rho \mathbf{x} \times \mathbf{b}\,\mathrm{d}v + \int_{\mathscr{S}} \mathbf{x} \times \mathbf{t}\,\mathrm{d}s. \tag{3.12}$$

Besides the *external force* $\mathbf{f}(\mathscr{B})$, there exist interactions between different parts of the body, resulting from the atomic or molecular interactions. To describe these *internal forces*, Cauchy made in 1822 the basic hypothesis that the interaction between two arbitrary parts of a body that have a common boundary may be replaced by a continuous distribution of surface forces acting on both sides of the common boundary. These forces referred per unit area, which have by hypothesis the same nature as the externai surface forces, are called *stress vectors*; they will be denoted by the same symbol $\mathbf{t}$ as the external surface forces. Accordingly, the balance equations (3.9) and (3.10) may be applied to *any* part $\mathscr{P}$ of the body.

Cauchy's assumption, which is also known as the *stress principle*, plays a fundamental role in continuum mechanics, because it allows the unified description of the internal forces, irrespective of the peculiar atomic structure of the body. To better understand its simplifying character, assume that a part $\mathscr{P}$ of $\mathscr{B}$ is cut out of the body; then, according to the stress principle, the action of the rest of the body on $\mathscr{P}$ could be replaced by surface forces acting on the boundary of $\mathscr{P}$ (Fig. 3.1). However, this is certainly not true whenever the action range $l$ of the internal forces, although small, cannot be neglected, e.g. in regions of high strain gradients. In such cases the concepts of boundary and stress vectors should be reconsidered. Namely, the boundary of $\mathscr{P}$ should be replaced by a "shell" or "boundary layer" of thickness $l$, containing a distribution of supplementary body forces that are necessary for preserving the form of $\mathscr{P}$. Clearly, this would result in a continuum non-local theory which stands closer to the lattice theory of crystals than the local theory considered in this chapter [1].

Let us explore now in more detail the consequences of the stress principle. We begin by noting that the stress vector depends not only on $\mathbf{x}$ but also on the orientation of the separation surface, i.e.[2]

$$\mathbf{t} = \mathbf{t}(\mathbf{x}, \mathbf{n}). \tag{3.13}$$

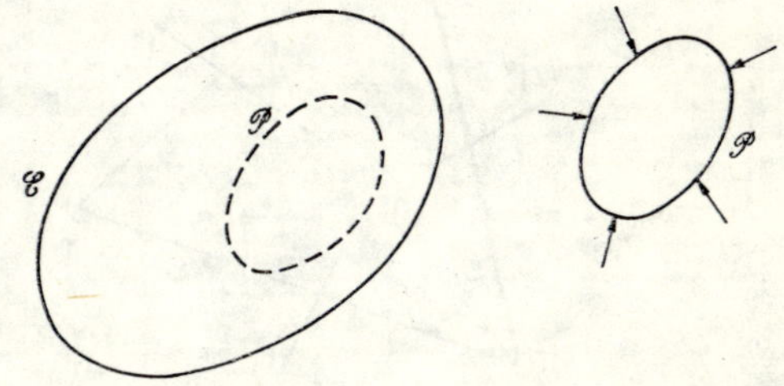


Fig. 3.1. On the definition of the stress vectors.

The set of all stress vectors $\mathbf{t}(\mathbf{x}, \mathbf{n})$ for $\mathbf{x}$ fixed and all $\mathbf{n}$ is called the *stress state around the point* $\mathbf{x}$.

To establish the dependence of $\mathbf{t}$ on $\mathbf{n}$, we apply the balance equation (3.11) to a tetrahedron which has three mutually orthogonal faces parallel to the co-ordinate planes of a Cartesian system of co-ordinates and intersecting at $\mathbf{x}$, and a fourth face with unit outward normal $\mathbf{n}$ (Fig. 3.2). Let $h$ be the height of the tetrahedron and $S$ the area of the oblique face $P_1P_2P_3$. Then the areas of the orthogonal faces are $Sn_1$, $Sn_2$, and $Sn_3$, respectively. Assume that $\rho\dot{\mathbf{v}}$ and $\rho\mathbf{b}$ are bounded and that $\mathbf{t}$ is a continuous function of both $\mathbf{x}$ and $\mathbf{n}$. Then, by the mean-value theorem of the integral calculus, we deduce from (3.11) that

$$S(n_1\mathbf{t}_1^* + n_2\mathbf{t}_2^* + n_3\mathbf{t}_3^* + \mathbf{t}^*) + \frac{1}{3}\, hS\rho\mathbf{K} = \mathbf{0}, \tag{3.14}$$

where $\mathbf{K}$ is a constant vector depending on the evaluation of the volume integrals in (3.11), and $\mathbf{t}_1^*$, $\mathbf{t}_2^*$, $\mathbf{t}_3^*$, and $\mathbf{t}^*$ are stress vectors applied in certain points of the corresponding faces of the tetrahedron. Dividing through by $S$ in (3.14) and letting $h \to 0$ for fixed $\mathbf{n}$, it follows that

$$\mathbf{t}(\mathbf{x}, \mathbf{n}) = -(\mathbf{t}_1 n_1 + \mathbf{t}_2 n_2 + \mathbf{t}_3 n_3), \tag{3.15}$$

where now the stress vectors $\mathbf{t}_1$, $\mathbf{t}_2$, $\mathbf{t}_3$, and $\mathbf{t}$ are calculated at $\mathbf{x}$. Equation (3.15) shows that the stress state around a point $\mathbf{x}$ of the body is completely determined by the stress vectors acting on three mutually orthogonal planes intersecting at $\mathbf{x}$.

---

[1] A first step towards such a continuum non-local theory can be taken by replacing the supplementary body forces by additional surface forces *and* double forces acting on the boundary of $\mathscr{P}$, as it is done for instance in the so-called theory of the materials of grade two. For the modelling of crystal defects using this more general approach see Teodosiu [332–334].

[2] For convenience we shall suppress the argument $t$ in what follows.

If now $P_1 \to P$, $n_1 \to 1$, and $n_2, n_3 \to 0$, then, by the continuity of **t** with respect to **n**, we have $\mathbf{t}_1 = \mathbf{t}(\mathbf{x}, -\mathbf{n})$, since the unit outward normal $-\mathbf{e}_1$ to $S_1$ coincides in the limit to $-\mathbf{n}$. Substituting the limiting values into (3.15) gives

$$\mathbf{t}(\mathbf{x}, -\mathbf{n}) = -\mathbf{t}(\mathbf{x}, \mathbf{n}). \tag{3.16}$$

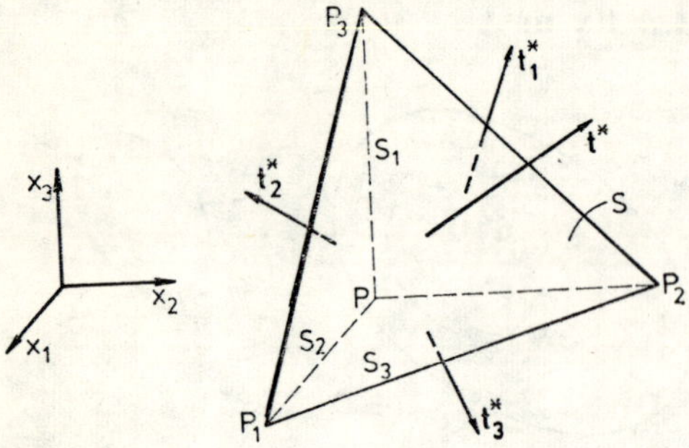


Fig. 3.2. Surface forces acting on a tetrahedron.

Clearly, this result does not depend on the limiting approach considered since the direction of the $x_1$-axis can be chosen arbitrarily. According to (3.16), the stress vectors acting at the same place on the two sides of an internal surface have the same magnitude, but are oppositely directed.

Let us choose now a Cartesian frame with unit vectors $\mathbf{e}_1, \mathbf{e}_2, \mathbf{e}_3$, and denote by $T_{km}$ the component parallel to the $x_k$-axis of the stress vector $-\mathbf{t}_m$, which acts on the positive side of the plane $x_m = \text{const}$ passing through **x** (Fig. 3.3), i.e.

$$\mathbf{t}_m(\mathbf{x}) = -T_{km}(\mathbf{x})\,\mathbf{e}_k. \tag{3.17}$$

Substituting (3.17) into (3.15), we obtain

$$t_k(\mathbf{x}, \mathbf{n}) = T_{km}(\mathbf{x})\,n_m. \tag{3.18}$$

Fig. 3.3. Positive stress components acting on the faces of a rectangular parallelepiped whose edges are parallel to the co-ordinate axes.

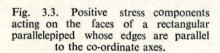

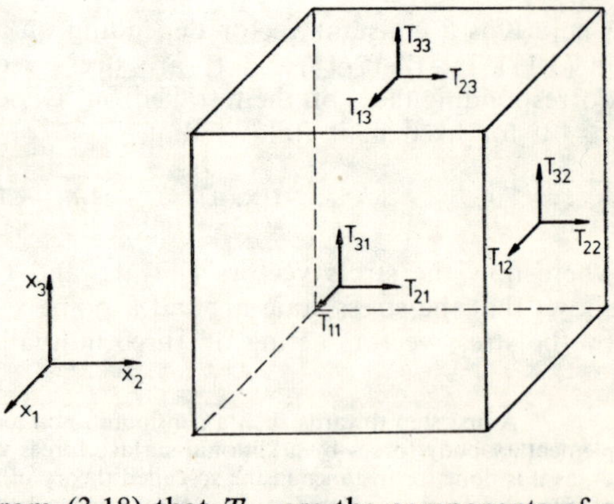


Since **t** and **n** are vectors, it is seen from (3.18) that $T_{km}$ are the components of a second-order tensor field **T**, which is called the *Cauchy stress tensor*. Moreover, as (3.18) is a tensor equation, it holds for any system of co-ordinates. In direct notation it reads

$$\mathbf{t}(\mathbf{x}, \mathbf{n}) = \mathbf{T}(\mathbf{x})\,\mathbf{n}, \tag{3.19}$$

showing that the stress tensor $\mathbf{T}(\mathbf{x})$ is the second-order tensor field (linear vector function) that assigns to each unit vector $\mathbf{n}$ the stress vector acting at $\mathbf{x}$ across the surface whose unit normal is $\mathbf{n}$.

It is important to note that when $\mathbf{t}(\mathbf{x}, \mathbf{n})$ is a prescribed traction on the boundary of the body, equation (3.19) represents a *traction boundary condition* for the determination of the stress tensor.

It can be also seen that the first subscript of $T_{km}$ identifies the component of the stress vector, while the second subscript identifies the plane [1]. The diagonal components ($k = m$) of the stress tensor are called *normal stresses*, while the off-diagonal components ($k \neq m$) are called *shear stresses*.

The components $T_{km}$ are considered positive or negative according as they have the directions indicated in Fig. 3.3 or opposite to them. In any case, by virtue of (3.16), the stress components acting on the negative sides of the rectangular parallelepiped shown in Fig. 3.3 will have senses opposite to those on the positive sides. It is easily seen that positive and negative normal stresses correspond to tensile and compressive tractions, respectively, whereas the algebraic sign of a shear stress has no intrinsic physical meaning.

Introducing now (3.19) into (3.11), we obtain the relation

$$\int_{\mathscr{V}} \rho \dot{\mathbf{v}}\, \mathrm{d}v = \int_{\mathscr{V}} \rho \mathbf{b}\, \mathrm{d}v + \int_{\mathscr{S}} \mathbf{T}\mathbf{n}\, \mathrm{d}s, \tag{3.20}$$

which, by making use of (2.3) and (1.53), may be rewritten as

$$\int_{\mathscr{V}} (\operatorname{div} \mathbf{T} + \rho \mathbf{b} - \rho \mathbf{a})\, \mathrm{d}v = \mathbf{0}. \tag{3.21}$$

Assuming that the integrand is a continuous function and recalling that $\mathscr{V}$ can be any material volume of $\mathscr{B}$, we deduce that (3.21) is equivalent to

$$\operatorname{div} \mathbf{T} + \rho \mathbf{b} = \rho \mathbf{a}, \qquad \frac{\partial T_{km}}{\partial x_m} + \rho b_k = \rho a_k. \tag{3.22}$$

This equation, which represents the local form of the balance equation of momentum, is called *Cauchy's first law of motion*.

Next, introducing (3.19) into (3.12), we have

$$\int_{\mathscr{V}} \rho \mathbf{x} \times \dot{\mathbf{v}}\, \mathrm{d}v = \int_{\mathscr{V}} \rho \mathbf{x} \times \mathbf{b}\, \mathrm{d}v + \int_{\mathscr{S}} \mathbf{x} \times (\mathbf{T}\mathbf{n})\, \mathrm{d}s. \tag{3.23}$$

---

[1] Some authors, e.g. Malvern [227], reverse this convention, using the first subscript for the vector component and the second subscript for the plane. Their stress tensor is then the transpose of the one defined here.

By making use again of the divergence theorem (1.53), the last integral in the right-hand side may be transformed as follows

$$\int_{\mathscr{S}} \mathbf{x} \times (\mathbf{Tn})\, ds = \mathbf{e}_k \int_{\mathscr{S}} \epsilon_{klm} x_l T_{mp}\, n_p\, ds = \mathbf{e}_k \int_{\mathscr{V}} \frac{\partial}{\partial x_p} (\epsilon_{klm} x_l\, T_{mp})\, dv =$$

$$= \mathbf{e}_k \int_{\mathscr{V}} \epsilon_{klm} T_{ml}\, dv + \int_{\mathscr{V}} (\mathbf{x} \times \operatorname{div} \mathbf{T})\, dv.$$

Substituting this result into (3.22) and rearranging terms, we find

$$\int_{\mathscr{V}} \mathbf{x} \times (\operatorname{div} \mathbf{T} + \rho \mathbf{b} - \rho \dot{\mathbf{v}})\, dv = \mathbf{e}_k \int_{\mathscr{V}} \epsilon_{klm} T_{ml}\, dv = 0. \tag{3.24}$$

By $(3.21)_1$ the first integral vanishes and hence, since $\mathscr{V}$ is arbitrary, we deduce that (3.24) is equivalent to

$$\epsilon_{klm} T_{ml} = 0. \tag{3.25}$$

Multiplying (3.25) by $\epsilon_{krs}$, summing with respect to $k$, and taking into account the identity (1.11), we finally obtain

$$\mathbf{T} = \mathbf{T}^T, \qquad T_{rs} = T_{sr}. \tag{3.26}$$

Thus, the local form (3.26) of the balance equation of the moment of momentum, which is called *Cauchy's second law of motion*, is equivalent to the assertion that. the stress tensor is symmetric, in the absence of body couples and couple stresses

### 3.3. The Piola-Kirchhoff stress tensors

We have seen that the stress vector $\mathbf{t}(\mathbf{x}, \mathbf{n})$ is the internal surface force acting at $\mathbf{x}$ per unit deformed area in the current configuration ($k$) across a surface with unit normal $\mathbf{n}$. Therefore, both $\mathbf{t}$ and the associated Cauchy stress tensor $\mathbf{T}$ are adequate for the spatial description. On the other hand, in non-linear elasticity theory, it is often convenient to use the material description, in order to solve problems in which the initial boundary of a body is deformed in a prescribed way, or the tractions keep their initial direction and magnitude per unit undeformed area in the reference configuration. There are two alternatives for such a material description, leading to the introduction of the so-called Piola-Kirchhoff stress tensors.

The first *Piola-Kirchhoff stress tensor*, $\mathbf{S}$, is defined by

$$\mathbf{SN} dS = \mathbf{t}\, ds = \mathbf{Tn}\, ds. \tag{3.27}$$

The first of these relations may be rewritten as

$$\mathbf{s} = \mathbf{SN}, \tag{3.28}$$

where $\mathbf{s} = \mathbf{t}\, \mathrm{d}s/\mathrm{d}S$. Hence, the first Piola-Kirchhoff stress tensor is the second-order tensor (linear vector function) that assigns to the unit normal $\mathbf{N}$ of the undeformed element of surface the actual force $\mathbf{t}\,\mathrm{d}s$ on the deformed surface element, but reckoned per unit undeformed area $\mathrm{d}S$.

With the aid of (2.25), we deduce from (3.27) that the stress tensors $\mathbf{T}$ and $\mathbf{S}$ are related by

$$\mathbf{T} = j\mathbf{S}\mathbf{F}^T, \qquad T_{km} = j\, S_{kp} F_{mp}, \tag{3.29}$$

where $j = J^{-1}$. Substituting $(3.29)_2$ into Cauchy's first law of motion $(3.22)_2$ leads to

$$\frac{\partial}{\partial x_m}(j\, S_{kp} F_{mp}) + \rho b_k = \rho a_k. \tag{3.30}$$

On the other hand, using the differentiation rule of a determinant and the chain rule of differentiation, and considering (2.7), (2.8), and (2.10), we successively obtain

$$\begin{aligned}\frac{\partial}{\partial x_m}(j\, F_{mp}) &= \frac{\partial j}{\partial x_m}\frac{\partial x_m}{\partial X_p} + j\frac{\partial F_{mp}}{\partial x_m} = \frac{\partial J^{-1}}{\partial X_p} + j\frac{\partial F_{mp}}{\partial X_k}\frac{\partial X_k}{\partial x_m} = \\ &= -\frac{1}{J}\frac{\partial F_{mk}}{\partial X_p} F^{-1}_{km} + j\frac{\partial F_{mp}}{\partial X_k} F^{-1}_{km} = 0,\end{aligned} \tag{3.31}$$

and hence, by (2.7) and (3.3),

$$\frac{\partial}{\partial x_m}(j S_{kp}\, F_{mp}) = j\, F_{mp}\frac{\partial S_{kp}}{\partial x_m} = \frac{\rho}{\rho_0}\frac{\partial S_{kp}}{\partial x_m}\frac{\partial x_m}{\partial X_p} = \frac{\rho}{\rho_0}\frac{\partial S_{kp}}{\partial X_p}.$$

Putting this result into (3.30), we arrive at the simplified equation

$$\operatorname{Div}\mathbf{S} + \rho_0\mathbf{b} = \rho_0\mathbf{a}, \qquad \frac{\partial S_{km}}{\partial X_m} + \rho_0 b_k = \rho_0 a_k. \tag{3.32}$$

Thus, Cauchy's first law of motion preserves its form when passing to material co-ordinates, provided that $\mathbf{T}$, $\rho$, and div be replaced by $\mathbf{S}$, $\rho_0$, and Div, respectively. On the contrary, Cauchy's second law of motion (3.26) combined with (3.29) leads to the relation

$$\mathbf{S}\mathbf{F}^T = \mathbf{F}\mathbf{S}^T, \tag{3.33}$$

which shows that the first Piola-Kirchhoff stress tensor is not, in general, a symmetric tensor. By introducing the *second Piola-Kirchhoff stress tensor*, $\mathbf{\Pi}$, defined by the relations

$$\mathbf{\Pi} = \mathbf{F}^{-1}\mathbf{S} = J\mathbf{F}^{-1}\,\mathbf{T}(\mathbf{F}^{-1})^T, \tag{3.34}$$

this difficulty is removed, since now (3.26) and (3.34) give

$$\mathbf{\Pi} = \mathbf{\Pi}^T, \tag{3.35}$$

showing that $\mathbf{\Pi}$ is symmetric whenever $\mathbf{T}$ is symmetric, i.e. in the absence of body couples and couple stresses. In exchange, Cauchy's first law of motion (3.32) assumes in terms of $\mathbf{\Pi}$ the more complicated form

$$\mathrm{Div}\,(\mathbf{F\Pi}) + \rho_0 \mathbf{b} = \rho_0 \mathbf{a}, \qquad \frac{\partial}{\partial X_m}(F_{kp}\Pi_{pm}) + \rho_0 b_k = \rho_0 a_k. \tag{3.36}$$

## 4. Thermodynamics of elastic deformation. Constitutive equations

For our purposes in continuum mechanics we shall always use as *thermodynamic system* a *closed system*, i.e. a part of the material universe not interchanging matter with its surroundings. Moreover, we assume that the only energy transfers to the system are by *mechanical work* done on the system by surface tractions and body forces, by *heat transfer* through the boundary, and possibly by distributed internal heat sources.

A *thermodynamic state variable* is any macroscopic quantity which characterizes the system, e.g. the temperature or the strain tensor. The set of the instantaneous values of all state variables at a given time is called the *thermodynamic state* of the system at that time. Clearly, the selection of the state variables depends on and implies a certain idealization of the system and of its evolution.

A thermodynamic state variable of a homogeneous system is called *extensive* or *intensive* according as it is proportional to or independent of the mass of the system. The density per unit mass of an extensive state variable is obviously an intensive one.

The passing of a system from a thermodynamic state into another one is called a *thermodynamic process*. A thermodynamic process is said to be *reversible* or *irreversible* according as the time-reversal of the external actions exerted on the system leads or not to a reversal of the process. A thermodynamic process is called *cyclic* if the final and initial states of the system carried through the process coincide.

### 4.1. The first law of thermodynamics

The first law of thermodynamics relates the work done on the system and the heat transfer into the system to the change in energy of the system.

Let us take as thermodynamic system an arbitrary part $\mathscr{P}$ of a body $\mathscr{B}$ and consider its evolution during the time interval $[t_0, t_M]$, where $t_0$ is the time at which the body occupied the reference configuration. Denote as usual by $\mathscr{V}$ the region occupied by $\mathscr{P}$ in the current configuration $(k)$ at time $t \in [t_0, t_M]$ and by $\mathscr{S}$ the boundary of $\mathscr{V}$.

The mechanical power input $P$ resulting from the surface tractions $\mathbf{t}$ and the body forces $\mathbf{b}$ acting on $\mathscr{P}$ at time $t$ is given by

$$P = \int_{\mathscr{S}} \mathbf{t}\cdot\mathbf{v}\,\mathrm{d}s + \int_{\mathscr{V}} \rho\mathbf{b}\cdot\mathbf{v}\,\mathrm{d}v. \tag{4.1}$$

The *heat input rate* $Q$ may be written in the analogous form

$$Q = \int_{\mathscr{S}} q \, \mathrm{d}s + \int_{\mathscr{V}} \rho r \, \mathrm{d}v, \tag{4.2}$$

where $q = q(\mathbf{x}, \mathbf{n}, t)$ is the *heat flux* input resulting from conduction through the surface $\mathscr{S}$ and measured per unit surface area, and $r = r(\mathbf{x}, t)$ is the *heat supply* per unit mass in $\mathscr{B}$ from the external world (possibly from a radiation field).

Generalizing from many experimental observations, it is found that

$$\oint (P + Q) \, \mathrm{d}t = 0 \tag{4.3}$$

for all *cyclic* thermodynamic processes. In addition, it is found that the integral $\int (P + Q) \, \mathrm{d}t$ calculated for any process carrying a homogeneous system is proportional to the total mass of the system. These experimental results imply the existence of an extensive state variable $E$, called the *total energy* of the system, such that

$$\dot{E} = P + Q. \tag{4.4}$$

Equation (4.4), which is named the *energy balance equation*, represents the mathematical form of the *first law of thermodynamics*.

The difference

$$U = E - K \tag{4.5}$$

between the total energy of the system and its *kinetic energy*

$$K = \frac{1}{2} \int_{\mathscr{V}} \rho \mathbf{v}^2 \, \mathrm{d}v \tag{4.6}$$

is called the *internal energy* of the system. Since $E$ and $K$ are extensive variables, $U$ must be an extensive variable, too. Consequently, we can write

$$U = \int_{\mathscr{V}} \rho \varepsilon \, \mathrm{d}v \tag{4.7}$$

where $\varepsilon$ is the *specific internal energy* per unit mass.

Introducing (4.1), (4.2), and (4.5–7) into (4.4), we obtain

$$\frac{\mathrm{d}}{\mathrm{d}t} \int_{\mathscr{V}} \left(\frac{1}{2} \mathbf{v}^2 + \varepsilon\right) \rho \, \mathrm{d}v = \int_{\mathscr{S}} (\mathbf{t} \cdot \mathbf{v} + q) \, \mathrm{d}s + \int_{\mathscr{V}} (\mathbf{b} \cdot \mathbf{v} + r) \, \rho \, \mathrm{d}v. \tag{4.8}$$

By applying (4.8) to a tetrahedron included in $\mathscr{V}$ which has three sides intersecting at $\mathbf{x}$ and parallel to the co-ordinate planes of a Cartesian frame and using a reasoning similar to that leading to the existence of the stress tensor, it can be shown that there exists a vector $\mathbf{q}(\mathbf{x}, t)$, called the *heat flux vector*, such that

$$q(\mathbf{x}, \mathbf{n}, t) = -\,\mathbf{q}(\mathbf{x}, t)\cdot\mathbf{n}. \tag{4.9}$$

The negative sign is needed because $\mathbf{q}\cdot\mathbf{n}$ is the *outward* heat flux per unit area of $\mathscr{S}$, whereas $q$ is the heat flux *input*.

On the other hand, by virtue of (3.19), we have

$$\mathbf{t}\cdot\mathbf{v} = (\mathbf{T}\mathbf{n})\cdot\mathbf{v} = T_{km}n_m v_k = (\mathbf{T}^T\mathbf{v})\cdot\mathbf{n}. \tag{4.10}$$

Substituting now (4.9) and (4.10) into (4.8) and making use of (1.52) and (3.6) leads to

$$\int_{\mathscr{V}} [\rho\mathbf{v}\cdot\mathbf{a} + \dot{\varepsilon} - \operatorname{div}(\mathbf{T}^T\mathbf{v}) + \operatorname{div}\mathbf{q} - \rho\,(\mathbf{b}\cdot\mathbf{v} + r)]\,dv = 0. \tag{4.11}$$

Remembering that $\mathscr{V}$ is arbitrary and assuming that the integrand in the left-hand side of (4.11) is a continuous function on $\mathscr{B}$, we deduce that (4.11) is equivalent to

$$\rho(\mathbf{v}\cdot\mathbf{a} + \dot{\varepsilon}) = \operatorname{div}(\mathbf{T}^T\mathbf{v}) - \operatorname{div}\mathbf{q} + \rho(\mathbf{b}\cdot\mathbf{v} + r). \tag{4.12}$$

+)

But

$$\operatorname{div}(\mathbf{T}^T\mathbf{v}) = \frac{\partial}{\partial x_m}(T_{km}v_k) = v_k\frac{\partial T_{km}}{\partial x_m} + T_{km}\frac{\partial v_k}{\partial x_m} = \mathbf{v}\cdot\operatorname{div}\mathbf{T} + \mathbf{T}\cdot\operatorname{grad}\mathbf{v},$$

and hence, considering also (3.22), equation (4.12) reduces to

$$\rho\dot{\varepsilon} = \mathbf{T}\cdot\operatorname{grad}\mathbf{v} - \operatorname{div}\mathbf{q} + \rho r. \tag{4.13}$$

This relation represents the *local form of the energy balance equation.*

### 4.2. The second law of thermodynamics

The first law of thermodynamics can be regarded as an expression of the interconvertibility of heat and work, provided that the total energy of the system remains constant. Therefore, this principle places no restriction on the direction of thermodynamic processes. On the contrary, the *second law of thermodynamics* introduces a severe discrimination between reversible and irreversible processes. According to this law there exists an extensive state variable $S$, called the *entropy* of the system, which satisfies the relation

$$\dot{S} \geqslant -\int_{\mathscr{S}} \frac{\mathbf{q}\cdot\mathbf{n}}{\theta}\,ds + \int_{\mathscr{V}} \frac{\rho r}{\theta}\,dv, \tag{4.14}$$

+) anschaulicher als (4.13)!

where $\theta(\mathbf{x}, t) > 0$ is the *absolute temperature*, the equality being valid for *reversible* processes and the inequality for *irreversible* processes.

The first integral in the right-hand side of (4.14) is the total entropy flux across $\mathscr{S}$ due to conduction, whereas the second integral is the entropy supplied per unit time into the interior of $\mathscr{P}$ from the external world (possibly from a radiation field). According to the inequality in (4.14), the rate of entropy increase is greater than the entropy input rate, thus implying internal entropy production in an irreversible process.

Denoting by $\eta(\mathbf{x}, t)$ the *specific entropy* per unit mass, we can write in view of (3.6)

$$\dot{S} = \frac{\mathrm{d}}{\mathrm{d}t} \int_{\mathscr{V}} \rho\eta \, \mathrm{d}v = \int_{\mathscr{V}} \rho\dot{\eta} \, \mathrm{d}v. \tag{4.15}$$

Substituting (4.15) into (4.14) and using the divergence formula (1.52) to transform the surface integral into a volume integral, we obtain

$$\int_{\mathscr{V}} \left[ \rho\dot{\eta} + \operatorname{div}\left(\frac{\mathbf{q}}{\theta}\right) - \frac{\rho r}{\theta} \right] \mathrm{d}v \geqslant 0. \tag{4.16}$$

Remembering that $\mathscr{V}$ is an arbitrary material volume and assuming that the integrand is a continuous function on $\mathscr{B}$, we deduce that (4.16) is equivalent to the relation

$$\rho\dot{\eta} + \operatorname{div}\left(\frac{\mathbf{q}}{\theta}\right) - \frac{\rho r}{\theta} \geqslant 0, \tag{4.17}$$

which is called the *Clausius-Duhem inequality*. Since

$$\operatorname{div}\left(\frac{\mathbf{q}}{\theta}\right) = \frac{1}{\theta} \operatorname{div}\mathbf{q} - \frac{1}{\theta^2} \mathbf{q}\cdot\operatorname{grad}\theta$$

and $\theta > 0$, (4.17) may be rewritten as

$$\rho\theta\dot{\eta} + \operatorname{div}\mathbf{q} - \rho r - \frac{1}{\theta}\mathbf{q}\cdot\operatorname{grad}\theta \geqslant 0. \tag{4.18}$$

Another form of this relation, which is particularly convenient for further applications, may be obtained by considering the local form (4.13) of the first law of thermodynamics. Namely, by solving (4.13) with respect to $\operatorname{div}\mathbf{q} - r$ and substituting the result obtained into (4.18), it follows that

$$-\rho\dot{\varepsilon} + \rho\theta\dot{\eta} + \mathbf{T}\cdot\operatorname{grad}\mathbf{v} - \frac{1}{\theta}\mathbf{q}\cdot\operatorname{grad}\theta \geqslant 0. \tag{4.19}$$

Finally, by introducing the *specific free energy* per unit mass $\psi(\mathbf{x}, t)$, defined by

$$\psi = \varepsilon - \eta\theta, \tag{4.20}$$

we deduce from (4.19) that

$$-\rho\dot{\psi} - \rho\eta\dot{\theta} + \mathbf{T}\cdot\operatorname{grad}\mathbf{v} - \frac{1}{\theta}\mathbf{q}\cdot\operatorname{grad}\theta \geqslant 0. \tag{4.21}$$

### 4.3. Thermoelastic materials. Elastic materials

The principles of mechanics and the laws of thermodynamics must be satisfied by any thermodynamic process, irrespective of the material of the body undergoing the process. On the other hand, the thermomechanical behaviour of a material is characterized by some supplementary relations between the state variables and possibly their histories. Such specific relations are called *constitutive equations*. Since in what follows we shall deal only with elastic materials, we devote the remaining of this section to the constitutive equations characterizing the elastic materials and their thermomechanical behaviour.

A *thermoelastic* material is defined by four constitutive equations giving the specific free energy $\psi$, the stress tensor $\mathbf{T}$, the specific entropy $\eta$, and the heat flux $\mathbf{q}$ at each material point $\mathbf{X}$ in terms of the deformation gradient $\mathbf{F}$, the absolute temperature $\theta$, and the temperature gradient $\mathbf{g} = \operatorname{Grad}\theta$ at $\mathbf{X}$

$$\psi = \hat{\psi}(\mathbf{F}, \theta, \mathbf{g}), \quad \mathbf{T} = \hat{\mathbf{T}}(\mathbf{F}, \theta, \mathbf{g}), \quad \eta = \hat{\eta}(\mathbf{F}, \theta, \mathbf{g}), \quad \mathbf{q} = \hat{\mathbf{q}}(\mathbf{F}, \theta, \mathbf{g}), \tag{4.22}$$

where the argument $\mathbf{X}$ has been suppressed for convenience. We assume that the *response* functions $\hat{\psi}, \hat{\mathbf{T}}, \hat{\eta}, \hat{\mathbf{q}}$ are continuously differentiable on their common domain.

Let us consider first the restrictions placed on the constitutive equations (4.22) by the second law of thermodynamics. Differentiating the first of these equations with respect to $t$ gives

$$\dot{\psi} = \frac{\partial\hat{\psi}}{\partial\mathbf{F}}\cdot\dot{\mathbf{F}} + \frac{\partial\hat{\psi}}{\partial\theta}\dot{\theta} + \frac{\partial\hat{\psi}}{\partial\mathbf{g}}\cdot\dot{\mathbf{g}}. \tag{4.23}$$

Next, by (2.7) and (2.2), we have

$$\dot{\mathbf{F}} = \operatorname{Grad}\mathbf{v} = (\operatorname{grad}\mathbf{v})\,\mathbf{F}. \tag{4.24}$$

Substituting (4.24) into (4.23), and the result obtained into (4.21), we find after rearranging terms the inequality

$$\left[\hat{\mathbf{T}}(\mathbf{F}^T)^{-1} - \rho\frac{\partial\hat{\psi}}{\partial\mathbf{F}}\right]\cdot\dot{\mathbf{F}} - \rho\left(\hat{\eta} + \frac{\partial\hat{\psi}}{\partial\theta}\right)\dot{\theta} - \rho\frac{\partial\hat{\psi}}{\partial\mathbf{g}}\cdot\dot{\mathbf{g}} - \frac{1}{\theta}\hat{\mathbf{q}}\cdot\operatorname{grad}\theta \geqslant 0. \tag{4.25}$$

On the other hand, it can be shown (see, e.g. Carlson [56], p. 303) that there exists at least a thermodynamic process which is compatible with the balance equations and such that $\dot{\mathbf{F}}, \dot{\theta}, \dot{\mathbf{g}}$ take arbitrarily assigned values at a given material point and at a given time. Consequently, the last inequality is fulfilled only if the terms multiplying $\dot{\mathbf{F}}, \dot{\theta}$, and $\dot{\mathbf{g}}$ vanish identically. It then follows that the constitutive equations (4.22) of the thermoelastic materials must assume the reduced form

$$\psi = \hat{\psi}(\mathbf{F}, \theta), \qquad \mathbf{T} = \rho \frac{\partial \hat{\psi}(\mathbf{F}, \theta)}{\partial \mathbf{F}} \mathbf{F}^T, \qquad \eta = -\frac{\partial \hat{\psi}(\mathbf{F}, \theta)}{\partial \theta}, \tag{4.26}$$

and the heat flux must obey the *heat conduction inequality*

$$\hat{\mathbf{q}}\,(\mathbf{F}, \theta, \mathbf{g}) \cdot \operatorname{grad} \theta \leqslant 0. \tag{4.27}$$

Clearly, the conditions (4.26) and (4.27) are also sufficient for the fulfilment of (4.25) and hence of (4.21). For the restrictions placed by the residual inequality (4.27) on the constitutive equation $(4.22)_4$ for the heat flux we refer to Carlson [56], p. 309. In particular, it can be shown by making use of (4.27) that the heat flux vanishes together with grad $\theta$.

By (4.20),

$$\dot{\varepsilon} = \dot{\psi} + \eta\dot{\theta} + \theta\dot{\eta},$$

while (4.23), (4.24), and (4.26) imply that

$$\dot{\psi} = \frac{1}{\rho} \mathbf{T} \cdot \operatorname{grad} \mathbf{v} - \eta\dot{\theta}, \tag{4.28}$$

and hence

$$\dot{\varepsilon} = \frac{1}{\rho} \mathbf{T} \cdot \operatorname{grad} \mathbf{v} + \theta\dot{\eta}. \tag{4.29}$$

Finally, by substituting (4.29) into (4.13), we obtain the *reduced form of the energy balance equation* for thermoelastic materials

$$\rho\theta\dot{\eta} = -\operatorname{div} \mathbf{q} + \rho r. \tag{4.30}$$

An alternative approach of the constitutive equations of thermoelastic materials is to choose as independent variable the' specific entropy $\eta$ instead of the absolute temperature $\theta$ and to replace the specific free energy $\psi$ by the specific internal energy $\eta$ as dependent variable, i.e. to start by adopting constitutive equations of the form

$$\varepsilon = \tilde{\varepsilon}\,(\mathbf{F}, \eta, \mathbf{g}), \quad \mathbf{T} = \tilde{\mathbf{T}}(\mathbf{F}, \eta, \mathbf{g}), \quad \theta = \tilde{\theta}\,(\mathbf{F}, \eta, \mathbf{g}), \quad \mathbf{q} = \tilde{\mathbf{q}}\,(\mathbf{F}, \eta, \mathbf{g}). \tag{4.31}$$

Then, by making use again of the Clausius-Duhem inequality and employing a similar reasoning as above, it can be proved that the first three constitutive equations (4.31) must assume the reduced form

$$\varepsilon = \tilde{\varepsilon}(\mathbf{F}, \eta), \qquad \mathbf{T} = \rho \frac{\partial \tilde{\varepsilon}(\mathbf{F}, \eta)}{\partial \mathbf{F}} \mathbf{F}^T, \qquad \theta = \frac{\partial \tilde{\varepsilon}(\mathbf{F}, \eta)}{\partial \eta}. \tag{4.32}$$

By comparing the constitutive equations $(4.26)_2$ and $(4.32)_2$, we see that, when ignoring the thermal variables $\theta$ and $\eta$, they both reduce to the same purely mechanical constitutive equation

$$\mathbf{T} = j \frac{\partial \tilde{W}(\mathbf{F})}{\partial \mathbf{F}} \mathbf{F}^T, \tag{4.33}$$

where $\tilde{W}$ is the so-called the *specific strain energy* per unit volume in the reference configuration and is taken equal to $\rho_0 \hat{\psi}$, respectively $\rho_0 \tilde{\varepsilon}$. Materials characterized by the constitutive equations (4.33) are called elastic materials [1].

The form of (4.33) may be further simplified, by requiring that the specific strain energy be invariant under a superimposed rigid motion of the body. Let

$$\mathbf{x}^* = \chi^*(\mathbf{X}, t) \tag{4.34}$$

be a motion of the body, differing from the real motion (2.1) by a rigid motion. By analogy with (2.17), we may write

$$\chi^*(\mathbf{X}, t) = \mathbf{c}(t) + \mathbf{Q}(t)(\chi(\mathbf{X}, t) - \mathbf{a}), \tag{4.35}$$

where $\mathbf{c}$ is a time-dependent vector, $\mathbf{Q}$ is a time-dependent orthogonal tensor, and $\mathbf{a}$ is a fixed position vector. The deformation gradient of the modified motion is

$$\mathbf{F}^* = \mathbf{QF} \tag{4.36}$$

and the corresponding finite strain tensor is

$$\mathbf{D}^* = \tfrac{1}{2}(\mathbf{F}^{*T}\mathbf{F}^{T} - \mathbf{1}) = \tfrac{1}{2}(\mathbf{F}^T\mathbf{Q}^T\mathbf{QF} - \mathbf{1}) = \mathbf{D}, \tag{4.37}$$

since $\mathbf{Q}^T\mathbf{Q} = \mathbf{1}$. The invariance condition stated above may be written now in the form

$$\tilde{W}(\mathbf{F}^*) = \tilde{W}(\mathbf{F}),$$

or

$$\tilde{W}(\mathbf{QF}) = \tilde{W}(\mathbf{F}). \tag{4.38}$$

---

[1] These materials are sometimes called *hyperelastic*, whereas the name elastic materials is preserved for the more general case when the dependence of $\mathbf{T}$ on $\mathbf{F}$ cannot be derived from a scalar potential like $\tilde{W}$. However, it was not possible up to now to find out elastic materials that are not hyperelastic, too.

(*) $j = J^{-1}$ (3.29), $J = \det F > 0$ (2.8)

This relation must be satisfied for every proper orthogonal tensor $\mathbf{Q}$ and every invertible tensor $\mathbf{F}$ in the domain of $\tilde{W}$. With the special choice $\mathbf{Q} = \mathbf{R}^T$, where $\mathbf{R}$ is the rotation tensor in the polar decomposition $(2.11)_1$, we obtain from (4.38) the *necessary* condition $\tilde{W}(\mathbf{F}) = \tilde{W}(\mathbf{U})$. Instead of $\mathbf{U}$ we may obviously use the tensor $\mathbf{C} = \mathbf{U}^2$, or the finite strain tensor $\mathbf{D} = \frac{1}{2}(\mathbf{C} - \mathbf{1})$. In the last case we have, say,

$$\tilde{W}(\mathbf{F}) = W(\mathbf{D}). \tag{4.39}$$

Clearly, by (4.37), the condition (4.39) is also *sufficient* for the invariance of the strain energy under superimposed rigid motions.

Finally, by introducing (4.39) into (4.33) and taking into account that

$$\frac{\partial \tilde{W}(\mathbf{F})}{\partial F_{km}} = \frac{\partial W(\mathbf{D})}{\partial D_{rs}} \frac{\partial D_{rs}}{\partial F_{km}} = \frac{1}{2}\left(\frac{\partial W(\mathbf{D})}{\partial D_{ms}} F_{ks} + \frac{\partial W(\mathbf{D})}{\partial D_{rm}} F_{kr}\right) = F_{kr} \frac{\partial W(\mathbf{D})}{\partial D_{rm}},$$

we obtain the relation

$$\mathbf{T} = j\,\mathbf{F} \frac{\partial W(\mathbf{D})}{\partial \mathbf{D}} \mathbf{F}^T, \qquad T_{km} = j\,\mathbf{F}_{kr} F_{ms} \frac{\partial W(\mathbf{D})}{\partial D_{rs}}, \tag{4.40}$$

which is the *non-linear constitutive equation of elastic materials*. The response function $W(\mathbf{D})$ will be also called the *strain-energy function*. Generally, the form of this function may depend on the particle $X$, as well as on the reference configuration. If it is possible to choose a reference configuration of the body $\mathscr{B}$ so that $W(\mathbf{D})$ is the same for all particles $X \in \mathscr{B}$, we say that the body is *homogeneous*.

By making use of the considerations which have led us to (4.40), it can be shown that the theory of elasticity, which primarily concerns the purely mechanical behaviour of the materials, may be also applied to isothermal and adiabatic processes.

A thermodynamic process is called *isothermal* when the temperature is uniform throughout the body and is time-independent ($\theta = \theta_0$). For such processes the constitutive equation (4.40) still holds provided that we set $W(\mathbf{D}) = \rho_0 \hat{\psi}(\mathbf{D}, \theta_0)$. A thermodynamic process may be satisfactorily approximated by an isothermal process when it is sufficiently slow for allowing the levelling of the temperatures of different parts of the body with the temperature $\theta_0$ of the surrounding medium (ideal exchange of heat). An isothermal process may also be considered as a limiting case of a real thermodynamic process when the conductivity of the material tends to infinity. In this case, the heat flux is no longer determined by a constitutive equation but by the energy equation (4.30) combined with the thermoelastic boundary conditions.

A thermodynamic process is called *adiabatic* when $\mathbf{q} = \mathbf{0}$, $r = 0$. The energy equation (4.30) gives in this case $\dot{\eta} = 0$, hence $\eta = \eta_0$ (const). Consequently, any adiabatic thermoelastic process is *isentropic*, too. The constitutive equation (4.40) still holds provided that we take $W(\mathbf{D}) = \rho_0 \tilde{\varepsilon}(\mathbf{D}, \eta_0)$. A thermodynamic process may be satisfactorily approximated by an adiabatic process when it is sufficiently rapid for preventing the heat exchange between the body and its surroundings and between different parts of the body (ideal thermal insulation). An adiabatic process may be also viewed as a limiting case of a real thermodynamic process when the conductivity of the material tends to zero.

The isothermal and the adiabatic processes are *ideal* processes, but they often provide a rather good approximation of real thermodynamic processes. Finally, it should be noticed that a thermoelastic material behaves like an elastic material throughout isothermal processes and like a different elastic material throughout adiabatic processes. In particular, the elastic constants, which will be considered below, assume different values in these two cases, the differences between them being, however, small.

Let us resume now the case of the elastic constitutive equation (4.40). Assume that the reference configuration is stress-free (natural state). Since in the reference configuration we have $\mathbf{F} = \mathbf{1}$, hence $\mathbf{D} = \mathbf{0}$ and $j = 1$, it follows from (4.40) that

$$\left.\frac{\partial W(\mathbf{D})}{\partial \mathbf{D}}\right|_{\mathbf{D}=0} = \mathbf{0}. \tag{4.41}$$

Moreover, since $W$ is defined (like $\hat{\psi}$ and $\tilde{\varepsilon}$) to within an additive constant, we may always take

$$W(\mathbf{0}) = 0. \tag{4.42}$$

Next, assuming that the strain energy can be expanded in a series of powers of $\mathbf{D}$ and taking into account (4.41) and (4.42), we obtain

$$W(\mathbf{D}) = \frac{1}{2!}\, c_{klmn} D_{kl} D_{mn} + \frac{1}{3!}\, C_{klmnrs} D_{kl} D_{mn} D_{rs} + \cdots, \tag{4.43}$$

where

$$c_{klmn} = \left.\frac{\partial^2 W(\mathbf{D})}{\partial D_{kl}\, \partial D_{mn}}\right|_{\mathbf{D}=0}, \qquad C_{klmnrs} = \left.\frac{\partial^3 W(\mathbf{D})}{\partial D_{kl}\, \partial D_{mn}\, \partial D_{rs}}\right|_{\mathbf{D}=0}. \tag{4.44}$$

The components of the fourth-order tensor $\mathbf{c}$ are called *second-order elastic constants*, while the components of the sixth-order tensor $\mathbf{C}$ are called *third-order elastic constants*. Clearly, by (4.44), these elastic constants must satisfy the symmetry conditions

$$\left.\begin{aligned} c_{klmn} &= c_{lkmn} = c_{klnm} = c_{mnkl}, \\ C_{klmnrs} &= C_{lkmnrs} = C_{klnmrs} = C_{klmnsr} = C_{mnklrs} = C_{rsmnkl} = C_{klrsmn}. \end{aligned}\right\} \tag{4.45}$$

To linearize the constitutive equation (4.40) of the elastic materials, we assume that $\varepsilon = \|\mathbf{H}\| \ll 1$ and neglect all terms of order $O(\varepsilon^2)$ or higher, as $\varepsilon \to 0$, taking also into account the results obtained in Sect. 2.6. We notice first that

$$\frac{\partial W(\mathbf{D})}{\partial D_{kl}} = c_{klmn} D_{mn} + \frac{1}{2} C_{klmnrs} D_{mn} D_{rs} + \cdots = c_{klmn} E_{mn} + O(\varepsilon^2), \quad (4.46)$$

and

$$j = [\det(\mathbf{1} + \mathbf{H})]^{-1} = 1 - H_{mm} + O(\varepsilon^2). \quad (4.47)$$

Substituting (2.28), (4.46), and (4.47) into (4.40), we find that, *to within an error of* $O(\varepsilon^2)$ *as* $\varepsilon \to 0$,

$$T_{kl} = c_{klmn} E_{mn}. \quad (4.48)$$

This constitutive equation, which is characteristic of *linear elastic* materials, is called *the generalized Hooke's law.* We assume in what follows that equations (4.48) may be solved with respect to $\mathbf{E}$, thus leading to

$$E_{kl} = s_{klmn} T_{mn}. \quad (4.49)$$

The components of the fourth-order tensor $\mathbf{s}$ are called *second-order elastic compliances.* It can be shown that they satisfy the relations

$$c_{klmn} s_{mnpr} = \frac{1}{2} (\delta_{kp} \delta_{lr} + \delta_{kr} \delta_{lp}). \quad (4.50)$$

To the same approximation as above it results from (3.29) and (3.34) that $\mathbf{\Pi} = \mathbf{S} = \mathbf{T}$, and hence, in the linear theory of elasticity, the Piola-Kirchhoff stress tensors coincide with the Cauchy stress tensor. Moreover, if we assume that $\mathbf{v}$ and grad $\mathbf{v}$ are of the order $O(\varepsilon)$, then the convective term (grad $\mathbf{v}$) $\mathbf{v}$ may be neglected in comparison with the first term in the right-hand side of (2.5), i.e.

$$\mathbf{a} = \frac{\partial \mathbf{v}}{\partial t} = \frac{\partial^2 \mathbf{u}}{\partial t^2}. \quad (4.51)$$

Consequently, the three forms (3.22), (3.32), and (3.36) of the first law of motion are indistinguishable and reduce to

$$\operatorname{div} \mathbf{T} + \rho_0 \mathbf{b} = \rho_0 \frac{\partial^2 \mathbf{u}}{\partial t^2}. \quad (4.52)$$

Finally, we note that in the linear theory of elasticity only the lowest-order terms in $\varepsilon$ are retained in the expression (4.43) of the specific strain energy, thus obtaining

$$W = W_2(\mathbf{E}) = \frac{1}{2} c_{klmn} E_{kl} E_{mn}. \quad (4.53)$$

Equations (4.48) and (4.52), together with the linearized kinematic equations given in Sects. 2.6 and 2.7, provide the field equations of the *linear* or *infinitesimal theory of elasticity.*

In order to investigate some *non-linear* phenomena in elastic materials subjected to large deformations, without increasing too much, however, the mathematical difficulties, it is customary to retain in the constitutive equation (4.40) besides the linear terms also the non-linear terms of lowest order, i.e. those of second order in $\varepsilon$. The effects arising from the presence of these non-linear terms are called *second-order elastic effects*. To obtain the corresponding constitutive equation we first note that, by $(4.46)_1$, $(2.30)_1$, and (2.33),

$$\frac{\partial W(\mathbf{D})}{\partial D_{kl}} = c_{klmn}(E_{mn} + \tfrac{1}{2} H_{pm}H_{pn}) + \tfrac{1}{2} C_{klmnrs}E_{mn}E_{rs} + 0(\varepsilon^3). \tag{4.54}$$

Next, by substituting (2.28), (4.47), and (4.54) into (4.40) and neglecting terms of third and higher order in $\varepsilon$ we obtain the required *constitutive equation of second-order elasticity*

$$\begin{aligned} T_{kl} = {} & c_{klmn}E_{mn}(1 - E_{pp}) + c_{plmn}H_{kp}\, E_{mn} + c_{kpmn}E_{mn}H_{lp} \\ & + \tfrac{1}{2}\, c_{klmn}H_{pm}H_{pn} + \tfrac{1}{2}\, C_{klmnrs}E_{mn}E_{rs}. \end{aligned} \tag{4.55}$$

Before closing this section, we remark that, in view of the symmetry conditions (4.45), it is advantageous to denote each pair of indexes of the elastic constants by a single index after *Voigt's convention* [372], namely

$$11 \sim 1, \quad 22 \sim 2, \quad 33 \sim 3, \quad 23 \sim 4, \quad 13 \sim 5, \quad 12 \sim 6. \tag{4.56}$$

According to this convention, we shall write

$$c_{klmn} = c_{KM}, \qquad C_{klmnrs} = C_{KMR}, \tag{4.57}$$

where the small subscripts range over the values 1, 2, 3, while the capital subscripts range over the values 1, 2, ..., 6. Moreover, the symmetry conditions (4.45) assume now the concise form

$$c_{KM} = c_{MK}, \qquad C_{KMR} = C_{MKR} = C_{RMK} = C_{KRM}. \tag{4.58}$$

There are several ways for extending Voigt's notation to the components of the tensors $\mathbf{T}, \mathbf{D}, \mathbf{E}$, and $\mathbf{s}$. The convention mostly used presently is that of Brugger [44]. According to this convention, the components $T_{kl}$, $D_{kl}$, $E_{kl}$, and $s_{klmn}$ are replaced by $T_K$, $D_K$, $E_K$, and $s_K$, respectively, after the rule

$$T_{kl} = T_K, \qquad 2D_{kl} = (1 + \delta_{kl})\, D_K, \qquad 2E_{kl} = (1 + \delta_{kl})\, E_K, \tag{4.59}$$

$$4s_{klmn} = (1 + \delta_{kl})\,(1 + \delta_{mn})\, s_{KM}, \tag{4.60}$$

in which the correspondence between the pairs of small subscripts and the capital subscripts is given by (4.56). In particular, the detailed form of $(4.59)_2$ is

$$D_{11} = D_1, \quad D_{22} = D_2, \quad D_{33} = D_3, \quad 2D_{23} = D_4, \quad 2D_{13} = D_5, \quad 2D_{12} = D_6. \tag{4.61}$$

Unlike other conventions, Brugger's notation has the advantage that it does not lead to the occurrence of new numerical coefficients either in the expression of the specific strain energy or in the constitutive equations. Namely, (4.43) may be rewritten as

$$W(\mathbf{D}) = \frac{1}{2!} c_{KM} D_K D_M + \frac{1}{3!} C_{KMR} D_K D_M D_R + \cdots \tag{4.62}$$

while the equations (4.48–50) of the linearized theory become

$$T_K = c_{KM} E_M, \quad E_K = s_{KM} T_M, \quad c_{MP} s_{PM} = \delta_{KM}, \tag{4.63}$$

where $K$, $M$, $P$ range over the values 1, 2, ···, 6.

Finally, by making use of the symmetry conditions (4.58), it can be shown that the maximum number of independent second-order elastic constants is 21, and that of independent third-order elastic constants is 56. We shall see in Sect. 5.2 that, whenever the elastic behaviour of the material exhibits certain symmetry properties, the number of independent elastic constants decreases accordingly.

# 5. Material symmetry

## 5.1. Material symmetry of elastic solids

All the considerations in the preceding section assumed a fixed reference configuration, the dependence of the constitutive equations and in particular of the function $W$ on this configuration being implied. We shall consider now this dependence in more detail. Let $\mathbf{D}$ and $\hat{\mathbf{D}}$ be the finite strain tensors associated with the deformations carrying a *homogeneous* elastic body from two different stress-free reference configurations $(K)$ and respectively $(\hat{K})$ into the current configuration $(k)$ (Fig. 5.1). Since the strain energy in the configuration $(k)$ does not depend on the choice of the reference configuration, we must obviously have

$$W_K(\mathbf{D}) = W_{\hat{K}}(\hat{\mathbf{D}}), \tag{5.1}$$

while the response functions $W_K$ and $W_{\hat{K}}$ may differ in general from each other.

Now, the question naturally arises: when are the functions $W_K$ and $W_{\hat{K}}$ identical? Or, in other words, what is the deformation bringing the material from a given configuration $(K)$ into another configuration $(\hat{K})$ such that the response of the material to any further deformation from both configurations be the same?

Experiments show that *elastic solids* have preferred configurations, such that any pure strain from a preferred configuration affects the subsequent behaviour of the material. Consequently, the gradient of the deformation relating the configurations $(K)$ and $(\hat{K})$ must be an orthogonal tensor $\mathbf{Q}$, since only in this case the corresponding finite strain tensor vanishes identically (cf. Sect. 2.3).

Denoting by $\mathbf{F}$ and $\hat{\mathbf{F}}$ the gradients of the deformations bringing the configurations $(K)$ and $(\hat{K})$ into the configuration $(k)$, respectively, we have $\mathbf{F} = \hat{\mathbf{F}}\,\mathbf{Q}$, and hence

$$\hat{\mathbf{D}} = \frac{1}{2}\,(\hat{\mathbf{F}}^T\hat{\mathbf{F}} - \mathbf{1}) = \frac{1}{2}\,(\mathbf{Q}\mathbf{F}^T\mathbf{F}\mathbf{Q}^T - \mathbf{1}) = \mathbf{Q}\mathbf{D}\mathbf{Q}^T,$$

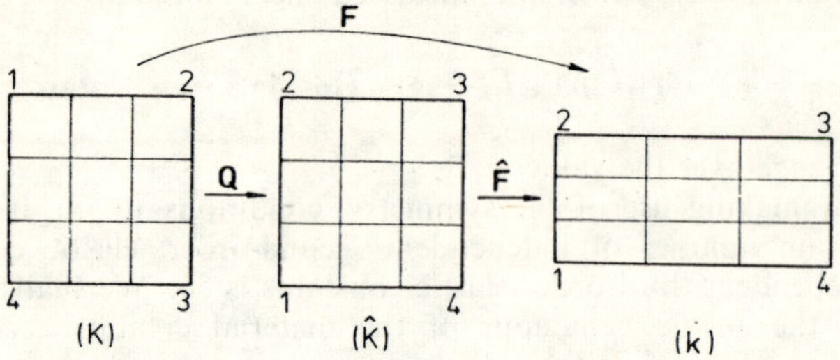


Fig. 5.1. Two different stress-free reference configurations $(K)$ and $(\hat{K})$, used to describe the deformation of an elastic body in the current configuration $(k)$.

since $\mathbf{Q}^T\mathbf{Q} = \mathbf{1}$. Substituting this result into (5.1) and assuming that the functions $W_K$ and $W_{\hat{K}}$ are identical, it results that

$$W_K(\mathbf{D}) = W_K(\mathbf{Q}\mathbf{D}\mathbf{Q}^T) \tag{5.2}$$

for every symmetric tensor $\mathbf{D}$. The set $\mathscr{G}$ of all orthogonal tensors $\mathbf{Q}$ that obey (5.2) for every symmetric tensor $\mathbf{D}$ is called the *symmetry group* or the *isotropy group* of the material [1].

A homogeneous elastic solid is called *isotropic* if there is at least one configuration such that the symmetry group $\mathscr{G}$ of the corresponding function $W$ coincide with the full orthogonal group; a configuration with this property is called an *undistorted state* of the material. On the contrary, when $\mathscr{G}$ is a proper subgroup of the orthogonal group, the elastic solid is called *aelotropic* or *anisotropic*.

By (5.2), when $\mathbf{Q}$ belongs to $\mathscr{G}$ then $-\mathbf{Q}$ belongs to $\mathscr{G}$, too. In fact, any symmetry group $\mathscr{G}$ can be represented as a direct product between the minimal symmetry group $\{\mathbf{1}, -\mathbf{1}\}$, consisting of the identity $\mathbf{1}$ and the inversion $-\mathbf{1}$, and another group, say $\mathscr{G}^+$, which consists only of proper orthogonal deformations, i.e. rotations ([358], Sect. 33). Consequently, the type of anisotropy is characterized by the type of the rotation group $\mathscr{G}^+$. In particular, an elastic solid is isotropic or anisotropic according as $\mathscr{G}^+$ equals the proper orthogonal group or is a proper subgroup of it.

To characterize the material symmetry of an elastic body it is sufficient to indicate the so-called *generators* of $\mathscr{G}^+$. These are defined as a set of elements of

[1] It is easily seen that if $\mathbf{Q}_1, \mathbf{Q}_2 \in \mathscr{G}$ then $\mathbf{Q}_1\mathbf{Q}_2 \in \mathscr{G}$, if $\mathbf{Q} \in \mathscr{G}$ then $\mathbf{Q}^{-1} = \mathbf{Q}^T \in \mathscr{G}$, and that $\mathbf{1} \in \mathscr{G}$, hence $\mathscr{G}$ is a group indeed. For inhomogeneous bodies, the symmetry group may generally vary from one material point to another.

$\mathscr{G}^+$ having the property that any element of $\mathscr{G}^+$ can be represented as a product of generators and, eventually, of their inverses.

After Coleman and Noll [78] we denote the generators of $\mathscr{G}^+$ by $\mathbf{R}_{\mathbf{n}}^{\varphi}$, which means the (proper) orthogonal tensor corresponding to a right-handed rotation through the angle $\varphi$, $0 < \varphi < 2\pi$, about an axis in the direction of the unit vector $\mathbf{n}$. We say that $\mathbf{n}$ is a *symmetry axis of order q* of the material, if $\mathscr{G}^+$ contains the rotation $\mathbf{R}_{\mathbf{n}}^{2\pi/q}$.

The anisotropic solids can be divided into twelve subsystems (subgroups of the proper orthogonal group) according to the symmetry of their elastic properties. The first eleven subsystems correspond to the thirty-two crystal classes. Table 5.1 shows the symbols of these classes introduced by Schoenflies [283] (column S) and by Hermann [160] and Mauguin [237] (column H—M). The fifth column of this table shows the generators of the corresponding rotation groups, where $\{\mathbf{i}, \mathbf{j}, \mathbf{k}\}$ denotes a right-handed orthonormal basis and $\mathbf{p} = (\mathbf{i} + \mathbf{j} + \mathbf{k})/\sqrt{3}$.

Smith and Rivlin [309] have proved that the specific strain energy $W(\mathbf{D})$ is invariant with respect to the rotations belonging to $\mathscr{G}^+$ if and only if it depends on $\mathbf{D}$ only through a certain number of scalars $I_1, I_2, \ldots, I_p$ of $\mathbf{D}$ that are invariant under $\mathscr{G}^+$, i.e. if

$$W = W(I_1, I_2, \ldots, I_p).$$

In table 5.1, the scalar invariants corresponding to each of the first eleven types of anisotropy are listed in the last column. $D_1, \ldots, D_6$ denote the Cartesian components of $\mathbf{D}$ with respect to the orthonormal basis $\{\mathbf{i}, \mathbf{j}, \mathbf{k}\}$; they are labelled according to Voigt's notation and assuming that the axes $x_1$, $x_2$, and $x_3$ of the co-ordinate system are chosen along the unit vectors $\mathbf{i}$, $\mathbf{j}$, and $\mathbf{k}$, respectively.

Smith and Rivlin [309] showed that when $W(\mathbf{D})$ is a polynomial in the components $D_1, \ldots, D_6$ it can be expressed as a polynomial in the invariant scalars $I_1, \ldots, I_p$. Moreover, Smith [310] has shown that these invariants satisfy certain algebraic relations, which he has subsequently used to obtain unique representations of polynomial strain-energy functions in terms of $I_1, \ldots, I_p$.

The last type of anisotropy, called *transverse isotropy*, is characterized by the property that all directions perpendicular to a certain direction, e.g. that of unit vector $\mathbf{k}$, are elastically equivalent. In this case the rotation group $\mathscr{G}^+$ consists of $\mathbf{1}$ and all rotations $\mathbf{R}_{\mathbf{k}}^{\varphi}$, $0 < \varphi < 2\pi$, about the axis determined by the unit vector $\mathbf{k}$. Transverse isotropy is appropriate to real materials having a laminated or bundled structure. As shown by Ericksen and Rivlin [104], the strain-energy function of an elastic solid which is transversely isotropic with respect to the $x_3$-axis must have the form

$$W = W(\mathrm{I}_{\mathbf{D}}, \mathrm{II}_{\mathbf{D}}, \mathrm{III}_{\mathbf{D}}, D_3, D_4^2 + D_5^2), \tag{5.3}$$

where

$$\mathrm{I}_{\mathbf{D}} = \operatorname{tr} \mathbf{D} = D_1 + D_2 + D_3, \tag{5.4}$$

$$\mathrm{II}_{\mathbf{D}} = \frac{1}{2}\,[(\operatorname{tr} \mathbf{D})^2 - \operatorname{tr} \mathbf{D}^2] = D_1D_2 + D_2D_3 + D_1D_3 - \frac{1}{4}\,(D_4^2 + D_5^2 + D_6^2), \tag{5.5}$$

$$\mathrm{III}_{\mathbf{D}} = \det \mathbf{D} = D_1D_2D_3 + \frac{1}{4}\,(D_4D_5D_6 - D_1D_4^2 - D_2D_5^2 - D_3D_6^2) \tag{5.6}$$

are the so-called *principal invariants* of $\mathbf{D}$.

*Table 5.1*

**Scalar invariants of $\mathscr{G}^{+}$**

| System | Sub-system | Class symbol S | Class symbol H – M | Generators | List of invariants |
|---|---|---|---|---|---|
| Triclinic | 1 | $C_1$, $C_i$ | 1, $\bar{1}$ | **1** | $D_1, D_2, D_3, D_4, D_5, D_6$ |
| Monoclinic | 2 | $C_s$, $C_2$, $C_{2h}$ | m, 2, $\frac{2}{m}$ | $\mathbf{R}_k^{\pi}$ | $D_1, D_2, D_3, D_6, D_4^2, D_5^2, D_4D_5$ |
| Rhombic | 3 | $C_{2v}$, $D_2$, $D_{2h}$ | 2mm, 222, mmm | $\mathbf{R}_i^{\pi}$, $\mathbf{R}_j^{\pi}$ | $D_1, D_2, D_3, D_4^2, D_5^2, D_6^2, D_4D_5D_6$ |
| Tetragonal | 4 | $C_4$, $S_4$, $C_{4h}$ | 4, $\bar{4}$, $\frac{4}{m}$ | $\mathbf{R}_k^{\pi/2}$ | $D_1 + D_2, D_3, D_4^2 + D_5^2, D_6^2, D_1D_2, D_6(D_1 - D_2),$ $D_4D_5(D_1 - D_2), D_4D_5D_6, D_6(D_4^2 - D_5^2), D_1D_4^2 + D_2D_5^2,$ $D_5D_6(D_4^2 - D_5^2), D_4^2D_5^2$ |
| | 5 | $D_{26}$, $C_{4v}$, $D_4$, $D_{4h}$ | $\bar{4}$2m, 4mm, 422, $\frac{4}{m}$mm | $\mathbf{R}_k^{\pi/2}$, $\mathbf{R}_i^{\pi}$ | $D_1 + D_2, D_3, D_4^2 + D_5^2, D_6^2, D_1D_2, D_4D_5D_6,$ $D_1D_4^2 + D_2D_5^2, D_4^2D_5^2$ |
| Cubic | 6 | T, $T_h$ | 23, m3 | $\mathbf{R}_i^{\pi}$, $\mathbf{R}_j^{\pi}$, $\mathbf{R}_p^{2\pi/3}$ | $D_1 + D_2 + D_3, D_1D_2 + D_2D_3 + D_3D_1, D_4^2 + D_5^2 + D_6^2,$ $D_1D_2D_3, D_4D_5D_6, D_1D_5^2 + D_2D_6^2 + D_3D_4^2,$ $D_1D_6^2 + D_2D_4^2 + D_3D_5^2, D_1D_3^2 + D_2D_1^2 + D_3D_2^2,$ $D_4^2D_5^2 + D_5^2D_6^2 + D_6^2D_4^2, D_1D_2D_6^2 + D_2D_3D_4^2 + D_3D_1D_5^2,$ $D_1D_2D_5^2 + D_2D_3D_6^2 + D_3D_1D_4^2, D_1D_5^2D_6^2 + D_2D_6^2D_4^2 + D_3D_4^2D_5^2,$ $D_1D_6^2D_4^2 + D_2D_4^2D_5^2 + D_3D_5^2D_6^2, D_4^2D_6^4 + D_5^2D_4^4 + D_6^2D_5^4$ |
| | 7 | $T_d$, O, $O_h$ | $\bar{4}$3m, 432, m3m | $\mathbf{R}_i^{\pi/2}$, $\mathbf{R}_j^{\pi/2}$, $\mathbf{R}_k^{\pi/2}$ | $D_1 + D_2 + D_3, D_1D_2 + D_2D_3 + D_3D_1, D_4^2 + D_5^2 + D_6^2,$ $D_1D_2D_3, D_4D_5D_6, (D_2 + D_3)D_4^2 + (D_3 + D_1)D_5^2 + (D_1 + D_2)D_6^2,$ $D_4^2D_5^2 + D_5^2D_6^2 + D_6^2D_4^2, D_1D_2D_6^2 + D_2D_3D_4^2 + D_3D_1D_5^2,$ $D_1D_5^2D_6^2 + D_2D_6^2D_4^2 + D_3D_4^2D_5^2$ |

| | | | | | |
|---|---|---|---|---|---|
| Hexagonal | 8 | $C_3$, $C_{3i}$ | 3, $\bar{3}$ | $R_k^{2\pi/3}$ | $D_3,\ D_1 + D_2,\ 4D_1D_2 - D_6^2,\ D_4^2 + D_5^2,\ (D_1 - D_2)D_5 - D_4D_6,$<br>$(D_1 - D_2)\ D_4 + D_5\ D_6,\ D_1\ [(D_1 + 3D_2)^2 - 3D_6^2],$<br>$D_5(D_5^2 - 3D_4^2),\ D_4(D_4^2 - 3D_5^2),$<br>$D_1D_5^2 + D_2D_4^2 + 2D_4\ D_5D_6,\ 3D_6(D_1 - D_2)^2 - D_6^3,$<br>$D_5[(D_1 + D_2)^2 + D_6^2 - 4D_2^2] - 4D_1D_4D_6,$<br>$D_4[(D_1 + D_2)^2 + D_6^2 - 4D_2^2] + 4D_1D_5D_6,$<br>$2(D_1 - D_2)\ D_4D_5 + D_6(D_4^2 - D_5^2)$ |
| | 9 | $C_{3v}$, $D_3$, $D_{3d}$ | 3m, 32, $\bar{3}$m | $R_i^\pi$, $R_k^{2\pi/3}$ | $D_3,\ D_1 + D_2,\ 4D_1D_2 - D_6^2,\ D_4^2 + D_5^2,\ (D_1 - D_2)\ D_4 + D_5D_6,$<br>$D_1[(D_1 + 3D_2)^2 - 3D_6^2],\ D_4(D_4^2 - 3D_5^2),$<br>$D_1D_5^2 + D_2D_4^2 + 2D_4D_5D_6,$<br>$D_4[(D_1 + D_2)^2 + D_6^2 - 4D_2^2] + 4D_1D_5D_6$ |
| | 10 | $C_6$, $C_{3h}$, $C_{6h}$ | 6, $\bar{6}$, $\frac{6}{m}$ | $R_k^{\pi/3}$ | $D_3,\ D_1 + D_2,\ 4D_1D_2 - D_6^2,\ D_4^2 + D_5^2,$<br>$D_1[(D_1 + 3D_2)^2 - 3D_6^2],\ D_1D_4^2 + D_2D_5^2 - 2D_4D_5D_6,$<br>$3D_6(D_1 - D_2)^2 - D_6^3,\ D_6(D_4^2 - D_5^2) + 2\ (D_1 - D_2)\ D_4\ D_5,$<br>$D_5^2[(D_1 + D_2)^2 - 4D_2^2 + D_6^2] -$<br>$-2D_1[(D_1 + 3D_2)\ (D_4^2 + D_5^2) - 4D_4D_5D_6],$<br>$D_4D_5[(D_1 + D_2)^2 - 4D_2^2 + D_6^2] + 4D_1D_6(D_4^2 - D_5^2),$<br>$D_1\ (3D_4^4 + D_5^4) + 2D_2D_5^2(3D_4^2 + D_5^2) - 8D_4D_5^3D_6,$<br>$D_6[(D_4^2 + D_5^2)^2 - 4D_4^2(D_4^2 - D_5^2)] - 4D_4D_5^3(D_1 - D_2),$<br>$D_5^2(3D_4^2 - D_5^2)^2,\ D_4D_5[3(D_4^2 - D_5^2)^2 - 4D_4^2D_5^2]$ |
| | 11 | $D_{3h}$, $C_{6v}$, $D_6$, $D_{6h}$ | $\bar{6}$m2, 6mm, 622, $\frac{6}{m}$mm | $R_i^\pi$, $R_k^{\pi/3}$ | $D_3,\ D_1 + D_2,\ 4D_1D_2 - D_6^2,\ D_4^2 + D_5^2,$<br>$D_1[(D_1 + 3D_2)^2 - 3D_6^2],\ D_1\ D_4^2 + D_2D_5^2 - 2D_4D_5D_6,$<br>$D_5^2[(D_1 + D_2)^2 - 4D_2^2 + D_6^2] -$<br>$-\ 2D_1[(D_1 + 3D_2)\ (D_4^2 + D_5^2) - 4D_4D_5D_6],$<br>$D_1(3D_4^4 + D_5^4) + 2D_2D_5^2(3D_4^2 + D_5^2) - 4D_4D_5^3D_6,\ D_5^2(3D_4^2 - D_5^2)^2$ |

O: orthotropic Bilder siehe Kittel, Einführung ... S. 37 - 42
rhombisch = orthorhombisch

When the symmetry group contains the reflection on a plane, that plane is called a *plane of elastic symmetry* of the material. A material is called *orthotropic* if it possesses three mutually perpendicular planes of elastic symmetry, i.e. if its symmetry group contains reflections on three mutually perpendicular planes. Such a triple of reflections is $-\mathbf{R}_{\mathrm{i}}^{\pi}$, $-\mathbf{R}_{\mathrm{j}}^{\pi}$, $-\mathbf{R}_{\mathrm{k}}^{\pi}$. Since $\mathbf{R}_{\mathrm{i}}^{\pi}\mathbf{R}_{\mathrm{j}}^{\pi} = \mathbf{R}_{\mathrm{k}}^{\pi}$ and $\mathbf{R}_{\mathrm{j}}^{\pi/2}\mathbf{R}_{\mathrm{j}}^{\pi/2} = \mathbf{R}_{\mathrm{k}}^{\pi}$, it follows that the crystals listed in table 5.1 under subsystems 3, 5, 6, and 7 are particular orthotropic materials.

Unlike anisotropic solids, the *isotropic* solids have no preferred directions in an undistorted state. For such materials, the strain energy may be expressed as a function of the principal invariants, i.e.

$$W = W(\mathrm{I_D}, \mathrm{II_D}, \mathrm{III_D}), \tag{5.7}$$

provided that the reference configuration is chosen as an undistorted state of the material.

Let us resume now the general case of an elastic solid, isotropic or anisotropic. If the strain-energy function $W$ is a known function of the scalar invariants $I_1, \ldots, I_p$, then the corresponding constitutive equation for the stress tensor results from $(4.40)_2$ as

$$T_{km} = \frac{1}{2} j\, F_{kr} F_{ms} \sum_{\alpha=1}^{p} \left( \frac{\partial I_\alpha}{\partial D_{rs}} + \frac{\partial I_\alpha}{\partial D_{sr}} \right) \frac{\partial W}{\partial I_\alpha}. \tag{5.8}$$

Since the calculation of the derivatives involved in (5.8) is generally rather tedious, we shall not give here the explicit form of the constitutive equation (5.8) for various types of anisotropy. However, we shall come back to this point below, when considering the constitutive equations of linear and second-order elasticity.

## 5.2. Second-order elastic constants

We will consider in this subsection the explicit form taken for each type of anisotropy by the first term,

$$W_2(\mathbf{D}) = \frac{1}{2} c_{KM} D_K D_M, \tag{5.9}$$

of the expansion (4.62) of the strain-energy function [1]. To this aim we shall retain from the last column of Table 5.1 only the invariants and their products that are of second degree in the components of $\mathbf{D}$. By comparing the dependence of $W_2$ on $D_K$ obtained in this way with (5.9), it is possible to derive the restrictions imposed by each type of material symmetry on the second-order elastic constants $c_{KM}$.

For the sake of simplicity we assume again that the body is homogeneous; for an inhomogeneous body all the considerations below are still valid at any fixed

---

[1] This first term of the expansion will play a special role in what follows, since, according to (4.53), the strain-energy function of a linear elastic material is $W_2(\mathbf{E})$.

point of the body, but the type of anisotropy and/or the values of the elastic constants may vary from one point to another.

The *triclinic* system corresponds to the lowest material symmetry. As shown in Table 5.1, the strain-energy function can be in this case an arbitrary function of $\mathbf{D}$. Consequently, a triclinic material has 21 independent second-order elastic constants, and $W_2(\mathbf{D})$ has the general form (5.9).

Materials belonging to the *monoclinic* system possess at any point an axis of symmetry of second order or a plane of elastic symmetry. Choosing the unit vector $\mathbf{k}$ along the axis of symmetry, respectively taking it perpendicular to the plane of elastic symmetry, the group $\mathscr{G}^+$ will have the generator $\mathbf{R}_{\mathbf{k}}^{\pi}$. Considering Table 5.1, we infer that the function $W_2(\mathbf{D})$ must have in this case the form

$$W_2 = W_2(D_1, D_2, D_3, D_4^2, D_5^2, D_4D_5). \tag{5.10}$$

Consequently, the quadratic form (5.9) cannot contain the terms $D_KD_4$ and $D_KD_5$, $K = 1, 2, 3, 6$, and hence

$$c_{14} = c_{24} = c_{34} = c_{46} = c_{15} = c_{25} = c_{35} = c_{56} = 0. \tag{5.11}$$

The materials belonging to the *rhombic* system possess at any point two mutually perpendicular planes of elastic symmetry. Choosing the unit vectors $\mathbf{i}$ and $\mathbf{j}$ of the orthonormal basis $\{\mathbf{i}, \mathbf{j}, \mathbf{k}\}$ perpendicular to these planes, the generators of $\mathscr{G}^+$ will be the rotations $\mathbf{R}_{\mathbf{i}}^{\pi}$ and $\mathbf{R}_{\mathbf{j}}^{\pi}$. From Table 5.1 we deduce that

$$W_2 = W_2(D_1, D_2, D_3, D_4^2, D_5^2, D_6^2), \tag{5.12}$$

and hence the quadratic form (5.9) cannot contain the terms $D_KD_M$ with $K = 1, 2, 3$ and $M = 4, 5, 6$. Consequently, the elastic constants $c_{KM}$ that have a single index equal to 4, 5, or 6 must vanish. In particular, it results that the plane perpendicular to the unit vector $\mathbf{k}$ is also a plane of elastic symmetry, and hence any linearly elastic material with rhombic symmetry is an orthotropic material, too.

The materials belonging to the *subsystem 4* of the *tetragonal* system have an axis of symmetry of fourth order. Taking the unit vector $\mathbf{k}$ along this axis, it follows from Table 5.1 that

$$W_2 = W_2(D_1 + D_2, D_3, D_4^2 + D_5^2, D_6^2, D_1D_2, D_1D_6 - D_2D_6). \tag{5.13}$$

Clearly, the conditions (5.11) must be satisfied since (5.13) is a particular case of (5.10). Moreover, by comparing (5.13) with (5.9), we see that

$$c_{11} = c_{22}, \quad c_{13} = c_{23}, \quad c_{44} = c_{55}, \quad c_{45} = c_{36} = 0, \quad c_{26} = -c_{16}. \tag{5.14}$$

The materials belonging to the *subsystem 5* of the *tetragonal* system possess, besides the axis of symmetry of fourth order, another axis of symmetry of second order, perpendicular to the former. Taking the unit vector $\mathbf{i}$ along the axis of symmetry of second order, it results that, in addition to (5.11) and (5.14), we must have $c_{16} = 0$.

Inspection of Table 5.1 reveals that the invariants of the first and second orders of both *subsystems 5 and 6* of the *cubic* system coincide and that in this case

$$W_2 = W_2(D_1 + D_2 + D_3,\ D_1D_2 + D_2D_3 + D_3D_1,\ D_4^2 + D_5^2 + D_6^2). \tag{5.15}$$

Since (5.15) is a particular case of (5.12), it follows that all conditions mentioned for the rhombic system must be fulfilled. Moreover, by comparing (5.15) with (5.9) we find

$$c_{11} = c_{22} = c_{33}, \quad c_{12} = c_{13} = c_{23}, \quad c_{44} = c_{55} = c_{66}. \tag{5.16}$$

The materials belonging to the *subsystem 8* of the *hexagonal* system have an axis of symmetry of third order. Taking the unit vector **k** along this axis and considering Table 5.1, we see that

$$\begin{aligned} W_2 = W_2(D_1 + D_2,\ D_3,\ 4D_1D_2 - D_6^2,\ D_4^2 + D_5^2,\\ D_1D_5 - D_2D_5 - D_4D_6,\quad D_1D_4 - D_2D_4 + D_5D_6). \end{aligned} \tag{5.17}$$

Comparing this expression with (5.9) yields

$$\left.\begin{aligned} &c_{11} = c_{22}, \quad c_{13} = c_{23}, \quad c_{44} = c_{55}, \quad c_{66} = \frac{1}{2}(c_{11} - c_{12}),\\ &c_{16} = c_{26} = c_{36} = c_{45} = 0, \quad c_{14} = -c_{24} = c_{56}, \quad c_{15} = -c_{25} = -c_{46}. \end{aligned}\right\} \tag{5.18}$$

From Table 5.1 it also follows that in the case of the *subsystem 9* of the *hexagonal* system, the strain-energy function no longer depends on the combination $D_1D_5 - D_2D_5 - D_4D_6$. Consequently, (5.18) must be supplemented by the conditions

$$c_{15} = c_{25} = c_{46} = 0. \tag{5.19}$$

The materials belonging to the *subsystems 10 and 11* of the *hexagonal* system possess an axis of symmetry of sixth order. Since in this case the strain-energy function

$$W_2 = W_2(D_1 + D_2,\ D_3,\ 4D_1D_2 - D_6^2,\ D_4^2 + D_5^2) \tag{5.20}$$

does not depend on $D_1D_4 - D_2D_4 + D_5D_6$ either, we must have

$$c_{14} = c_{24} = c_{56} = 0, \tag{5.21}$$

besides (5.18) and (5.19).

The strain-energy function of a material with *transverse isotropy* has the form (5.3). Accordingly, the quadratic form (5.9) can contain only the combinations

$$\left.\begin{aligned} &\mathrm{I}_{\mathbf{D}}^2 = (D_1 + D_2 + D_3)^2, \quad D_3\mathrm{I}_{\mathbf{D}} = D_3(D_1 + D_2 + D_3), \quad D_3^2,\\ &\mathrm{II}_{\mathbf{D}} = D_1D_2 + D_2D_3 + D_3D_1 - \frac{1}{4}(D_4^2 + D_5^2 + D_6^2), \quad D_4^2 + D_5^2, \end{aligned}\right\} \tag{5.22}$$

and hence the only non-zero second-order elastic constants are

$$c_{11} = c_{22}, \quad c_{33}, \quad c_{13} = c_{23}, \quad c_{44} = c_{55}, \quad c_{66} = \frac{1}{2}(c_{11} - c_{12}). \tag{5.23}$$

Therefore, linear elastic materials with transverse isotropy have the same constitutive equations as materials belonging to subsystems 10 and 11 of the hexagonal system. Thus, an axis of symmetry of sixth order assures the transverse isotropy of a linear elastic material with respect to that axis.

Table 5.2 summarizes the above results concerning the second-order elastic constants for all types of material symmetry. For conciseness, the symbol $c$ of the elastic constants has been omitted, only the subscripts of the non-zero independent elastic constants being listed for each type of material symmetry. The order number of the subsystem is written on the first line, and the corresponding number of independent second-order elastic constants is written in brackets on the second line of the table. Table 5.3 shows a compilation of experimental values of adiabatic (isentropic) second-order elastic constants of various single crystals.

*Table 5.2*

**Independent second-order elastic constants for various types of material symmetry**

| Triclinic | Monoclinic | Rhombic | Tetragonal | | Cubic | Hexagonal | | | Isotropic |
|---|---|---|---|---|---|---|---|---|---|
| 1 | 2 | 3 | 4 | 5 | 6,7 | 8 | 9 | 10,11 | — |
| (21) | (13) | (9) | (7) | (6) | (3) | (7) | (6) | (5) | (2) |
| 11 | 11 | 11 | 11 | 11 | 11 | 11 | 11 | 11 | $\lambda + 2\mu$ |
| 12 | 12 | 12 | 12 | 12 | 12 | 12 | 12 | 12 | $\lambda$ |
| 13 | 13 | 13 | 13 | 13 | 13 | 13 | 13 | 13 | $\lambda$ |
| 14 | 0 | 0 | 0 | 0 | 0 | 14 | 14 | 0 | 0 |
| 15 | 0 | 0 | 0 | 0 | 0 | 15 | 0 | 0 | 0 |
| 16 | 16 | 0 | 16 | 0 | 0 | 0 | 0 | 0 | 0 |
| 22 | 22 | 22 | 11 | 11 | 11 | 11 | 11 | 11 | $\lambda + 2\mu$ |
| 23 | 23 | 23 | 13 | 13 | 12 | 13 | 13 | 13 | $\lambda$ |
| 24 | 0 | 0 | 0 | 0 | 0 | −14 | −14 | 0 | 0 |
| 25 | 0 | 0 | 0 | 0 | 0 | −15 | 0 | 0 | 0 |
| 26 | 26 | 0 | −16 | 0 | 0 | 0 | 0 | 0 | 0 |
| 33 | 33 | 33 | 33 | 33 | 11 | 33 | 33 | 33 | $\mu$ |
| 34 | 0 | 0 | 0 | 0 | 0 | 0 | 0 | 0 | 0 |
| 35 | 0 | 0 | 0 | 0 | 0 | 0 | 0 | 0 | 0 |
| 36 | 36 | 0 | 0 | 0 | 0 | 0 | 0 | 0 | 0 |
| 44 | 44 | 44 | 44 | 44 | 44 | 44 | 44 | 44 | $\mu$ |
| 45 | 45 | 0 | 0 | 0 | 0 | 0 | 0 | 0 | 0 |
| 46 | 0 | 0 | 0 | 0 | 0 | −15 | 0 | 0 | 0 |
| 55 | 55 | 55 | 44 | 44 | 44 | 44 | 44 | 44 | $\mu$ |
| 56 | 0 | 0 | 0 | 0 | 0 | 14 | 14 | 0 | 0 |
| 66 | 66 | 66 | 66 | 66 | 44 | $\frac{1}{2}(11-12)$ | $\frac{1}{2}(11-12)$ | $\frac{1}{2}(11-12)$ | $\mu$ |

*Table 5.3*

**Experimental values of adiabatic second-order elastic constants in GPa**
**Cubic system, subsystem 7, crystal class $O_h$ (m3m)**

| Material | $c_{11}$ | $c_{12}$ | $c_{44}$ | Temp. | Source | Structure type |
|---|---|---|---|---|---|---|
| Ag | 122.2 | 90.7 | 45.4 | room | [161] | |
| Al | 106.43 | 60.35 | 28.21 | 300 K | [135] | |
| Au | 192.9 | 163.8 | 41.5 | room | [161] | $A_1$: f.c.c. |
| Cu | 166.1 | 119.9 | 75.6 | room | [161] | |
| Ni | 250.8 | 150.0 | 123.5 | 300 K | [4] | |
| Pb | 49.66 | 42.31 | 14.98 | 296 K | [248] | |
| Th | 75.3 | 48.9 | 47.8 | 300 K | [6] | |
| Cr | 350.0 | 67.8 | 100.8 | 298 K | [34] | |
| $\alpha$−Fe | 230.1 | 134.6 | 116.6 | 300 K | [147] | |
| K | 3.71 | 3.15 | 1.88 | 295 K | [312] | |
| Li | 13.42 | 11.30 | 8.89 | 298 K | [305] | |
| Mo | 463 | 161 | 109 | 300 K | [86] | $A_2$: b.c.c. |
| Na | 7.69 | 6.47 | 4.34 | 299 K | [236] | |
| Nb | 246.5 | 134.5 | 28.73 | 300 K | [33] | |
| Ta | 266.8 | 161.1 | 82.49 | 300 K | [417] | |
| V | 230.98 | 120.17 | 43.76 | 300 K | [35] | |
| W | 523.27 | 204.53 | 160.72 | 300 K | [417] | |
| $\alpha$−CuZn (55.1% Cu) | 119.0 | 102.3 | 174.4 | room | [7] | |
| C | 1079 | 124 | 578 | 298 K | [243] | |
| Ge | 128.528 | 48.260 | 66.799 | 298 K | [241] | $A_4$: Diamond |
| Si | 165.773 | 63.924 | 79.619 | 298 K | [241] | |
| AgBr | 56.10 | 32.70 | 7.24 | 300 K | [217] | |
| AgCl | 59.85 | 36.11 | 6.24 | 300 K | [217] | |
| KBr | 34.68 | 5.80 | 5.07 | 298 K | [303] | |
| KCl | 40.69 | 7.11 | 6.31 | 298 K | [303] | |
| KF | 64.80 | 16.00 | 12.52 | 300 K | [234] | |
| KI | 27.71 | 4.36 | 3.73 | 298 K | [18] | |
| LiBr | 39.20 | 18.90 | 18.85 | 300 K | [235] | |
| LiCl | 48.99 | 22.23 | 24.89 | 299 K | [272] | |
| LiF | 113.97 | 47.67 | 63.64 | room | [96] | |
| LiI | 29.07 | 14.21 | 14.07 | 295 K | [239] | |
| NaBr | 40.37 | 10.13 | 10.15 | 298 K | [181] | $B_1$: NaCl |
| NaCl | 49.36 | 12.88 | 12.78 | 300 K | [136] | |
| NaF | 97.00 | 23.80 | 28.22 | 300 K | [247] | |
| NaI | 30.35 | 9.15 | 7.42 | 298 K | [18] | |
| RbBr | 31.630 | 4.672 | 3.840 | 298 K | [66] | |
| RbCl | 36.589 | 6.153 | 4.753 | 298 K | [66] | |
| RbF | 55.09 | 14.49 | 92.39 | 300 K | [76] | |
| RbI | 25.730 | 3.776 | 2.790 | 298 K | [66] | |
| CaO | 223 | 59 | 81 | 298 K | [315] | |
| MgO | 296.64 | 95.08 | 155.81 | 298 K | [65] | |
| SrO | 173 | 45 | 56 | 298 K | [315] | |
| CsBr | 30.63 | 8.07 | 7.50 | room | [304] | |
| | 30.77 | 8.27 | 7.60 | 286 K | [64] | |
| CsCl | 36.64 | 8.82 | 8.04 | room | [304] | $B_2$: CsCl |
| | 36.83 | 8.93 | 8.17 | 286 K | [64] | |
| CsI | 24.46 | 6.61 | 6.29 | room | [304] | |
| | 24.62 | 6.59 | 6.44 | 286 K | [64] | |

Table 5.3. (continued)

**Cubic system, subsystem 7, crystal class $O_h$(m3m)**

| Material | $c_{11}$ | $c_{12}$ | $c_{44}$ | Temp. | Source | Structure type |
|---|---|---|---|---|---|---|
| $BaF_2$ | 89.48 | 38.54 | 24.95 | 295 K | [134] | $C_1$ : $CaF_2$ |
| $CaF_2$ | 164.94 | 44.62 | 33.80 | 295.5 K | [164] | |
| YFe–g | 269 | 107.7 | 76.4 | room | [100] | Garnet |

**Cubic system, subsystem 7, crystal class $T_d(\bar{4}3m)$**

| GaAs | 119.04 | 53.84 | 59.52 | room | [94] | |
| | 118.77 | 53.72 | 59.44 | room | [242] | $B_3$ : $\beta$ – ZnS |
| InSb | 67.00 | 36.49 | 30.19 | room | [95] | |

**Hexagonal system, subsystem 11, crystal class $D_{6h}$(6/mm), structure type $A_3$: h.c.p.**

| Material | $c_{11}$ | $c_{12}$ | $c_{13}$ | $c_{33}$ | $c_{44}$ | Temp. | Source |
|---|---|---|---|---|---|---|---|
| Be | 288.8 | 20.1 | 4.7 | 354.2 | 154.9 | 298 K | [278] |
| Cd | 115.2 | 39.72 | 40.53 | 51.22 | 20.25 | 300 K | [125] |
| Co | 307.1 | 165.0 | 102.7 | 358.1 | 75.5 | 298 K | [241] |
| Er | 86.3 | 30.5 | 22.7 | 85.5 | 28.1 | 298 K | [274] |
| Mg | 59.40 | 25.61 | 21.44 | 61.60 | 16.40 | 300 K | [306] |
| Ti | 40.80 | 35.4 | 29.0 | 52.80 | 7.26 | 300 K | [114] |
| Y | 77.9 | 28.5 | 21.0 | 76.9 | 24.31 | 300 K | [311] |
| Zn | 163.68 | 36.4 | 53.0 | 63.47 | 38.79 | 295 K | [3] |
| Zr | 143.68 | 73.04 | 65.88 | 165.17 | 32.14 | 298 K | [116] |
| $Ag_2Al$ | 141.5 | 84.7 | 74.6 | 168.5 | 34.08 | 298 K | [62] |

**Hexagonal system, subsystem 11, crystal class $C_{6v}$(6mm), structure type $B_4$: α-ZnS**

| BeO | 460.6 | 126.5 | 88.48 | 491.6 | 147.7 | 298 K | [77] |
| CdS | 84.31 | 52.08 | 45.67 | 91.83 | 14.58 | 300 K | [133] |
| CdSe | 74.90 | 46.09 | 39.26 | 84.51 | 13.15 | 298 K | [77] |
| α–ZnS | 123.4 | 58.5 | 45.5 | 139.6 | 28.85 | 298 K | [59] |

**Hexagonal system, subsystem 9, crystal class $D_{3d}$ (3m)**

| Material | $c_{11}$ | $c_{12}$ | $c_{13}$ | $c_{14}$ | $c_{33}$ | $c_{34}$ | Temp. | Source |
|---|---|---|---|---|---|---|---|---|
| Bi | 63.5 | 24.7 | 24.5 | 72.3 | 38.1 | 11.3 | 301 K | [176] |

**Tetragonal system, subsystem 5, crystal class $D_{4h}$(4/mmm), structure type $A_5$:$TiO_2$**

| Material | $c_{11}$ | $c_{12}$ | $c_{13}$ | $c_{33}$ | $c_{44}$ | $c_{66}$ | Temp | Source |
|---|---|---|---|---|---|---|---|---|
| β–Sn | 72.0 | 58.5 | 37.4 | 88.0 | 21.9 | 24.0 | 301 K | [176] |
| $TiO_2$ | 271.4 | 178.0 | 149.6 | 484.0 | 124.4 | 194.8 | 298 K | [228] |

**Rhombic system, subsystem 3, crystal class $D_{2h}$(mmm), T=298 K**

| Mat. | $c_{14}$ | $c_{12}$ | $c_{13}$ | $c_{22}$ | $c_{23}$ | $c_{33}$ | $c_{44}$ | $c_{55}$ | $c_{66}$ | Source |
|---|---|---|---|---|---|---|---|---|---|---|
| α–U | 214.74 | 46.94 | 21.77 | 198.57 | 107.91 | 267.11 | 124.44 | 73.42 | 74.33 | [139] |

Finally, in the case of an *isotropic* material, the quadratic form $W_2(\mathbf{D})$ depends only on $\mathrm{I}^2_{\mathbf{D}}$ and $\mathrm{II}_{\mathbf{D}}$. Therefore, we may write

$$W_2(\mathbf{D}) = \frac{\lambda + 2\mu}{2}\,\mathrm{I}^2_{\mathbf{D}} - 2\mu\,\mathrm{II}_{\mathbf{D}} = \frac{\lambda}{2}\,(D_{mm})^2 + \mu D_{km} D_{mk}, \tag{5.24}$$

where $\lambda$ and $\mu$ are *Lamé's constants*[1]. With Voigt's notation, (5.24) becomes

$$W_2(\mathbf{D}) = \frac{\lambda + 2\mu}{2}\,(D_1^2 + D_2^2 + D_3^2) + \lambda\,(D_1 D_2 + D_1 D_3 + D_2 D_3) + \\ + \frac{\mu}{2}\,(D_4^2 + D_5^2 + D_6^2) \tag{5.25}$$

and comparing this expression with (5.9) it results that the only non-zero second-order elastic constants of an isotropic material are

$$c_{11} = c_{22} = c_{33} = \lambda + 2\mu, \quad c_{12} = c_{13} = c_{23} = \lambda, \quad c_{44} = c_{55} = c_{66} = \mu.$$

As already mentioned in Sect. 4.4, in the linear elasticity, the strain-energy function is given by $W_2(\mathbf{E})$ and the constitutive equation is given by $(4.63)_1$. By using the results above, and in particular Table 5.2, it is a simple matter to write the explicit form of this constitutive equation for each type of material symmetry. In particular, by comparing $(5.24)_2$ to (4.53), we deduce that the tensor $\mathbf{c}$ of the second-order elastic constants has the components

$$c_{klmn} = \lambda\,\delta_{kl}\delta_{mn} + \mu\,(\delta_{km}\delta_{ln} + \delta_{kn}\delta_{lm}). \tag{5.26}$$

Finally, by substituting (5.26) into (4.48), we obtain the *constitutive equation of an isotropic linear elastic material*

$$\mathbf{T} = \lambda\,(\mathrm{tr}\,\mathbf{E})\,\mathbf{1} + 2\mu\mathbf{E}, \qquad T_{kl} = \lambda E_{mm}\delta_{kl} + 2\mu E_{kl}. \tag{5.27}$$

## 5.3. Higher-order elastic constants

We shall see in chapter III that the solving of non-linear elastic problems requires the knowledge of higher-order elastic constants and in the first place that of the third-order elastic constants. That is why we will devote most of this subsection to a closer examination of third-order elastic constants for various types of material symmetry. To this end we make use of a similar reasoning as before, starting from the results of Smith and Rivlin indicated in Table 5.1 and retaining for each

---

[1] It can be shown that $W_2(\mathbf{D})$ is positive definite if and only if $\mu > 0$, $3\lambda + 2\mu > 0$.

type of anisotropy only the invariants and their products that are of third degree in the components of **D**. The expressions thus obtained are subsequently compared with the terms of third-order degree in the expansion (4.62) of the strain-energy function,

$$W_3(\mathbf{D}) = \frac{1}{3!} C_{KMR} D_K D_M D_R \tag{5.28}$$

and this leads to the restrictions imposed on the third-order elastic constants $C_{KMR}$ by each type of material symmetry.

Since this method has been repeatedly used in the preceding subsection, we confine ourselves to illustrating its application for materials with highest cubic symmetry and for isotropic materials.

In the case of *subsystem 7* of the *cubic* system, $W_3(\mathbf{D})$ may depend only on the following combinations of the invariants listed in the last column of Table 5.1

$$\left.\begin{aligned} &(D_1 + D_2 + D_3)^3, \quad (D_1 + D_2 + D_3)(D_1D_2 + D_2D_3 + D_3D_1), \\ &D_1D_2D_3, \ D_4D_5D_6, \ (D_1 + D_2 + D_3)(D_4^2 + D_5^2 + D_6^2), \\ &(D_2 + D_3) D_4^2 + (D_3 + D_1) D_5^2 + (D_1 + D_2) D_6^2. \end{aligned}\right\} \tag{5.29}$$

By comparing (5.29) to (5.28), it results that the only non-zero third-order elastic constants are

$$\left.\begin{aligned} &C_{111} = C_{222} = C_{333}, \ C_{112} = C_{113} = C_{122} = C_{223} = C_{133} = C_{233}, \ C_{123}, \\ &C_{144} = C_{255} = C_{366}, \ C_{155} = C_{166} = C_{244} = C_{266} = C_{344} = C_{355}, \ C_{456}. \end{aligned}\right\} \tag{5.30}$$

Therefore, a material with highest cubic symmetry has six independent third-order elastic constants.

Table 5.4 shows the results obtained by a similar reasoning for all other types of anisotropy. They coincide with those derived in a different way by Fumi and Hearmon (see Hearmon [159], where a different notation is used, however, for the components of the tensors **C** and **D**). For conciseness the symbol $C$ is again omitted in Table 5.4, only the indexes of the non-zero elastic constants being listed. The number of independent third-order elastic constants is written in brackets under the order number of each system. Table 5.5 shows a compilation of experimental values of the adiabatic (isentropic) third-order elastic constants for various single crystals.

In the case of *isotropic* materials, $W_3(\mathbf{D})$ may contain only the products of third degree in the components of **D** of the principal invariants $\mathrm{I}_\mathbf{D}$, $\mathrm{II}_\mathbf{D}$, and $\mathrm{III}_\mathbf{D}$, i.e.

$$\begin{aligned} \mathrm{I}_\mathbf{D}^3 &= (D_1 + D_2 + D_3)^3, \\ \mathrm{I}_\mathbf{D}\,\mathrm{II}_\mathbf{D} &= (D_1 + D_2 + D_3)\,[D_1D_2 + D_2D_3 + D_3D_1 - \tfrac{1}{4}(D_4^2 + D_5^2 + D_6^2)], \\ \mathrm{III}_\mathbf{D} &= D_1D_2D_3 + \tfrac{1}{4}(D_4D_5D_6 - D_1D_4^2 - D_2D_5^2 - D_3D_6^2). \end{aligned}$$

*Table 5.5*

**Experimental values of adiabatic third-order elastic constants in GPa for cubic materials[1], at room temperature**

| Material | $C_{111}$ | $C_{112}$ | $C_{123}$ | $C_{144}$ | $C_{166}$ | $C_{456}$ | Source | Structure type |
|---|---|---|---|---|---|---|---|---|
| Cu | −1271 | −814 | −50 | −3 | −780 | −95 | [161] | |
| Ag | −843 | −529 | 189 | 56 | −637 | 83 | [161] | $A_1$: f.c.c. |
| Al | −1076 | −315 | 36 | −23 | −340 | −30 | [350] | |
| Au | −1729 | −922 | −233 | −13 | −648 | −12 | [161] | |
| Ge | −710 | −389 | −18 | −23 | −292 | −53 | [241] | |
| | −681 | −363 | −9 | 9 | −306 | −43 | [92] | $A_4$: Diamond |
| Si | −825 | −451 | −64 | 12 | −310 | −64 | [241] | |
| | −744 | −418 | 2 | 29 | −315 | −70 | [93] | |
| LiF | −1423 | −264 | 15.6 | 85 | −273 | 94 | [96] | |
| KCl | −701 | −22.4 | 13.3 | 12.7 | −24.5 | 11.8 | [63] | |
| | −726 | −24 | 11 | 23.0 | −26.0 | 16.0 | [96] | |
| NaCl | −880 | −57.1 | 28.4 | 25.8 | −61.1 | 27.1 | [63] | $B_1$: NaCl |
| | −823 | 20.0 | 53.0 | 23.0 | −61 | 20 | [140] | |
| | −863.6 | −49.6 | 9.3 | 7.1 | −58.7 | 13.2 | [326] | |
| | −843 | −50.0 | 46.0 | 29.0 | −60 | 26.0 | [96] | |
| MgO | −4895 | −95 | −69.0 | 113 | −659 | 147 | [32] | |
| $BaF_2$ | −584 | −299 | −206 | −121 | −88.9 | −27.1 | [134] | $C_1$: $CaF_2$ |
| YFe−g | −2330 | −717 | −33 | −148 | −306 | −97 | [100] | Garnet |
| GaAs | −675 | −402 | −4 | −70 | −320 | −69 | [94] | |
| | −622 | −387 | −57 | 2 | −269 | −39 | [242] | $B_3$: $\beta$−ZnS |
| InSb | −314 | −210 | −48 | 9 | −118 | 0.2 | [95] | |

[1] All materials listed in this table belong to subsystem 7 and they pertain to the crystal class $O_h$ (m3m), except GaAs and InSb that belong to the crystal class $T_d$ ($\bar{4}3m$).

Comparing these expressions with (5.28), we see that the six independent third-order elastic constants of subsystem 7 must be related by three supplementary equations, namely

$$C_{144} = (C_{112} - C_{123})/2, \quad C_{155} = (C_{111} - C_{112})/4, \quad C_{456} = (C_{111} - 3C_{112} + 2C_{123})/8. \tag{5.31}$$

By choosing after Toupin and Bernstein [355] as independent elastic constants

$$C_{123} = \nu_1, \quad C_{144} = \nu_2, \quad C_{456} = \nu_3, \tag{5.32}$$

we obtain from (5.31) the expressions of the other three elastic constants

$$C_{111} = \nu_1 + 6\nu_2 + 8\nu_3, \quad C_{112} = \nu_1 + 2\nu_2, \quad C_{155} = \nu_2 + 2\nu_3.$$

It can be shown that the tensor **C** of an isotropic material has the Cartesian components

$$C_{klmnrs} = \nu_1 \delta_{kl}\delta_{mn}\delta_{rs} + \nu_2 \{\delta_{kl}(\delta_{mr}\delta_{ns} + \delta_{ms}\delta_{nr}) + \delta_{mn}(\delta_{kr}\delta_{ls} + \delta_{ks}\delta_{lr}) + \\ + \delta_{rs}(\delta_{km}\delta_{ln} + \delta_{kn}\delta_{lm})\} + \nu_3 \{\delta_{km}(\delta_{lr}\delta_{ns} + \delta_{ls}\delta_{nr}) + \delta_{ln}(\delta_{kr}\delta_{ms} + \\ + \delta_{ks}\delta_{mr}) + \delta_{kn}(\delta_{lr}\delta_{ms} + \delta_{ls}\delta_{mr}) + \delta_{lm}(\delta_{kr}\delta_{ns} + \delta_{ks}\delta_{nr})\}. \tag{5.33}$$

Finally, by substituting (5.26) and (5.33) into (4.55), we deduce that the *constitutive equation of second-order elasticity for isotropic materials* is

$$T_{kl} = \lambda E_{mm}\delta_{kl} + 2\mu E_{kl} + \left\{\frac{\lambda}{2} H_{mn}H_{mn} + \left(\frac{\nu_1}{2} - \lambda\right)(E_{mm})^2 + \nu_2 E_{mn}E_{mn}\right\}\delta_{kl} + \\ + 2(\lambda - \mu + \nu_2) E_{mm}E_{kl} + \mu H_{km}H_{lm} + 4(\mu + \nu_3) E_{km}E_{ml}. \tag{5.34}$$

Going a step further in the expansion (4.43) of the strain-energy function, we consider the terms of fourth-order degree in the components of the finite strain tensor, $W_4(\mathbf{D}) = (1/24)\, L_{ijklmnrs} D_{ij} D_{kl} D_{mn} D_{rs}$, where

$$L_{ijklmnrs} = \left.\frac{\partial^4 W(\mathbf{D})}{\partial D_{ij}\,\partial D_{kl}\,\partial D_{mn}\,\partial D_{rs}}\right|_{\mathbf{D}=0}$$

are the *fourth-order elastic constants*. The number of independent fourth-order elastic constants for all crystal classes and for the isotropic case have been obtained independently by Markenscoff [450] and by Brendel [402]. The latter author has also developed computer programs that allow to obtain the dependence relations between the elastic constants of $n$'th order, as well as the independent and the zero constants for each crystal class.

Fourth-order elastic constants are used to describe higher-order non-linearities occurring in such phenomena as generation of higher-harmonics in finite-amplitude waves, pressure dependence of elastic constants at higher pressures, temperature dependence of the second-order elastic constants, and shock waves in solids that can sustain large compressions.

The most precise method for the determination of higher-order elastic constants is based on the accurate measurement of the velocity of small-amplitude waves superposed on homogeneously prestressed media (Thurston and Brugger [481], Markenscoff [449]). The third- and fourth-order elastic constants are related, respectively, to the first and second derivative of the wave velocity with respect to the initially applied stress, both taken at the zero stress.

It is interesting to note that some of the experimental results available to date (see, e.g. Chang and Barsch [66, 406, 407], Graham [421]) show that partial contractions $C_{ijklmm}$ and $L_{ijklmmrr}$ of the third- and fourth-order elastic constants are about 10 to 25 times, respectively 200 to 500 times, larger than the corresponding second-order elastic constants. This clearly illustrates the rather slow convergence of the expansion (4.43), at least in the cases investigated so far.

## 5.4. Transformation of elastic constants under a change of co-ordinates

Until now we have constantly assumed that the Cartesian axes of co-ordinates $x_1, x_2, x_3$ are taken, respectively, along the unit vectors **i, j, k** that are associated with the preferred directions of the elastic material, in order to take the maximum advantage of the material symmetry for simplifying the form of the constitutive equations. An alternative approach, which proves to be particularly useful in continuum mechanics, is to choose a co-ordinate frame that exploits the geometric symmetry of the body and/or the symmetry of stress state. In such situations it is necessary to know the rules governing the change of the elastic constants under a transformation of co-ordinates.

For the sake of convenience, in what follows, the unit vectors **i, j, k,** which are associated with the preferred directions of the material in the reference configuration, will be denoted by $\mathbf{e}_1$, $\mathbf{e}_2$, $\mathbf{e}_3$, respectively. Let $D_{kl}$, $c_{klmn}$, $C_{klmnrs}$ be the components of the tensors $\mathbf{D}, \mathbf{c}, \mathbf{C}$ in the bases $\mathbf{e}_k\mathbf{e}_l$, $\mathbf{e}_k\mathbf{e}_l\mathbf{e}_m\mathbf{e}_n$, and $\mathbf{e}_k\mathbf{e}_l\mathbf{e}_m\mathbf{e}_n\mathbf{e}_r\mathbf{e}_s$, respectively, and $D'_{kl}$, $c'_{klmn}$, $C'_{klmnrs}$ the components of the same tensors in the bases $\mathbf{e}'_k\mathbf{e}'_l$, $\mathbf{e}'_k\mathbf{e}'_l\mathbf{e}'_m\mathbf{e}'_n$, and $\mathbf{e}'_k\mathbf{e}'_l\mathbf{e}'_m\mathbf{e}'_n\mathbf{e}'_r\mathbf{e}'_s$, respectively. We denote, as in Sect. 1.1, by $q_{kp}$'s the direction cosines of the unit vectors $\mathbf{e}_k$ with respect to the unit vectors $\mathbf{e}'_p$, i.e.

$$q_{kp} = \cos(\mathbf{e}_k, \mathbf{e}'_p) = \mathbf{e}_k \cdot \mathbf{e}'_p. \tag{5.35}$$

We have then, by $(1.35)_1$,

$$D_{kl} = q_{kp} q_{lr} D'_{pr}. \tag{5.36}$$

By making use of Voigt's notation, the last relation gives for typical components of **D**

$$D_1 = q_{11}^2 D'_1 + q_{12}^2 D'_2 + q_{13}^2 D'_3 + q_{12}q_{13}D'_4 + q_{13}q_{11}D'_5 + q_{11}q_{12}D'_6,$$

$$D_4 = 2q_{21}q_{31}D'_1 + 2q_{22}q_{32}D'_2 + 2q_{23}q_{33}D'_3 + (q_{22}q_{33} + q_{23}q_{32})\,D'_4 + \\ + (q_{23}q_{31} + q_{21}q_{33})\,D'_5 + (q_{21}q_{32} + q_{22}q_{31})\,D'_6$$

and, in general,

$$D_K = Q_{KP} D'_P, \tag{5.37}$$

where the transformation matrix $Q = [Q_{KP}]$ is given by

$$Q = \begin{bmatrix} q_{11}^2 & q_{12}^2 & q_{13}^2 & q_{12}q_{13} & q_{13}q_{11} & q_{11}q_{12} \\ q_{21}^2 & q_{22}^2 & q_{23}^2 & q_{22}q_{23} & q_{23}q_{21} & q_{21}q_{22} \\ q_{31}^2 & q_{32}^2 & q_{33}^2 & q_{32}q_{33} & q_{33}q_{31} & q_{31}q_{32} \\ 2q_{21}q_{31} & 2q_{22}q_{32} & 2q_{23}q_{33} & q_{22}q_{33}+q_{23}q_{32} & q_{23}q_{31}+q_{21}q_{33} & q_{21}q_{32}+q_{22}q_{31} \\ 2q_{31}q_{11} & 2q_{32}q_{12} & 2q_{33}q_{13} & q_{32}q_{13}+q_{33}q_{12} & q_{33}q_{11}+q_{31}q_{13} & q_{31}q_{12}+q_{32}q_{11} \\ 2q_{11}q_{21} & 2q_{12}q_{22} & 2q_{13}q_{23} & q_{12}q_{23}+q_{13}q_{22} & q_{13}q_{21}+q_{11}q_{23} & q_{11}q_{22}+q_{12}q_{21} \end{bmatrix}. \tag{5.38}$$

Since the strain-energy function $W(\mathbf{D})$ must be invariant under a change of co-ordinate frame, it follows from (4.62) that

$$W(\mathbf{D}) = \frac{1}{2!}\, c_{KM} D_K D_M + \frac{1}{3!}\, C_{KMR} D_K D_M D_R + \dots$$

$$= \frac{1}{2!}\, c'_{PT} D'_P D'_T + \frac{1}{3!}\, C'_{PTV} D'_P D'_T D'_V + \dots$$

Substituting (5.37) into this relation and equating coefficients of $D'_P D'_T$ and $D'_P D'_T D'_V$ leads to the *transformation rule of the elastic constants of second and third order under a change of co-ordinates*[1]

$$c'_{PT} = Q_{KP} Q_{MT} c_{KM}, \qquad C'_{RTV} = Q_{KP} Q_{MT} Q_{RV} C_{KMR}. \tag{5.39}$$

Although the application of these transformation rules requires the previous calculation of the matrix $Q = [Q_{KM}]$ by (5.38), it is generally more convenient than the use of the direct transformation rules of the components $c_{klmn}$ and $C_{klmnrs}$ by means of the formulae given in Sect. 1.1 and of the matrix $q = [q_{km}]$.

# 6. Linear theory of elasticity

## 6.1. Fundamental field equations

The linearization of the kinematic equations and of the elastic constitutive equations has been done in Sects. 2.6 and 4.3, respectively. We have also remarked that in the linear theory the spatial co-ordinates of the particles can be identified with their material co-ordinates when calculating the gradients of scalar, vector, or tensor fields. Consequently, we shall adopt in the linear theory of elasticity the simplified notation

$$\frac{\partial(\cdot)}{\partial X_m} = \frac{\partial(\cdot)}{\partial x_m} = (\cdot)_{,m}.$$

Moreover, by using Lagrange's theorem, it may be shown that, if $\|(\mathrm{Grad}\,\boldsymbol{\Phi})\mathbf{u}\| \ll \|\boldsymbol{\Phi}\|$ at any point $\mathbf{X}$ of the body, where $\boldsymbol{\Phi}$ is an arbitrary vector or tensor field of class $C^1$ and $\mathbf{u}$ is the displacement vector, then $\boldsymbol{\Phi}(\mathbf{X}, t) \approx \boldsymbol{\Phi}(\mathbf{x}, t)$. In the linear theory of elasticity it is assumed that all fields occurring in the for-

---

[1] It is interesting to note that $D_K$, $c_{KM}$, and $C_{KMR}$ follow, under a change of co-ordinate frame, the transformation rules of the Cartesian components of a vector, a second-order tensor, and a third-order tensor, respectively. It should be remembered, however, that capital Latin subscripts range over the values 1, 2, ..., 6, and hence the dimensions of the corresponding vector and tensor spaces increase accordingly.

mulation of the boundary-value problems satisfy this condition; consequently, no distinction is being made between material and spatial co-ordinates of the particles.

In what follows we shall consider mostly applications of linear elastostatics [1] to the modelling of crystal defects. Therefore, we recollect below for convenience the *basic field equations of linear elastostatics*, namely the kinematic equations

$$E_{km} = \frac{1}{2}(u_{k,m} + u_{m,k}), \tag{6.1}$$

the equilibrium equations

$$T_{km,m} + f_k = 0, \tag{6.2}$$

and the constitutive equations

$$T_{kl} = c_{klmn} E_{mn}, \tag{6.3}$$

or

$$E_{kl} = s_{klmn} T_{mn}. \tag{6.4}$$

We recall that **u** is the displacement vector, **E** is the infinitesimal strain tensor, **T** is the stress tensor, $\mathbf{f} = \rho_0 \mathbf{b}$ is the body force per unit volume, **c** is the tensor of second-order elastic constants, and **s** is the tensor of second-order elastic compliances. We assume throughout that the body is *homogeneous*, i.e. the tensors **c** and **s** do not depend on **x**.

For *isotropic* bodies, (6.3) reduces to

$$T_{kl} = \lambda E_{mm} \delta_{kl} + 2\mu E_{kl}, \tag{6.5}$$

where $\lambda$ and $\mu$ are Lamé's constants. By contraction, (6.5) leads to

$$T_{mm} = 3K E_{mm}, \tag{6.6}$$

where

$$K = \lambda + \frac{2\mu}{3} \tag{6.7}$$

is the *bulk modulus*. In view of (6.6), we can solve equations (6.5) with respect to $E_{kl}$, thus obtaining

$$E_{kl} = \frac{1}{2\mu}\left(T_{kl} - \frac{\nu}{1+\nu} T_{mm}\delta_{kl}\right), \tag{6.8}$$

where

$$\nu = \frac{\lambda}{2(\lambda + \mu)} \tag{6.9}$$

is *Poisson's ratio*.

[1] Linear elastostatics deals with the equilibrium of linear elastic bodies.

As already shown in Sect. 2.7, the components of the infinitesimal strain tensor **E** must satisfy the compatibility equations

$$-\in_{ikl}\in_{jmn}E_{ln,km}=0, \tag{6.10}$$

which can be also written as

$$\Delta E_{kl}-E_{lm,km}-E_{km,lm}+E_{mm,kl}=0. \tag{6.11}$$

We have also seen in Sect. 4.3 that the strain energy density of a linear elastic material may be expressed as a quadratic form in the components of **E**, namely

$$W=W_2(\mathbf{E})=\frac{1}{2}\,c_{klmn}E_{kl}E_{mn}. \tag{6.12}$$

In view of (6.3) and (6.4), this relation may be rewritten in the alternative forms

$$W=\frac{1}{2}\,T_{km}E_{km}=\frac{1}{2}\,s_{klmn}T_{kl}T_{mn}. \tag{6.13}$$

In the *isotropic* case, the strain-energy function takes the form (5.24) with **D** replaced by **E**, i.e.

$$W=W_2(\mathbf{E})=\frac{\lambda}{2}\,(E_{mm})^2+\mu E_{km}E_{km}. \tag{6.14}$$

Throughout the remainder of this chapter we assume that **f** is a continuous vector field on $\overline{\mathscr{V}}=\mathscr{V}\cup\mathscr{S}$, where $\mathscr{V}$ is the region occupied by the elastic body in the current configuration and $\mathscr{S}$ denotes its boundary. We also assume that $\mathscr{S}$ is the union of a finite number of non-intersecting closed surfaces that are two-sided and piecewise smooth.

By an *admissible state* we mean an ordered array $\mathscr{s}=[\mathbf{u},\mathbf{E},\mathbf{T}]$, where **u** is of class $C^2$ on $\overline{\mathscr{V}}$, while **E** and **T** are of class $C^2$ on $\overline{\mathscr{V}}$. An admissible state that satisfies equations (6.1—3) is called an *elastic state* corresponding to the body force **f**. Clearly, by (6.1—3), when **u** is the displacement field of an elastic state, the regularity conditions adopted for **u** imply those assumed for **E**, **T**, and **f**.

By virtue of (3.18), the surface traction **t** on $\mathscr{S}$ corresponding to the stress tensor **T** is

$$\mathbf{t}(\mathbf{x})=\mathbf{T}(\mathbf{x})\mathbf{n}(\mathbf{x}), \tag{6.15}$$

where $\mathbf{n}(\mathbf{x})$ is the outward unit normal to $\mathscr{S}$ at **x**. We call the pair [**f**, **t**] the *external force system* for the elastic state $\mathscr{s}$.

The following two theorems are consequences of the equations of equilibrium (6.2).

**Theorem of work and energy.** *If the elastic state* [**u**, **E**, **T**] *corresponds to the external force system* [**f**, **t**] *then*

$$\mathscr{W}=\int_{\mathscr{V}}W\,\mathrm{d}v=\frac{1}{2}\int_{\mathscr{V}}\mathbf{T}\cdot\mathbf{E}\,\mathrm{d}v=\frac{1}{2}\int_{\mathscr{S}}\mathbf{t}\cdot\mathbf{u}\,\mathrm{d}s+\frac{1}{2}\int_{\mathscr{V}}\mathbf{f}\cdot\mathbf{u}\,\mathrm{d}v. \tag{6.16}$$

*Proof.* By (6.13) and (6.1), we have

$$\mathscr{W} = \frac{1}{2}\int_{\mathscr{V}} T_{kl}E_{kl}\,dv = \frac{1}{2}\int_{\mathscr{V}} T_{kl}u_{k,l}dv = \frac{1}{2}\int_{\mathscr{V}} \{(T_{kl}u_k)_{,l} - u_kT_{kl,l}\}\,dv.$$

Now, by making use of (1.52) and taking into account (6.2) and (6.15), we obtain (6.16) and the theorem is proved.

**Betti's reciprocal theorem** [25]. *Let* [**u**, **E**, **T**] *and* [**u***, **E***, **T***] *be two elastic states corresponding to the external force systems* [**f**, **t**] *and* [**f***, **t***], *respectively. Then*

$$\int_{\mathscr{V}} \mathbf{T}\cdot\mathbf{E}^*\,dv = \int_{\mathscr{S}} \mathbf{t}\cdot\mathbf{u}^*ds + \int_{\mathscr{V}} \mathbf{f}\cdot\mathbf{u}^*\,dv = \int_{\mathscr{V}} \mathbf{T}\cdot\mathbf{E}^*\,dv = \tag{6.17}$$

$$= \int_{\mathscr{S}} \mathbf{t}^*\cdot\mathbf{u}\,ds + \int_{\mathscr{V}} \mathbf{f}^*\cdot\mathbf{u}\,dv.$$

*Proof.* By using an analogous reasoning as in the proof of the preceding theorem it is easy to see that

$$\int_{\mathscr{V}} \mathbf{T}\cdot\mathbf{E}^*\,dv = \int_{\mathscr{S}} \mathbf{t}\cdot\mathbf{u}^*\,ds + \int_{\mathscr{V}} \mathbf{f}\cdot\mathbf{u}^*\,dv,$$

$$\int_{\mathscr{V}} \mathbf{T}^*\cdot\mathbf{E}\,dv = \int_{\mathscr{S}} \mathbf{t}^*\cdot\mathbf{u}\,ds + \int_{\mathscr{V}} \mathbf{f}^*\cdot\mathbf{u}\,dv.$$

On the other hand, the symmetry of **c** implies that $\mathbf{T}\cdot\mathbf{E}^* = \mathbf{T}^*\cdot\mathbf{E}$ and this completes the proof.

## 6.2. Boundary-value problems of linear elastostatics

The field equations (6.1 − 3) of linear elastostatics constitute a system of 15 scalar equations with 15 unknowns: six components of the stress tensor, six components of the strain tensor, and three components of the displacement vector. A boundary-value problem is the problem of finding solutions of the field equations that satisfy certain *boundary conditions.*

The main boundary-value problems occurring in linear elastostatics are of the following three types:

1. *The displacement boundary-value problem.* The displacement vector is prescribed on the boundary of the body, i.e.

$$\mathbf{u} = \mathbf{u}^\circ \quad \text{on } \mathscr{S}. \tag{6.18}$$

2. *The traction boundary-value problem.* The surface traction is prescribed on the boundary of the body, i.e.

$$\mathbf{Tn} = \mathbf{t}^\circ \quad \text{on } \mathscr{S}. \tag{6.19}$$

3. *The mixed boundary-value problem.* The displacement vector is prescribed on a part $\mathscr{S}_1$ of the boundary and the surface traction is prescribed on the complementary part $\mathscr{S}_2$ of $\mathscr{S}(\mathscr{S}_1 \cup \mathscr{S}_2 = \mathscr{S}, \mathscr{S}_1$ and $\mathscr{S}_2$ have no common interior points), i.e.

$$\mathbf{u} = \mathbf{u}^\circ \quad \text{on } \mathscr{S}_1, \qquad \mathbf{Tn} = \mathbf{t}^\circ \quad \text{on } \mathscr{S}_2. \tag{6.20}$$

Clearly, the *displacement boundary conditions* (6.18) or $(6.20)_1$ correspond to constraining the boundary $\mathscr{S}$ of the body, or a part $\mathscr{S}_1$ of it, to assume a given shape in the deformed configuration, whereas the *traction boundary conditions* (6.19) or $(6.20)_2$ correspond to prescribing the loading on the surface $\mathscr{S}$ of the body, or a part $\mathscr{S}_2$ of it, respectively.

Any solution of a boundary-value problem in statics must be such that the total force and the total torque acting on the body in the configuration of equilibrium vanish:

$$\int_{\mathscr{S}} \mathbf{t}\, ds + \int_{\mathscr{V}} \mathbf{f}\, dv = \mathbf{0}, \qquad \int_{\mathscr{S}} \mathbf{x} \times \mathbf{t}\, ds + \int_{\mathscr{V}} \mathbf{x} \times \mathbf{f}\, dv = \mathbf{0}.$$

We shall consider now shortly the uniqueness question appropriate to the boundary-value problems of elastostatics formulated above [1].

**Kirchhoff's uniqueness theorem** [179]. *If the elastic body is simply-connected and the density of its strain energy $W = W_2(\mathbf{E})$ is positive definite, then*

(i) *the displacement boundary-value problem has at most one solution*;

(ii) *the mixed boundary-value problem has at most one solution*;

(iii) *two solutions of the traction boundary-value problem differ by an infinitesimal rigid displacement.*

*Proof.* Consider first the mixed boundary conditions (6.20), which include as particular cases the displacement and the traction boundary conditions ($\mathscr{S}_2 = \emptyset$ and $\mathscr{S}_1 = \emptyset$, respectively). Let $\mathbf{u}'$ and $\mathbf{u}''$ be two displacement fields that satisfy equations (6.1—3) and the boundary conditions (6.20). We denote by $\mathbf{E}', \mathbf{E}''$ the

---

[1] The extension of the uniqueness theorems to the dynamic case may be done without great difficulty provided that the boundary conditions be supplemented by initial conditions of the type $\mathbf{u}(\mathbf{x}, t_0) = \tilde{\mathbf{u}}(\mathbf{x})$, $\dot{\mathbf{u}}(\mathbf{x}, t_0) = \tilde{\mathbf{v}}(\mathbf{x})$ for $\mathbf{x} \in \overline{\mathscr{V}}$, and the time-variation of the functions $\mathbf{u}^\circ$ and $\mathbf{t}^\circ$ be equally prescribed (see, e.g. Sokolnikoff [313], Sect. 27).

infinitesimal strain tensors and by $\mathbf{T}', \mathbf{T}''$ the stress tensors corresponding to the displacement vectors $\mathbf{u}'$ and $\mathbf{u}''$, respectively. Let

$$\mathbf{u} = \mathbf{u}' - \mathbf{u}'', \quad \mathbf{E} = \mathbf{E}' - \mathbf{E}'', \quad \mathbf{T} = \mathbf{T}' - \mathbf{T}''. \tag{6.21}$$

From (6.1 – 3), (6.20), and (6.21) it follows that

$$E_{km} = \frac{1}{2}(u_{k,m} + u_{m,k}), \quad T_{km,m} = 0, \quad T_{kl} = c_{klmn}E_{mn} \text{ in } \mathscr{V} \tag{6.22}$$

and that

$$\mathbf{u} = \mathbf{0} \text{ on } \mathscr{S}_1, \quad \mathbf{t} = \mathbf{0} \text{ on } \mathscr{S}_2. \tag{6.23}$$

By (6.23) we have also $\mathbf{u} \cdot \mathbf{t} = 0$ on $\mathscr{S}$, since $\mathscr{S}_1$ and $\mathscr{S}_2$ are complementary subsets of $\mathscr{S}$. Substituting this result into (6.16) and taking into account that the stress field $\mathbf{T}$ corresponds, by $(6.22)_2$, to zero body forces, it results that

$$\int_{\mathscr{V}} W \,\mathrm{d}v = 0. \tag{6.24}$$

Consequently, since $W = W_2(\mathbf{E})$ is by hypothesis a positive definite quadratic form in the components of $\mathbf{E}$, we deduce that $\mathbf{E} = \mathbf{0}$ and, by $(6.22)_3$, that $\mathbf{T} = \mathbf{0}$, too. Moreover, as already mentioned in Sect. 2.7, $\mathbf{E} = \mathbf{0}$ implies that $\mathbf{u}$ is a rigid displacement. By (6.21), we conclude that $\mathbf{E}' = \mathbf{E}''$, $\mathbf{T}' = \mathbf{T}''$ in $\mathscr{V}$ and that $\mathbf{u}'$ and $\mathbf{u}''$ differ by a rigid displacement. In the case of the displacement boundary conditions, we must have $\mathbf{u} = \mathbf{0}$ on $\mathscr{S}$ or a part of it, and hence $\mathbf{u}$ must vanish identically. On the other hand, the solution of the traction boundary-value problem, if any, is unique to within a rigid displacement.

Let us consider now a *multiply-connected* elastic body $\mathscr{B}$. We have seen in Sect. 2.8 that it is possible to produce in such a body a *state of self-stress*, i.e. a non-zero stress state corresponding to vanishing external forces. Assume that the body $\mathscr{B}$ occupies an $n$-tuply connected region $\mathscr{V}$ and let $S_1, S_2, \ldots, S_{n-1}$ be $n - 1$ cuts rendering $\mathscr{V}$ simply-connected. Let the faces of each cut be relatively displaced by a small *rigid* displacement, and the opposing faces of the cuts be joined, by removal or insertion, if necessary, of a thin sheet of matter of the same kind as that forming the original body. Then the body will be again multiply-connected, but in a state of self-strain, called a Volterra dislocation. As shown in Sect. 2.8, the displacement $\mathbf{u}$ may be defined as a single-valued vector field of class $C^3$ on $\mathscr{V} \setminus \bigcup_{k=1}^{n-1} S_k$, the jumps of $\mathbf{u}$ across the surfaces $S_k$ being equal to the relative rigid displacements of the cut faces, whereas the infinitesimal strain field $\mathbf{E}$ corresponding to $\mathbf{u}$ is continuous across the surfaces $S_k$, and the extension (by continuity) of $\mathbf{E}$ to $\mathscr{V}$ is of class $C^2$. Alternatively, the displacement may be considered as a multiple-valued vector field of class $C^3$ on $\mathscr{V}$ with cyclic constants given again by the translation and rotation vectors of the relative rigid displacements of the cut faces.

For a multiply-connected body, Kirchhoff's uniqueness theorem must be replaced by the following theorem, which we give without proof.

**Volterra's uniqueness theorem** [373]. *Suppose that an elastic body $\mathscr{B}$ occupies an n-tuply connected region $\mathscr{V}$, that its strain-energy density $W = W_2(\mathbf{E})$ is a positive definite function, and that the jumps of the displacement vector* **u** *across $n-1$ cuts transforming $\mathscr{V}$ into a simply-connected region are prescribed. Then*

(i) *the displacement boundary-value problem has at most one solution;*

(ii) *the mixed boundary-value problem has at most one solution;*

(iii) *two solutions of the traction boundary-value problem differ by an infinitesimal rigid displacement.*

Finally, let us consider the case when the region $\mathscr{V}$ occupied by the elastic body is *infinite*, but its boundary $\mathscr{S}$ consists of a finite number of closed, bounded, piecewise smooth, and non-intersecting surfaces. Besides the boundary conditions (6.18), (6.19), or (6.20) on $\mathscr{S}$, we consider also the following alternative *complementary conditions*:

$$(C_1)\quad \lim_{\|\mathbf{x}\|\to\infty} \|\mathbf{u}(\mathbf{x}) - \hat{\mathbf{u}}\| = 0,$$

$$(C_2)\quad \lim_{\|\mathbf{x}\|\to\infty} \|\mathbf{T}(\mathbf{x}) - \hat{\mathbf{T}}\| = 0, \qquad \int_{\mathscr{S}} \mathbf{t}\,ds = \hat{\mathbf{f}}, \qquad \int_{\mathscr{S}} \mathbf{x}\times\mathbf{t}\,ds = \hat{\mathbf{m}},$$

$$(C_3)\quad \lim_{\|\mathbf{x}\|\to\infty} \|\mathbf{T}(\mathbf{x}) - \hat{\mathbf{T}}\| = 0, \qquad \int_{\mathscr{S}} \mathbf{t}\,ds = \hat{\mathbf{f}}, \ \lim_{\|\mathbf{x}\|\to\infty} \|\boldsymbol{\omega}(\mathbf{x}) - \hat{\boldsymbol{\omega}}\| = 0,$$

where $\hat{\mathbf{u}}$, $\hat{\mathbf{f}}$, $\hat{\mathbf{m}}$, and $\hat{\boldsymbol{\omega}}$ are constant vectors, $\hat{\mathbf{T}}$ is a constant symmetric second-order tensor, and $\boldsymbol{\omega} = -\frac{1}{2}\operatorname{curl}\mathbf{u}$ is the infinitesimal rotation vector. With the above notation we can formulate the following uniqueness theorem, which holds for the solutions $\mathbf{u}(\mathbf{x})$ that are uniform and of class $C^2$ in $\bar{\mathscr{V}} = \mathscr{V} \cup \mathscr{S}$.

**Bézier's uniqueness theorem** [26]. *Suppose that the strain-energy density $W = W_2(\mathbf{E})$ of the body $\mathscr{B}$ is positive definite and that $\mathscr{B}$ occupies an infinite region $\mathscr{V}$, whose boundary $\mathscr{S}$ consists of a finite number of closed, bounded, piecewise smooth, and non-intersecting surfaces. Then:*

(i) *the displacement, the traction, and the mixed boundary-value problems have at most one solution that satisfies the complementary condition* $(C_1)$;

(ii) *the displacement and the mixed boundary-value problems have at most one solution that satisfies the complementary conditions* $(C_2)$ *or* $(C_3)$;

(iii) *two solutions of the traction boundary-value problem differ by an infinitesimal rigid displacement if they satisfy the complementary condition* $(C_2)$, *and by an infinitesimal rigid translation if they satisfy the complementary condition* $(C_3)$.

For a proof of this theorem in the isotropic case we refer to Fichera [115], Gurtin and Sternberg [149], and Gurtin [150], Sect. 50. An analogous theorem holds for the plane problem of linear elastostatics (see Muskhelishvili [254], Sect. 41).

## 6.3. Stress and displacement formulations of the boundary-value problems

The *stress formulation* of the boundary-value problem is generally used in conjunction with the traction boundary condition (6.19). This formulation can be obtained in the following way. By substituting (6.4) into (6.10), we obtain the *compatibility conditions in terms of stresses*

$$-\epsilon_{ikl}\epsilon_{jmn}s_{lnpq}T_{pq,km} = 0.$$

In the *isotropic* case, these equations take a much simpler form. Namely, by introducing (6.8) into (6.11) and considering (6.2), it follows that

$$\Delta T_{kl} + \frac{1}{1+\nu}(T_{mm,kl} - \nu\delta_{kl}\Delta T_{mm}) + f_{k,l} + f_{l,k} = 0,$$

wherefrom it results by contraction

$$\Delta T_{mm} = -\frac{1+\nu}{1-\nu} f_{m,m}.$$

Combining the last two relations yields the *Beltrami-Michell compatibility equations*[1]

$$\Delta T_{kl} + \frac{1}{1+\nu} T_{mm,kl} + f_{k,l} + f_{l,k} + \frac{\nu}{1-\nu}\delta_{kl} f_{m,m} = 0. \qquad (6.25)$$

Besides the boundary conditions (6.19), the six unknown stresses must satisfy the three equilibrium equations (6.2) and the six compatibility conditions (6.25), i.e. nine field equations; however, equations (6.25) represent only three independent conditions, as already noticed in Sect. 2.7.

In the absence of body forces [2], the equilibrium equations (6.2) can be identically satisfied, in view of $(1.51)_2$, by using *Beltrami's solution* [23]

$$\mathbf{T} = \operatorname{inc} \boldsymbol{\chi}, \qquad T_{ij} = -\epsilon_{ikl}\epsilon_{jmn}\chi_{ln,km}, \qquad (6.26)$$

where $\boldsymbol{\chi}$ is the (symmetric) *stress function tensor*.

It can be shown (see Schaefer [282] and Gurtin [150], Sect. 17) that Beltrami's solution is *complete*, i.e. any stress field admits a representation as a Beltrami solution, if either the body is simply-connected or it is multiply-connected, but the resultant force and the resultant moment vanish on each closed surface in $\bar{\mathscr{V}}$ (in particular, on each closed surface of the boundary)[3]. On the other hand, if the body is simply-connected, the six distinct scalar stress functions $\chi_{km}$ may be subjected to three supplementary conditions, provided that these conditions be *admissible*, in the sense of not restricting the generality of the possible stress states. On using as before a Cartesian frame, and putting $\chi_{11} = \chi_{22} = \chi_{33}$ into (6.26), *Morera's solution* results [251]; alternatively, setting $\chi_{12} = \chi_{23} = \chi_{31} = 0$, one obtains *Maxwell's solution* [238]; both these sets of supplementary conditions can be shown to be admissible.

Any general solution of the equilibrium equations in terms of stress functions must still satisfy the compatibility conditions (6.25) and the boundary conditions (6.19). While the uncoupling of equations (6.25) by stress functions can be done

---

[1] Derived by Beltrami [23] for $\mathbf{t} = \mathbf{0}$, and by Michell [246] in the general case.

[2] The general case may be reduced to this particular one, by finding out a particular solution of the equations of equilibrium [282], [331].

[3] A stress field having this property is called *self-equilibrated*.

by a suitable choice of the admissible supplementary conditions (see Sect. 18.2), the simultaneous uncoupling of the boundary conditions is rarely possible.

The *displacement formulation* of the boundary-value problems is generally used in conjunction with the boundary conditions (6.18) or (6.20), i.e. when the displacement is prescribed on the boundary of the body or on a part of it. However, it can be also applied, without much difficulty, in the case of the traction boundary-value problem.

By substituting (6.1) into (6.3), and the result obtained into (6.2), we obtain

$$c_{klmn}u_{m,nl} + f_k = 0, \qquad k = 1, 2, 3. \tag{6.27}$$

The boundary-value problems of linear elastostatics can now be formulated as follows: Find a class $C^2$ vector field $\mathbf{u}(\mathbf{x})$ that satisfies the equations (6.27) in $\mathscr{V}$ and the boundary conditions (6.18), or (6.19), or (6.20) on $\mathscr{S}$. These boundary-value problems can be simplified to a certain extent by solving equations (6.27) in terms of displacement potentials that are solutions of simpler field equations. We illustrate below this procedure, limiting ourselves for the sake of simplicity to the *isotropic* case. Substituting (5.26) into (6.27) yields *Navier's equations*

$$(\lambda + \mu)u_{m,mk} + \mu\Delta u_k + f_k = 0. \tag{6.28}$$

These equations can be solved for instance by setting

$$u_k = \Phi_k - \frac{1}{4(1-\nu)}(x_m\Phi_m + \Phi_0)_{,k}, \quad k = 1, 2, 3,$$

where $\mathbf{\Phi}$ is a vector field, and $\Phi_0$ is a scalar field, both of class $C^3$ in $\mathscr{V}$, and that satisfy the equations of Poisson type

$$\mu\Delta\Phi_0 = x_m f_m, \qquad \mu\Delta\Phi_k = -f_k, \qquad k = 1, 2, 3.$$

The above solution of the field equations in terms of the potentials $\Phi_0$ and $\mathbf{\Phi}$ is called the *Papkovitch-Neuber representation*, since Papkovitch [263] and Neuber [259] have proved independently that this solution is complete, i.e. every sufficiently regular solution of Navier's equations admits a representation of this form [1].

## 6.4. Green's tensor function of an infinite elastic medium

In this subsection we consider an elastic body occupying the entire space $\mathscr{E}$, referred to a rectangular Cartesian system of co-ordinates $x_k$.

We call *fundamental singular solution* or *Green's tensor function* of the infinite elastic medium the second-order tensor field $\mathbf{G}(\mathbf{x})$ with the following properties [2]:

(i) For any point of $\mathscr{E}$ with position vector $\mathbf{x} \neq \mathbf{0}$ and for each $p = 1, 2, 3$, the displacement field $u_k^{(p)}(\mathbf{x}) = G_{kp}(\mathbf{x})$ defines a (regular) elastic state corresponding

---

[1] For the admissible supplementary conditions that may be imposed on the Papkovitch-Neuber potentials we refer to Eubanks and Sternberg [113].

[2] As shown by Sternberg and Eubanks [321], properties (i)–(iii) uniquely characterize $\mathbf{G}(\mathbf{x})$.

to zero body force. In particular, by (6.27),

$$c_{ijkl}G_{kp,lj}(\mathbf{x}) = 0, \quad i, p = 1, 2, 3. \tag{6.29}$$

(ii) $\mathbf{G}(\mathbf{x})$ is a homogeneous function of degree $-1$ in $x_k$. In particular, we have

$$\mathbf{G}(\mathbf{x}) = O(r^{-1}), \quad \mathbf{T}^{(p)}(\mathbf{x}) = O(r^{-2}) \quad \text{as } r \to 0 \text{ and also as } r \to \infty, \tag{6.30}$$

where $r = \|\mathbf{x}\|$, and $T_{ij}^{(p)} = c_{ijkl}G_{kp,l}$ are the components of the stress tensor corresponding to the displacement $\mathbf{u}^{(p)}$.

(iii) For all $\eta > 0$ and $p = 1, 2, 3$,

$$\int_{\Sigma_\eta} \mathbf{T}^{(p)}\mathbf{n}\,\mathrm{d}s = \mathbf{e}_p, \quad \int_{\Sigma_\eta} T_{ij}^{(p)}n_j\,\mathrm{d}s = \delta_{ip}, \tag{6.31}$$

where $\Sigma_\eta$ is the sphere with radius $\eta$ centred at the origin, and $\mathbf{n}$ is the inward unit normal to $\Sigma_\eta$.

Equation (6.31) shows that the resultant of the stress vectors corresponding to the displacement $u_k^{(p)}(\mathbf{x}) = G_{kp}(\mathbf{x})$ and acting on any sphere centred at the origin equals $\mathbf{e}_p$. That is why $G_{kp}(\mathbf{x})$ is also said to be the component in the direction of the $x_k$-axis of the displacement produced by a *unit concentrated force* acting at the origin and directed along the $x_p$-axis. Since the elastic medium occupies the entire space, it may be seen that a unit concentrated force acting at an arbitrary point with position vector $\mathbf{x}'$ and directed along the $x_p$-axis produces a displacement field $u_k^{(p)}(\mathbf{x}) = G_{kp}(\mathbf{x} - \mathbf{x}')$. Finally, it results that an arbitrary concentrated force $\mathbf{P}$ acting at $\mathbf{x}'$ produces the displacement field

$$\mathbf{u}(\mathbf{x}) = \mathbf{G}(\mathbf{x} - \mathbf{x}')\mathbf{P}, \quad u_k(\mathbf{x}) = G_{kp}(\mathbf{x} - \mathbf{x}')P_p. \tag{6.32}$$

Of a special interest is the differential equations satisfied by the Green's tensor function in the sense of the theory of distributions. Let $\varphi(\mathbf{x})$ be an arbitrary function of class $C^\infty$ and of bounded support on $\mathscr{E}$. According to the definition of the derivatives of a distribution[1] we have

$$(c_{ijkl}G_{kp,lj}(\mathbf{x}), \varphi(\mathbf{x})) = -(c_{ijkl}G_{kp,l}(\mathbf{x}), \varphi_{,j}(\mathbf{x})) = (c_{ijkl}G_{kp}(\mathbf{x}), \varphi_{,jl}(\mathbf{x})). \tag{6.33}$$

Denote by $\Omega_r$ the exterior domain bounded by the sphere $\Sigma_r$ of radius $r$ and the centre at the origin. Integrating by parts twice, taking into account that $\varphi$ vanishes together with all its derivatives for sufficiently large values of $r$, we successively obtain

$$\int_{\Omega_r} c_{ijkl}G_{kp}(\mathbf{x})\,\varphi_{,jl}(\mathbf{x})\,\mathrm{d}v = \int_{\Omega_r} (c_{ijkl}G_{kp}(\mathbf{x})\varphi_{,j}(\mathbf{x}))_{,l}\,\mathrm{d}v - \int_{\Omega_r} c_{ijkl}G_{kp,l}(\mathbf{x})\varphi_{,j}(\mathbf{x})\,\mathrm{d}v =$$

$$= \int_{\Sigma_r} c_{ijkl}G_{kp}(\mathbf{x})\varphi_{,j}(\mathbf{x})n_l\mathrm{d}s - \int_{\Omega_r} (c_{ijkl}G_{kp,l}(\mathbf{x})\varphi(\mathbf{x}))_{,j}\mathrm{d}v + \int_{\Omega_r} c_{ijkl}\,G_{kp,lj}(\mathbf{x})\,\varphi(\mathbf{x})\,\mathrm{d}v$$

---

[1] For the basic results of the theory of distributions used below see, e.g. Gelfand and Shilov [137] or Kecs and Teodorescu [178].

and hence, considering also (6.29),

$$\int_{\Omega_r} c_{ijkl} G_{kp}(\mathbf{x})\, \varphi_{,jl}(\mathbf{x})\, \mathrm{d}v = \int_{\Sigma_r} c_{ijkl} G_{kp}(\mathbf{x}) \varphi_{,j}(\mathbf{x}) n_l \mathrm{d}s - \int_{\Sigma_r} c_{ijkl} G_{kp,l}(\mathbf{x})\, \varphi(\mathbf{x})\, n_j \mathrm{d}s.$$

Next, by making use of the mean theorem of the integral calculus and taking into consideration (6.30) and (6.31), it results that

$$\lim_{r\to 0} \int_{\Sigma_r} c_{ijkl} G_{kp}(\mathbf{x}) \varphi_{,j}(\mathbf{x}) n_l \mathrm{d}s = 0,$$

$$\lim_{r\to 0} \int_{\Sigma_r} c_{ijkl} G_{kp,l}(\mathbf{x}) \varphi(\mathbf{x}) n_j\, \mathrm{d}s = \lim_{r\to 0} \int_{\Sigma_r} T_{ij}^{(p)} n_j \varphi\, \mathrm{d}s = \delta_{ip}\varphi(\mathbf{0}),$$

and hence

$$\lim_{r\to 0} \int_{\Omega_r} c_{ijkl} G_{kp}(\mathbf{x}) \varphi_{,jl}(\mathbf{x})\, \mathrm{d}v = -\delta_{ip}\varphi(\mathbf{0}).$$

Combining this result with (6.33), we conclude that the regular functionals associated to $c_{ijkl}G_{kp,lj}(\mathbf{x})$ on the regions $\Omega_r$ tend to $-\delta_{ip}\delta(\mathbf{x})$ as $r \to 0$, where $\delta(\mathbf{x})$ is Dirac's distribution. Consequently, the components of the distribution associated to $\mathbf{G}(\mathbf{x})$ satisfy the system of equations [1]

$$c_{ijkl} G_{kp,lj}(\mathbf{x}) + \delta_{ip}\delta(\mathbf{x}) = 0, \quad i, p = 1, 2, 3. \tag{6.34}$$

Assume now that the elastic medium is subjected to the action of a body force $\mathbf{f}(\mathbf{x})$ of class $C^1$ in $\mathscr{E}$ and that satisfies the condition

$$\mathbf{f}(\mathbf{x}) = O(r^{-3}) \quad \text{as } r \to \infty. \tag{6.35}$$

Making use of the properties of the convolution and considering (6.34) we may write

$$f_i(\mathbf{x}) = \delta_{ip}\delta(\mathbf{x}-\mathbf{x}') * f_p(\mathbf{x}') = -c_{ijkl} G_{kp,lj}(\mathbf{x}-\mathbf{x}') * f_p(\mathbf{x}')$$

$$= -c_{ijkl}\, \{G_{kp}(\mathbf{x}-\mathbf{x}') * f_p(\mathbf{x}')\}_{,lj},$$

where the derivatives are taken with respect to $x_l$ and $x_j$. By comparing this relation with (6.27), we deduce that

$$u_k(\mathbf{x}) = G_{kp}(\mathbf{x}-\mathbf{x}') * f_p(\mathbf{x}') \tag{6.36}$$

---

[1] The Green's tensor function is sometimes defined as the particular solution of (6.34), in the sense of the theory of distributions, that vanishes at infinity. The definition adopted above has the advantage that it needs not the regularization of the solution of (6.34).

is the unique solution of (6.27) that satisfies the supplementary condition

$$\lim_{r\to 0} \|\mathbf{u}(\mathbf{x})\| = 0. \tag{6.37}$$

Finally, by taking into account the way in which the distribution $\mathbf{G}(\mathbf{x})$ has been generated, as well as the continuity of the convolution, (6.36) may be rewritten as

$$u_k(\mathbf{x}) = \int_{\mathscr{E}} G_{kp}(\mathbf{x}-\mathbf{x}') f_p(\mathbf{x}')\,\mathrm{d}V', \tag{6.38}$$

the convergence of this integral being granted by the conditions $(6.30)_1$ and (6.35).

The properties (i) and (ii) imply that the partial derivatives of $n$'th order $G_{kp,k_1\ldots k_n}$ of $\mathbf{G}(\mathbf{x})$ define certain displacement fields that are homogeneous functions of degree $-n+1$ of the co-ordinates $x_k$ and such that the corresponding elastic states are regular for all points of $\mathscr{E}$ except the origin. As shown by Fredholm [123], these functions play in the anisotropic elasticity theory the same role as the spherical harmonics of negative degree in the potential theory. Namely, it can be proved that if $\mathbf{u}(\mathbf{x})$ satisfies the equations of equilibrium with null body forces,

$$c_{ijkl}\, u_{k,lj} = 0, \tag{6.39}$$

outside a sphere $\Sigma_{r_0}$ of radius $r_0$ and with centre at the origin, then each component $u_k(\mathbf{x})$ of the displacement field may be written as a sum of two series: a power series and a series of derivatives of Green's function. These expansions, which are analogous to Laurent's series, are uniformly convergent in any closed region that is exterior to the ball bounded by $\Sigma_{r_0}$[1].

Let us consider now in more detail the physical significance of the partial derivatives of first order of $\mathbf{G}(\mathbf{x})$. We can obviously write

$$G_{kp,m}(\mathbf{x}) = \lim_{h\to 0} \frac{1}{h}\{G_{kp}(\mathbf{x}) - G_{kp}(\mathbf{x}-h\mathbf{e}_m)\}. \tag{6.40}$$

Consequently, the elastic state associated to the displacement field

$$u_k(\mathbf{x}) = G_{kp,m}(\mathbf{x}) \tag{6.41}$$

is the limiting value as $h \to 0$ of a sum of two elastic states: the first corresponds to the concentrated force $\mathbf{e}_p/h$ acting at the origin; the second corresponds to a concentrated load $-\mathbf{e}_p/h$ acting at the point with position vector $h\,\mathbf{e}_m$ (Fig. 6.1). A straightforward calculation shows that the resultant of the stress vectors acting on any sphere $\Sigma_\eta$ with radius $\eta$ and centre at the origin is zero, while their resulting couple equals $-\mathbf{e}_m \times \mathbf{e}_p$. Following the terminology introduced by Love [222],

[1] See also Bézier [26]. An analogous theorem that is valid in the isotropic case has been proved by Kelvin as early as 1863.

Sect. 132, we say that the elastic state corresponding to the displacement (6.41) is produced by a *unit double force*, which is statically equivalent to a *directed concentrated couple* [1] or to **0** according as $m \neq p$ or $m = p$. In the latter case we say that the singularity at the origin is a *unit double force without moment.*

Fig. 6.1. On the definition of a unit double force by a limiting process.

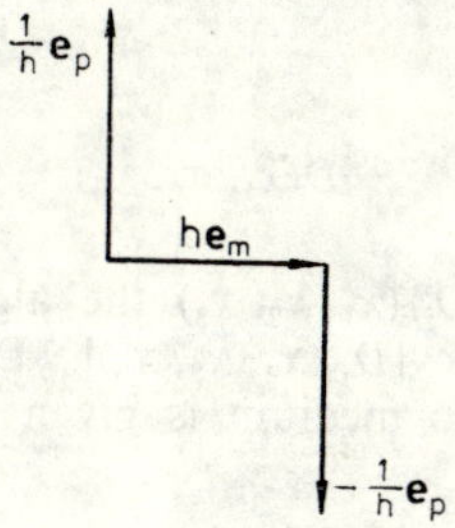


The elastic state corresponding to the displacement field

$$u_k(\mathbf{x}) = G_{km,m}(\mathbf{x}), \tag{6.42}$$

which is produced by three mutually orthogonal unit double forces without moment acting at the origin, is called a *centre of compression*, whereas the elastic state corresponding to the opposite of (6.42) is called a *centre of dilatation.*

As has been shown by Sternberg and Eubanks [321], the part of (6.30) concerning the behaviour of $\mathbf{G}$ as $r \to 0$ is indispensable for a unique characterization of the singular elastic state produced by a unit concentrated force. In fact, this condition eliminates the possibility of superimposing self-equilibrated singular elastic states, such as those produced by double forces without moment; indeed, it is apparent from the reasoning above that such states correspond to displacement fields of the order $O(r^{-n})$, $n \geqslant 2$, as $r \to \infty$, and hence do not satisfy (6.30).

Green's tensor functions are particularly important for the modelling of crystal defects, since they correspond to singular elastic states. That is why we will consider in the following in more detail the most powerful methods of determining Green's tensor functions for various types of material symmetry, namely Fredholm's method and the method of Fourier transformation.

*Fredholm's method.* The Green's tensor function of an infinite isotropic elastic medium was determined by Kelvin [351] in 1848. Later on, Fredholm [123] deduced the form of $\mathbf{G}(\mathbf{x})$ in terms of the roots of a sextic algebraic equation, for an elastic medium with general anisotropy. We cannot follow here the rather intricate reasoning of Fredholm and content ourselves, therefore, with explaining his result. Putting

$$D_{ik}(\mathbf{x}) = D_{ik}(x_1, x_2, x_3) = c_{ijkl}x_jx_l, \tag{6.43}$$

---

[1] For a detailed discussion of the elastic states produced by double forces in an isotropic elastic medium we refer to Gurtin [150], Sect. 51 and to Kecs and Teodorescu [178], Chap. 5 and Sect. 10.1.

the equations of equilibrium (6.39) may be rewritten in the symbolic form

$$D_{ik}\left(\frac{\partial}{\partial x_1}, \frac{\partial}{\partial x_2}, \frac{\partial}{\partial x_3}\right) u_k = 0. \tag{6.44}$$

Let

$$D(x_1, x_2, x_3) = \det\left[D_{ik}(x_1, x_2, x_3)\right] \tag{6.45}$$

and denote by $D_{ik}^*(x_1, x_2, x_3)$ the algebraic complement of $D_{ik}(x_1, x_2, x_3)$ in the symmetric matrix $[D_{ik}(x_1, x_2, x_3)]$. Then, Green's tensor function of an infinite anisotropic elastic medium is given by

$$G_{kp}(\mathbf{x}) = \operatorname{Re}\left\{\frac{1}{2\pi i}\sum_{\nu=1}^{3}\frac{D_{kp}^*(\xi_\nu, \eta_\nu, 1)}{x_1\dfrac{\partial D}{\partial \eta}(\xi_\nu, \eta_\nu, 1) - x_2\dfrac{\partial D}{\partial \xi}(\xi_\nu, \eta_\nu, 1)}\right\}, \tag{6.46}$$

where $\xi_\nu, \eta_\nu$ are the roots of the system of equations

$$D(\xi, \eta, 1) = 0, \quad \xi x_1 + \eta x_2 + x_3 = 0, \tag{6.47}$$

and the sum in the right-hand side of (6.46) is extended to the three roots with $\operatorname{Im} \xi_\nu > 0$, which are assumed to be simple. By eliminating $\eta$ between equations (6.47), it results that $\xi_\nu$ are the roots of the sextic algebraic equation

$$D\left(\xi, -\frac{\xi x_1 + x_3}{x_2}, 1\right) = a_0\xi^6 + a_1\xi^5 + a_2\xi^4 + a_3\xi^3 + a_4\xi^2 + a_5\xi + a_6 = 0, \tag{6.48}$$

whereas $(6.47)_2$ yields $\eta_\nu = -(\xi_\nu x_1 + x_3)/x_2$.

It can be shown that equation (6.48) has real coefficients; moreover, by introducing the spherical co-ordinates $r, \theta, \varphi$, it may be seen that these coefficients do not depend on $r$. Consequently, it results that the function $\mathbf{G}(\mathbf{x})$ determined by (6.46) is indeed a homogeneous function of degree $-1$ in the co-ordinates $x_k$, and we may write

$$G_{kp}(\mathbf{x}) = r^{-1}H_{kp}(\theta, \varphi). \tag{6.49}$$

As shown by Gebbia [126], the roots of equation (6.48) can be obtained in closed form only for isotropic materials and for hexagonal crystals. The expression of Green's tensor function for materials with hexagonal symmetry has been independently derived by Lifshits and Rozentsveig [216] and by Kröner [186]; later on, the same result has been reobtained in a different way by Willis [381].

Starting from Fredholm's formula, Mann, v. Jan, and Seeger [229] evaluated numerically for copper the components of Green's tensor function as well as its

(*) $G_{kp}(\underline{x}) = G_{pk}(\underline{x})$, because $D^*_{kp} = D^*_{pk}$

derivatives corresponding to unit double forces without moment directed along the axes of the cubic lattice. Lie and Koehler [215] performed a similar calculation for Al, Cu, and Li. In order to obtain the derivatives of $G_{kp}$, they fitted the function $H_{kp}(\theta, \varphi)$ in (6.49) to a truncated double Fourier series

$$H_{kp}(\theta, \varphi) = A_{kpqr} \cos(F_{kpq}\theta) \cos(G_{kpr}\, \varphi), \tag{6.50}$$

where $A_{kpqr}$ are the Fourier coefficients, and the $F_{kpq}$'s and $G_{kpr}$'s are some known polynomials of first degree in $q$, respectively $r$, depending on $k$ and $p$, while the summation is performed only over the subscripts $q$ and $r$. It should be mentioned, however, that the error introduced by the subsequent differentiation term by term of this series increases rapidly with the order of differentiation. Similar techniques have been used by Bullough, Norgett, and Webb [51].

More recently, Meissner [244] substantially improved the Fredholm technique, by deriving explicit formulae for calculating the coefficients $a_k$ of the sextic polynomial in (6.48), as well as the algebraic complements $D^*_{kp}$, and by elaborating programmes for the numerical calculation of Green's tensor function for materials belonging to the rhombic system and for the general anisotropic case. He also worked out programmes allowing a very precise evaluation of the coefficients $A_{kpqr}$ of the double Fourier series (6.50) for rhombic crystals and applied them to α-uranium. Meissner's results yield accuracies of at least 0.01% for $G_{kp}$ and 0.1% for its first order derivatives.

*The Fourier transform method.* The Fourier transform of (6.34) is

$$c_{jlms} k_l k_s \tilde{G}_{mp}(\mathbf{k}) = \delta_{jp}, \tag{6.51}$$

where $\mathbf{k}$ is the Fourier wave vector; $\tilde{\mathbf{G}}(\mathbf{k})$ is the Fourier transform of $\mathbf{G}(\mathbf{x})$ and is given by the integral

$$\tilde{\mathbf{G}}(\mathbf{k}) = \int_{\mathscr{E}} \mathbf{G}(\mathbf{x}) e^{i\mathbf{k}\cdot\mathbf{x}} dv, \tag{6.52}$$

whose convergence for $\|\mathbf{x}\| \to 0$ is again assured by the first condition (6.30).

By making use of the notation (6.43) we may rewrite (6.51) in the tensor form

$$\mathbf{D}(\mathbf{k})\, \tilde{\mathbf{G}}(\mathbf{k}) = \mathbf{1}, \tag{6.53}$$

wherefrom it results that

$$\tilde{\mathbf{G}}(\mathbf{k}) = \mathbf{D}^{-1}(\mathbf{k}), \tag{6.54}$$

with $\mathbf{D}^{-1} = \mathbf{D}^*/D$, and $D$ given by (6.45). Now, by the Fourier inversion theorem and taking into account that $\mathbf{G}(\mathbf{x})$ is a real-valued function, we obtain from (6.54)

$$\mathbf{G}(\mathbf{x}) = \frac{1}{8\pi^3} \operatorname{Re} \int_{\tilde{\mathscr{E}}} \tilde{\mathbf{G}}(\mathbf{k})\, e^{-i\mathbf{k}\cdot\mathbf{x}}\, d\tilde{v} = \frac{1}{8\pi^3} \operatorname{Re} \int_{\tilde{\mathscr{E}}} \mathbf{D}^{-1}(\mathbf{k})\, e^{-i\mathbf{k}\cdot\mathbf{x}}\, d\tilde{v}, \tag{6.55}$$

where $\tilde{\mathscr{E}}$ is the phase space, and $d\tilde{v}$ is the volume element in $\tilde{\mathscr{E}}$.

In the *isotropic* case, by introducing (5.26) into (6.51), we obtain

$$(\lambda + \mu)k_j k_m \tilde{G}_{mp}(\mathbf{k}) + \mu k^2 \tilde{G}_{jp}(\mathbf{k}) = \delta_{jp},$$

where $k = \|\mathbf{k}\|$. Multiplying both sides of this equation by $k_j$ and summing over $j$ yields

$$(\lambda + 2\mu)\, k^2 k_m \tilde{G}_{mp}(\mathbf{k}) = k_p.$$

By eliminating $k_m G_{mp}(\mathbf{k})$ between the last two relations and using (6.9), we deduce that

$$\tilde{G}_{jp}(\mathbf{k}) = \frac{1}{2\mu}\left(\frac{2\delta_{jp}}{k^2} - \frac{1}{1-\nu}\frac{k_j k_p}{k^4}\right). \tag{6.56}$$

Next, from the relation [1]

$$\frac{1}{\pi^2}\int_{\tilde{\mathscr{E}}} \frac{1}{k^4}\, \mathrm{e}^{-i\mathbf{k}\cdot\mathbf{x}}\, \mathrm{d}\tilde{v} = -r,$$

where $r = \|\mathbf{x}\|$, it results by differentiation that

$$\frac{1}{\pi^2}\int_{\tilde{\mathscr{E}}} \frac{k_j k_p}{k^4}\, \mathrm{e}^{-i\mathbf{k}\cdot\mathbf{x}}\, \mathrm{d}\tilde{v} = r_{,jp}, \qquad \frac{1}{\pi^2}\int_{\tilde{\mathscr{E}}} \frac{1}{k^2}\, \mathrm{e}^{-i\mathbf{k}\cdot\mathbf{x}}\, \mathrm{d}\tilde{v} = r_{,pp}. \tag{6.57}$$

Substituting now (6.56) into (6.55) and considering (6.57), we obtain

$$G_{jp}(\mathbf{x}) = \frac{1}{16\,\pi\mu(1-\nu)}\,[2(1-\nu)\delta_{jp} r_{,mm} - r_{,jp}]. \tag{6.58}$$

Finally, by taking into account that $r = (x_1^2 + x_2^2 + x_3^2)^{1/2}$, we deduce that

$$G_{jp}(\mathbf{x}) = \frac{1}{16\,\pi\mu(1-\nu)}\,\frac{1}{r}\left[(3-4\nu)\delta_{jp} + \frac{x_j x_p}{r^2}\right], \tag{6.59}$$

and hence *Green's tensor function for an infinite isotropic elastic medium* is

$$\mathbf{G}(\mathbf{x}) = \frac{1}{16\pi\mu(1-\nu)}\,\frac{1}{r}\left[(3-4\nu)\mathbf{1} + \frac{\mathbf{x}\mathbf{x}}{r^2}\right]. \tag{6.60}$$

It can be easily verified that this function has indeed the properties (i)—(iii) given at the beginning of this subsection.

---

[1] See, e.g. Jones [175], p. 222.

Although the relation $(6.55)_2$ has been derived by Zeilon [390] as early as 1911, it has not been effectively used until much later. Lifshits and Rozentsveig [216] and Leibfried [213] employed a perturbation method in order to determine first-order contributions of cubic anisotropy to the Green's tensor function (6.60) and to the dilatation produced by a dilatation centre. The same problem was reconsidered more recently by Barnett [14], who obtained various representations of the cubic Green's tensor components as power series in the *anisotropy factor* for cubic materials

$$H = 1 - \frac{c_{11} - c_{12}}{c_{44}}.$$

It should be noticed, however, that such expansions have a rather limited range of applicability, since $H$ takes sufficiently low values only for a small number of cubic crystals, such as aluminium and diamond.

From (6.43) and (6.53) it is easily seen that $\mathbf{D}(\mathbf{k})$ and $\mathbf{D}^{-1}(\mathbf{k})$ are homogeneous functions of degree 2 and $-2$ in $k_i$, respectively. As shown by Kröner [186], this property can be exploited to obtain expansions of Green's tensor components as series of surface spherical harmonics. This method has been applied by Mann, v. Jan and Seeger [229], who calculated numerically the coefficients of the expansions for copper and compared the results obtained with those given by Fredholm's formula. Bross [42] used a similar procedure to derive expansions for Green's tensor components in terms of cubic harmonics.

As shown by Barnett [16], a considerable progress in the numerical calculation of Green's tensor function and of its derivatives can be achieved by transforming the triple improper integrals in (6.55) into ordinary line integrals about the unit circle in a plane orthogonal to $\mathbf{x}$. To obtain this transformation we first write (6.55) in the form

$$\mathbf{G}(\mathbf{x}) = \frac{1}{8\pi^3} \int_{\tilde{\mathscr{E}}} \frac{1}{k^2} \mathbf{D}^{-1}(\boldsymbol{\xi}) \cos(kr\boldsymbol{\xi}\cdot\boldsymbol{\rho})\, d\tilde{v}, \tag{6.61}$$

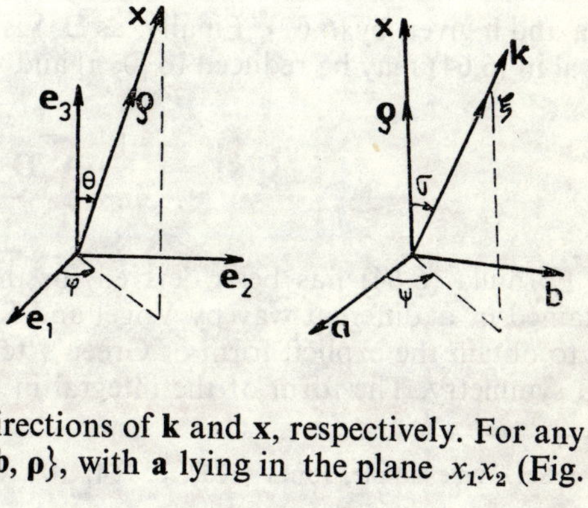


Fig. 6.2. Orthogonal frames used for the calculation of Green's tensor function.

where $\boldsymbol{\xi}$ and $\boldsymbol{\rho}$ are unit vectors in the directions of $\mathbf{k}$ and $\mathbf{x}$, respectively. For any fixed $\mathbf{x}$, we choose an orthogonal frame $\{\mathbf{a}, \mathbf{b}, \boldsymbol{\rho}\}$, with $\mathbf{a}$ lying in the plane $x_1x_2$ (Fig. 6.2).

It can be shown [16] that the components of the unit vectors **a**, **b**, **ρ** with respect to the basis $\mathbf{e}_1$, $\mathbf{e}_2$, $\mathbf{e}_3$, to which $c_{ijkl}$ are referred, are given by

$$\left.\begin{array}{lll} a_1 = \sin\varphi, & a_2 = -\cos\varphi, & a_3 = 0, \\ b_1 = \cos\theta\cos\varphi, & b_2 = \cos\theta\sin\varphi, & b_3 = -\sin\theta, \\ \rho_1 = \sin\theta\cos\varphi, & \rho_2 = \sin\theta\sin\varphi, & \rho_3 = \cos\theta, \end{array}\right\} \tag{6.62}$$

where $\varphi$ and $\theta$ are the angular spherical co-ordinates associated to $\{\mathbf{e}_k\}$. We shall calculate now the integral (6.61) by making use of the spherical co-ordinates $k$, $\sigma$, $\psi$ associated to the new basis $\{\mathbf{a}, \mathbf{b}, \boldsymbol{\rho}\}$. Since

$$\mathrm{d}\tilde{v} = k^2 \sin\sigma\, \mathrm{d}k\, \mathrm{d}\sigma\, \mathrm{d}\psi, \quad \boldsymbol{\xi}\cdot\boldsymbol{\rho} = \cos\sigma,$$

we obtain from (6.61)

$$\mathbf{G}(\mathbf{x}) = \frac{1}{8\pi^3}\int_0^{2\pi} \mathrm{d}\psi \int_0^{\pi} \mathbf{D}^{-1}(\boldsymbol{\xi}) \sin\sigma\, \mathrm{d}\sigma \int_0^{\infty} \cos(kr\cos\sigma)\, \mathrm{d}k. \tag{6.63}$$

On the other hand, we have[1]

$$\int_0^{\infty} \cos(k\cos\sigma)\, \mathrm{d}k = \pi\,\delta(\cos\sigma) = \frac{\pi}{\sin\sigma}\,\delta(\sigma - \pi/2),$$

and hence (6.63) reduces to

$$\mathbf{G}(\mathbf{x}) = \frac{1}{8\pi^2 r}\int_0^{2\pi} \mathbf{D}^{-1}[\boldsymbol{\xi}(\psi)]\, \mathrm{d}\psi, \tag{6.64}$$

where the integrand must be calculated for $\sigma = \pi/2$, i.e. in the plane defined by the unit vectors **a** and **b**. Hence we must take in the right-hand side of (6.64)

$$\boldsymbol{\xi}(\psi) = \mathbf{a}\cos\psi + \mathbf{b}\sin\psi,$$

with **a** and **b** given by (6.62). Finally, as $\mathbf{D}^{-1}$ is an even function of $\boldsymbol{\xi}$, the integration interval in (6.64) may be reduced to $[0, \pi]$ and we obtain

$$\mathbf{G}(\mathbf{x}) = \frac{1}{4\pi^2 r}\int_0^{\pi} \mathbf{D}^{-1}[\boldsymbol{\xi}(\psi)]\, \mathrm{d}\psi. \tag{6.65}$$

Formula (6.64) has been derived for the first time by Synge [327] and later reobtained in a different way by Vogel and Rizzo [371]. It has been used by Willis [381] to obtain the explicit form of Green's tensor function for materials with hexagonal symmetry. The form of the integral in (6.65) is very well suited to rapid and

[1] See, for example, Jones [175], p. 254.

accurate numerical integration by standard Romberg procedures and has been successfully used by Barnett and Swanger [15] for calculations of the energy of straight dislocations in anisotropic media. Moreover, Barnett [16] has shown that the first two derivatives of the Green's tensor function can be calculated by similar integrals, namely

$$G_{ip,s}(\mathbf{x}) = \frac{1}{4\pi^2 r^2}\int_0^\pi (\xi_s F_{ip} - \rho_s D_{ip}^{-1})\,\mathrm{d}\psi, \tag{6.66}$$

$$G_{ip,sm}(\mathbf{x}) = \frac{1}{4\pi^2 r^3}\int_0^\pi [2\rho_s\rho_m D_{ip}^{-1} - 2(\xi_s\rho_m + \xi_m\rho_s)]F_{ip} + \xi_s\xi_m A_{ip}]\,\mathrm{d}\xi, \tag{6.67}$$

where

$$F_{ip} = c_{jrnq} D_{ij}^{-1} D_{np}^{-1}(\xi_r\rho_q + \xi_q\rho_r),$$

$$A_{ip} = c_{jrnq}[(\xi_r\rho_q + \xi_q\rho_r)(F_{ij}D_{rp}^{-1} + F_{np}D_{ij}^{-1}) - D_{ij}^{-1}D_{np}^{-1}\rho_q\rho_r].$$

As pointed out by Willis [381] and Barnett [16], Fredholm's formula (6.46) can be obtained from (6.65) by the substitution $y = e^{i\psi}$, which converts the integrals over $\psi$ into line integrals about the circle $|y| = 1$, and by using subsequently the residue theorem to evaluate the line integrals in terms of the roots of a sextic polynomial occurring in the integrand. Formula (6.65) has the advantage that it holds even when this polynomial has multiple roots, e.g. in the isotropic case. Moreover, the integrands in (6.65—67) have no singularities, and hence Green's tensor function and its partial derivatives of the first and second orders can be easily calculated using standard numerical techniques. Thus, the errors occurring in former variants of Fredholm's method when differentiating truncated double Fourier series are completely avoided. Equations (6.65—67) have been applied by Barnett [16] to Cu and by Meissner [244] to α—U. The accuracy obtained by using a Romberg integration scheme was in both cases between 0.1 and 0.01%.

The considerations above show that the numerical calculation of the Green's tensor function of an infinite elastic medium with general anisotropy by Fredholm's method, as well as by the Fourier transform method, has been reduced at present to the application of some standard programmes.

Finally, we mention that for boundary-value elastic problems that are independent of one co-ordinate, Green's tensor function is known in finite form for the general anisotropic case (Eshelby, Read, and Shockley [109], Stroh [324]). This stimulated a series of investigations concerning the expression of three-dimensional Green's tensor functions in terms of the angular derivatives of two-dimensional Green's tensor functions (Indenbom and Orlov [169], [170], Malén [224], Malén and Lothe [225]).

## 6.5. Concentrated loads. Integral representation of solutions to concentrated load problems

In the previous subsection we have introduced the notions of concentrated force and associated singular elastic state for an *infinite* elastic medium. We will consider now the concepts of concentrated force and Green's tensor function in the case of a *finite* elastic body $\mathscr{V}$, bounded by a surface $\mathscr{S}$. Let $\mathbf{P}$ be a vector-valued function whose domain $\mathscr{D}$ is a finite set of points of $\overline{\mathscr{V}} = \mathscr{V} \cup \mathscr{S}$. Interpreting $\mathbf{P}$ as a *system of concentrated loads* we say that $[\mathbf{u}, \mathbf{E}, \mathbf{T}]$ is a *singular elastic state* corresponding to the external force system $[\mathbf{f}, \mathbf{t}, \mathbf{P}]$ if

(i) $[\mathbf{u}, \mathbf{E}, \mathbf{T}]$ is a (regular) elastic state on $\overline{\mathscr{V}} \setminus \mathbf{D}$ corresponding to the external force system $[\mathbf{f}, \mathbf{t}]$.

(ii) For each $\mathbf{x}' \in \mathscr{D}$, we have

$$\mathbf{u}(\mathbf{x}) = O(r^{-1}), \quad \mathbf{T}(\mathbf{x}) = O(r^{-2}) \quad \text{as } r = \|\mathbf{x} - \mathbf{x}'\| \to 0. \tag{6.68}$$

(iii) For each $\mathbf{x}' \in \mathscr{D}$,

$$\lim_{\eta \to 0} \int_{\mathscr{V} \cap \Sigma_\eta(\mathbf{x}')} \mathbf{T}\mathbf{n} \, \mathrm{d}s = \mathbf{P}(\mathbf{x}'), \tag{6.69}$$

where $\Sigma_\eta(\mathbf{x}')$ is the sphere of radius $\eta$ and centre at $\mathbf{x}'$, and $\mathbf{n}$ is the inward unit normal to $\Sigma_\eta(\mathbf{x}')$.

For singular elastic states holds the following generalization of Betti's theorem, due to Turteltaub and Sternberg [362].

**Reciprocal theorem for singular elastic states.** *Let* $\mathbf{P}$ *and* $\mathbf{P}^*$ *be systems of concentrated loads with disjoint domains* $\mathscr{D}$ *and* $\mathscr{D}^*$. *If* $[\mathbf{u}, \mathbf{E}, \mathbf{T}]$ *and* $[\mathbf{u}^*, \mathbf{E}^*, \mathbf{T}^*]$ *are singular elastic states corresponding to the external force systems* $[\mathbf{f}, \mathbf{t}, \mathbf{P}]$ *and* $[\mathbf{f}^*, \mathbf{t}^*, \mathbf{P}^*]$, *respectively, then*

$$\begin{aligned}\int_{\mathscr{V}} \mathbf{T} \cdot \mathbf{E}^* \, \mathrm{d}v &= \int_{\mathscr{S}} \mathbf{t} \cdot \mathbf{u}^* \, \mathrm{d}s + \int_{\mathscr{V}} \mathbf{f} \cdot \mathbf{u}^* \mathrm{d}v + \sum_{\mathbf{x}' \in \mathscr{D}} \mathbf{P}(\mathbf{x}') \cdot \mathbf{u}^*(\mathbf{x}') = \\ &= \int_{\mathscr{V}} \mathbf{T}^* \cdot \mathbf{E} \, \mathrm{d}v = \int_{\mathscr{S}} \mathbf{t}^* \cdot \mathbf{u} \, \mathrm{d}s + \int_{\mathscr{V}} \mathbf{f}^* \cdot \mathbf{u} \, \mathrm{d}v + \sum_{\mathbf{x}' \in \mathscr{D}^*} \mathbf{P}^*(\mathbf{x}') \cdot \mathbf{u}(\mathbf{x}').\end{aligned} \tag{6.70}$$

The proof of this theorem is based on the application of Betti's reciprocal theorem for a domain that is obtained from $\mathscr{V}$ by eliminating disjoint balls centred at the points of $\mathscr{D}$ and $\mathscr{D}^*$, and of a sufficiently small radius $\eta$. Then, letting $\eta \to 0$ and making use of the properties (i)—(iii) in the definition above yields (6.70). When some of the points of $\mathscr{D}$ and/or $\mathscr{D}^*$ belong to $\mathscr{S}$, the surface integrals in (6.70) are to be interpreted as Cauchy principal values.

The reciprocal theorem for singular elastic states is still valid in the case of an *infinite* media with *finite* boundary provided that

$$\mathbf{u}(\mathbf{x}),\ \mathbf{u}^*(\mathbf{x}) = O(r^{-1});\ \mathbf{T}(\mathbf{x}),\ \mathbf{T}^*(\mathbf{x}) = O(r^{-2});\ \mathbf{f}(\mathbf{x}),\ \mathbf{f}^*(\mathbf{x}) = O(r^{-3}) \tag{6.71}$$

as $r = \|\mathbf{x}\| \to \infty$.

Next, we introduce after Turteltaub and Sternberg [362] the notion of Green's tensor function for the boundary-value problems corresponding to the boundary conditions (6.18—20) in the presence of concentrated loads. To this end we need the following

**Lemma** (Gurtin [150], p. 185). *Let* **f** *and* **m** *be two vectors. Then there exists a unique rigid displacement field* **w(x)** *that satisfies the system of equations*

$$\int_{\mathscr{S}} \mathbf{w}\, ds = \mathbf{f}, \quad \int_{\mathscr{S}} (\mathbf{x} - \mathbf{c}) \times \mathbf{w}\, ds = \mathbf{m}, \tag{6.72}$$

*where* **c** *is the position vector of the centroid of* $\mathscr{S}$. *This solution is given by*

$$\mathbf{w}(\mathbf{x}) = \mathbf{w}^\circ + \boldsymbol{\omega}^\circ \times (\mathbf{x} - \mathbf{c}), \quad \mathbf{w}^\circ = \frac{1}{a}\mathbf{f}, \quad \boldsymbol{\omega}^\circ = \mathbf{I}^{-1}\mathbf{m}, \tag{6.73}$$

*where* $a$ *is the area of* $\mathscr{S}$, *and* **I** *is the centroidal inertia tensor, whose components are the moments of inertia of* $\mathscr{S}$ *with respect to the principal axes of* $\mathscr{S}$ *passing through its centroid.*

A second-order tensor field with components denoted by $\hat{G}_{kp}(\mathbf{x}; \mathbf{x}')$ is called *Green's tensor function of the region* $\mathscr{V}$ provided

(i) The elastic displacement

$$\hat{u}_k^{(p)}(\mathbf{x}; \mathbf{x}') = \hat{G}_{kp}(\mathbf{x}; \mathbf{x}') \tag{6.74}$$

and the corresponding stress tensor

$$\hat{T}_{ij}^{(p)}(\mathbf{x}; \mathbf{x}') = c_{ijkl}\hat{G}_{kp,l}(\mathbf{x}; \mathbf{x}') \tag{6.75}$$

represent the singular elastic state corresponding to vanishing body forces and to a concentrated load $\mathbf{e}_p$ acting at $\mathbf{x}'$.

(ii) If $\mathscr{S}_1$ is not empty (displacement or mixed boundary-value problem), then

$$\hat{\mathbf{u}}^{(p)} = \mathbf{0} \text{ on } \mathscr{S}_1, \quad \hat{\mathbf{T}}^{(p)}\mathbf{n} = \mathbf{0} \text{ on } \mathscr{S}_2; \tag{6.76}$$

if $\mathscr{S}_1$ is empty (traction boundary-value problem), then

$$\hat{\mathbf{T}}^{(p)}\mathbf{n} = \mathbf{w} \text{ on } \mathscr{S}_2, \tag{6.77}$$

where **w** is given by (6.73) with $\mathbf{f} = -\mathbf{e}_p$ and $\mathbf{m} = -(\mathbf{x}' - \mathbf{c}) \times \mathbf{e}_p$.

Substituting (6.77) into (6.72) it may be easily shown that the boundary condition (6.77) insures that balance of forces and moments are satisfied when $\mathscr{S}_1$ is empty.

The Green's tensor function defined above depends not only on the material, as in the case of the infinite elastic medium, but also on the region occupied by the elastic body, as well on the boundary conditions. On the other hand it is obvious that it admits the decomposition

$$\hat{G}_{kp}(\mathbf{x}; \mathbf{x}') = G_{kp}(\mathbf{x} - \mathbf{x}') + u_k^{(p)}(\mathbf{x}; \mathbf{x}'), \tag{6.78}$$

where $\mathbf{G}(\mathbf{x} - \mathbf{x}')$ is the Green's tensor function of the infinite elastic medium, and $\tilde{\mathbf{u}}^{(p)}(\mathbf{x}; \mathbf{x}')$ is the (regular) displacement field corresponding to vanishing body and concentrated forces and such that $\mathbf{G}(\mathbf{x}; \mathbf{x}')$ satisfies the boundary conditions (6.76) or (6.77).

We say that an integrable vector field $\mathbf{u}(\mathbf{x})$ on $\mathscr{S}$ is *normalized* if

$$\int_{\mathscr{S}} \mathbf{u}\, ds = \mathbf{0}, \quad \int_{\mathscr{S}} (\mathbf{x} - \mathbf{c}) \times \mathbf{u}\, ds = \mathbf{0}. \tag{6.79}$$

Given a solution $\mathbf{u}$ of the traction problem, the field $\mathbf{u} + \mathbf{w}$ with $\mathbf{w}$ rigid is also a solution. On the other hand, according to the above lemma, there exists a unique rigid displacement $\mathbf{w}$ such that $\mathbf{u} + \mathbf{w}$ is normalized. Therefore, we may always assume, without loss in generality, that the solutions of the traction problem are normalized.

The following theorem gives an integral representation of the solution to boundary-value problems of linear elastostatics in terms of Green's tensor function.

**Integral representation theorem**[1]. *Let* $\mathbf{u}(\mathbf{x})$ *be the solution of one of the boundary-value problems* (6.18—20) *corresponding to the external force system* $[\mathbf{f}, \mathbf{t}, \mathbf{P}]$, *and assume that* $\mathbf{u}$ *is normalized if* $\mathscr{S}_1$ *is empty. Then for any* $\mathbf{x}' \in \mathscr{V} \setminus \mathscr{D}$

$$u_p(\mathbf{x}') = -\int_{\mathscr{S}_1} (\hat{\mathbf{T}}^{(p)}\mathbf{n}) \cdot \mathbf{u}^\circ\, ds + \int_{\mathscr{S}_2} \hat{\mathbf{u}}^{(p)} \cdot \mathbf{t}^\circ\, ds + \int_{\mathscr{V}} \hat{\mathbf{u}}^{(p)} \cdot \mathbf{f}\, dv + \sum_{\mathbf{x} \in \mathscr{D}} \hat{\mathbf{u}}^{(p)}(\mathbf{x}; \mathbf{x}') \cdot \mathbf{P}(\mathbf{x}), \tag{6.80}$$

*where* $\hat{\mathbf{u}}^{(p)}$ *and* $\hat{\mathbf{T}}^{(p)}$ *are given by* (6.74) *and* (6.75).

*Proof.* First assume that $\mathscr{S}_1$ is not empty. By making use of the reciprocal theorem (6.70) and taking into account (6.76), we obtain

$$\int_{\mathscr{S}_1} (\hat{\mathbf{T}}^{(p)}\mathbf{n}) \cdot \mathbf{u}^\circ ds + u_p(\mathbf{x}') = \int_{\mathscr{S}_2} \mathbf{t}^\circ \cdot \hat{\mathbf{u}}^{(p)} ds + \int_{\mathscr{V}} \mathbf{f} \cdot \hat{\mathbf{u}}^{(p)}\, dv + \sum_{\mathbf{x} \in \mathscr{D}} \mathbf{P}(\mathbf{x}) \cdot \hat{\mathbf{u}}^{(p)}(\mathbf{x}; \mathbf{x}'),$$

---

[1] This theorem was given in the *isotropic* case by Lauricella [209] for the displacement and traction boundary-value problems without concentrated loads and by Turteltaub and Sternberg [362] for the traction problem with concentrated loads. Fredholm [123] derived the representation formula (6.80) in the *anisotropic* case for the displacement and mixed boundary-value problems without concentrated loads.

and the theorem is proved. On the other hand, if $\mathscr{S}_1$ is empty, then from (6.77), (6.73), and (6.79), we find that

$$\int_{\mathscr{S}} (\hat{\mathbf{T}}^{(p)}\mathbf{n}) \cdot \mathbf{u}\, ds = \mathbf{w}^{\circ} \cdot \int_{\mathscr{S}} \mathbf{u}\, ds + \boldsymbol{\omega}^{\circ} \cdot \int_{\mathscr{S}} (\mathbf{x} - \mathbf{c}) \times \mathbf{u}\, ds = 0, \qquad (6.81)$$

and the reciprocal theorem leads again to (6.80).

The advantage of using Green's tensor functions is that, after solving the particular boundary-value problem whose solution is $\tilde{\mathbf{u}}^{(p)}(\mathbf{x}; \mathbf{x}')$, the general solution corresponding to any other boundary-value problem may be obtained by quadratures, provided that the subboundaries $\mathscr{S}_1$ and $\mathscr{S}_2$ remain unchanged.

By applying the integral representation formula (6.80) to the singular elastic state corresponding to the displacement $\hat{u}_p^{(k)}(\mathbf{x}'; \mathbf{x})$, it results that Green's tensor function has the symmetry property

$$\hat{G}_{kp}(\mathbf{x}; \mathbf{x}') = \hat{G}_{kp}(\mathbf{x}'; \mathbf{x}). \qquad (6.82)$$

Finally, by taking into account (6.74) and (6.82), it can be shown that the representation formula (6.80) generalizes the relations (6.32) and (6.38) established above for the infinite elastic medium.

CHAPTER II

# THE LINEAR ELASTIC FIELD OF SINGLE DISLOCATIONS

## 7. The elastic model of a single dislocation

### 7.1. Introduction of the dislocation concept

The X-rays experiments made by Max von Laue in 1912 have definitely proved the atomistic and periodic nature of crystalline substances. It was reckoned by then that the structure plays a determining part in the physical and mechanical behaviour of such materials. However, most of the natural and artificial crystalline materials are polycrystals, i.e. they consist of randomly oriented single crystals and have isotropic macroscopic properties. This fact has somehow delayed the interpretation of the behaviour of polycrystalline materials in terms of the phenomena taking place inside the individual grains. It was the artificial growth of single crystals that has opened new prospects to the understanding of the correlation between the structure and the properties of crystalline materials. It has subsequently been proved for instance that the plastic deformation of metals takes place along certain preferred planes, called *glide planes*, and along certain preferred directions within these planes, called *glide directions*. As a rule, the glide planes have the maximum atomic density, and the glide directions have the closest atomic package.

One of the problems the physicists have been most concerned with from the very beginning of their studies on single crystals has been the explanation of the experimental value of the *yield stress*, i.e. of the stress level at which plastic deformation begins. Indeed, lattice calculations done by Frenkel, Polanyi, and Schmid between 1926 and 1929 led to theoretical values of the yield stress 100 to 1000 times higher than the experimental ones. However, these calculations assumed the crystalline structure to be perfect and the crystalline planes to glide along each other as a whole, like playing cards, hypotheses which proved later to be unsuitable to real materials. Almost in the same period it has been recognized that crystalline defects play a fundamental role in all phenomena taking place with material transport, e.g. in plastic deformation. Thus, Prandtl and Dehlinger succeeded as early as 1928 in explaining anelastic and recrystallization phenomena by using defect models very similar to what is presently called a dislocation.

In 1934, Orowan [262], Taylor [330], and Polanyi [271] imagined for the first time, independently of each other, the model of a linear crystalline defect called *edge dislocation*. Here is the explanation given by Taylor for the formation of an edge dislocation in an ideal crystal acted on by a shear stress (Fig. 7.1 a). When the shear stress attains a certain critical value, a glide step appears at the surface of the crystal, while a supplementary atomic half-plane, whose boundary is the *dislocation line*, occurs inside the crystal (Fig. 7.1. b). The subsequent glide propagation may be conceived as the dislocation motion through the crystal. At each stage of the plastic deformation, the dislocation line separates the region of

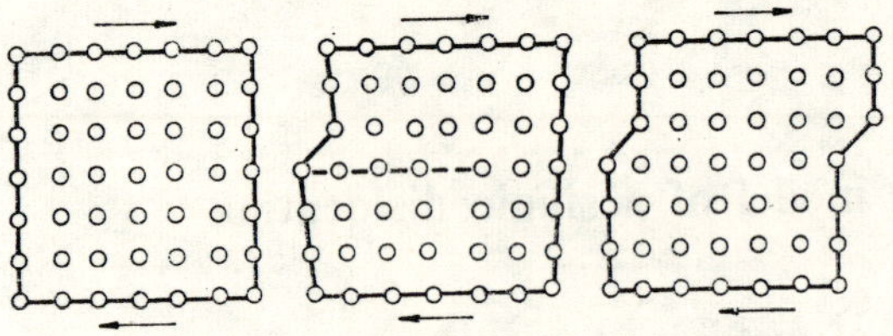


Fig. 7.1. Taylor's model of the atomic positions (a) before, (b) during, and (c) after the passage of an edge dislocation across a cubic lattice.

the glide plane on which the glide already took place from the one on which glide has not yet occurred. When the dislocation leaves the crystal, the crystalline structure resumes its initial regularity, but the two parts of the crystal separated by the glide plane preserve a relative displacement equal to one atomic spacing (Fig. 7.1 c). The glide lines occurring at the surface of a deformed single crystal are the result of a large number of dislocations emerging at the crystal surface along the boundary of the same glide plane.

Assuming that every crystal contains a large number of grown-in dislocations, Taylor was able to calculate the yield stress as the necessary stress to move a dislocation through the elastic field of all other dislocations, thus obtaining an evaluation in satisfactory agreement with the experimental result. On the other hand, Taylor noticed that the elastic field of the dislocations immobilized inside the crystal by various obstacles hinders the further motion of the gliding dislocations. Thus, the slip can proceed only under the action of an increasing applied stress, a phenomenon called *work-hardening* or *strain-hardening*. Moreover, by assuming that the number of the immobilized dislocations increases proportionally to the amount of glide, Taylor inferred that the flow stress should increase after a parabolic law, in agreement with the general aspect of the stress-plastic strain curve for an f.c.c. metal and a sufficiently high initial dislocation density.

Taylor's theory, based upon the hypothesis of the step-by-step propagation of plastic glide, succeeded in giving a first qualitative as well as quantitative explanation of the process of plastic deformation by means of the motion and interaction of dislocations. Subsequently, several other aspects concerning the origin of dislocations, the mechanism of their multiplication during plastic deformation, and the characteristic stages of work-hardening for various types of single crystals have been elucidated. It is interesting to note that most of these studies were of theoretical and predictive nature; it was only by 1950 that dislocation lines could be directly observed by making use of the electron microscope [1].

## 7.2. The Burgers vector

In a real crystal, dislocations generally occur as closed lines called *dislocation loops*, or as lines ending at the surface of the crystal.

A dislocation is characterized by its line and by the elementary glide vector associated with the dislocation, the so-called *Burgers vector*. The first correct definition of the Burgers vector was given by Frank [122] in 1951. We shall explain this definition in the case of an edge dislocation in a crystal with primitive cubic lattice; however, it is valid for an arbitrary curvilinear dislocation line and for an arbitrary crystalline lattice. The left side of Fig. 7.2 shows a perfect crystal and the right side a distorted crystal containing an edge dislocation. To define the Burgers vector we proceed as follows. Choose an arbitrary positive sense on the dislocation line $L$ and denote by $\mathbf{l}$ the unit vector tangent to $L$ at a current point and pointing in the positive direction. Describe within the distorted crystal a closed atomic circuit $PQ$, called the *Burgers circuit*, directed clockwise when looking down along the

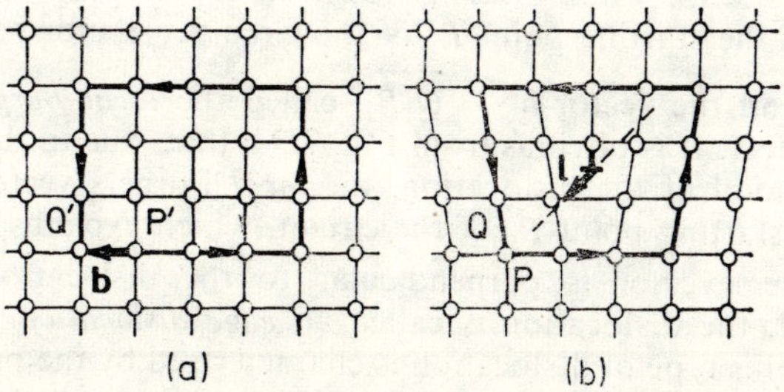


Fig. 7.2. Burgers circuits used to define the true Burgers vector **b** in a crystal. (a) Perfect lattice. (b) Crystal with an edge dislocation.

positive sense on $L$ (Fig. 7.2 b). Then, repeat the atomic circuit in the same sense within the perfect crystal, thus obtaining a closure failure (Fig. 7.2 a). The vector closing the last circuit and directed from the starting point $P'$ to the final point $Q'$ of the circuit is called the *true Burgers vector* [2] and is denoted by **b**. From this

[1] For various theories of plastic deformation based on the laws of motion, multiplication, and interaction of crystal defects see Cottrell [84], Seeger [286], Kronmüller [196], Zarka [387—389], Teodosiu [335, 344], Bullough [50], Perzyna [268, 269], Teodosiu and Sidoroff [345], where further references on this subject can be also found.

[2] There is no generally accepted convention for the sense of **b**. The convention adopted by us is known as the *SF/RH* rule, since **b** is directed from the *starting* point of the Burgers circuit to its *finish* whereas the Burgers circuit appears *right-handed* with respect to the positive sense chosen on the dislocation line. The same convention has been used for instance by Burgers [54], Read [275], Seeger [286], J. Weertman and J. R. Weertman [379], Nabarro [258], and more recently by Kosevich [440] and Gairola [418]. For further comments concerning the conventions used in various standard books on dislocation theory see de Wit [384] and Hirth and Lothe [162], p. 22.

definition it follows that **b** is always a vector of the perfect lattice. Moreover, it is easily seen that the true Burgers vector does not depend on the point chosen along the dislocation line [1], and hence it can be really considered as a characteristic of the dislocation.

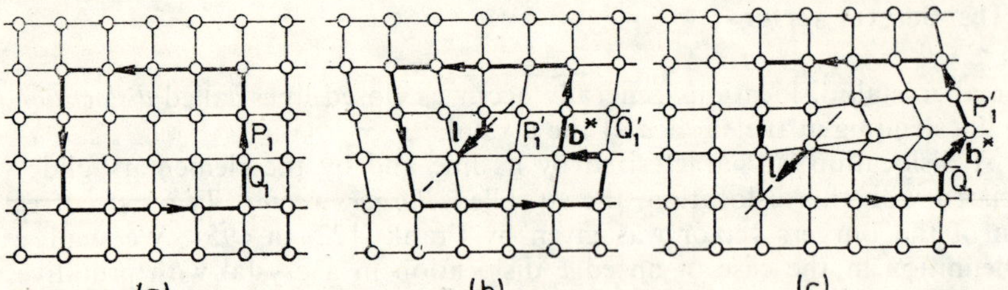


Fig. 7.3. Burgers circuits used to define the local Burgers vector **b*** in a crystal. (a) Perfect lattice. (b) Crystal with an edge dislocation. (c) Crystal with a screw dislocation.

To define the Burgers vector we may proceed the other way round. Namely, we can choose a closed atomic circuit $P_1Q_1$ within the perfect crystal (Fig. 7.3 a), such that the corresponding circuit in the distorted crystal encircle the dislocation line in a clockwise sense when looking down along the positive sense on $L$. The final point $Q_1'$ and the starting point $P_1'$ of the circuit in the distorted crystal do no longer coincide, and the vector $\mathbf{b}^* = \overrightarrow{Q_1'P_1'}$, called the *local Burgers vector*, defines now the lattice defect. It is obvious from Fig. 7.3 b that, due to the lattice distortion in the neighbourhood of the dislocation, the *local* Burgers vector does depend on the choice of the starting point $P_1$ of the circuit, which explains its name [2].

If the Burgers vector is perpendicular to the dislocation line ($\mathbf{b} \perp \mathbf{l}$), as shown in Fig. 7.2, the dislocation is called an *edge dislocation*. Inspection of Fig. 7.2 b reveals that this type of dislocation is characterized by the presence of a supplementary atomic half-plane. The vectors **b** and **l** determine the *glide plane*, whereas the vector **b** defines the *glide direction* associated with the dislocation motion. The position of the edge dislocation is marked by the symbol $\perp$ or $\top$, the horizontal line showing the direction of the glide plane and the vertical one the position of the supplementary atomic half-plane situated above or below the glide plane, respectively.

If the Burgers vector is parallel to the dislocation line ($\mathbf{b} \| \mathbf{l}$), the dislocation is called a *screw dislocation*, due to the resemblance of the atomic planes distorted by the dislocation to the spiral ridge of a screw of axis **l** and pitch $\|\mathbf{b}\|$. This type of lattice defect has been first imagined by Burgers [54] in 1939. Fig. 7.3 c shows a

---

[1] More precisely, **b** does not depend on the choice of the Burgers circuit as long as it surrounds the same dislocation line. In particular, the true Burgers vector is independent of the lattice distortion.

[2] Inspection of Figs. 7.3 b and 7.2 a reveals that the local Burgers vector **b*** may be considered as the true Burgers vector applied at $P_1$ and deformed together with the lattice. This relation will be given a more quantitative form in the following subsection.

Burgers circuit and the corresponding local Burgers vector in the case of a screw dislocation. The position of a screw dislocation is marked by the symbol ⊗, respectively ⊙, according as the Burgers vector is directed towards or out of the figure, that is according as the crystal planes build a right-handed or a left-handed screw with respect to the positive sense chosen on the dislocation line.

The edge and the screw dislocations are merely special types of dislocations. In the general case of a curvilinear dislocation, the vector **b** is still constant, but **l** varies along the dislocation line, which means that various segments of the dislocation line may be of different type. If the angle between **b** and **l** is not 0° or 90° the dislocation segment is said to be of a *mixed type*.

Since a dislocation line is the boundary between a region which has slipped and another region which has not slipped, it is intuitively obvious that it cannot end within an otherwise perfect crystal region. Thus, a dislocation line must be either a closed line, or a line terminated at a free surface, another dislocation line, an inclusion, a grain boundary, or some other defect. For instance, if one attempts to end the dislocation shown in Fig. 7.2 b by completing the supplementary lower lattice half-plane of the edge dislocation by an upper half-plane, one finds that this is possible only with the introduction of another edge dislocation with its line perpendicular to the initial dislocation (Hirth and Lothe [162], p. 23).

If the Burgers circuit surrounds more than one dislocation, and if it appears clockwise when looking down the positive sense chosen on each dislocation line, then the corresponding Burgers vector equals the sum of the Burgers vectors of all dislocations encircled by the circuit. Thus, the Burgers vectors of the individual dislocations can be summed up to obtain the *resultant* Burgers vector of a group of dislocations. This property will be used in Sect. 17 in order to extend the concept of Burgers vector to continuous distributions of dislocations.

## 7.3. Simulation of crystal dislocations by Volterra dislocations

As shown in Figs. 7.2 and 7.3, any dislocation produces a lattice deformation, which decreases with increasing distance from the dislocation line. In order to evaluate the deformation of a dislocated crystal it is advantageous to consider the crystal as a linear elastic continuum, at least at sufficiently large distances from the dislocation line. This approach must be given up, however, when considering the region of the crystal close to the dislocation line; indeed, in this highly distorted region, which is called the *dislocation core* and amounts to a few atomic spacings around the dislocation line, even the non-linear elasticity theory proves to be inappropriate [1].

Since dislocations and the accompanying lattice deformations can persist in the unloaded state of a body, the stresses produced by dislocations are called *self-stresses* or *residual stresses*.

---

[1] We shall come back to this point in Sect. 16, where several methods will be presented for studying the dislocation core by combining the continuum elastic model with the atomic model of the dislocation.

In his pioneering work, Taylor [330] realized the possibility of evaluating such self-stresses by the linear elasticity theory and approximated the long-range stress field of an edge dislocation by the linear elastic stress field corresponding to a Volterra dislocation in an isotropic hollow cylinder. Actually, this solution was available ever since 1907, long before the dislocation was considered as a crystal defect. Indeed, Volterra [373] determined the elastic state that occurs in a hollow circular isotropic cylinder when this is subjected to the following operations (cf.

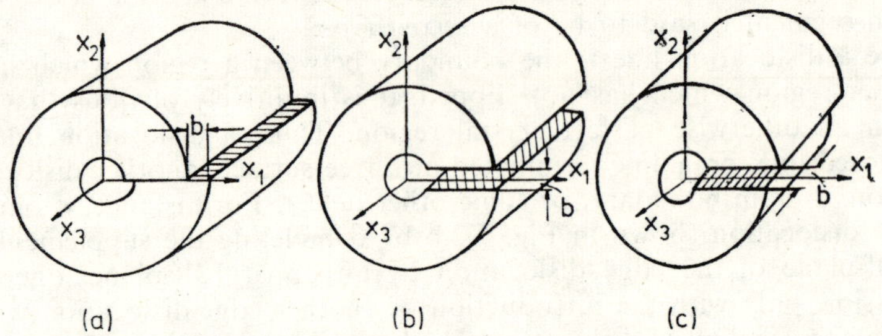


Fig. 7.4. Volterra dislocations of translational type in a hollow cylinder.

Sect. 2.8). First, the doubly-connected region occupied by the elastic body is rendered simply-connected by cutting it along a smooth surface joining the bounding cylindrical surfaces, e.g. a plane passing through the axis of the cylinder, and having that axis for an edge (Fig. 7.4). Next, one face of the cut is displaced with respect to the other by a small *rigid* displacement. Then, the opposing faces of the cut are joined, by removing or inserting, if necessary, a thin layer of material of the same kind as that of the cylinder. Finally, the external forces that have acted on the cylinder during these operations are removed. The body thus deformed will be, in general, in a state of self-stress [1].

Let us choose a Cartesian system of co-ordinates as shown in Fig. 7.4, with the cut taken as the $x_1x_3$-plane. The rigid relative displacement of the opposite faces of the cut may be decomposed into three translations along and three rotations around the axes of co-ordinates. It is easily seen that the deformation resulted after a *rigid translation* of the cut faces can be used to simulate a *crystal dislocation* (Fig. 7.4 a, b, c). Indeed, by removing a thin cylinder corresponding to the core of the edge dislocation shown in Fig. 7.3 b one obtains a configuration of the crystal which is similar to that of the dislocated cylinder in Fig. 7.4 a. Analogously, the screw dislocation in Fig. 7.3 c corresponds to the Volterra dislocation of the cylinder shown in Fig. 7.4 c. Finally, the deformation of the cylinder in Fig. 7.4 b can be produced in a crystal by inserting a supplementary atomic half-plane $x_1x_2$, and hence corresponds to an edge dislocation with the glide plane $x_2x_3$.

A Volterra dislocation obtained after *rigidly rotating* the faces of the cut corresponds to a defect which is presently named a *disclination*. It seems, however,

---

[1] Such deformations of multiply-connected bodies, which are presently called Volterra dislocations, have been named by Volterra "distorsioni". The name "dislocation" is due to Love [222], Sect. 156 A.

that no disclination can appear in a crystalline lattice, on account of the high self-energy required [1]. Therefore, we shall limit ourselves in the following to considering only Volterra dislocations of translational type.

The simulation of crystal dislocations by means of Volterra dislocations in an elastic continuum may be easily generalized to arbitrary dislocation loops. Assume that an elastic body $\mathscr{B}$ occupies a simply-connected region $\tilde{\mathscr{V}}$ of boundary $\tilde{\mathscr{S}}$ in

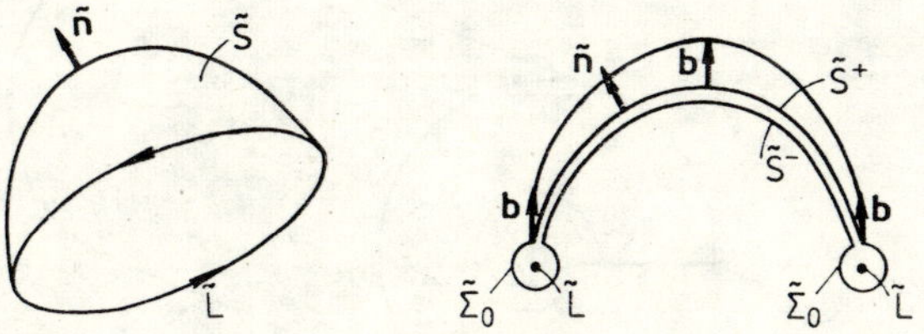


Fig. 7.5. Simulation of a dislocation loop $\tilde{L}$ by a Volterra dislocation.

the natural state, and let $\tilde{L}$ be a smooth closed line in $\tilde{\mathscr{V}}$ and $\tilde{S}$ a smooth and two-sided surface bounded by $\tilde{L}$ (Fig. 7.5). Arbitrarily choose a positive sense on $\tilde{L}$ and denote by $\tilde{\mathbf{n}}$ the unit normal to $\tilde{S}$ directed according to the right-hand rule with respect to the positive sense on $\tilde{L}$. Exclude a thin tube of boundary $\tilde{\Sigma}_0$ around the dislocation line and cut the elastic body along the surface $\tilde{S}$, so as to render it again simply-connected. Translate the positive cut face $\tilde{S}^+$, into which $\tilde{\mathbf{n}}$ points, by a vector $\mathbf{b}$ with respect to the negative face $\tilde{S}^-$. Finally, add or remove material, and join the two faces of the cut, thus re-establishing the continuity of the body. The result is a dislocation loop of line $\tilde{L}$ and true Burgers vector $\mathbf{b}$, the tube inside $\tilde{\Sigma}_0$ playing the role of the dislocation core. The only difference against the case of the straight dislocation considered above is that now the surface $\tilde{S}$ may be closed within the elastic body, as a consequence of the line $\tilde{L}$ being also closed inside the body.

As already mentioned, dislocations cannot end within an otherwise perfect region of a crystal. This property becomes obvious when simulating crystal dislocations by Volterra dislocations. Indeed, assume that a simply-connected body contains a finite dislocation line terminating within the body. Clearly, by cutting out a thin tube corresponding to the dislocation core the connectivity of the body does not change. On the other hand, cf. Sect. 2.8, Volterra dislocations are not possible in a simply-connected body and this indirectly proves that crystal dislocations cannot end within a perfect crystal region (cf. also Nabarro [258], p. 13).

---

[1] In exchange, disclinations may be used for modelling defects occurring in polymers, in inhomogeneous magnetoelastic fields, or in the flux lines of the magnetic field within a superconducting material (see, e.g. Anthony [5], Anthony and Kröner [393], Kröner and Anthony [444]).

For the dislocation lines ending at the free surface of the body, we shall use the following convention for the orientation of the cut surface. We consider the dislocation line $\tilde{L}$ and the cut surface as imaginarily continued outside the body until $\tilde{L}$ becomes a closed line. Then, the choice of the unit normal $\tilde{\mathbf{n}}$, of the positive and negative sides of $\tilde{S}$, and the generation of the corresponding Volterra dislocation may proceed like in the case of a dislocation loop. This convention is illustrated in Fig. 7.6 for an edge dislocation lying in the axis of a circular cylinder.

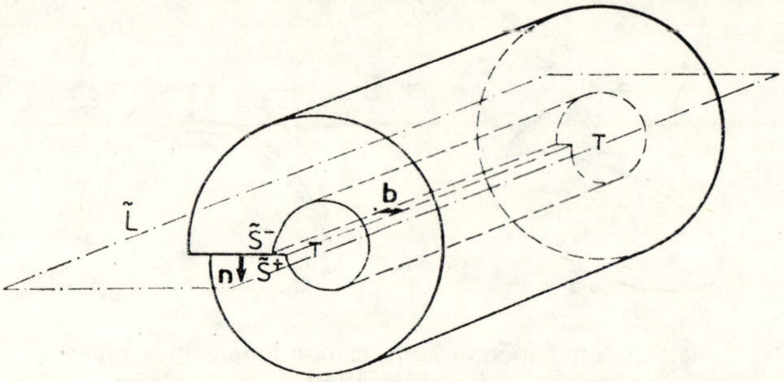


Fig. 7.6. Illustration of the convention used for the orientation of the cut surface of a Volterra dislocation, when the corresponding crystal dislocation ends at the free surface of the crystal.

The connection between crystal dislocations and Volterra dislocations is rendered more explicit by defining Burgers vectors of the latter with the help of line integrals whose integration paths are similar to Burgers circuits in a crystal. Let us denote by $(K)$ and $(k)$ the configurations of a simply-connected body $\mathscr{B}$ in the natural and the dislocated state, and let $\mathbf{X}$ and $\mathbf{x}$ be the position vectors of a current material point in the configurations $(K)$ and $(k)$, respectively. Denote by $\tilde{\mathscr{V}}$ the region occupied by the elastic body in the configuration $(K)$, by $\tilde{\mathscr{S}}$ the boundary of $\tilde{\mathscr{V}}$, and by $\tilde{\mathscr{V}}_0$ the doubly-connected region obtained after excluding the dislocation core by cutting out a thin tube of boundary $\tilde{\Sigma}_0$ around the dislocation line. Let $\tilde{S}$ be a smooth and two-sided barrier connecting the surfaces $\tilde{\Sigma}_0$ and $\tilde{\mathscr{S}}$ and rendering $\tilde{\mathscr{V}}_0$ simply-connected. To simplify the following discussion, we will again think of $\tilde{L}$ as being an edge dislocation lying in the axis of a cylinder $\tilde{\mathscr{V}}$; moreover, we shall assume that $\tilde{S}$ is a plane cut passing through $\tilde{L}$, the positive and negative faces of which are defined as shown in Fig. 7.6. However, the basic relations given below are general, not restricted to this particular example.

In the simply-connected region $\tilde{\mathscr{V}}_0 \setminus \tilde{\mathscr{S}}$ the deformation is uniquely defined. Let $\mathscr{V}_0$ and $\mathscr{S}$ be the images of $\tilde{\mathscr{V}}_0$ and $\tilde{\mathscr{S}}$, respectively, in the configuration $(k)$. Assuming as usual that the deformation $\chi\colon \tilde{\mathscr{V}}_0 \setminus \tilde{S} \to \mathscr{V}_0 \setminus S$ is one-to-one and of class $C^3$, we can write, with the notation in Sect. 2.5,

$$\mathbf{x} = \chi(\mathbf{X}) = \mathbf{X} + \mathbf{u}(\mathbf{X}), \tag{7.1}$$

$$\mathbf{F} = \mathbf{1} + \mathbf{H}, \tag{7.2}$$

where **u**(**X**) is the displacement field, and

$$\mathbf{F}(\mathbf{X}) = \text{Grad}\,\boldsymbol{\chi}(\mathbf{X}), \quad \mathbf{H}(\mathbf{X}) = \text{Grad}\,\mathbf{u}(\mathbf{X}). \tag{7.3}$$

Alternatively, we may describe the deformation in terms of the positions assumed by the material points in the deformed configuration ($k$). Denoting by $\boldsymbol{\chi}^{-1} : \mathscr{V}_0 \setminus S \to \tilde{\mathscr{V}}_0 \setminus \tilde{S}$ the inverse of the mapping (7.1), we have

$$\mathbf{X} = \boldsymbol{\chi}^{-1}(\mathbf{x}) = \mathbf{x} - \mathbf{u}(\mathbf{x}), \tag{7.4}$$

$$\mathbf{F}^{-1}(\mathbf{x}) = \mathbf{1} - \text{grad}\,\mathbf{u}(\mathbf{x}), \tag{7.5}$$

where

$$\mathbf{F}^{-1}(\mathbf{x}) = \text{grad}\,\boldsymbol{\chi}^{-1}(\mathbf{x}). \tag{7.6}$$

Clearly, **u**(**X**) and **u**(**x**) must be considered as different functions, expressing the displacement vector field in terms of the positions of the material points in the configurations ($K$) and ($k$), respectively, which are related in their turn by (7.1) and (7.4).

Let now $C$ be a closed curve in $\mathscr{V}_0$, which encircles the dislocation line $L$ in a right-handed sense and intersects $S^+$ and $S^-$ in the points $P$ and $Q$, respectively (Fig. 7.7 b). Then, the curve $C'$ corresponding to $C$ through (7.4) will encircle $\tilde{\Sigma}_0$

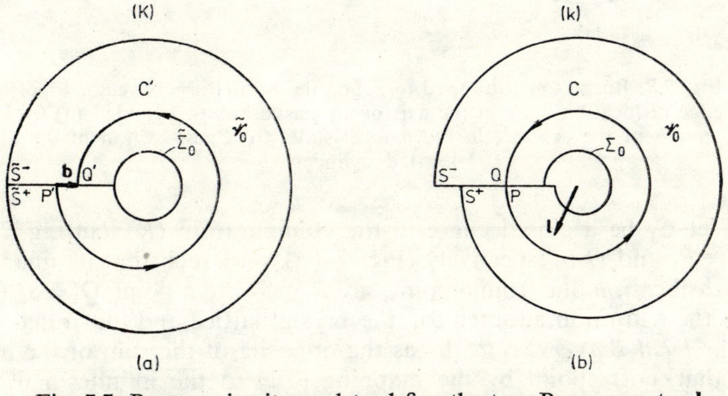


Fig. 7.7. Burgers circuits used to define the true Burgers vector **b** of an edge dislocation lying in the axis of an elastic hollow cylinder. (a) Cross-section of the cylinder in the natural state. (b) Cross-section of the dislocated cylinder.

from a point, say $P' \in \tilde{S}^+$, to a point $Q' \in \tilde{S}^-$ (Fig. 7.8 a). By analogy with the definition adopted for the crystal lattice and illustrated in Fig. 7.2, we define the *true Burgers vector* **b** as the sum of the infinitesimal vectors d**X** that correspond through the mapping (7.4) to the infinitesimal vectors d**x** taken along $C$. Thus

$$\mathbf{b} = \overrightarrow{P'Q'} = \oint_C \mathrm{d}\boldsymbol{\chi}^{-1}(\mathbf{x}) = -\oint_C \mathrm{d}\mathbf{u}(\mathbf{x}) = \mathbf{u}(P) - \mathbf{u}(Q),$$

since $\oint_C d\mathbf{x} = 0$. This relation may be rewritten as

$$\mathbf{u}^+(\mathbf{x}) - \mathbf{u}^-(\mathbf{x}) = \mathbf{b} \quad \text{on } S, \tag{7.7}$$

where $\mathbf{x}$ is the position vector of a current point on $S$, whereas $\mathbf{u}^+(\mathbf{x})$ and $\mathbf{u}^-(\mathbf{x})$ denote the limiting values of the displacement field $\mathbf{u}(\mathbf{x})$ on $S^+$ and $S^-$, respectively. Since the faces of the undeformed cut $\tilde{S}$ have been relatively displaced by a rigid translation, the true Burgers vector $\mathbf{b}$ is independent of the choice of $P$, as it should be.

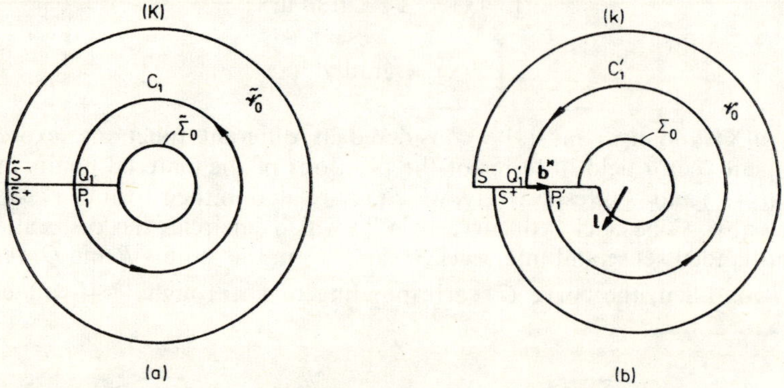


Fig. 7.8. Burgers circuits used to define the local Burgers vector **b*** of an edge dislocation lying in the axis of an elastic hollow cylinder. (a) Cross-section of the cylinder in the natural state. (b) Cross-section of the dislocated cylinder.

Next, let $C_1$ be a closed curve in the configuration $(K)$, cutting $\tilde{S}^+$ and $\tilde{S}^-$ in the points $P_1$ and $Q_1$, respectively (Fig. 7.8 a), and such that its image $C_1'$ in $(k)$ encircle the dislocation line from a point, say $P_1' \in S^+$, to a point $Q_1' \in S^-$ (Fig. 7.8 b). Similarly to the definition adopted for the crystal lattice and illustrated in Fig. 7.3, we define the *local Burgers vector* $\mathbf{b}^*$ as the opposite of the sum of the infinitesimal vectors $d\mathbf{x}$ that correspond by the mapping (7.1) to the infinitesimal vectors $d\mathbf{X}$ on $C_1$. Thus

$$\mathbf{b}^* = \overrightarrow{Q_1'P_1'} = -\oint_{C_1} d\boldsymbol{\chi}(\mathbf{X}) = -\oint_{C_1} d\mathbf{u}(\mathbf{X}) = \mathbf{u}(P_1) - \mathbf{u}(Q_1),$$

since $\oint_{C_1} d\mathbf{X} = 0$. This equation may be rewritten as

$$\mathbf{u}^+(\mathbf{X}) - \mathbf{u}^-(\mathbf{X}) = \mathbf{b}^*(\mathbf{X}) \quad \text{on } \tilde{S}, \tag{7.8}$$

where $\mathbf{X}$ is the position vector of a current point on $\tilde{S}$, whereas $\mathbf{u}^+(\mathbf{X})$ and $\mathbf{u}^-(\mathbf{X})$ denote the limiting values of the displacement field $\mathbf{u}(\mathbf{X})$ on $\tilde{S}^+$ and $\tilde{S}^-$, respectively.

Clearly, the local Burgers vector $\mathbf{b}^*$ depends on the strain around $P_1$, since both $P_1$ and $Q_1$ lie on the deformed surface of the cut.

It is now easy to establish the relation between the true and local Burgers vectors associated with a Volterra dislocation. To this end, let us denote by $\mathbf{X}$ and $\mathbf{X} + \mathbf{b}$ the position vectors of the points $P'$ and $Q'$, respectively. Since their images $P$ and $Q$ in $(k)$ coincide, we have

$$\chi^+(\mathbf{X}) = \chi^-(\mathbf{X} + \mathbf{b}) \quad \text{on } \tilde{S}, \tag{7.9}$$

whence, by (7.1),

$$\mathbf{u}^+(\mathbf{X}) - \mathbf{u}^-(\mathbf{X} + \mathbf{b}) = \mathbf{b} \quad \text{on} \quad \tilde{S}. \tag{7.10}$$

Next, neglecting terms of the order $O(b^2)$, where $b = \|\mathbf{b}\|$ is the magnitude of the true Burgers vector, and taking into account $(7.3)_2$, we may write

$$\mathbf{u}^+(\mathbf{X}) - \mathbf{u}^-(\mathbf{X}) - \mathbf{H}^-(\mathbf{X})\,\mathbf{b} = \mathbf{b} \quad \text{on } \tilde{S}.$$

Finally, by making use of (7.8) and (7.2), we deduce that

$$\mathbf{b}^*(\mathbf{X}) = \mathbf{F}^-(\mathbf{X})\mathbf{b}, \tag{7.11}$$

which is the desired result (Teodosiu [337], vol. 1). Alternatively, by replacing $\mathbf{X}$ with $\mathbf{X} - \mathbf{b}$ in (7.9) and using a similar reasoning as above, it may be shown that, to within terms of second order in $b$, we have

$$\mathbf{b}^*(\mathbf{X}) = \mathbf{F}^+(\mathbf{X})\mathbf{b}, \tag{7.12}$$

and hence [1]

$$[\mathbf{F}^+(\mathbf{X}) - \mathbf{F}^-(\mathbf{X})]\mathbf{b} = O(b^2). \tag{7.13}$$

We have already remarked that the local Burgers vector depends on the lattice deformation, and hence on the choice of the starting point of the Burgers circuit. Equations (7.11) and (7.12) give now a quantitative form to this dependence. Actually, they show that the local Burgers vector at $\mathbf{X}$ results by deforming together with the elastic body a material vector equal to the true Burgers vector $\mathbf{b}$ and applied at $\mathbf{X}$.

In the particular example considered above the cut $\tilde{S}$ was straight and its faces were rigidly displaced in their own plane. Consequently, it was possible to assume that they are rejoined with perfect fit, i.e. without adding or removing material. When this is not the case, the correct definition of the mapping $\chi$ requires a more sophisticated discussion. However, it may be shown (Teodosiu and Soós [479]) that the basic relations (7.7), (7.8), (7.10), and (7.11) derived above are still valid in the general case of an arbitrary dislocation loop and of an arbitrary cut.

---

[1] Generally, one assumes *a priori* that $\mathbf{F}$ is of class $C^1$ in $\mathcal{V}_0$. Then, of course, the left-hand side of (7.13) vanishes identically on any cut surface $\tilde{S}$ [479].

## 7.4. Linear elastostatics of single dislocations

At sufficiently large distances from the dislocation line, say 5 to 10 atomic spacings, it may be assumed that $\|\mathbf{H}\| \ll 1$ and thus one may apply the linear theory of elasticity. Then, by virtue of (7.2) and (7.11), we may identify the local Burgers vector $\mathbf{b}^*$ with the true Burgers vector $\mathbf{b}$ and the position vectors of the material points in the configuration $(k)$ with those in the reference configuration $(K)$. Consequently, neglecting terms of the order $O(b^2)$ or higher, (7.7) and (7.11) yield

$$\mathbf{u}^+(\mathbf{x}) - \mathbf{u}^-(\mathbf{x}) = \mathbf{b} = \mathbf{b}^*, \tag{7.14}$$

where $\mathbf{x}$ denotes the position vector of a current point on a cut $S = \tilde{S}$ transforming $\mathscr{V}_0 = \tilde{\mathscr{V}}_0$ into a simply-connected region, whereas $\mathbf{u}^+(\mathbf{x})$ and $\mathbf{u}^-(\mathbf{x})$ are the limiting values of the displacement vector field $\mathbf{u}(\mathbf{x})$ on the positive and negative face of the cut, respectively [1].

The basic field equations are (cf. Sect. 6.1)

$$\mathbf{E} = \operatorname{sym} \operatorname{grad} \mathbf{u}, \qquad E_{km} = \frac{1}{2}(u_{k;m} + u_{m,k}), \tag{7.15}$$

$$\operatorname{div} \mathbf{T} = \mathbf{0}, \qquad T_{km,m} = 0, \tag{7.16}$$

$$T_{kl} = c_{klmn} E_{mn}, \tag{7.17}$$

where $\mathbf{E}$ is the strain tensor, $\mathbf{T}$ is the Cauchy stress tensor, and $\mathbf{c}$ is the tensor of the second-order elastic constants. According to the theory of Volterra dislocations (Sect. 2.8), we shall assume that the strain field $\mathbf{E}$ is continuous across $S$ and its extension (by continuity) to $\mathscr{V}_0$ is of class $C^2$.

The solution of the above field equations must also fulfil certain boundary conditions. For instance, when the surface tractions are prescribed on the external boundary $\mathscr{S}$ of $\mathscr{V}_0$ and on the boundary $\Sigma_0$ of the dislocation core, the boundary condition reads

$$\mathbf{T}\mathbf{n} = \mathbf{t}^*, \qquad T_{km} n_m = t_k^* \quad \text{on } \mathscr{S} \cup \Sigma_0, \tag{7.18}$$

where $\mathbf{t}^*$ is the surface traction and $\mathbf{n}$ is the outward unit normal to the boundary of $\mathscr{V}_0$. More sophisticated boundary conditions can be also considered, e.g. the tractions prescribed on $\mathscr{S}$ and the displacements on $\Sigma_0$. The uniqueness of the solution of such boundary-value problems is generally covered by Volterra's uniqueness theorem for multiply-connected regions (see Sect. 6.2).

---

[1] As shown in Sect. 2.8, it is open to us to consider the displacement field either as single-valued and discontinuous in the simply-connected region $\mathscr{V}_0 \setminus S$, with a jump equal to $\mathbf{b}$ across the barrier $S$, or as multiple-valued and continuous in the doubly-connected region $\mathscr{V}_0$, with a vector cyclic constant $\mathbf{b}$. However, the former point of view will be constantly adopted throughout this book, since it corresponds better to the physical way crystal dislocations are generated.

If we are interested merely in calculating the self-stresses produced by dislocations, we can take $\mathbf{t}^* = \mathbf{0}$ on the external boundary $\mathscr{S}$ of the body [1]. On the contrary, the boundary $\Sigma_0$ of the tube used to isolate the dislocation line is acted on by forces arising from the dislocation core, which can be determined only by a combined continuum and atomistic calculation (see Sect. 16). It may be shown, however, that terms corresponding to these forces in the stress field decay much more rapidly with increasing distance from the dislocation line than those depending only on $L$ and $\mathbf{b}$, the latter characterizing thus the long-range stress field of the dislocation. That is why, many of the calculations done in the elastic theory of dislocations assume in general that $\mathbf{t}^* = \mathbf{0}$ on $\Sigma_0$ or even ignore altogether the boundary conditions on $\Sigma_0$.

We shall devote the remaining part of this chapter to the calculation of the linear elastic field and of the linear elastic interactions of stationary and moving dislocations. Non-linear effects in the elastic field of dislocations will be considered in Chapters III and IV.

### 7.5. Somigliana dislocations

Volterra dislocations require the continuity of the strain components and of their partial derivatives of first and second orders across the dislocation cut. We have seen that this condition implies that the relative displacement of the cut faces be rigid; moreover, as shown in Sect. 2.8, Volterra dislocations are possible only in multiply-connected bodies.

However, in order to re-establish the continuity of a cut body it is not even necessary to require the continuity of the strain or stress tensor across the cut. Actually, as shown by Somigliana [475, 476], it is sufficient that the tractions acting on the cut faces be in equilibrium at any point of the cut.

Let $\mathscr{B}$ be an elastic body of arbitrary connectivity, occupying a region $\mathscr{V}$ of boundary $\mathscr{S}$, and let $S$ denote a regular surface, which is contained in $\mathscr{V}$ or has a part of its boundary on $\mathscr{S}$. Assume that $\mathscr{B}$ is cut along $S$, then a thin sheet of material of the same kind as that of $\mathscr{B}$ is introduced or removed, and the continuity of the body is re-established, leaving an arbitrary discontinuity of the displacement across $S$, restricted only by the equilibrium and boundary conditions. The resulted state of self-strain is called a *Somigliana dislocation* [2].

Suppose that the displacement discontinuity across $S$ is sufficiently small to allow the application of linear elasticity. Then, arbitrarily choosing a positive and a negative face of $S$, we may write the jump conditions across the cut under the form

$$\mathbf{u}^+(\mathbf{x}) - \mathbf{u}^-(\mathbf{x}) = \mathbf{g}(\mathbf{x}), \tag{7.19}$$

$$\mathbf{t}^+(\mathbf{x}) + \mathbf{t}^-(\mathbf{x}) = \mathbf{0}, \tag{7.20}$$

where the superscripts + and − denote the limiting values of the corresponding fields on the positive and negative face of $S$, respectively, $\mathbf{x}$ is the position vector of an arbitrary point of $S$, and $\mathbf{g}(\mathbf{x})$ denotes the prescribed jump of $\mathbf{u}$ across $S$.

The boundary-value problem associated with conditions (7.19) and (7.20) is slightly more complicated than the mixed boundary-value problem of linear elasticity. Indeed, while the cut is open, the tractions *or* the displacements may be prescribed at all points of the boundary, including the faces of the cut. On the other hand, when the cut is closed, the six scalar equations corresponding to (7.19) and (7.20) replace a set of three equations on each side of the cut.

---

[1] For a straight dislocation lying in an infinite elastic medium we shall require that stresses vanish as $r^{-1}$ as $r \to \infty$, where $r$ is the distance from the dislocation line.

[2] The possibility of such states of self-strain has been recognized by Somigliana as early as 1905, i.e. immediately after the publication of the first notes by Volterra (cf. V. Volterra and E. Volterra [485], p. 13).

Somigliana dislocations in an isotropic hollow cylinder have been thoroughly studied by Yoffe [448, 489], who also applied them to discuss the structure of the dislocation core by a physically realistic non-linear model . Bogdanoff [400] pointed out the contribution of the discontinuities allowed in the first-order derivatives of the strain field, whereas Ju [434] treated plane problems corresponding to Somigliana dislocations with straight or logarithmic spiral cuts by means of complex variable techniques. Somigliana dislocations in an anisotropic elastic medium and their application to the determination of second-order effects in the elastic field of dislocations have been recently considered by Teodosiu [478] and by Teodosiu and Soós [479] (see also Sects. 10.6 and 14).

# 8. Straight dislocations in isotropic media

Even in polycrystalline materials, the influence of the elastic field of dislocations is significant mostly within the grains, which frequently exhibit a high anisotropy, thus limiting the applicability of the isotropic approximation. However, owing to the extreme simplicity of the solutions of isotropic elasticity, they have been almost exclusively used in the first thirty years of dislocation theory and are still being widely employed.

That is why we will consider in this section the case of straight dislocations lying in isotropic elastic cylinders or in an infinite isotropic elastic medium. We shall make use almost exclusively of complex-variable techniques, which allow a unitary and systematic solution of boundary-value problems.

## 8.1. Edge dislocation in an elastic cylinder

Consider an edge dislocation whose line $L$ has infinite length and coincides with the axis of an isotropic elastic circular cylinder of radius $R$. Choose the dislocation

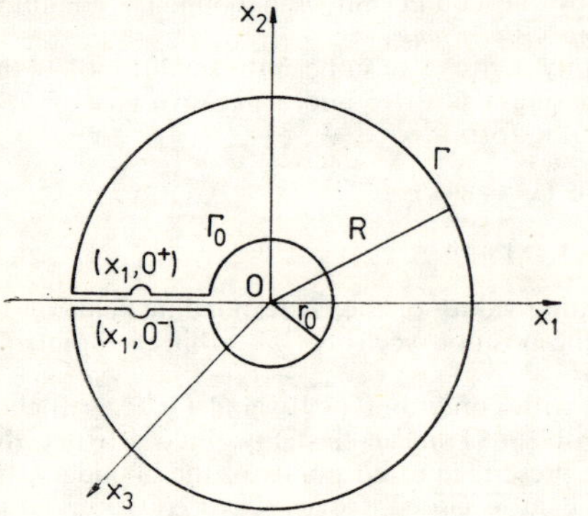


Fig. 8.1. Cut along the strip $x_2 = 0$, $-R \leqslant x_1 \leqslant -r_0$, used to define a single-valued displacement field around a straight dislocation lying along the axis of an elastic cylinder, taken as $x_3$-axis.

line as $x_3$-axis, and the direction of the Burgers vector as $x_1$-axis of a rectangular Cartesian system of co-ordinates (Fig. 8.1).

Let us apply the linear theory of elasticity outside an infinite circular cylindrical surface $\Sigma_0$ of axis $x_3$ and radius $r_0 < R$. We assume that the surface tractions acting on $\Sigma_0$ and on the outer boundary of the cylinder are known[1] and do not depend on $x_3$. Then the displacement vector field must be also independent of $x_3$ and its direction must be parallel to the $x_1x_2$-plane, i.e.

$$u_1 = u_1(x_1, x_2), \qquad u_2 = u_2(x_1, x_2), \qquad u_3 = 0. \tag{8.1}$$

The elastic state of the cylinder, characterized by (8.1), is said to be a *state of plane strain.*

We shall also make use of the cylindrical co-ordinates $\rho, \theta, z$ related to the Cartesian co-ordinates $x_1, x_2, x_3$ by

$$x_1 = \rho \cos \theta, \qquad x_2 = \rho \sin \theta, \qquad x_3 = z, \tag{8.2}$$

where $\theta \in (-\pi, \pi]$ is the polar angle in the $x_1x_2$-plane, measured in a clockwise sense when looking down along the $x_3$-axis.

Let $\Gamma_0$ and $\Gamma$ be the circles situated in the $x_1x_2$-plane, with centre at the origin and of radius $r_0$, and respectively $R$, and let $\Delta$ be the region between $\Gamma_0$ and $\Gamma$. We consider $\mathbf{u}$ as being single-valued and of class $C^3$ in the simply-connected region obtained from $\Delta$ by eliminating its points belonging to the negative $x_1$-axis (Fig. 8.1). Then the component $u_2$ of the displacement vector will be continuous across the cut $x_2 = 0$, $-R \leqslant x_1 \leqslant -r_0$, while the component $u_1$ will have a jump across this cut, given by

$$u_1(x_1, 0^+) - u_1(x_1, 0^-) = -b, \qquad -R \leqslant x_1 \leqslant -r_0, \tag{8.3}$$

where $b$ is the magnitude of the true Burgers vector.

Substituting (8.1) into $(7.15)_2$, it follows that the infinitesimal strain tensor has the non-zero components

$$E_{11} = \frac{\partial u_1}{\partial x_1}, \qquad E_{22} = \frac{\partial u_2}{\partial x_2}, \qquad E_{12} = \frac{1}{2}\left(\frac{\partial u_1}{\partial x_2} + \frac{\partial u_2}{\partial x_1}\right), \tag{8.4}$$

whereas $E_{13} = E_{23} = E_{33} = 0$. Next, by introducing (8.4) into (6.5), we find that the non-zero components of the stress tensor are

$$\left.\begin{aligned} T_{11} &= \lambda\left(\frac{\partial u_1}{\partial x_1} + \frac{\partial u_2}{\partial x_2}\right) + 2\mu\frac{\partial u_1}{\partial x_1}, \\ T_{22} &= \lambda\left(\frac{\partial u_1}{\partial x_1} + \frac{\partial u_2}{\partial x_2}\right) + 2\mu\frac{\partial u_2}{\partial x_2}, \\ T_{12} &= \mu\left(\frac{\partial u_1}{\partial x_2} + \frac{\partial u_2}{\partial x_1}\right), \end{aligned}\right\} \tag{8.5}$$

$$T_{33} = \lambda\left(\frac{\partial u_1}{\partial x_1} + \frac{\partial u_2}{\partial x_2}\right) = \nu(T_{11} + T_{22}), \tag{8.6}$$

[1] As already mentioned above, the surface forces exerted on $\Sigma_0$ from the dislocation core may be obtained only by a combined atomic and continuum calculation, which in its turn requires the solving of the boundary-value problem formulated below.

whereas $T_{13} = T_{23} = 0$. Since the components of the stress tensor are also independent of $x_3$, the first two equilibrium equations (7.16) assume the reduced form

$$\frac{\partial T_{11}}{\partial x_1} + \frac{\partial T_{12}}{\partial x_2} = 0, \qquad \frac{\partial T_{12}}{\partial x_1} + \frac{\partial T_{22}}{\partial x_2} = 0, \tag{8.7}$$

and the third one is identically satisfied.

Since we have assumed that $t_3^* = t_z^* = 0$ for $\rho = r_0$ and $\rho = R$, we shall consider only the physical components $t_\rho^*$ and $t_\theta^*$ of the surface tractions acting on the bounding cylindrical surfaces of the elastic body. In addition, we shall suppose that these components may be expanded in complex Fourier series of the polar angle $\theta$, i.e.[1]

$$t_\rho^* + i t_\theta^* = \begin{cases} \sum_{k=-\infty}^{\infty} t_k^{(1)} \, \mathrm{e}^{ik\theta} & \text{for } \rho = r_0 \\ \sum_{k=-\infty}^{\infty} t_k^{(2)} \, \mathrm{e}^{ik\theta} & \text{for } \rho = R. \end{cases} \tag{8.8}$$

Since both the dislocation core and the remaining part of the elastic cylinder are in equilibrium, the total force and the total couple acting on each of the surfaces $\rho = r_0$ and $\rho = R$ vanish, and this implies that

$$\int_0^{2\pi} (t_\rho^* \cos\theta - t_\theta^* \sin\theta) \, \mathrm{d}\theta = 0, \qquad \int_0^{2\pi} (t_\rho^* \sin\theta + t_\theta^* \cos\theta) \, \mathrm{d}\theta = 0,$$
$$\int_0^{2\pi} t_\theta^* \, \mathrm{d}\theta = 0. \tag{8.9}$$

The first two of these conditions may be written in the equivalent complex form

$$\int_0^{2\pi} (t_\rho^* + i t_\theta^*) \, \mathrm{e}^{i\theta} \, \mathrm{d}\theta = 0. \tag{8.10}$$

Introducing (8.8) into (8.9) and (8.10), yields

$$t_{-1}^{(1)} = t_{-1}^{(2)} = 0, \qquad \mathrm{Im}\, t_0^{(1)} = \mathrm{Im}\, t_0^{(2)} = 0. \tag{8.11}$$

---

[1] For more generality we have assumed in (8.8) that the outer cylindrical surface is acted by surface tractions, too. When this is not the case, one should take $t_k^{(2)} = 0$ for all integers $k$.

On the other hand, by making use of $(1.64)_4$ and (1.73), it may be shown that

$$T_{\rho\rho} + iT_{\rho\theta} = \frac{1}{2}[T_{11} + T_{22} + i(T_{11} - T_{22} + 2i\, T_{12})e^{-2i\theta}],$$

and hence the boundary conditions (7.18) can be written in the equivalent complex form

$$\frac{1}{2}[T_{11} + T_{22} + i(T_{11} - T_{22} + 2i\, T_{12})e^{-2i\theta}] = t_\rho^* + it_\theta^* \quad \text{for } \rho = r_0 \text{ and } \rho = R, \tag{8.12}$$

the right-hand side of this equation being given by (8.8).

The easiest way to solve the above formulated boundary-value problem is based on the representation of the elastic solution with the aid of complex potentials (see also Muskhelishvili [254], Sect. 30—32 and 45, and Gurtin [150], Sect. 47). Let us first introduce the complex variables

$$z = x_1 + i\,x_2, \qquad \bar{z} = x_1 - ix_2, \tag{8.13}$$

the *complex displacement*

$$U = u_1 + i\,u_2, \tag{8.14}$$

and the *complex stresses*

$$\Theta = T_{11} + T_{22}, \qquad \Phi = T_{11} - T_{22} + 2iT_{12}. \tag{8.15}$$

By using the relations

$$\frac{\partial}{\partial z} = \frac{1}{2}\left(\frac{\partial}{\partial x_1} - i\frac{\partial}{\partial x_2}\right), \qquad \frac{\partial}{\partial \bar{z}} = \frac{1}{2}\left(\frac{\partial}{\partial x_1} + i\frac{\partial}{\partial x_2}\right), \tag{8.16}$$

we deduce from (8.14) that

$$\begin{aligned} \frac{\partial U}{\partial z} &= \frac{1}{2}\left[\frac{\partial u_1}{\partial x_1} + \frac{\partial u_2}{\partial x_2} - i\left(\frac{\partial u_1}{\partial x_2} - \frac{\partial u_2}{\partial x_1}\right)\right], \\ \frac{\partial U}{\partial \bar{z}} &= \frac{1}{2}\left[\frac{\partial u_1}{\partial x_1} - \frac{\partial u_2}{\partial x_2} + i\left(\frac{\partial u_1}{\partial x_2} + \frac{\partial u_2}{\partial x_1}\right)\right]. \end{aligned} \tag{8.17}$$

Next, by using (8.13—17), we derive the equivalent complex form of the *jump condition* (8.3)

$$U(x_1, 0^+) - U(x_1, 0^-) = -b, \qquad -R \leqslant x_1 \leqslant -r_0, \tag{8.18}$$

of the *constitutive equations* (8.5)

$$\Theta = 2(\lambda + \mu)\left(\frac{\partial U}{\partial z} + \frac{\partial \bar{U}}{\partial \bar{z}}\right), \qquad \Phi = 4\mu\,\frac{\partial U}{\partial \bar{z}}, \tag{8.19}$$

of the *equilibrium equations* (8.7)

$$\frac{\partial \Theta}{\partial \bar{z}} + \frac{\partial \Phi}{\partial z} = 0, \tag{8.20}$$

and of the *boundary conditions* (8.12)

$$\tfrac{1}{2}(\Theta + \Phi e^{-2i\theta}) = t_\rho^* + it_\theta^* \quad \text{for } \rho = r_0 \text{ and } \rho = R. \tag{8.21}$$

Equation (8.20) may be identically satisfied by putting

$$\Phi = -4\frac{\partial^2 F}{\partial \bar{z}^2}, \qquad \Theta = 4\frac{\partial^2 F}{\partial z\, \partial \bar{z}}, \tag{8.22}$$

where $F$ is a real-valued function of class $C^4$, which is called *Airy's stress function.* The function $F$ must also satisfy the Beltrami-Michell compatibility equations (6.25). To derive their complex equivalent for the state of plane strain, we directly eliminate $U$ between equations (8.19), thus obtaining

$$(\lambda + \mu)\frac{\partial^2 \Theta}{\partial z\, \partial \bar{z}} - 2\mu\left(\frac{\partial^2 \Phi}{\partial z^2} + \frac{\partial^2 \bar{\Phi}}{\partial \bar{z}^2}\right) = 0,$$

wherefrom, by (8.22), it results that [1]

$$\frac{\partial^4 F}{\partial z^2\, \partial \bar{z}^2} = 0 \qquad \text{in } \Delta_0. \tag{8.23}$$

By successively integrating this equation with respect to $z$ and $\bar{z}$, and taking into account that $F$ and $\dfrac{\partial^2 F}{\partial z\, \partial \bar{z}}$ are real-valued functions, we obtain

$$\frac{\partial^2 F}{\partial z\, \partial \bar{z}} = \operatorname{Re} \varphi'(z), \qquad \frac{\partial F}{\partial \bar{z}} = \tfrac{1}{2}\{\varphi(z) + z\overline{\varphi'(z)} + \overline{\psi(z)}\}, \tag{8.24}$$

$$F(z, \bar{z}) = \operatorname{Re}\{\bar{z}\varphi(z) + \chi(z)\}, \tag{8.25}$$

---

[1] It is easily seen from (8.16) that

$$4\frac{\partial^2}{\partial z \partial \bar{z}} = \frac{\partial^2}{\partial x_1^2} + \frac{\partial^2}{\partial x_2^2},$$

and hence (8.23) implies that $F$ is a biharmonic function in $\Delta_0$.

where $\varphi(z)$ and $\psi(z)$ are arbitrary analytic functions of $z$ in $\Delta_0$ and $\chi(z) = \int_0^z \psi(z)\mathrm{d}z$. Substituting $(8.24)_2$ into (8.22) yields the representation of complex stresses in terms of the complex potentials $\varphi(z)$ and $\psi(z)$

$$\left.\begin{aligned} T_{11} + T_{22} = \Theta = 4\mathrm{Re}\,\varphi'(z), \\ T_{11} - T_{22} + 2iT_{12} = \Phi = -2\{z\overline{\varphi''(z)} + \psi'(z)\}. \end{aligned}\right\} \tag{8.26}$$

Next, we have from (8.19) and (8.26)

$$\left.\begin{aligned} 2\mu \frac{\partial U}{\partial \bar{z}} = -\, z\overline{\varphi''(z)} - \overline{\psi'(z)}, \\ \frac{\partial U}{\partial z} + \frac{\partial \bar{U}}{\partial \bar{z}} = \frac{1}{\lambda + \mu}\{\varphi'(z) + \overline{\varphi'(z)}\}. \end{aligned}\right\} \tag{8.27}$$

Integrating the first of these equations with respect to $\bar{z}$ gives

$$2\mu\, U(z, \bar{z}) = z\overline{\varphi'(z)} - \overline{\psi(z)} + \eta(z), \tag{8.28}$$

where $\eta(z)$ is an arbitrary analytic function of $z$. Introducing (8.28) into $(8.27)_2$ and making use of (6.9), we find

$$\eta'(z) + \overline{\eta'(z)} = \frac{\lambda + 3\mu}{\lambda + \mu}\{\varphi'(z) + \overline{\varphi'(z)}\} = (3 - 4\nu)\{\varphi'(z) + \overline{\varphi'(z)}\},$$

wherefrom, by integration, it results that [1]

$$\eta(z) = (3 - 4\nu)\varphi(z) + 2\mu(\omega_0 iz + u_0 + iv_0), \tag{8.29}$$

where $\omega_0$, $u_0$, and $v_0$ are arbitrary real constants. Finally, by substituting (8.29) into (8.28), we obtain

$$2\mu U(z, \bar{z}) = (3 - 4\nu)\varphi(z) - z\overline{\varphi'(z)} - \overline{\psi(z)} + 2\mu(\omega_0 iz + u_0 + iv_0). \tag{8.30}$$

The expression $\omega_0 iz + u_0 + iv_0$ is an infinitesimal complex rigid displacement, composed of an infinitesimal translation of components $u_0$, $v_0$ and of an infinitesimal rotation of angle $\omega_0$ around the $x_3$-axis. Clearly, this expression could be included

[1] In deriving (8.29) we have taken into account that the imaginary part of an analytic function whose real part vanishes must be a constant.

in the arbitrary functions $\varphi(z)$ and $\overline{\psi(z)}$. However, we prefer to preserve the form (8.30) of the complex displacement and to impose the supplementary conditions

$$\varphi(0) = 0, \qquad \psi(0) = 0, \qquad \operatorname{Im}\varphi'(0) = 0, \tag{8.31}$$

which exhaust the arbitrariness in the choice of the functions $\varphi(z)$ and $\psi(z)$ corresponding to a given elastic state [254].

We also notice that from $(2.39)_2$, (8.1), $(8.17)_1$, and (8.30) it follows that the only non-zero component of the elastic rotation vector is

$$\omega_3 = -\frac{1}{2}\left(\frac{\partial u_1}{\partial x_2} - \frac{\partial u_2}{\partial x_1}\right) = \operatorname{Im}\frac{\partial U}{\partial z} = \frac{2(1-\nu)}{\mu}\operatorname{Im}\varphi'(z) + \omega_0. \tag{8.32}$$

The relations (8.26) and (8.30) give *Kolosov's representation* of the solution of the plane strain problem of linear elasticity in terms of the complex potentials $\varphi(z)$ and $\psi(z)$.

Substituting (8.26) and (8.8) into (8.21) yields

$$\varphi'(z) + \overline{\varphi'(z)} - \mathrm{e}^{-2i\theta}\{z\overline{\varphi''(z)} + \overline{\psi'(z)}\} = \begin{cases} \sum\limits_{k=-\infty}^{\infty} t_k^{(1)}\,\mathrm{e}^{ik\theta} & \text{for } z = r_0\,\mathrm{e}^{i\theta} \\ \sum\limits_{k=-\infty}^{\infty} t_k^{(2)}\,\mathrm{e}^{ik\theta} & \text{for } z = R\,\mathrm{e}^{i\theta}. \end{cases} \tag{8.33}$$

The boundary-value problem may be given now the following formulation: *Find the functions $\varphi(z)$ and $\psi(z)$ that are analytic in $\Delta_0$ and continuous in $\overline{\Delta}_0 = \Delta_0 \cup \Gamma_0 \cup \Gamma$ and that satisfy the jump condition* (8.18) *and the boundary conditions* (8.33).

Since the stress components and the elastic rotation are continuous across the negative $x_1$-axis, equations (8.26) and (8.32) imply that the analytic functions $\varphi'(z)$ and $\psi'(z)$ must be continuous and single-valued in $\Delta$ and hence they can be expanded in Laurent power series of $z$

$$\varphi'(z) = \sum_{k=-\infty}^{\infty} a_k z^k, \qquad \psi'(z) = \sum_{k=-\infty}^{\infty} b_k z^k. \tag{8.34}$$

Termwise integration of these series gives

$$\varphi(z) = a_{-1}\ln z + \sum_{\substack{k=-\infty \\ k\neq -1}}^{\infty} \frac{a_k\, z^{k+1}}{k+1}, \qquad \psi(z) = b_{-1}\ln z + \sum_{\substack{k=-\infty \\ k\neq -1}}^{\infty} \frac{b_k z^{k+1}}{k+1}, \tag{8.35}$$

the constants of integration being zero owing to the first two conditions (8.31). It is well known (see, e.g. Knopp [436]) that one may choose a single-valued determination of the multiple-valued function $\ln z$ by introducing a suitable cut in the $z$-plane. For example, by choosing the cut $x_2 = 0,\ x_1 \leqslant 0$, we can take

$$\ln z = \ln|z| + i \arg z, \tag{8.36}$$

where [1]

$$|z| = \rho = \sqrt{x_1^2 + x_2^2},$$

$$\arg z = \theta = \begin{cases} \operatorname{cotan}^{-1}(x_1/x_2) & \text{for } x_2 > 0 \\ 0 & \text{for } x_2 = 0,\, x_1 > 0 \\ \operatorname{cotan}^{-1}(x_1/x_2) - \pi & \text{for } x_2 < 0. \end{cases} \tag{8.37}$$

According to this definition, the limiting values of $\arg z$ on the upper and lower faces of the cut are $\pi$ and $-\pi$, respectively. Consequently, by introducing (8.34) and (8.35) into (8.30), and the result obtained into the jump condition (8.18), it follows that

$$(3 - 4\nu)a_{-1} + \bar{b}_{-1} = -\, i\mu b/\pi. \tag{8.38}$$

Next, by substituting (8.34) into (8.33) and equating coefficients of $e^{ik\theta}$ for $k = 0,\ \pm 1, \pm 2, \ldots,$ we obtain for $k = 0$:

$$2a_0 - \bar{b}_{-2} r_0^{-2} = t_0^{(1)}, \qquad 2a_0 - \bar{b}_{-2} R^{-2} = t_0^{(2)}, \tag{8.39}$$

since $\operatorname{Im} a_0 = 0$ on account of $(8.31)_3$; for $k = -1$, considering also $(8.11)_1$:

$$a_{-1} - \bar{b}_{-1} = 0; \tag{8.40}$$

for $k = 1$:

$$a_1 r_0 + 2\bar{a}_{-1} r_0^{-1} - \bar{b}_{-3} r_0^{-3} = t_1^{(1)}, \quad a_1 R + 2\bar{a}_{-1} R^{-1} - \bar{b}_{-3} R^{-3} = t_1^{(2)}; \tag{8.41}$$

for $k = \pm 2, \pm 3, \ldots$:

$$\left.\begin{aligned} a_k r_0^k + \bar{a}_{-k} r_0^{-k}(1 + k) - \bar{b}_{-k-2} r_0^{-k-2} &= t_k^{(1)}, \\ a_k R^k + \bar{a}_{-k} R^{-k}(1 + k) - \bar{b}_{-k-2} R^{-k-2} &= t_k^{(2)}. \end{aligned}\right\} \tag{8.42}$$

---

[1] This definition of the logarithm has the advantage of being also valid for real values of the argument and of giving $\arg \bar{z} = -\arg z$. In addition, the cut chosen for $\ln z$ coincides with that adopted above to make the displacement single-valued.

From (8.38—40) we deduce that

$$a_0 = \frac{R^2 t_0^{(2)} - r_0^2 t_0^{(1)}}{2(R^2 - r_0^2)}, \qquad b_{-2} = \frac{r_0^2 R^2(\bar{t}_0^{(2)} - \bar{t}_0^{(1)})}{R^2 - r_0^2}, \qquad a_{-1} = \bar{b}_{-1} = \frac{i\mu b}{4\pi(1-\nu)}. \tag{8.43}$$

Next, introducing $(8.43)_3$ into (8.41) yields

$$\left.\begin{aligned} a_1 &= \frac{i\mu b}{2\pi(1-\nu)(R^2+r_0^2)} + \frac{R^3 t_1^{(2)} - r_0^3 t_1^{(1)}}{R^4 - r_0^4}, \\ b_{-3} &= \frac{i\mu b R^2 r_0^2}{2\pi(1-\nu)(R^2+r_0^2)} + \frac{R^3 r_0^3(r_0 t_1^{(2)} - R t_1^{(1)})}{R^4 - r_0^4}, \end{aligned}\right\} \tag{8.44}$$

and from (8.42) it results that

$$\left.\begin{aligned} a_k &= \frac{(R^{2-2k} - r_0^{2-2k})(R^{k+2} t_k^{(2)} - r_0^{k+2} t_k^{(1)}) - (1+k)(R^2 - r_0^2)(R^{2-k}\bar{t}_{-k}^{(2)} - r_0^{2-k}\bar{t}_{-k}^{(1)})}{(R^{2k+2} - r_0^{2k+2})(R^{2-2k} - r_0^{2-2k}) - (1-k^2)(R^2 - r_0^2)^2}, \\ b_{-k-2} &= r_0^{2k+2}\bar{a}_k + (1+k) r_0^2 a_{-k} - r_0^{k+2}\bar{t}_k^{(1)}, \end{aligned}\right\} \tag{8.45}$$

for $k = \pm 2, \pm 3, \ldots$ Equations (8.43—45) determine all coefficients occurring in the expansions (8.35) of the complex potentials $\varphi(z)$ and $\psi(z)$. Thus, the boundary-value problem formulated above is completely solved. Indeed, (8.26) and (8.30) give now the stress and the displacement components, the latter being determined, as was to be expected, to within an infinitesimal rigid displacement.

It should be noticed that in obtaining the solution for an elastic cylinder of infinite length, it was tacitly assumed that the state of plane strain is maintained by the surface tractions

$$t_3^* = \pm T_{33}(x_1, x_2), \tag{8.46}$$

with $T_{33}$ given by (8.6), acting on the bases $x_3 = +\infty$, respectively $x_3 = -\infty$, of the cylinder. In order to use this solution for an elastic cylinder whose length, although finite, is large with respect to the radius of its cross section and whose bases are free of surface tractions, some correcting terms, arising from the condition that the resultant force and couple acting on the bases be zero, must be introduced. This procedure will be detailed in the next subsection for the case of a screw dislocation.

When both surfaces $\rho = r_0$ and $\rho = R$ are free of traction, i.e. $t_k^{(1)} = t_k^{(2)} = 0$ for all $k$, $(8.43)_3$ and (8.44) yield

$$\left.\begin{aligned} a_{-1} &= \bar{b}_{-1} = \frac{i\mu b}{4\pi(1-\nu)}, \qquad a_1 = \frac{i\mu b}{2\pi(1-\nu)(R^2 + r_0^2)}, \\ b_{-3} &= \frac{i\mu b R^2 r_0^2}{2\pi(1-\nu)(R^2+r_0^2)}, \end{aligned}\right\} \tag{8.47}$$

and all other coefficients $a_k$, $b_k$ vanish. It results then from (8.35), (8.26), and (8.30), by putting $u_0 = v_0 = \omega_0 = 0$, that

$$\varphi(z) = a_{-1}\ln z + \frac{a_1 z^2}{2}, \qquad \psi(z) = b_{-1}\ln z - \frac{b_{-3}}{2z^2},$$

$$2\mu U = (3-4\nu)\left(a_{-1}\ln z + \frac{a_1 z^2}{2}\right) - \bar{a}_{-1}\frac{z}{\bar{z}} - \bar{a}_1 z\bar{z} - \bar{b}_{-1}\ln\bar{z} + \frac{\bar{b}_{-3}}{2\bar{z}^2},$$

$$T_{11} + T_{22} = 4\,\mathrm{Re}\left(\frac{a_{-1}}{z} + a_1 z\right),$$

$$T_{11} - T_{22} + 2iT_{12} = -2\left(\bar{a}_1 z - \frac{\bar{a}_{-1}z}{\bar{z}^2} + \frac{\bar{b}_{-1}}{\bar{z}} + \frac{\bar{b}_{-3}}{2\bar{z}^2}\right).$$

Finally, by taking into account that

$$T_{\rho\rho} + T_{\theta\theta} = T_{11} + T_{22}, \qquad T_{\rho\rho} - T_{\theta\theta} + 2iT_{\rho\theta} = (T_{11} - T_{22} + 2iT_{12})\mathrm{e}^{-2i\theta},$$

we obtain from the above relations and (8.47) the displacement components

$$\left.\begin{aligned} u_1 &= -\frac{b}{2\pi}\left\{\theta + \frac{1}{4(1-\nu)}\left[1 + \frac{(3-4\nu)\rho^2}{R^2+r_0^2} - \frac{R^2 r_0^2}{\rho^2(R^2+r_0^2)}\right]\sin 2\theta\right\}, \\ u_2 &= \frac{b}{8\pi(1-\nu)}\left\{2(1-2\nu)\ln\rho + \frac{2\rho^2}{R^2+r_0^2} + \right. \\ &\qquad \left. + \left[1 + \frac{(3-4\nu)\rho^2}{R^2+r_0^2} - \frac{R^2 r_0^2}{\rho^2(R^2+r_0^2)}\right]\cos 2\theta\right\}, \end{aligned}\right\} \tag{8.48}$$

and the physical components in cylindrical co-ordinates of the stress tensor

$$\left.\begin{aligned}
T_{\rho\rho} &= \frac{\mu b}{2\pi(1-\nu)}\left(\frac{1}{\rho} - \frac{\rho}{R^2 + r_0^2} - \frac{R^2 r_0^2}{R^2 + r_0^2}\frac{1}{\rho^3}\right)\sin\theta,\\
T_{\theta\theta} &= \frac{\mu b}{2\pi(1-\nu)}\left(\frac{1}{\rho} - \frac{3\rho}{R^2 + r_0^2} + \frac{R^2 r_0^2}{R^2 + r_0^2}\frac{1}{\rho^3}\right)\sin\theta,\\
T_{\rho\theta} &= \frac{\mu b}{2\pi(1-\nu)}\left(-\frac{1}{\rho} + \frac{\rho}{R^2 + r_0^2} + \frac{R^2 r_0^2}{R^2 + r_0^2}\frac{1}{\rho^3}\right)\cos\theta,\\
T_{zz} &= \nu(T_{\rho\rho} + T_{\theta\theta}), \quad T_{\rho z} = T_{z\rho} = 0.
\end{aligned}\right\} \tag{8.49}$$

It is generally admitted that the linear dimensions of the body containing the dislocations are much larger than the range of the elastic field produced by dislocations; this assumption comes in our case to letting $R \to \infty$. Moreover, since the boundary conditions on the surface $\rho = r_0$ of the dislocation core are unknown without a simultaneous atomic calculation, the terms of order $O(\rho^{-3})$ in the stress components and of order $O(\rho^{-2})$ in the displacement components, which arise from satisfying these boundary conditions, are frequently neglected as $\rho \to \infty$. With these approximations, (8.48) and (8.49) yield the simplified relations

$$\left.\begin{aligned}
u_1 &= -\frac{b}{2\pi}\left[\theta + \frac{\sin 2\theta}{4(1-\nu)}\right],\\
u_2 &= \frac{b}{8\pi(1-\nu)}\left[2(1-2\nu)\ln\rho + \cos 2\theta\right],
\end{aligned}\right\} \tag{8.50}$$

$$\left.\begin{aligned}
T_{\rho\rho} &= T_{\theta\theta} = \frac{1}{2\nu}T_{zz} = \frac{\mu b}{2\pi(1-\nu)}\frac{\sin\theta}{\rho},\\
T_{\rho\theta} &= -\frac{\mu b}{2\pi(1-\nu)}\frac{\cos\theta}{\rho}, \quad T_{\rho z} = T_{z\rho} = 0,
\end{aligned}\right\} \tag{8.51}$$

which are used, especially when looking for a rough evaluation of the effects produced by the elastic field of an edge dislocation. Equations (8.51) show that the stress field has a rather long range, for its components decrease merely as $\rho^{-1}$ as $\rho \to \infty$. This is one of the reasons why the theory of elasticity has been so successful in modelling crystal dislocations.

Substituting (8.51) into (6.8), we obtain the physical components of the strain tensor in cylindrical co-ordinates

$$\left.\begin{aligned} E_{\rho\rho} &= E_{\theta\theta} = \frac{b(1-2\nu)}{4\pi(1-\nu)}\frac{\sin\theta}{\rho}, \\ E_{\rho\theta} &= -\frac{b}{4\pi(1-\nu)}\frac{\cos\theta}{\rho}, \\ E_{zz} &= E_{\rho z} = E_{\theta z} = 0. \end{aligned}\right\} \tag{8.52}$$

Finally, from (8.51), (8.52), and $(6.13)_1$, it follows that the strain energy density produced by the edge dislocation is

$$W = \frac{\mu b^2}{8\pi^2(1-\nu)^2}\frac{1-2\nu\sin^2\theta}{\rho^2}. \tag{8.53}$$

Consequently, the *strain energy stored per unit length* of the edge dislocation in an infinite elastic medium between the surfaces $\rho = r_0$ and $\rho = R$ is

$$w = \int_0^1 \mathrm{d}z \int_{r_0}^R \rho \mathrm{d}\rho \int_0^{2\pi} W \mathrm{d}\theta = \frac{\mu b^2}{4\pi(1-\nu)} \ln\frac{R}{r_0}. \tag{8.54}$$

When the energy per unit dislocation length of the dislocation core, say $w_0$, is also taken into account, the *total energy per unit length of the edge dislocation* is given by

$$w_t = \frac{\mu b^2}{4\pi(1-\nu)} \ln\frac{R}{r_0} + w_0. \tag{8.55}$$

In passing from (8.54) to (8.55) we have neglected the contribution of the tractions acting on the surface $\Sigma_0$ to the elastic strain energy density $W$, since it decreases at least as $\rho^{-2}$ when $\rho \to \infty$.

It is apparent from (8.54) that the strain energy grows to infinity as $R \to \infty$; this shows once again how important is the strain energy to the total free energy of the dislocation. Although $R$ takes a finite value for any crystal, the strain energy is relatively high. So, for metals, (8.54) yields values of about 5 to 10 eV for the strain energy $wb$ per atomic plane crossing an edge dislocation. This result completely eliminates the possibility of thermal generation of dislocations; in other words, dislocations do not correspond to a state of thermodynamic equilibrium of the crystal [50].

## 8.2. Screw dislocation in an elastic cylinder

Let us consider now a screw dislocation whose line $L$ is *infinite* and coincides with the axis of an isotropic elastic circular cylinder of radius $R$ (Fig. 8.1). We make use of the same notation as for the edge dislocation, but we introduce from the very beginning the cylindrical co-ordinates $\rho, \theta, z$ defined by (8.2). Due to the symmetry of the problem, the strain and stress components must be independent of $z$ and $\theta$, and the displacement vector must be parallel to the dislocation line and independent of $z$, i.e.

$$u_z = u_z(\rho, \theta), \qquad u_\rho = u_\theta = 0. \tag{8.56}$$

On the same symmetry grounds, the tractions exerted by the dislocation core must reduce to a radial pressure that is independent of $z$. For the sake of simplicity we shall assume, however, that both cylindrical surfaces $\rho = r_0$ and $\rho = R$ are free of tractions [1], i.e.

$$T_{\rho\rho} = T_{\rho\theta} = T_{\rho z} = 0 \qquad \text{for } \rho = r_0 \text{ and } \rho = R. \tag{8.57}$$

In contradistinction to the case of the edge dislocation, the displacement components $u_\rho$ and $u_\theta$ are now continuous across the cut $\theta = \pi$, $-R \leqslant \rho \leqslant -r_0$, while $u_z$ has a jump across this cut, given by

$$u_z(\rho, \pi) - u_z(\rho, -\pi) = -b, \qquad -R \leqslant \rho \leqslant -r_0, \tag{8.58}$$

where $b$ is the magnitude of the true Burgers vector. Clearly, this condition is satisfied if we take in (8.56)

$$u_z = -\frac{b\theta}{2\pi}, \qquad \theta \in (-\pi, \pi]. \tag{8.59}$$

Introducing (8.59) into (1.75) and (1.76), and then the result obtained into (6.5), we deduce that the only non-zero components of the displacement gradient **H**, of the infinitesimal strain tensor **E**, and of the stress tensor **T** are

$$H_{z\theta} = -\frac{b}{2\pi\rho}, \quad E_{\theta z} = E_{z\theta} = -\frac{b}{4\pi\rho}, \quad T_{\theta z} = T_{z\theta} = -\frac{\mu b}{2\pi\rho}. \tag{8.60}$$

By taking into consideration (1.77), it may be seen that the equilibrium equations (7.16) and the boundary conditions (8.57) are identically satisfied. Consequently, by virtue of Volterra's uniqueness theorem, we conclude that (8.59) and (8.60) give the desired solution of the boundary-value problem.

By $(6.13)_1$, the strain energy density produced by the screw dislocation is

$$W = \tfrac{1}{2}(E_{\theta z} T_{\theta z} + E_{z\theta} T_{z\theta}) = \frac{\mu b^2}{8\pi^2\rho^2}. \tag{8.61}$$

---

[1] The solution corresponding to a non-zero pressure acting from the dislocation core will be derived in Sect. 14.3 by superposing effects.

By substituting (8.61) into $(8.54)_1$, we find that the *strain energy stored per unit length of the screw dislocation* in the elastic cylinder of radii $r_0$ and $R$ is

$$w = \frac{\mu b^2}{4\pi} \ln \frac{R}{r_0}, \tag{8.62}$$

while the *total energy per unit length of the screw dislocation* is $w_t = w + w_0$, where $w_0$ denotes as above the contribution of the dislocation core. As $0 < \nu < 0.5$, it is easily seen, by comparing (8.62) to $(8.54)_2$, that the strain energy of a screw dislocation is smaller than that of an edge dislocation.

When the screw dislocation lies along the axis of an elastic cylinder of *finite* length $l$, the above solution should be corrected in order to assure that the bases of the cylinder are free of tractions. However, we shall content ourselves to require the vanishing of the resultant force and couple of the surface forces acting on the ends of the cylinder. According to Saint-Venant's principle [1], the solution obtained in this way will be correct at distances larger than about $2R$ from each basis, which is quite satisfactory when $l \gg R$.

It is easily verified that the tractions corresponding to the shear stress $(8.60)_3$ on the bases of the cylinder, namely

$$\mathbf{t} = \mp \frac{\mu b}{2\pi\rho} \mathbf{e}_\theta \quad \text{for} \quad z = \pm \frac{l}{2}, \qquad \rho \in [r_0, R],$$

have a vanishing resultant force on each basis, but produce the torque

$$M_z = \int_0^{2\pi} \mathrm{d}\theta \int_{r_0}^{R} t_\theta \, \rho^2 \, \mathrm{d}\rho = -\frac{\mu b(R^2 - r_0^2)}{2} \tag{8.63}$$

on the upper basis, and $-M_z$ on the lower basis of the cylinder. As long as the shear stresses acting on the ends of the dislocation core are not known, we can extend the distribution of shear stresses $(8.60)_3$ up to the dislocation line, which comes to take $r_0 = 0$ in (8.63). Consequently, we shall superimpose on the elastic state (8.59), (8.60) obtained for the infinite cylinder the elastic state produced by the torques $\pm\mu b \, R^2/2$ acting on the bases $z = \pm l/2$ of a cylinder of finite length $l$, namely [2]

$$u_\theta = \frac{b\rho z}{\pi R^2}, \quad E_{\theta z} = \frac{b\rho}{2\pi R^2}, \quad T_{\theta z} = \frac{\mu b \rho}{\pi R^2}. \tag{8.64}$$

---

[1] Saint-Venant's principle asserts that a system of loads acting on the plane ends of a cylindrical body and having zero resultant force and couple at each end produces a stress field that is negligibly small away from the ends. For an analytic substantiation of this principle, see Toupin [356] or Gurtin [150], Sects. 54–56 a.

[2] This elementary solution may be found in any standard book on linear elasticity (see, e.g. Timoshenko and Goodier [353], or Solomon [314], p. 239).

By superposing the elastic states (8.59), (8.60), and (8.64), we find that

$$\left.\begin{aligned} u_\theta &= \frac{b\rho z}{\pi R^2}, \quad u_z = -\frac{b\theta}{2\pi}, \\ E_{\theta z} &= -\frac{b}{4\pi\rho}\left(1 - \frac{2\rho^2}{R^2}\right), \quad T_{\theta z} = -\frac{\mu b}{2\pi\rho}\left(1 - \frac{2\rho^2}{R^2}\right). \end{aligned}\right\} \tag{8.65}$$

It is interesting to note that the correction (8.64) leads to a twist per unit length equal to $b/\pi R^2$. This is the so-called *Eshelby twist*, which has been observed in thin, long whiskers containing a single screw dislocation (Hirth and Lothe [162], p. 61).

Finally, we notice that the elastic states corresponding to edge and screw dislocations in an infinite isotropic elastic medium are "uncoupled", in the sense that the components of the fields **u**, **E**, and **T** which are non-zero for an edge dislocation vanish for the screw dislocation and conversely. This remark allows to derive at once the elastic state produced by a *mixed dislocation* whose Burgers vector makes an angle $\beta$ with the positive direction of the dislocation line, by simply replacing $b$ with $b \sin \beta$ in the elastic state produced by an edge dislocation, with $b \cos \beta$ in that produced by a screw dislocation, and summing up the results thus obtained. In particular, we deduce from (8.54) and (8.62) that the *strain energy stored per unit length of a mixed dislocation* between the cylindrical surfaces $\rho = r_0$ and $\rho = R$ in an infinite isotropic elastic medium is

$$w = \frac{\mu b^2}{4\pi}\left(\cos^2 \beta + \frac{\sin^2\beta}{1 - \nu}\right) \ln \frac{R}{r_0}. \tag{8.66}$$

## 8.3. Influence of the boundaries on the isotropic elastic field of straight dislocations

We have considered so far only dislocations lying in the axis of a circular elastic cylinder. Edge and screw dislocations whose lines are parallel to but do not coincide with the axis of an elastic cylinder have been studied by Dietze [88], who determined also the elastic field of straight dislocations parallel to the boundary of an elastic half-space or to the faces of an infinite elastic plate (cf. Seeger [286], Sect. 66).

A problem frequently encountered in various applications is the determination of the elastic state produced by a dislocation near a free boundary, a grain boundary, or a bimetallic interface. In order to fulfil the boundary conditions on such surfaces one has to supplement the solution corresponding to the infinite elastic medium by additional terms whose weight increases with decreasing the distance separating the dislocation from the boundary. The derivative of the dislocation strain energy with respect to the distance between the dislocation and the boundary, taken with opposite sign, is by definition the (attractive or repulsive) force exerted by the boundary on the dislocation.

Head [154, 156] has shown that an edge dislocation situated near the interface between two semi-infinite media with different elastic properties will be attracted by the interface when it lies within the more rigid half-space.

Special attention has been also given to the interaction between the surface coating of an elastic half-space and an edge (Conners [80], Weeks, Dundurs, and Stippes [375]) or screw dislocation (Head [155], Chou [71]), as well as to the interaction between a straight dislocation and a partially bonded bimetallic interface (Tamate [328], Tamate and Kurihara [329]). The elastic field of an edge dislocation situated near or inside a circular inclusion has been obtained by Dundurs and Mura [98] and, respectively, by Dundurs and Sendecky [99]; Dundurs [412] has given a general review of this and related work, while List [446] succeeded to give a unified treatment of these problems by making use of complex-variable techniques.

Finally, the stress field of an edge dislocation near an elliptical hole in an isotropic medium has been investigated by Vitek [482], who has considered also the important limiting case when one axis of the ellipse is reduced to zero, leaving a dislocation in the neighbourhood of a crack [483]. This last problem has been also treated for a straight dislocation of mixed type by Hirth and Wagoner [428] and by Rice and Thompson [464].

For a comprehensive and critical review of the solutions to boundary problems associated with the elastic field of dislocations, we refer to a recent article by Eshelby [416].

# 9. Dislocation loops in isotropic media

## 9.1. Displacements and stresses produced by dislocation loops in an infinite isotropic elastic medium

As shown in Sect. 7.3, a dislocation loop of line $L$ and true Burgers vector $\mathbf{b}$ can be simulated in a linear elastic body by a Volterra dislocation in the following way. First eliminate the dislocation core by surrounding the dislocation line with a toroidal hole of boundary $\Sigma_0$, and cut the body along a smooth and two-sided surface $S$ bounded by $L$. Arbitrarily choose a positive sense on $L$ and denote by $\mathbf{n}$ the unit normal to $S$ that is right-handed with respect to this positive sense (Fig. 9.1). Translate the positive face $S^+$ of the cut (into which points $\mathbf{n}$) by a vector $\mathbf{b}$ relatively to the negative face $S^-$. Finally, add or remove material, if necessary, and re-establish the continuity of the body by joining the faces of the cut. Denoting by $\mathbf{u}^+(x)$, and respectively $\mathbf{u}^-(x)$, the limiting values of the displacement vector field on $S^+$ and $S^-$, we have

$$\mathbf{u}^+(\mathbf{x}) - \mathbf{u}^-(\mathbf{x}) = \mathbf{b}. \tag{9.1}$$

Let us first suppose that the elastic continuum is *infinite* and let $\mathbf{G}$ be Green's tensor function of the elastic medium (cf. Sect. 6.5). Denote by $\mathbf{u}(\mathbf{x})$ the displacement field and by $\mathbf{T}(\mathbf{x})$ the stress field produced by the dislocation. By making use of the reciprocal theorem (6.70) for the singular elastic state produced by the unit force $\mathbf{e}_p$ acting at $\mathbf{x}$ and the elastic state generated by the dislocation, we obtain

$$\int_{S^-} (\mathbf{T}\,\mathbf{n})\cdot\mathbf{u}^{(p)}\,\mathrm{d}s - \int_{S^+} (\mathbf{T}\mathbf{n})\cdot\mathbf{u}^{(p)}\,\mathrm{d}s = \int_{S^-} (\mathbf{T}^{(p)}\mathbf{n})\cdot\mathbf{u}^-\,\mathrm{d}s -$$

$$- \int_{S^+} (\mathbf{T}^{(p)}\mathbf{n})\cdot\mathbf{u}^+\,\mathrm{d}s + u_p(\mathbf{x}).$$

In deriving this relation we have neglected the tractions acting on $\Sigma_0$ from the dislocation core and we have also taken into account that the outward unit normal to the boundary of the elastic medium is $-\mathbf{n}$ on $S^+$ and $\mathbf{n}$ on $S^-$. Since $\mathbf{u}^{(p)}$ is continuous across $S$, the left-hand side of the last relation vanishes and we find, by virtue of (9.1),

$$u_p(\mathbf{x}) = \int_{S^+} \mathbf{b}\cdot[\mathbf{T}^{(p)}(\mathbf{x}' - \mathbf{x})\,\mathbf{n}(\mathbf{x}')]\,\mathrm{d}s'. \tag{9.2}$$

Finally, by putting $n_j(\mathbf{x}')\,\mathrm{d}s = \mathrm{d}s'_j$ and taking into consideration that

$$T_{ij}^{(p)}(\mathbf{x}' - \mathbf{x}) = c_{ijkl}G_{kp,l'}(\mathbf{x}' - \mathbf{x}) = -c_{ijkl}G_{kp,l}(\mathbf{x} - \mathbf{x}'),$$

we may rewrite (9.2) as

$$u_p(\mathbf{x}) = -\int_{S^+} b_i c_{ijkl}\,G_{kp,l}(\mathbf{x} - \mathbf{x}')\,\mathrm{d}s'_j. \tag{9.3}$$

The formula (9.3) has been obtained by Volterra [373] in the isotropic case and by Burgers [54] in the anisotropic case.

When the elastic medium occupies a *finite* region $\mathscr{V}$ and we are interested to determine the elastic state produced by a dislocation, we must add to the displacement field (9.3) a regular elastic displacement field corresponding to the tractions $-\mathbf{Tn}$ applied on the boundary of the body. In case Green's tensor function $\hat{\mathbf{G}}(\mathbf{x};\mathbf{x}')$ for the region $\overline{\mathscr{V}}$ is known, the normalized displacement field produced by the dislocation may be directly derived, according to Sect. 6.5, by the formula

$$u_p(\mathbf{x}) = \int_{S^+} b_i c_{ijkl}\,\hat{G}_{kp,l'}(\mathbf{x}':\mathbf{x})\,\mathrm{d}s'_j. \tag{9.4}$$

Resuming now the case of the *infinite* medium, we notice that, if we choose another surface $\hat{S}$ passing through $L$ (Fig. 9.1) and repeat the operations already used for generating the dislocation loop, then, denoting the corresponding elastic displacement components by $\hat{u}_p(\mathbf{x})$ and making use of (1.53), it results from (9.3) that

$$u_p(\mathbf{x}) - \hat{u}_p(\mathbf{x}) = -\int_{S^+} b_i c_{ijkl} G_{kp,l}(\mathbf{x}-\mathbf{x}')\,ds'_j +$$

$$+\int_{\hat{S}^+} b_i c_{ijkl} G_{kp,l}(\mathbf{x}-\mathbf{x}')\,ds'_j = -\int_{S^+\cup\hat{S}^-} b_i c_{ijkl} G_{kp,l}(\mathbf{x}-\mathbf{x}')\,ds'_j$$

$$= \int_V b_i c_{ijkl} G_{kp,lj}(\mathbf{x}-\mathbf{x}')\,dV',$$

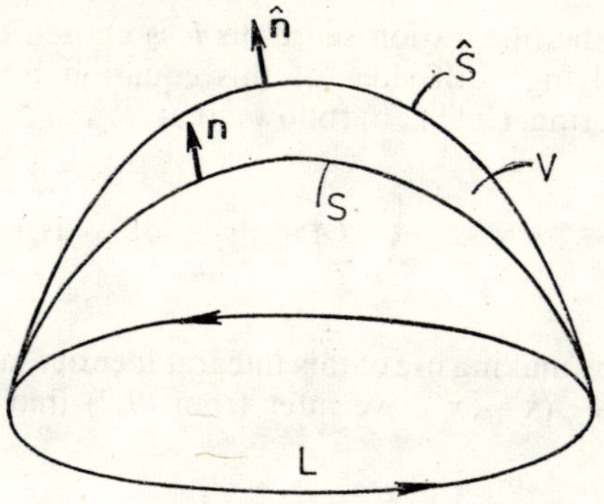


Fig. 9.1. Two different cuts, $S$ and $\hat{S}$, used to generate the same dislocation loop $L$.

where $V$ is the region between $S$ and $\hat{S}$. Hence, by (6.34),

$$\mathbf{u}(\mathbf{x}) - \hat{\mathbf{u}}(\mathbf{x}) = \begin{cases} -\mathbf{b} & \text{if} \quad \mathbf{x} \in V \\ \mathbf{0} & \text{if} \quad \mathbf{x} \notin V. \end{cases} \tag{9.5}$$

This result may be easily understood on taking into account that the jump condition (9.1) is satisfied on both $S$ and $\hat{S}$. Moreover, as noted by Leibfried [213], the displacement fields $\mathbf{u}(\mathbf{x})$ and $\hat{\mathbf{u}}(\mathbf{x})$ differ by a rigid translation that vanishes outside $V$. Consequently, the strains and stresses corresponding to these fields and extended by continuity to $S$ and $\hat{S}$ coincide in the whole space.

From the reasoning above it follows that the strains and stresses produced by a dislocation are independent of the choice of the cut, being fully determined by the line $L$ and the Burgers vector $\mathbf{b}$. We shall give this statement a more mathematical form by expressing the dislocation strains and stresses as line integrals taken along $L$. To this end, we first derive the gradient of the displacement field (9.3), which has the components

$$H_{pr}(\mathbf{x}) = -\int_{S^+} b_i c_{ijkl} G_{kp,lr}(\mathbf{x}-\mathbf{x}')\,ds'_j. \tag{9.6}$$

In order to transform the surface integral in the right-hand side of (9.6) into a line integral, we note that, in view of (6.39), we may write

$$H_{pr}(\mathbf{x}) = -\int_{S^+} b_i c_{ijkl} [G_{kp,lr}(\mathbf{x}-\mathbf{x}')\, ds'_j - G_{kp,lj}(\mathbf{x}-\mathbf{x}')\, ds'_r] \tag{9.7}$$

for any $\mathbf{x} \notin S$. On the other hand, by applying Stokes' formula to the Cartesian components of an arbitrary vector or tensor field $\mathbf{A}(\mathbf{x})$, we obtain

$$\int_{S^+} \epsilon_{nqt} A_{\dots,q}\, ds_t = -\oint_L A_{\dots}\, dx_n, \tag{9.8}$$

where the integration sense on $L$ is chosen clockwise when looking down along $\mathbf{n}$. Multiplying both sides of this equation by $\epsilon_{nrj}$, summing with respect to $n$, and considering (1.11), it follows that

$$\int_{S^+} (A_{\dots,r}\, ds_j - A_{\dots,j}\, ds_r) = \oint_L \epsilon_{njr} A_{\dots}\, dx_n. \tag{9.9}$$

Next, by making use of this integral identity, and taking into account that $\mathbf{G}_{,r}(\mathbf{x}-\mathbf{x}') = -\mathbf{G}_{,r'}(\mathbf{x}-\mathbf{x}')$, we infer from (9.7) that

$$H_{pr}(\mathbf{x}) = \oint_L \epsilon_{njr} b_i c_{ijkl} G_{kp,l}(\mathbf{x}-\mathbf{x}')\, dx'_n. \tag{9.10}$$

This formula has been derived for the general anisotropic case by Mura [253] in 1963.

Finally, remembering that the infinitesimal strain tensor $\mathbf{E}$ is the symmetric part of the displacement gradient $\mathbf{H}$, we deduce from (9.10) and (7.17) the stresses produced by the dislocation loop:

$$T_{ij}(\mathbf{x}) = \oint_L c_{ijpr} \epsilon_{mtr} b_q c_{qtkl} G_{kp,l}(\mathbf{x}-\mathbf{x}')\, dx'_m. \tag{9.11}$$

The remaining part of this subsection will be devoted to the isotropic case.

### 9.2. Burgers' formula

For *isotropic* media we see from (6.58) that

$$G_{kp}(\mathbf{x}-\mathbf{x}') = \frac{1}{16\pi\mu(1-\nu)} [2(1-\nu)\delta_{kp} R_{,mm} - R_{,kp}], \tag{9.12}$$

where $R = \|\mathbf{x} - \mathbf{x}'\|$. Next, by (5.26) and (6.9), we obtain after some calculation

$$b_i c_{ijkl} G_{kp,l}(\mathbf{x} - \mathbf{x}') = \frac{1}{8\pi(1 - \nu)} [\nu b_j R_{,mmp} + $$

$$+ (1 - \nu)(b_p R_{,mmj} + b_i \delta_{jp} R_{,mmi}) - b_i R_{,ijp}]. \tag{9.13}$$

Substituting this result into (9.3) and rearranging terms yields

$$u_p(\mathbf{x}) = -\frac{1}{8\pi} \int_{S^+} b_p R_{,mmj} \, ds_j' - \frac{1}{8\pi} \int_{S^+} (b_j R_{,mmj} \, ds_p' - b_j R_{,mmp} \, ds_j') -$$

$$- \frac{1}{8\pi(1 - \nu)} \int_{S^+} (b_j R_{,pmm} \, ds_j' - b_j R_{,pmj} \, ds_m'). \tag{9.14}$$

Next, by transforming the last two integrals in the right-hand side with the aid of (9.9), we find that

$$u_p(\mathbf{x}) = -\frac{1}{8\pi} \int_{S^+} b_p R_{,mmj} \, ds_j' - \frac{1}{8\pi} \oint_L \in_{ijp} b_j R_{,mm} \, dx_i' -$$

$$- \frac{1}{8\pi(1 - \nu)} \oint_L \in_{imj} b_j R_{,mp} \, dx_i'.$$

Finally, by taking into account that

$$R_{,i} = \frac{X_i}{R}, \quad R_{,mp} = \frac{\delta_{mp}}{R} - \frac{X_m X_p}{R^3}, \quad R_{,mm} = \frac{2}{R}, \tag{9.15}$$

$$\int_{S^+} R_{,mmj} \, ds_j' = -2 \int_{S^+} \frac{X_j \, ds_j'}{R^3} = -2\Omega, \tag{9.16}$$

where $X_i = x_i - x_i'$, and $\Omega$ denotes the solid angle under which $S^+$ is seen from the point with position vector $\mathbf{x}$, we deduce from the last relation that

$$u_p(\mathbf{x}) = \frac{b_p \Omega}{4\pi} + \frac{1}{4\pi} \oint_L \frac{\in_{pji} b_j \, dx_i'}{R} + \frac{1}{8\pi(1 - \nu)} \left( \oint_L \frac{\in_{ijm} b_j X_m \, dx_i'}{R} \right)_{,p} . \tag{9.17}$$

This formula, which may be rewritten in direct notation as

$$\mathbf{u}(\mathbf{x}) = \frac{\mathbf{b}\Omega}{4\pi} + \frac{1}{4\pi}\oint_L \frac{\mathbf{b}\times \mathrm{d}\mathbf{x}'}{R} + \frac{1}{8\pi(1-\nu)}\,\mathrm{grad}\left(\oint_L \frac{(\mathbf{b}\times \mathbf{R})\cdot \mathrm{d}\mathbf{x}'}{R}\right), \tag{9.18}$$

has been obtained by Burgers [54] in 1939.

Since the magnitude of the solid angle under which is seen the surface $S^+$ depends only on the boundary $L$ of the surface, it results from (9.18) that the displacement field $\mathbf{u}(\mathbf{x})$ is completely determined by the dislocation line $L$ and the Burgers vector $\mathbf{b}$. Moreover, since the integrals in (9.18) are single-valued functions of $\mathbf{x}$, and the solid angle varies by $-4\pi$ when the point $\mathbf{x}$ encircles the dislocation line in the positive sense of $C$ (Fig. 9.1), it follows that the displacement field (9.18) satisfies indeed the required jump condition (9.1).

### 9.3. The formula of Peach and Koehler

To obtain the displacement gradient and the stresses produced by a dislocation loop in an isotropic medium, we first replace (9.13) into (9.10) and obtain after rearranging terms

$$H_{pr}(\mathbf{x}) = \frac{b}{8\pi}\oint_L \in_{njr}[(b_p R_{,j} - b_j R_{,p})_{,mm} + b_i\delta_{jp} R_{,mmi} + \frac{1}{1-\nu}(b_j R_{,m} - b_m R_{,j})_{,mp}]\,\mathrm{d}x'_n.$$

On the other hand, by virtue of (1.11), we have

$$\in_{njr}(b_p R_{,j} - b_j R_{,p}) = \in_{njr}\in_{kpj}\in_{kst} b_s R_{,t}$$

$$= \in_{kst}(\delta_{rk}\delta_{np} - \delta_{rp}\delta_{nk}) b_s R_{,t} = (\in_{rst}\delta_{np} - \in_{nst}\delta_{pr}) b_s R_{,t}$$

and an analogous calculation gives

$$\oint_L \in_{njr}(b_j R_{,m} - b_m R_{,j})_{,mp}\,\mathrm{d}x'_n = \oint_L b_s(\in_{nst}R_{,tpr} - \in_{rst}R_{,npt})\,\mathrm{d}x'_n$$

$$= \oint_L b_s\in_{nst}R_{,tpr}\mathrm{d}x'_n - \oint_L \mathrm{d}(b_s\in_{rst}R_{,pt}) = \oint_L b_s\in_{nst}R_{,tpr}\,\mathrm{d}x'_n.$$

By taking into account these transformations, the expression of the displacement gradient becomes

$$H_{pr}(\mathbf{x}) = \frac{1}{8\pi} \oint_L b_s [(\epsilon_{rst}\delta_{np} - \epsilon_{nst}\,\delta_{pr} + \epsilon_{npr}\delta_{st})\, R_{,mmt} + \\ + \frac{1}{1-\nu} \epsilon_{nst} R_{,tpr}]\, dx'_n. \tag{9.19}$$

Substituting now (9.19) into (7.15), we obtain the components of the infinitesimal strain tensor

$$E_{pr}(\mathbf{x}) = \frac{1}{8\pi} \oint_L b_s \left[ \left( \frac{1}{2} \epsilon_{rst}\,\delta_{np} + \frac{1}{2} \epsilon_{pst}\delta_{nr} - \epsilon_{nst}\delta_{pr} \right) R_{,mmt} + \\ + \frac{1}{1-\nu} \epsilon_{nst}\, R_{,tpr} \right] dx'_n,$$

where from, by contraction, we derive the dilatation

$$E_{pp}(\mathbf{x}) = -\frac{1-2\nu}{8\pi(1-\nu)} \oint_L b_s \epsilon_{nst}\, R_{,mmt}\, dx'_n.$$

Finally, by introducing the last two relations into (6.5) and considering (6.9), we find the stresses generated by the dislocation loop

$$T_{pr}(\mathbf{x}) = \frac{\mu b_s}{4\pi} \oint_L \left[ \frac{1}{2} R_{,mmt} (\epsilon_{rst}\, dx'_p + \epsilon_{pst}\, dx'_r) + \\ + \frac{1}{1-\nu} \epsilon_{nst}\, (R_{,tpr} - \delta_{pr}\, R_{,mmt})\, dx'_n \right]. \tag{9.20}$$

Formula (9.20) has been derived by Peach and Koehler [265] in 1950, by differentiating Burgers' formula (see also de Wit [385]), in a somewhat more explicit form than that given above and which could be found by substituting (9.15) into (9.20).

## 9.4. Planar dislocation loops

Most of the results available in the literature on curved dislocations concern planar dislocation loops. If **n** denotes the unit normal to the loop plane, then the dislocation loop is said to be a *glide loop* or a *prismatic loop*, according as the Burgers vector **b** is parallel or perpendicular to **n**. Irregular-shaped glide loops are frequently generated during plastic deformation, by gradual expansion of small loops originated, e.g. by a Frank-Read mechanism. Loops of prismatic type may be formed by precipitation of vacancies or interstitial atoms which arise as a result of quenching or irradiation.

The elastic field of a circular glide dislocation loop has been calculated by Keller and quoted by Kröner [190], and that of a circular prismatic loop has been deduced by Kroupa [200]. Their results have been re-analyzed by Marcinkowski and Sree Harsha [232], who corrected some errors in [190] and undertook a detailed numerical analysis of the variation of the stress field around a circular dislocation glide loop. The stress field of a planar elliptical dislocation loop of arbitrary Burgers vector has been recently determined by Mastrojannis, Mura, and Keer [451], who published the explicit expressions of the in-plane values of the dislocation stress field.

A very efficient method for determining the elastic field of planar dislocation loops is to first calculate the solution corresponding to an infinitesimal rectangular dislocation loop and then to integrate the result obtained over the surface bounded by the loop. This method has been largely used by Kroupa [200—202] (see also Hirth and Lothe [162], p. 128). The elastic field of an infinitesimal dislocation loop is also of intrinsic interest, for it provides a good approximation to the long-range elastic field of a finite dislocation loop of arbitrary shape at sufficiently large distances from the loop.

There exists an extensive literature concerning the dislocation loops lying in an isotropic elastic half-space. Thus, Steketee [318] has expressed in an integral form the displacements produced by a dislocation loop in an isotropic elastic half-space, and Baštecká [21] has determined the stresses generated by a circular dislocation prismatic loop lying in a plane parallel to the boundary of the half-space. The case of an infinitesimal dislocation loop of arbitrary orientation in an elastic half-space has been independently treated by Tikhonov [352] and by Bacon and Groves [12]. Their results have been extended by Vagera [363] to dislocation loops situated near the boundary between two different elastic half-spaces, a configuration used, e.g. for modelling the interaction between a dislocation and a grain boundary.

## 10. Straight dislocations in anisotropic media

As already mentioned, even in polycrystalline materials, the elastic field of dislocations plays a significant role mostly within the grains, which are single crystals and frequently highly anisotropic. This explains the continuously increased interest in anisotropic elastic solutions to dislocation problems, which has led in the last ten years to substantial analytic and numerical results. It should be noted that anisotropic elasticity does not provide only quantitative corrections to the isotropic solutions, but may also change qualitatively the predictions based on isotropic theory.

Two-dimensional solutions concerning infinite straight dislocations in anisotropic media have been obtained as early as 1953 by Eshelby, Read, and Shockley

[109], and by Seeger and Schöck [285], their ideas being subsequently developed by Stroh [323]. It is striking that all these researches completely ignored the essential developments of the theory of anisotropic elasticity brought about by Lekhnitsky [210, 211] and later by Green [144, 145]; the ways opened by the results of the last authors for the elastic simulation of crystal defects are still insufficiently exploited.

## 10.1. Generalized plane strain of an anisotropic elastic body

Consider an anisotropic elastic body $\mathscr{B}$ referred to a rectangular Cartesian system of co-ordinates $x_k$ and assume that the displacement vector does not depend on one of the co-ordinates, say $x_3$. Thus

$$u_k = u_k(x_1, x_2), \quad k = 1, 2, 3. \tag{10.1}$$

The elastic state corresponding to this displacement field is called after Lekhnitski [210] a *state of generalized plane strain* [1].

Assuming that $\mathscr{B}$ is free of body forces and taking into account that (10.1) implies the stress components being also independent of $x_3$, we infer that the equilibrium equations (7.16) take the reduced form

$$T_{k1,1} + T_{k2,2} = 0, \quad k = 1, 2, 3. \tag{10.2}$$

Substituting (10.1) into $(7.15)_2$ and using the notation $(4.59)_3$, it follows that

$$\left.\begin{aligned} E_1 = u_{1,1}, \quad E_2 = u_{2,2}, \quad E_3 = 0, \quad E_4 = u_{3,2}, \\ E_5 = u_{3,1}, \quad E_6 = u_{1,2} + u_{2,1}. \end{aligned}\right\} \tag{10.3}$$

Next, putting $E_3 = 0$ in the third equation $(4.63)_2$, we have

$$T_3 = - \sum_{M \neq 3} (s_{3M}/s_{33})\, T_M$$

and, introducing this result into the other five equations $(4.63)_2$, we find that

$$E_K = S_{KM} T_M, \tag{10.4}$$

where

$$S_{KM} = s_{KM} - s_{K3} s_{M3}/s_{33}. \tag{10.5}$$

Inspection of (4.63), (10.5), and (10.6) reveals that $S_{K3} = S_{3M} = 0$ and that the $5 \times 5$ matrix $S_{KM}$, $K, M = 1, 2, 4, 5, 6$ is reciprocal to the matrix obtained by omitting the third row and the third column of the matrix $c_{KM}$.

---

[1] Throughout this section small Latin and Greek indices range over the values 1, 2, 3, and capital Latin indices over the values 1,2, . . ., 6. The summation convention over a twice repeated small or capital Latin index will be always implied, whereas eventual summation over Greek indices will be explicitly indicated.

As shown by Lekhnitsky [210], any solution of equations (10.2—4) can be represented in terms of three complex potentials as[1]

$$\left.\begin{aligned} u_1 &= 2\,\mathrm{Re} \sum_{\alpha=1}^{3} A_{1\alpha} f_\alpha(z_\alpha) - \omega_3^0\, x_2 + u_1^0, \\ u_2 &= 2\,\mathrm{Re} \sum_{\alpha=1}^{3} A_{2\alpha} f_\alpha(z_\alpha) + \omega_3^0 x_1 + u_2^0, \\ u_3 &= 2\,\mathrm{Re} \sum_{\alpha=1}^{3} A_{3\alpha} f_\alpha(z_\alpha) + u_3^0, \end{aligned}\right\} \tag{10.6}$$

$$T_{k1} = -\,2\,\mathrm{Re} \sum_{\alpha=1}^{3} p_\alpha L_{k\alpha} f_\alpha'(z_\alpha), \quad T_{k2} = 2\,\mathrm{Re} \sum_{\alpha=1}^{3} L_{k\alpha} f_\alpha'(z_\alpha), \quad k = 1, 2, 3, \tag{10.7}$$

where $u_1^0$, $u_2^0$, $u_3^0$, and $\omega_3$ are arbitrary real constants, and $f_\alpha(z_\alpha)$ denotes, for each $\alpha = 1, 2, 3$, an analytic function of the complex variable

$$z_\alpha = x_1 + p_\alpha x_2, \qquad \mathrm{Im}\, p_\alpha > 0.$$

The quantities $A_{k\alpha}$, $L_{k\alpha}$, and $p_\alpha$ depend only on the elastic constants and on the orientation of the $x_k$-axes. They may be calculated by using the following steps:

(i) Find the reduced elastic compliances $S_{KM}$, by using (10.5), where $s_{KM}$ are the elastic compliances with respect to the $x_k$-axes.

(ii) Determine the polynomials

$$\left.\begin{aligned} l_2(p) &= S_{55}p^2 - 2S_{45}\,p + S_{44}, \\ l_3(p) &= S_{15}p^3 - (S_{14} + S_{56})\,p^2 + (S_{25} + S_{46})\,p - S_{24}, \\ l_4(p) &= S_{11}p^4 - 2S_{16}p^3 + (2S_{12} + S_{66})\,p^2 - S_{26}\,p + S_{22}. \end{aligned}\right\} \tag{10.8}$$

(iii) Solve the sextic equation

$$l(p) = l_2(p)\, l_4(p) - l_3^2(p) = 0, \tag{10.9}$$

and label the roots with positive imaginary parts, $p_1, p_2, p_3$, such that [2]

$$l_2(p_1) \neq 0, \quad l_2(p_2) \neq 0, \quad l_4(p_3) \neq 0. \tag{10.10}$$

---

[1] For a detailed discussion of the completeness of this representation and of its connection with previous work on dislocation theory see Teodosiu and Nicolae [339]. More general results concerning the completeness of the solutions of systems of differential equations have been given by Lopatinsky [218].

[2] As shown by Lekhnitsky [210], if the strain energy function is positive definite, equation (10.9) admits three pairs of complex conjugate roots. We assume throughout that the roots with positive imaginary parts, $p_1$, $p_2$, $p_3$, are simple. Multiple roots seem to have little physical significance, except the isotropic case ($p_1 = p_2 = p_3 = i$), which is best treated separately. Moreover, it can be proved (Teodosiu and Nicolae [339]) that conditions (10.10) can always be fulfilled when $p_1$, $p_2$, and $p_3$ are simple.

(iv) Form the matrix

$$[L_{k\alpha}] = \begin{bmatrix} -p_1 & -p_2 & -p_3\lambda_3 \\ 1 & 1 & \lambda_3 \\ -\lambda_1 & -\lambda_2 & -1 \end{bmatrix}, \tag{10.11}$$

where

$$\lambda_1 = -\frac{l_3(p_1)}{l_2(p_1)}, \quad \lambda_2 = -\frac{l_3(p_2)}{l_2(p_2)}, \quad \lambda_3 = -\frac{l_3(p_3)}{l_4(p_3)}. \tag{10.12}$$

(v) Calculate the coefficients $A_{k\alpha}$ by the formulae

$$\left.\begin{aligned} A_{1\alpha} &= S_{11}p_\alpha^2 - S_{16}p_\alpha + S_{12} + \lambda_\alpha(S_{15}p_\alpha - S_{14}), \\ A_{2\alpha} &= \{S_{12}p_\alpha^2 - S_{26}p_\alpha + S_{22} + \lambda_\alpha(S_{25}p_\alpha - S_{24})\}/p_\alpha, \\ A_{3\alpha} &= \{S_{14}p_\alpha^2 - S_{46}p_\alpha + S_{24} + \lambda_\alpha(S_{45}p_\alpha - S_{44})\}/p_\alpha \end{aligned}\right\} \tag{10.13}$$

for $\alpha = 1, 2$ and

$$\left.\begin{aligned} A_{13} &= \lambda_3(S_{11}p_3^2 - S_{16}p_3 + S_{12}) + S_{15}p_3 - S_{14}, \\ A_{23} &= \{\lambda_3(S_{12}p_3^2 - S_{26}p_3 + S_{22}) + S_{25}p_3 - S_{24}\}/p_3, \\ A_{33} &= \{\lambda_3(S_{14}p_3^2 - S_{46}p_3 + S_{24}) + S_{45}p_3 - S_{44}\}/p_3. \end{aligned}\right\} \tag{10.14}$$

As shown by Stroh [323], the coefficients $A_{k\alpha}$ and $L_{k\alpha}$ satisfy the orthogonality conditions [1]

$$\left.\begin{aligned} A_{k\alpha}L_{k\beta} + A_{k\beta}L_{k\alpha} &= 0 \quad \text{for any } \alpha, \beta = 1, 2, 3,\ \alpha \neq \beta, \\ A_{k\alpha}\bar{L}_{k\beta} + \bar{A}_{k\beta}L_{k\alpha} &= 0 \quad \text{for any } \alpha, \beta = 1, 2, 3. \end{aligned}\right\} \tag{10.15}$$

Equations (10.6) and (10.7), with $L_{k\alpha}$ given by (10.11) and $A_{k\alpha}$ given by (10.13) and (10.14), express Lekhnitsky's representation of the generalized plane strain in terms of the complex potentials $f_\alpha(z_\alpha)$, $\alpha = 1, 2, 3$. Unlike the representation obtained by Eshelby, Read, and Shockley [109], this representation does not depend on the solution of algebraic systems and, in this respect, is as explicit as that given by Willis [383] (see Sect. 11.2). Moreover, it has the advantage over Willis' representation of not making use of Green's functions for infinite media, being thus applicable for solving boundary-value problems for finite anisotropic bodies as well. Finally, in comparison with Stroh's solutions, Lekhnitsky's representation has the advantage of being valid for the general anisotropic case.

[1] An elegant proof of these relations, based on Betti's reciprocal theorem, has been given by Malén and Lothe [225].

Finally, we notice that the equilibrium equations (10.2) of the generalized plane strain are identically satisfied [109] if we set

$$T_{k1} = -\Phi_{k,2}, \quad T_{k2} = \Phi_{k,1}, \quad k = 1, 2, 3, \tag{10.16}$$

where $\Phi_1$, $\Phi_2$, $\Phi_3$ are stress functions of class $C^2$. The functions $\Phi_1$ and $\Phi_2$ may be expressed in terms of Airy's stress function $F$ as

$$\Phi_1 = -F_{,2}, \quad \Phi_2 = F_{,1} \tag{10.17}$$

and must satisfy the consistency condition $\Phi_{1,2} + \Phi_{2,1} = 0$.

It may be shown [323, 339] that the stress functions can be represented in terms of the complex potentials $f_\alpha(z_\alpha)$ in the form

$$\Phi_k = 2 \operatorname{Re} \sum_{\alpha=1}^{3} L_{k\alpha} f_\alpha(z_\alpha), \quad k = 1, 2, 3. \tag{10.18}$$

Clearly, by introducing (10.18) into (10.16), we recover the expression (10.7) of the stress components. Equations (10.6) and (10.18) give an equivalent complete representation of the generalized plane strain.

## 10.2. **Straight dislocation in an infinite anisotropic elastic medium**

There exists an extensive literature concerning the anisotropic elastic field of straight dislocations for various crystals and dislocation orientations. For a detailed discussion of the cases when the elastic solution, including the roots of equation (10.4), may be analytically obtained, we refer to the articles of Eshelby, Read, and Shockley [109], Seeger and Schöck [285], Head [158], Duncan and Kuhlmann-Wilsdorf [97], Chou and Michell [74], as well as to the books by Hirth and Lothe [162], Sect. 13, and Steeds [317], Sect. 3.

In what follows, we shall expound the solution obtained by Teodosiu, Nicolae, and Paven [342] for an arbitrarily oriented straight dislocation lying in an infinite anisotropic medium, under consideration of the core boundary conditions. Since this solution makes use of Lekhnitsky's representation, no restrictions have to be imposed either on the anisotropy of the material, or on the dislocation character. Moreover, the solution is found when either tractions or displacements are prescribed on the core boundary. At the end of subsection 10.4 we shall give the numerical values of the parameters $A_{k\alpha}$, $L_{k\alpha}$, $p_\alpha$ entering Lekhnitsky's representation, for some typical crystals belonging to cubic and hexagonal systems, and for almost all dislocation orientations that are energetically possible in these crystals.

Consider a straight dislocation lying in an infinite anisotropic elastic medium, and take the positive direction of the dislocation line as $x_3$-axis of a rectangular Cartesian system of co-ordinates. We apply the linear theory of elasticity outside a circular cylindrical surface of radius $r_0$ and axis $x_3$, say $\Sigma_0$, considered as boundary of the dislocation core. Let us denote by $\Gamma_0$ the intersection line of $\Sigma_0$ with the $x_1x_2$-plane, and by $\Delta$ the region outside $\Gamma_0$ within this plane (Fig. 10.1).

The elastic medium outside $\Sigma_0$ is obviously subjected to a state of generalized plane strain. Consequently, as shown in the previous subsection, the displacement and stress components may be represented by (10.6) and (10.7), respectively, in

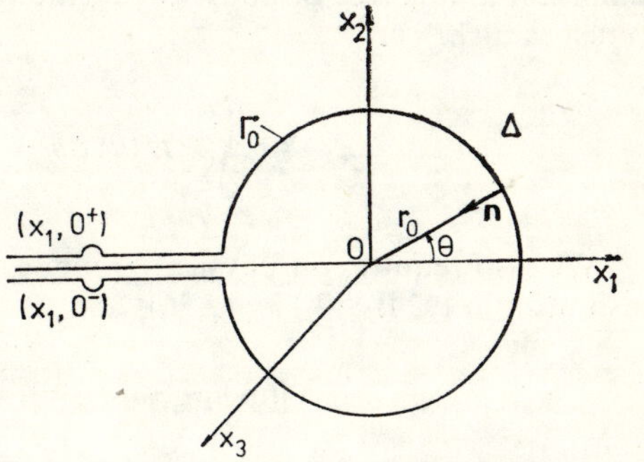


Fig. 10.1. Cut along the half-plane $x_2=0$, $x_1 \leqslant -r_0$, used to define a single-valued displacement field around a straight dislocation lying along the $x_3$-axis in an infinite elastic medium.

terms of three complex potentials $f_\alpha(z_\alpha)$, $\alpha = 1, 2, 3$, which are analytic functions of their arguments. As already mentioned, the parameters $A_{k\alpha}$, $L_{k\alpha}$, and $p_\alpha$ occurring in this representation depend only on the elastic constants and on the orientation of the $x_k$-axes.

In the case of a single dislocation the displacement vector can be considered as a single-valued function with a prescribed jump on an arbitrary cut connecting $\Gamma_0$ with infinity, or as a continuous but multiple-valued function in $\Delta$, with a prescribed cyclic constant around the dislocation line. In the following we adopt the first approach; more precisely, we assume that the displacement vector is single-valued and of class $C^3$ in the region obtained from $\Delta$ by removing its points belonging to the negative $x_1$-axis (Fig. 10.1), and that it is discontinuous across the cut $x_2 = 0$, $x_1 \leqslant -r_0$, its jump across the cut being given by

$$u_k(x_1, 0^+) - u_k(x_1, 0^-) = -b_k, \quad k = 1, 2, 3, \tag{10.19}$$

where **b** is the true Burgers vector of the dislocation.

We shall consider two types of boundary-value problems [1]. In the first case, we assume that the tractions acting on $\Sigma_0$ from the dislocation core are prescribed and that they can be approximated by some smooth functions, say $t_k^*(\theta)$, where $\theta \in (-\pi, \pi]$ is the polar angle in the $x_1x_2$-plane, measured clockwise when looking

[1] The physical significance of these boundary-value problems becomes clear when using semidiscrete methods to study the dislocation core (see Sect. 16). If the displacements of the atoms inside $\Sigma_0$ are known from an atomistic calculation, then the corresponding "strains", "stresses", and "tractions" may be calculated by using an interpolation technique and the stress-strain relations of linear elasticity. Alternatively, if we use the non-linear theory of elasticity outside $\Sigma_0$, and find the solution by solving a series of linear problems, then the displacements at $\Sigma_0$ or the tractions acting on $\Sigma_0$ are known at each step from the previous step of approximation.

down the $x_3$-axis. The stress tensor must then satisfy the boundary conditions

$$T_{k1}n_1 + T_{k2}n_2 = t_k^*(\theta) \quad \text{on } \Gamma_0, \quad k = 1, 2, 3, \tag{10.20}$$

where $\mathbf{n}$ is the inward unit normal to $\Gamma_0$. Moreover, since the dislocation core is in equilibrium and is free of body forces, the resultant force of the tractions acting on $\Sigma_0$ must vanish, i.e.

$$P_k = r_0 \int_{-\pi}^{\pi} t_k^*(\theta)\,\mathrm{d}\theta = 0, \qquad k = 1, 2, 3. \tag{10.21}$$

We also require, on physical grounds, that the stress and the elastic rotations vanish at infinity. By (7.17), (2.36), and (2.37), this implies that

$$\lim_{\rho \to 0} u_{k,m} = 0, \qquad k, m = 1, 2, 3, \tag{10.22}$$

where $\rho = \sqrt{x_1^2 + x_2^2}$.

From Bézier's uniqueness theorem given in Sect. 6.2, it follows that the boundary conditions (10.20) and (10.22), together with the jump condition (10.19), uniquely determine the elastic solution to within an infinitesimal rigid translation [1].

Let us transform now the boundary conditions by using the complex representation of the solution given at the beginning of this section. From Fig. 10.1 it is apparent that on $\Gamma_0$

$$\left.\begin{aligned} &x_1 = r_0 \cos\theta, \quad x_2 = r_0 \sin\theta, \quad z_\alpha = r_0(\cos\theta + p_\alpha \sin\theta), \\ &n_1 = -\cos\theta = -\frac{1}{r_0}\frac{\mathrm{d}x_2}{\mathrm{d}\theta}, \quad n_2 = -\sin\theta = \frac{1}{r_0}\frac{\mathrm{d}x_1}{\mathrm{d}\theta}. \end{aligned}\right\} \tag{10.23}$$

On the other hand, since $z_\alpha = x_1 + p_\alpha x_2$, it follows that

$$\frac{\partial}{\partial x_1} f_\alpha(z_\alpha) = f'_\alpha(z_\alpha), \qquad \frac{\partial}{\partial x_2} f_\alpha(z_\alpha) = p_\alpha f'_\alpha(z_\alpha),$$

and hence, by (10.7) and (10.23), we deduce that

$$T_{k1}n_1 + T_{k2}n_2 = \frac{2}{r_0} \operatorname{Re} \sum_{\alpha=1}^{3} L_{k\alpha} \frac{\mathrm{d}}{\mathrm{d}\theta} f_\alpha(r_0 \cos\theta + p_\alpha r_0 \sin\theta).$$

[1] In fact, Bézier's theorem concerns the case when $b_k = 0$ in (10.19), but its extension to the case $b_k \neq 0$ under the conditions of regularity imposed on the solution is straightforward.

Consequently, the boundary conditions (10.20) become

$$\frac{2}{r_0}\operatorname{Re}\sum_{\alpha=1}^{3} L_{k\alpha}\frac{\mathrm{d}}{\mathrm{d}\theta}f_\alpha(r_0\cos\theta + p_\alpha r_0\sin\theta) = t_k^*(\theta), \qquad k = 1, 2, 3. \tag{10.24}$$

Considering (10.7), we see that conditions (10.22) may be fulfilled by setting $\omega_3^0 = 0$ and requiring that

$$\lim_{\rho\to\infty}|f_\alpha'(z_\alpha)| = 0, \qquad \alpha = 1, 2, 3. \tag{10.25}$$

Moreover, (10.7) assume in this case the simplified form

$$u_k = 2\operatorname{Re}\sum_{\alpha=1}^{3} A_{k\alpha}f_\alpha(z_\alpha) + u_k^0, \qquad k = 1, 2, 3. \tag{10.26}$$

Denote by $\Delta_\alpha$ and $\bar{\Delta}_\alpha$ the regions corresponding to $\Delta$ and $\bar{\Delta} = \Delta \cup \Gamma_0$, respectively, in the $z_\alpha$-plane by the transformation $z_\alpha = x_1 + p_\alpha x_2$, for each $\alpha = 1, 2, 3$. The *traction boundary-value problem* can now be formulated in the following form: *Find three functions $f_1(z_1)$, $f_2(z_2)$, $f_3(z_3)$ that are analytic in $\Delta_1$, $\Delta_2$, $\Delta_3$, and continuous in $\bar{\Delta}_1$, $\bar{\Delta}_2$, $\bar{\Delta}_3$, respectively, and that satisfy the boundary conditions* (10.21), (10.24), (10.25), *and the jump conditions* (10.19).

Alternatively, when the displacements, rather than the tractions are prescribed on $\Gamma_0$, and when they can be approximated by some smooth functions, say $u_k^*(\theta)$, the boundary conditions (10.20) have to be replaced by

$$u_k = u_k^*(\theta) \quad \text{on} \quad \Gamma_0, \qquad k = 1, 2, 3. \tag{10.27}$$

Accordingly, when using the complex representation (10.26) of the displacement field, this *displacement boundary-value problem* can be formulated as follows: *Find three functions $f_1(z_1)$, $f_2(z_2)$, $f_3(z_3)$ that are analytic in $\Delta_1$, $\Delta_2$, $\Delta_3$, and continuous in $\bar{\Delta}_1$, $\bar{\Delta}_2$, $\bar{\Delta}_3$, respectively, and that satisfy the boundary conditions* (10.21), (10.25), (10.27), *and the jump conditions* (10.19).

The main difficulty raised by the solving of the boundary-value problems formulated above is that the images of $\Gamma_0$ in the $z_\alpha$-planes are no longer circles. To avoid this complication, we introduce after Lekhnitsky [211] new complex variables $\zeta_\alpha$, $\alpha = 1, 2, 3$, defined by

$$z_\alpha = (1 - ip_\alpha)\frac{\zeta_\alpha}{2} + (1 + ip_\alpha)\frac{r_0^2}{2\zeta_\alpha}. \tag{10.28}$$

Remembering that $\operatorname{Im} p_\alpha > 0$, $\alpha = 1, 2, 3$, it may be shown that the singularities of the transformation (10.28), i.e. the points where $z_\alpha'(\zeta_\alpha)$ vanishes, lie inside the circle $\Gamma_0$ and that the reciprocal transformation is

$$\zeta_\alpha = \frac{z_\alpha}{1 - ip_\alpha}\left[1 + \sqrt{1 - \frac{r_0^2(1 + p_\alpha^2)}{z_\alpha^2}}\right], \tag{10.29}$$

where the determination with positive real part of the square root in the right-hand side should be chosen. The transformations (10.28) and (10.29) establish, for each $\alpha = 1, 2, 3$, a one-to-one correspondence between the points situated on or outside the circle $\Gamma_0$ and their images in the $\zeta_\alpha$-plane. A direct calculation shows that

$$|\zeta_\alpha| \to \infty \quad \text{as} \quad \rho \to \infty \tag{10.30}$$

and that

$$\zeta_\alpha = r_0 e^{i\theta} \text{ if and only if } z = r_0 e^{i\theta}, \tag{10.31}$$

with $\theta \in (-\pi, \pi]$, which also implies that the circle $\Gamma_0$ is invariant to the transformation (10.28).

Let us put now

$$f_\alpha(z_\alpha(\zeta_\alpha)) \equiv \varphi_\alpha(\zeta_\alpha). \tag{10.32}$$

By virtue of (10.30) and (10.31), the traction boundary-value problem becomes: *Find three functions* $\varphi_\alpha(\zeta_\alpha)$, $\alpha = 1, 2, 3$, *that are analytic in the regions* $|\zeta_\alpha| > r_0$ *and continuous for* $|\zeta_\alpha| \geqslant r_0$, $\alpha = 1, 2, 3$, *respectively, and that satisfy the boundary conditions*

$$\frac{2}{r_0} \operatorname{Re} \sum_{\alpha=1}^{3} L_{k\alpha} \frac{d}{d\theta} \varphi_\alpha(r_0 e^{i\theta}) = t_k^*(\theta), \qquad k = 1, 2, 3, \tag{10.33}$$

$$\lim_{|\zeta_\alpha| \to \infty} |\varphi'_\alpha(\zeta_\alpha)| = 0, \qquad \alpha = 1, 2, 3, \tag{10.34}$$

*the condition* (10.21), *as well as the jump condition* (10.19). Indeed, (10.34) follows from (10.30) and (10.25), by taking into account that

$$f'_\alpha(z_\alpha) = \varphi'_\alpha(\zeta_\alpha)\,\zeta'_\alpha(z_\alpha) = \frac{\varphi'_\alpha(\zeta_\alpha)}{1 - ip_\alpha}\left\{1 + \frac{1}{\sqrt{1 - r_0^2(1 + p_\alpha^2)/z_\alpha^2}}\right\},$$

and hence $|\varphi'_\alpha(\zeta_\alpha)| \to 0$ as $\rho \to \infty$ and $|f'_\alpha(z_\alpha)| \to 0$. For further use, we note that, by (10.29), the last relation can be rewritten as

$$f'_\alpha(z_\alpha) = \varphi'_\alpha(\zeta_\alpha)\,\frac{\zeta_\alpha}{z_\alpha}\,\frac{1}{\sqrt{1 - r_0^2(1 + p_\alpha^2)/z_\alpha^2}}, \tag{10.35}$$

where $\zeta_\alpha$ should be considered as a function of $z_\alpha$ on the right-hand side.

Since $\varphi'_\alpha(\zeta_\alpha)$ is analytic and single-valued for $|\zeta_\alpha| > r_0$, it can be developed in a Laurent series in this region. Moreover, in view of (10.34), this series may contain only negative powers of $\zeta_\alpha$. Hence

$$\varphi'_\alpha(\zeta_\alpha) = \frac{D_\alpha}{2\pi i \zeta_\alpha} + \sum_{m=2}^{\infty} b_{\alpha m}\, \zeta_\alpha^{-m}, \qquad \alpha = 1, 2, 3, \tag{10.36}$$

where $D_\alpha$ and $b_{\alpha m}$ are arbitrary complex constants. Integrating term by term this series and neglecting additive constants, which can be included in $u_k^0$, we obtain

$$\varphi_\alpha(\zeta_\alpha) = \frac{D_\alpha}{2\pi i}\left\{\ln\frac{z_\alpha}{r_0} + \ln\frac{1 + \sqrt{1 - r_0^2(1 + p_\alpha^2)/z_\alpha^2}}{1 - ip_\alpha}\right\} + \sum_{m=1}^{\infty} a_{\alpha m}\, \zeta_\alpha^{-m}, \tag{10.37}$$

where

$$a_{\alpha m} = -\frac{b_{\alpha, m+1}}{m}, \qquad \alpha = 1, 2, 3; \qquad m = 1, 2, \ldots$$

It can be shown that, by cutting the $x_1x_2$-plane along the ray $x_2 = 0, x_1 \leqslant -r_0$, both logarithmic terms in (10.37) become single-valued functions and may be calculated by using the same formulae (8.36), (8.37) as for $\ln z$. Indeed, on the cut $\operatorname{Im} z_\alpha = 0$, $\operatorname{Re} z_\alpha \leqslant 0$, which is used to define a single-valued branch of $\ln z_\alpha$, we have, by $(10.23)_3$, $\sin\theta = 0$, $\cos\theta \leqslant 0$, and hence this cut coincides with the negative $x_1$-axis, like in the case of $\ln z$. Moreover, it can be shown by a direct calculation that the expression inside the braces in (10.37) assumes the value $i\theta$ on $\Gamma_0$. In particular, it results that

$$\varphi_\alpha(r_0 e^{i\theta}) = \frac{D_\alpha \theta}{2\pi} + \sum_{m=1}^{\infty} a_{\alpha m}\, r_0^{-m}\, e^{-im\theta}. \tag{10.38}$$

It is worth noting that writing simply $\ln(\zeta_\alpha/r_0)$ instead of the expression within the braces in (10.37), would have required the introduction of several cuts in the $x_1x_2$-plane, corresponding to the cuts $\operatorname{Im}\zeta_\alpha = 0$, $\operatorname{Re}\zeta_\alpha \leqslant 0$ used for calculating $\ln\zeta_\alpha$, as well as of different additive constants in (10.37), in order to assure the continuity of $u_k(x_1, x_2)$ across these cuts [1].

Consider now the boundary conditions (10.33) and assume that the functions $t_k^*(\theta)$, $k = 1, 2, 3$, defined in the interval $(-\pi, \pi]$ and periodically continued on the whole real axis, can be developed in Fourier series. Then, taking also into account (10.21), we can write

$$t_k^*(\theta) = 2 \operatorname{Re} \sum_{m=1}^{\infty} f_{km}\, e^{im\theta}, \qquad k = 1, 2, 3, \tag{10.39}$$

---

[1] See, e.g. Granzer [142], where such a procedure has been used in a similar case.

where

$$f_{km} = \frac{1}{2\pi} \int_{-\pi}^{\pi} t_k(\theta)\, e^{-im\theta}\, d\theta, \qquad k = 1, 2, 3;\; m = 1, 2, \ldots \tag{10.40}$$

Substituting (10.38) and (10.39) into (10.33), we obtain

$$\frac{1}{r_0} \operatorname{Re} \sum_{\alpha=1}^{3} L_{k\alpha} \left( \frac{D_\alpha}{2\pi} - i \sum_{m=1}^{\infty} m\, a_{\alpha m} r_0^{-m}\, e^{-im\theta} \right) = \operatorname{Re} \sum_{m=1}^{\infty} f_{km}\, e^{im\theta}, \qquad k = 1, 2, 3.$$

Equating now coefficients of like powers of $e^{i\theta}$ yields

$$2 \operatorname{Re} \sum_{\alpha=1}^{3} L_{k\alpha} D_\alpha = 0, \qquad k = 1, 2, 3, \tag{10.41}$$

$$a_{\alpha m} = \frac{i}{m} \overline{L_{\alpha k}^{-1} f_{km}}\, r_0^{m+1}, \quad \alpha = 1, 2, 3; \quad m = 1, 2, \ldots, \tag{10.42}$$

where $[L_{\alpha k}^{-1}]$ denotes the reciprocal matrix [1] of $[L_{k\alpha}]$.

Finally, as

$$u_k = 2 \operatorname{Re} \sum_{\alpha=1}^{3} A_{k\alpha} \varphi_\alpha(\zeta_\alpha) + u_k^0, \tag{10.43}$$

the jump conditions (10.19) give, considering (10.38),

$$2 \operatorname{Re} \sum_{\alpha=1}^{3} A_{k\alpha} D_\alpha = -b_k, \qquad k = 1, 2, 3. \tag{10.44}$$

As shown by Stroh [324], the three complex constants $D_\alpha$ can be determined by solving the system of six real linear algebraic equations (10.41) and (10.44), with the aid of the orthogonality relations (10.15). Indeed, multiplying (10.41) by $A_{k\beta}$, (10.44) by $L_{k\beta}$, summing up for $k = 1, 2, 3$, and adding the two relations obtained, it results, in view of (10.15), that [2]

$$D_\alpha = -\frac{L_{k\alpha}\, b_k}{2\, A_{m\alpha}\, L_{m\alpha}}, \qquad \alpha = 1, 2, 3. \tag{10.45}$$

---

[1] As shown by Stroh [323], if the roots $p_1, p_2, p_3$ are simple, the matrices $L_{k\alpha}$ and $A_{k\alpha}$ are non-singular.

[2] We recall that summation is to be performed in (10.45) only over repeated Latin indices.

By (10.42) and (10.45), the expressions (10.37) of the complex potentials $\varphi_\alpha(\zeta_\alpha)$ are fully determined. They can be written in a still more explicit form by considering (10.40) and calculating the matrix $[L_{\alpha k}^{-1}]$. Namely, it results from (10.11) that

$$[L_{\alpha k}^{-1}] = \frac{1}{L}\begin{bmatrix} \lambda_2\lambda_3 - 1 & \lambda_2\lambda_3 p_3 - p_2 & \lambda_3(p_3 - p_2) \\ 1 - \lambda_2\lambda_3 & p_1 - \lambda_1\lambda_3 p_3 & \lambda_3(p_1 - p_3) \\ \lambda_1 - \lambda_2 & \lambda_1 p_2 - \lambda_2 p_1 & p_2 - p_1 \end{bmatrix}, \tag{10.46}$$

where

$$L = \det\,[L_{k\alpha}] = p_1 - p_2 + \lambda_2\lambda_3(p_3 - p_1) + \lambda_1\lambda_3(p_2 - p_3).$$

Introducing now (10.42) into (10.37), we find

$$\varphi_\alpha(\zeta_\alpha) = \frac{D_\alpha}{2\pi i}\left\{\ln\frac{z_\alpha}{r_0} + \ln\frac{1 + \sqrt{1 - r_0^2(1 + p_\alpha^2)/z_\alpha^2}}{1 - ip_\alpha}\right\} + $$

$$+ ir_0 \sum_{m=1}^{\infty} \frac{1}{m} L_{\alpha k}^{-1} \bar{f}_{km}\left(\frac{r_0}{\zeta_\alpha}\right)^m.$$

Substituting this result into (10.43) directly yields the displacement field. The constants $u_k^0$ can be determined by prescribing the displacement vector of an arbitrary material point of the elastic medium. Finally, the stress components are given by (10.7), where the functions $f'_\alpha(z_\alpha)$, as determined by (10.35) and (10.37), must be replaced by

$$f'_\alpha(z_\alpha) = \frac{1}{z_\alpha\sqrt{1 - r_0^2(1 + p_\alpha^2)/z_\alpha^2}}\left\{\frac{D_\alpha}{2\pi i} - ir_0\sum_{m=1}^{\infty} L_{\alpha k}^{-1}\bar{f}_{km}\left(\frac{r_0}{\zeta_\alpha}\right)^m\right\}. \tag{10.47}$$

We pass now to the solution of the *displacement boundary-value problem.* Since conditions (10.21), (10.25), and (10.19) are common to both boundary-value problems, the functions $\varphi_\alpha(\zeta_\alpha)$ must have the same form (10.37) with $D_\alpha$ determined by (10.45).

Consider now condition (10.27). Since the atomic calculation of the dislocation core is done using a jump condition similar to (10.19), we obviously have

$$u_k^*(\pi) - u_k^*(-\pi) = -b_k, \quad k = 1, 2, 3. \tag{10.48}$$

Let us put

$$u_k^*(\theta) = -\frac{b_k\theta}{2\pi} + \tilde{u}_k(\theta), \tag{10.49}$$

with $\tilde{u}_k(\pi) = \tilde{u}_k(-\pi)$. Assuming that $\tilde{u}_k(\theta)$, as defined by (10.49) for $\theta \in (-\pi, \pi]$ and periodically continued on the whole real axis, can be developed in a Fourier series, we obtain

$$u_k^*(\theta) = -\frac{b_k\theta}{2\pi} + \mathrm{d}_{k0} + 2\,\mathrm{Re}\sum_{m=1}^{\infty} \mathrm{d}_{km}\,\mathrm{e}^{im\theta}, \qquad (10.50)$$

where

$$\mathrm{d}_{km} = \frac{1}{2\pi}\int_{-\pi}^{\pi} \tilde{u}_k(\theta)\,\mathrm{e}^{-im\theta}\,\mathrm{d}\theta, \qquad k = 1, 2, 3; \quad m=0, 1, 2, \ldots$$

Introducing now (10.42) and (10.50) into (10.27), and taking into account (10.38) and (10.44), it follows that

$$u_k^0 + 2\,\mathrm{Re}\sum_{\alpha=1}^{3} A_{k\alpha}\sum_{m=1}^{\infty} a_{\alpha m}\,r_0^{-m}\,\mathrm{e}^{-im\theta} = d_{k0} + 2\,\mathrm{Re}\sum_{m=1}^{\infty} \mathrm{d}_{km}\,\mathrm{e}^{im\theta}, \qquad k = 1, 2, 3.$$

Next, equating like powers of $\mathrm{e}^{i\theta}$ gives

$$u_k^0 = d_{k0}, \qquad k = 1, 2, 3, \qquad (10.51)$$

$$r_0^{-m}\sum_{\alpha=1}^{3} A_{k\alpha}a_{\alpha m} = \bar{d}_{km}, \qquad k = 1, 2, 3; \quad m = 1, 2, \ldots \qquad (10.52)$$

and, by solving (10.52) with respect to $a_{\alpha m}$, we obtain

$$a_{\alpha m} = r_0^m\,A_{\alpha k}^{-1}\bar{d}_{km}, \quad \alpha = 1, 2, 3;\ m = 1, 2, \ldots, \qquad (10.53)$$

where $[A_{\alpha k}^{-1}]$ denotes the reciprocal matrix of the (non-singular) matrix $[A_{k\alpha}]$. Finally, substituting (10.53) into (10.37), we deduce the expressions of the complex potentials $\varphi_\alpha(\zeta_\alpha)$, $\alpha = 1, 2, 3$:

$$\varphi_\alpha(\zeta_\alpha) = \frac{D_\alpha}{2\pi i}\left\{\ln\frac{z_\alpha}{r_0} + \ln\frac{1+\sqrt{1-r_0^2(1+p_\alpha^2)/z_\alpha^2}}{1-ip_\alpha}\right\} +$$

$$+\sum_{m=1}^{\infty} A_{\alpha k}^{-1}\,\bar{d}_{km}\left(\frac{r_0}{\zeta_\alpha}\right)^m, \qquad (10.54)$$

where $D_\alpha$ are given by (10.45). The displacement and stress components are again explicitly given by (10.43) and (10.7), by using (10.51), (10.35), and (10.54).

### 10.3. Neglecting the core boundary conditions

The solutions obtained in the previous subsection become considerably simpler when terms arising from satisfying the boundary conditions on $\Sigma_0$ are neglected. Indeed, both traction and displacement boundary-value problems amount in this case to the fulfilment of the same conditions (10.21), (10.25), and (10.19). It then results

$$f_\alpha(z_\alpha) = \frac{D_\alpha}{2\pi i} \ln z_\alpha, \quad \alpha = 1, 2, 3, \tag{10.55}$$

and we deduce from (10.6) and (10.7):

$$u_k = \frac{1}{\pi} \operatorname{Im} \sum_{\alpha=1}^{3} A_{k\alpha} D_\alpha \ln z_\alpha + u_k^0, \tag{10.56}$$

$$T_{k1} = -\frac{1}{\pi} \operatorname{Im} \sum_{\alpha=1}^{3} \frac{p_\alpha L_{k\alpha} D_\alpha}{z_\alpha}, \quad T_{k2} = \frac{1}{\pi} \operatorname{Im} \sum_{\alpha=1}^{3} \frac{L_{k\alpha} D_\alpha}{z_\alpha}, \tag{10.57}$$

where $D_\alpha$ are given by (10.45). It is easily seen that the stress and strain components decrease as $\rho^{-1}$ when $\rho \to \infty$, like in the isotropic case.

We will determine now the *strain energy stored per unit dislocation length* in an infinite elastic medium between the surfaces $\rho = r_0$ and $\rho = R$. Since a calculation similar with that performed in Sect. 8.1 would be rather tedious in the anisotropic case, we prefer to use a somewhat different reasoning on the lines of Stroh [323]. First, by applying the theorem of work and energy to an elastic cylinder of unit length and bounded by the cylindrical surfaces $\rho = r_0$, $\rho = R$, and the plane cut $x_2 = 0$, $-R \leqslant x_1 \leqslant -r_0$, we deduce that

$$w = \frac{1}{2} \int_C t_k u_k \, dl = \frac{1}{2} \int_C T_{km} n_m u_k \, dl,$$

where $C$ is the union $C = \Gamma_0 \cup \Gamma \cup AB \cup A'B'$, $\Gamma$ is the circle of radius $R$ and centred at the origin in the $x_1x_2$-plane, and $\mathbf{n}$ is the outward unit normal to $C$ in this plane (Fig. 10.2).

Since $\mathbf{n}$ equals $-\mathbf{e}_2$ on $AB$ and $\mathbf{e}_2$ on $A'B'$, the last relation yields

$$w = \frac{1}{2} \oint_{\Gamma_0} t_k u_k \, dl + \frac{1}{2} \oint_{\Gamma} t_k u_k \, dl + \frac{1}{2} \int_{-R}^{-r_0} T_{k2}(x_1, 0)\{u_k(x_1, 0^-) - u_k(x_1, 0^+)\} \, dx_1,$$

wherefrom, by (10.1), it follows that

$$w = \frac{1}{2}\oint_{\Gamma_0} t_k u_k \, dl + \frac{1}{2}\oint_{\Gamma} t_k u_k \, dl + \frac{1}{2} b_k \int_{-R}^{-r_0} T_{k2}(x_1, 0) \, dx_1. \tag{10.58}$$

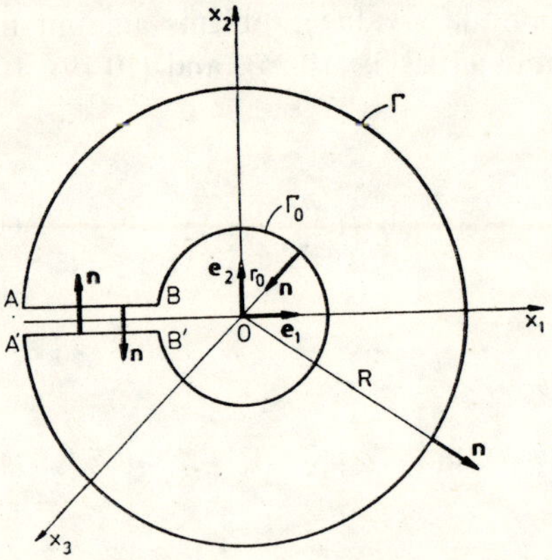


Fig. 10.2. On the calculation of the strain energy stored per unit dislocation length.

Next, taking into consideration (10.56) and (10.57), we see that on $\Gamma_0$, hence for $z_\alpha = r_0(\cos\theta + p_\alpha \sin\theta)$, we have $u_k = c_k \ln r_0 + g_k(\theta)$, $\mathbf{n} = \mathbf{e}_1 \cos\theta + \mathbf{e}_2 \sin\theta$, $t_k = h_k(\theta)/r_0$, where $c_k$ are some constants, and $g_k(\theta)$ and $h_k(\theta)$ do not depend on $r_0$. Consequently, considering also (10.21),

$$\oint_{\Gamma_0} t_k u_k \, dl = c_k(\ln r_0)\int_{-\pi}^{\pi} h_k(\theta)\, d\theta + \int_{-\pi}^{\pi} h_k(\theta)\, g_k(\theta)\, d\theta = \int_{-\pi}^{\pi} h_k(\theta) g_k(\theta)\, d\theta.$$

A similar calculation for the integral along $\Gamma$, taking into account that $\mathbf{n} = -\mathbf{e}_1 \cos\theta - \mathbf{e}_2 \sin\theta$ on $\Gamma$, leads to an opposite value, and hence the sum of the first two integrals in (10.58) vanishes. Next, by substituting $(10.57)_2$ into (10.58) gives

$$w = \frac{b_k}{2\pi} \operatorname{Im}\left(\sum_{\alpha=1}^{3} L_{k\alpha} D_\alpha\right) \int_{-R}^{-r_0} \frac{dx_1}{x_1} = -\frac{b_k}{2\pi} \operatorname{Im}\left(\sum_{\alpha=1}^{3} L_{k\alpha} D_\alpha\right) \ln \frac{R}{r_0}.$$

On the other hand, (10.41) implies that the number $\sum_{\alpha=1}^{3} L_{k\alpha} D_\alpha$ is pure imaginary and hence, in view of (10.45), the last relation may be rewritten as

$$w = \frac{Kb^2}{4\pi} \ln \frac{R}{r_0}, \tag{10.59}$$

where

$$Kb^2 = \frac{1}{i} \sum_{\alpha=1}^{3} \frac{L_{r\alpha} L_{s\alpha} b_r b_s}{A_{m\alpha} L_{m\alpha}}. \tag{10.60}$$

The expression (10.59) of $w$ has been first derived by Foreman [120], who also calculated the values of $w$ for various dislocation orientations in crystals with cubic and hexagonal symmetry. Clearly, the constant $K$, which is also called the *energy factor*, depends not only on the elastic constants and of the dislocation orientation, but also on the magnitude and on the direction of the Burgers vector. By comparing (10.59) with (8.62) and (8.54), we conclude that in the isotropic case the energy factor takes the value $\mu$ for a screw dislocation and $\mu/(1-\nu)$ for an edge dislocation.

We close this discussion by specializing the above results to the case where *the dislocation line is a two-fold symmetry axis or*, equivalently for elasticity, when *it is perpendicular to a reflection plane* [1]. Then, by (5.11), we have

$$c_{14} = c_{24} = c_{34} = c_{46} = c_{15} = c_{25} = c_{35} = c_{56} = 0, \tag{10.61}$$

and the reciprocal $[S_{KM}]$ of the matrix $[c_{KM}]$, $K, M = 1, 2, 4, 5, 6$, has the elements

$$\left.\begin{aligned} &S_{11} = (c_{22}c_{66} - c_{26}^2)/d, \qquad && S_{12} = (c_{16}c_{26} - c_{12}c_{66})/d,\\ &S_{16} = (c_{12}c_{26} - c_{16}c_{22})/d, && S_{22} = (c_{11}c_{66} - c_{16}^2)/d,\\ &S_{26} = (c_{11}c_{26} - c_{12}c_{16})/d, && S_{66} = (c_{11}c_{22} - c_{12}^2)/d,\\ &\quad S_{44} = c_{55}/d', \quad S_{45} = -c_{45}/d', && S_{55} = c_{44}/d',\\ &\quad S_{14} = S_{24} = S_{46} = S_{15} = S_{25} = S_{56} = 0, \end{aligned}\right\} \tag{10.62}$$

with the notation

$$d = c_{66}(c_{11}c_{22} - c_{12}^2) + 2c_{12}c_{16}c_{26} - c_{11}c_{26}^2 - c_{22}c_{16}^2, \quad d' = c_{44}c_{55} - c_{45}^2.$$

If we require that the strain energy function $W$ be positive definite, the quantities $d$ and $d'$, which are actually principal minors of the matrix $[c_{KM}]$, must be strictly positive.

From $(10.8)_2$ it follows now that $l_3(p)$ vanishes identically, and hence (10.9) may be decomposed into the quartic equation

$$l_4(p) \equiv S_{11}p^4 - 2S_{16}p^3 + (2S_{12} + S_{66})p^2 - 2S_{26}p + S_{22} = 0, \tag{10.63}$$

whose roots with positive imaginary parts will be denoted by $p_1$ and $p_2$, and the quadratic equation

$$l_2(p) \equiv S_{55}p^2 - 2S_{45}p + S_{44} = 0, \tag{10.64}$$

[1] This case has been first considered by Seeger and Schöck [285]. A complete discussion of the symmetry cases when analytic results are available or merely possible may be found in the book by Steeds [317] and in the review article by Steeds and Willis [477].

which gives

$$p_3 = (-c_{45} + i\sqrt{d'})/c_{44}. \tag{10.65}$$

Next, (10.12) yields

$$\lambda_1 = \lambda_2 = \lambda_3 = 0,$$

and hence (10.11) becomes

$$[L_{k\alpha}] = \begin{bmatrix} -p_1 & -p_2 & 0 \\ 1 & 1 & 0 \\ 0 & 0 & -1 \end{bmatrix}. \tag{10.66}$$

On the other hand, from (10.13) and (10.14) it follows that

$$A_{1\alpha} = S_{11}p_\alpha^2 - S_{16}p_\alpha + S_{12}, \quad A_{2\alpha} = S_{12}p_\alpha - S_{26} + S_{22}/p_\alpha, \quad A_{3\alpha}=0, \quad \alpha=1,2, \tag{10.67}$$

$$A_{13} = A_{23} = 0, \qquad A_{33} = S_{45} - S_{44}/p_3 = i\sqrt{d'}. \tag{10.68}$$

It may be easily seen now that, in the case considered, the elastic state produced by the dislocation consists of a pure screw and a pure edge part, like in the isotropic case. Indeed, let us consider first an *edge dislocation* with components of the Burgers vector

$$b_1 = b, \qquad b_2 = b_3 = 0. \tag{10.69}$$

From (10.45) and (10.66—68) we obtain

$$D_1 = -\frac{p_1 b}{2(p_1 A_{11} - A_{21})}, \qquad D_2 = -\frac{p_2 b}{2(p_2 A_{12} - A_{22})}, \qquad D_3 = 0, \tag{10.70}$$

whereas (10.56) and (10.57) yield

$$u_3 = 0, \qquad T_{13} = T_{23} = 0, \tag{10.71}$$

$$u_\beta = \frac{1}{\pi} \operatorname{Im} \sum_\alpha A_{\beta\alpha} D_\alpha \ln(x_1 + p_\alpha x_2) + u_\beta^0, \tag{10.72}$$

$$\left.\begin{aligned} T_{11} &= \frac{1}{\pi} \operatorname{Im} \sum_\alpha \frac{p_\alpha^2 D_\alpha}{x_1 + p_\alpha x_2}, \qquad T_{12} = -\frac{1}{\pi} \operatorname{Im} \sum_\alpha \frac{p_\alpha D_\alpha}{x_1 + p_\alpha x_2}, \\ T_{22} &= \frac{1}{\pi} \operatorname{Im} \sum_\alpha \frac{D_\alpha}{x_1 + p_\alpha x_2}, \end{aligned}\right\} \tag{10.73}$$

where $\sum_\alpha$ denotes summation with respect to $\alpha$ over the integers 1 and 2. Next, by $(4.63)_1$, (10.3), and $(10.71)_1$, we have

$$T_{33} = c_{13}u_{1,1} + c_{23}u_{2,2} + c_{36}(u_{1,2} + u_{2,1}), \tag{10.74}$$

wherefrom, by virtue of (10.72), it follows that

$$T_{33} = \frac{1}{\pi} \operatorname{Im} \left\{ \sum_\alpha [A_{1\alpha}(c_{13} + p_\alpha c_{36}) + A_{2\alpha}(p_\alpha c_{23} + c_{36})] \frac{D_\alpha}{x_1 + p_\alpha x_2} \right\}. \tag{10.75}$$

Finally, from (10.60) and (10.66), we deduce that the energy factor of the edge dislocation is

$$K = \frac{1}{i} \sum_\alpha \frac{p_\alpha^2}{A_{2\alpha} - p_\alpha A_{1\alpha}}. \tag{10.76}$$

We shall resume the case of the edge dislocation in Sect. 14.2, when considering the solution obtained in a different way by Teodosiu and Nicolae [338]. For a more explicit form of the above relations, which is valid, however, only when $c_{16} = c_{26} = 0$, see Hirth and Lothe [162], pp. 422—425.

Let us consider now a *screw dislocation* with the components of the Burgers vector

$$b_1 = b_2 = 0, \quad b_3 = b.$$

From (10.45) and (10.66—68) we obtain in this case

$$D_1 = D_2 = 0, \qquad D_3 = -\frac{b}{2A_{33}} = ib\sqrt{d'}, \tag{10.77}$$

and from (10.56), (10.57), and (10.74), we deduce that

$$u_1 = u_2 = 0, \quad T_{11} = T_{22} = T_{12} = T_{33} = 0, \tag{10.78}$$

$$u_3 = -\frac{b}{\pi} \arg(x_1 + p_3 x_2) + u_3^0, \tag{10.79}$$

$$\left.\begin{aligned} T_{13} &= \frac{\sqrt{d'}}{\pi} \operatorname{Re} \frac{p_3}{x_1 + p_3 x_2} = -\frac{b\sqrt{d'}}{\pi} \frac{c_{45}x_1 - c_{55}x_2}{c_{44}x_1^2 - 2c_{45}x_1x_2 + c_{55}x_2^2}, \\ T_{23} &= -\frac{b\sqrt{d'}}{\pi} \operatorname{Re} \frac{1}{x_1 + p_3 x_2} = -\frac{b\sqrt{d'}}{\pi} \frac{c_{44}x_1 - c_{45}x_2}{c_{44}x_1^2 - 2c_{45}x_1x_2 + c_{55}x_2^2}. \end{aligned}\right\} \tag{10.80}$$

Finally, by making use of (10.60) and (10.66), we infer that the energy factor of the screw dislocation is

$$K = \sqrt{d'}. \tag{10.81}$$

From the considerations above it is easily seen that the pure edge and the pure screw dislocations have complementary non-zero displacement and stress components. Hence the elastic state produced by a mixed straight dislocation can be derived by simply adding the elastic states corresponding to its edge and screw components.

## 10.4. Numerical results

As shown by (10.56), (10.57), and (10.60), the quantities $p_\alpha$, $A_{k\alpha}$, and $L_{k\alpha}$, $k$, $\alpha = 1, 2, 3$, completely characterize the principal singularity of the elastic field of a straight dislocation in a medium with general anisotropy. For this reason, a special program has been elaborated by Teodosiu, Nicolae, and Paven [340, 342], for calculating these quantities. The program was applied to 60 different crystals[1] belonging to the cubic and hexagonal systems, and to almost all cases of straight dislocations occurring in these crystals[2], by using the experimental values given in Table 5.3 of the adiabatic second-order elastic constants at room temperature with respect to the standard crystallographic axes $x_1^0$, $x_2^0$, $x_3^0$ shown in Fig. 10.3 a and 10.6 a (cf. also Mantea *et al.* [230]).

Tables 10.1 and 10.2 concern the dislocations considered for various crystal lattices. The orientations of the axes $x_1$, $x_2$, $x_3$ that were used to describe the elastic field of the dislocation, chosen in each case with the $x_3$-axis along the dislocation line, have been labeled from 1 to 6, the directions of the $x_k$-axes being indicated in Table 10.1. The direction of the dislocation line, the glide plane, and the Burgers vector are shown in Table 10.2 and are illustrated in Figs. 10.3–10.6.

*Table 10.1*

**Crystallographic orientations of the co-ordinate axes used to describe the elastic field of straight dislocations**

| Orientation label | Directions of the co-ordinate axes | | |
|---|---|---|---|
| | $x_1$ | $x_2$ | $x_3$ |
| 1 | $[110]$ | $[\bar{1}10]$ | $[001]$ |
| 2 | $[10\bar{1}]$ | $[010]$ | $[101]$ |
| 3 | $[\bar{1}10]$ | $[\bar{1}\bar{1}1]$ | $[112]$ |
| 4 | $[\bar{1}2\bar{1}]$ | $[\bar{1}01]$ | $[111]$ |
| 5 | $[2\bar{1}\bar{1}0]$ | $[0001]$ | $[0\bar{1}10]$ |
| 6 | $[10\bar{1}0]$ | $[0001]$ | $[1\bar{2}10]$ |

[1] Namely Ag, Al, Au, Cu, Ni, Pb, Th, Cr, $\alpha$–Fe, K, Li, Mo, Na, Nb, Ta, V, W, CuZn, AgBr, AgCl, the 16 alkali halides, CaO, MgO, SrO, C, Si, Ge, Be, Cd, Co, Er, Mg, Tl, Y, Zr, $Ag_2Al$, BeO, CdS, CdSe, and $\alpha$–ZnS.

[2] For hexagonal crystals, however, only the basal glide has been considered.

*Table 10.2.*

**Dislocations considered in numerical calculations**

| Crystal system | Crystal lattice | Orient. label | Glide system | Disl. line | Glide plane | Burgers vector | Dislocation type |
|---|---|---|---|---|---|---|---|
| Cubic | f.c.c. | 2 | ⟨110⟩{111} | [101] | $(11\bar{1})$ | $\frac{1}{2}$ [101] | screw |
| Cubic | f.c.c. | 2 | ⟨110⟩{111} | [101] | $(11\bar{1})$ | $\frac{1}{2}$ $[10\bar{1}]$ | sessile edge |
| Cubic | f.c.c. | 2 | ⟨110⟩{111} | [101] | $(11\bar{1})$ | $\frac{1}{2}$ $[1\bar{1}0]$ | 60° mixed |
| Cubic | f.c.c. | 3 | ⟨112⟩{111} | [112] | $(11\bar{1})$ | $\frac{1}{2}$ $[\bar{1}10]$ | edge |
| Cubic | f.c.c. | 3 | ⟨112⟩{111} | [112] | $(11\bar{1})$ | $\frac{1}{2}$ [011] | 30° mixed |
| Cubic | b.c.c. | 2 | ⟨110⟩{112} | [101] | $(12\bar{1})$ | $\frac{1}{2}$ $[1\bar{1}\bar{1}]$ | edge |
| Cubic | b.c.c. | 2 | ⟨110⟩{110} | [101] | $(10\bar{1})$ | $\frac{1}{2}$ [111] | 35°16′ mixed |
| Cubic | b.c.c. | 3 | ⟨112⟩{110} | [112] | $(1\bar{1}0)$ | $\frac{1}{2}$ $[\bar{1}\bar{1}1]$ | edge |
| Cubic | b.c.c. | 4 | ⟨111⟩{110}<br>⟨111⟩{112} | [111] | $(1\bar{1}0)$<br>$(2\bar{1}\bar{1})$ | $\frac{1}{2}$ [111] | screw |
| Cubic | rock salt | 1 | ⟨100⟩{110} | [001] | $(1\bar{1}0)$ | [110] | edge |
| Cubic | rock salt | 2 | ⟨110⟩{110} | [101] | $(10\bar{1})$ | [101] | screw |
| Cubic | rock salt | 4 | ⟨111⟩{110} | [111] | $(01\bar{1})$ | [101] | 35°16′ mixed |
| Cubic | diamond | 2 | ⟨110⟩{111} | [101] | $(11\bar{1})$ | $\frac{1}{2}$ [101] | screw |
| Cubic | diamond | 2 | ⟨110⟩{111} | [101] | $(11\bar{1})$ | $\frac{1}{2}$ $[1\bar{1}0]$ | 60° mixed |
| Hexagonal | h.c.p. | 5 | $\langle 0\bar{1}10\rangle(0001)$ | $[0\bar{1}10]$ | (0001) | $\frac{1}{3}$ $[2\bar{1}\bar{1}0]$ | edge |
| Hexagonal | h.c.p. | 5 | $\langle 0\bar{1}10\rangle(0001)$ | $[0\bar{1}10]$ | (0001) | $\frac{1}{3}$ $[1\bar{2}10]$ | 30° mixed |
| Hexagonal | h.c.p. | 6 | $\langle 1\bar{2}10\rangle(0001)$ | $[1\bar{2}10]$ | (0001) | $\frac{1}{3}$ $[1\bar{2}10]$ | screw |

To illustrate the numerical results obtained we give here only the values of the quantities $p_\alpha$, $A_{k\alpha}$, and $L_{k\alpha}$, calculated for dislocations in four typical crystals Cu, $\alpha$–Fe, NaCl, and Zn, having an f.c.c., b.c.c., rock-salt, and h.c.p. lattice, respectively. Crystals belonging to the diamond structure do not exhibit qualitative differences as concerns the values of $p_\alpha$, $A_{k\alpha}$, and $L_{k\alpha}$ (cf. also

Table 10.2). Only non-zero coefficients $A_{k\alpha}$ and $L_{k\alpha}$ are listed. Moreover since always

$$L_{11} = -p_1, \quad L_{12} = -p_2, \; L_{21} = L_{22} = -L_{33} = 1,$$

these values are not mentioned. The quantities $p_\alpha$ and $L_{k\alpha}$ are non-dimensional, whereas $A_{k\alpha}$ are expressed in units of $(\text{Mbar})^{-1} = 10^{-11}\text{m}^2/\text{N}$.

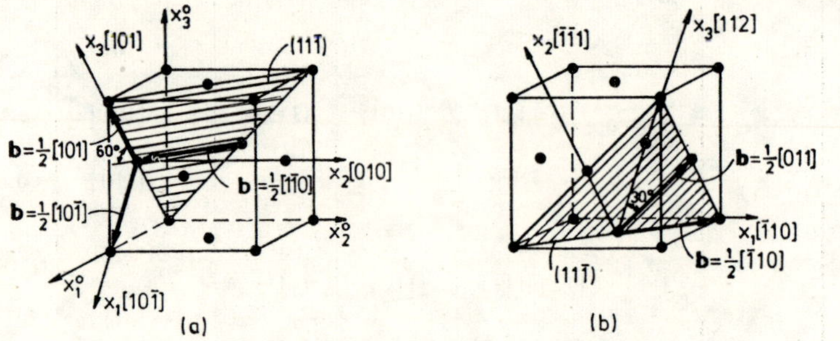


Fig. 10.3. Dislocations in an f.c.c. crystal. (a) The screw, the sessile edge, and the 60° mixed dislocations (orientation 2 of the $x_k$-axes). The standard crystallographic axes of the cubic lattice are denoted by $\overset{0}{x_1}$, $\overset{0}{x_2}$, and $\overset{0}{x_3}$. (b) The edge and the 30° mixed dislocations (orientation 3 of the $x_k$-axes).

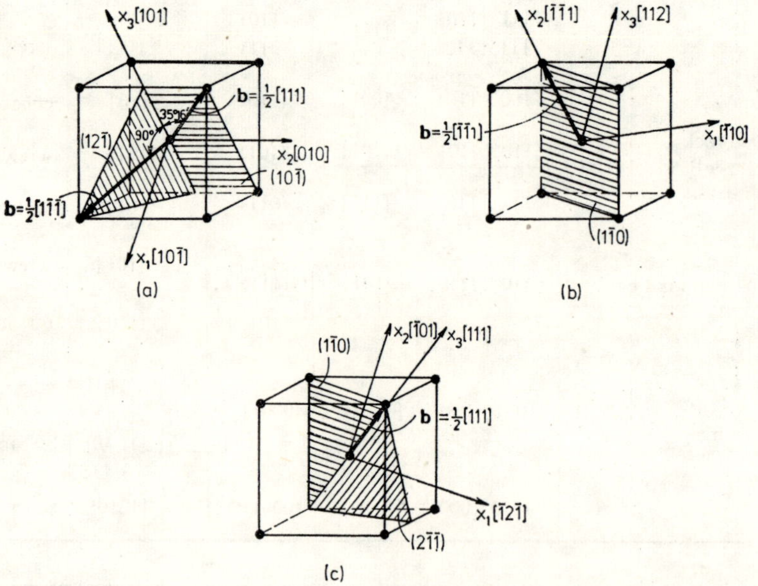


Fig. 10.4. Dislocations in a b.c.c. crystal. (a) The [110] edge and the 35°16′ mixed dislocation (orientation 2 of the $x_k$-axes). (b) The [112] edge dislocation (orientation 3 of the $x_k$-axes). (c) The screw dislocation (orientation 4 of the $x_k$-axes).

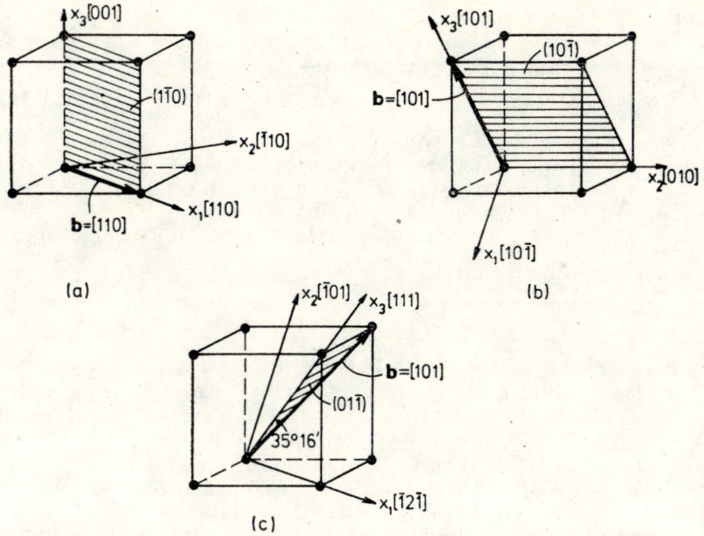


Fig. 10.5. Dislocations in a rock-salt crystal. (a) The edge dislocation (orientation 1 of the $x_k$-axes). (b) The screw dislocation (orientation 2 of the $x_k$-axes). (c) The 35°16′ mixed dislocation (orientation 4 of the $x_k$-axes).

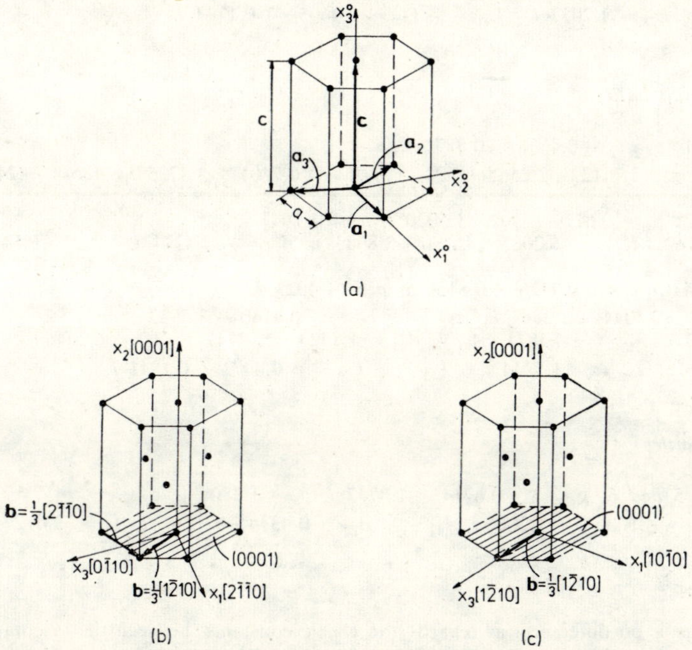


Fig. 10.6. Basal glide dislocations in an h.c.p. crystal. (a) Standard crystallographic axes, $x_1^0$, $x_2^0$, $x_3^0$, and crystallographic vectors, $\mathbf{a}_1$, $\mathbf{a}_2$, $\mathbf{a}_3$, $\mathbf{c}$, of the h.c.p. lattice. (b) The edge and the 30° mixed dislocations (orientation 5 of the $x_k$-axes). (c) The screw dislocation (orientation 6 of the $x_k$-axes).

*Cu (f.c.c. lattice)*

*Orientation* 2: $p_{1,2} = \mp 0.7056 + 0.8058\, i,\ p_3 = 0.5528\, i,$
$A_{11} = \bar{A}_{12} = -0.6614 - 0.8611\, i,\ A_{21} = -\bar{A}_{22} = -0.2273 - 1.1406\, i,\ \ A_{33} = 2.3929\, i.$

*Orientation* 3: $p_1 = 0.3906\ i,\ \ p_{2,3} = \pm 0.5845 + 1.4813\ i,$
$A_{11} = -0.2359,\ A_{12} = -1.6130 + 0.8382\, i,\ A_{13} = 1.0265 + 1.3844\ i,$
$A_{21} = -1.3017\ i,\ A_{22} = 0.0552 - 0.6557\ i,\ A_{23} = -0.2253 + 0.5818\ i,$
$A_{31} = 0.7085\, i,\ A_{32} = 1.1312 - 1.7939\ i,\ A_{33} = 0.2190 + 1.9987\ i,$
$L_{13} = -1.1134 + 1.0197\ i,\ L_{23} = -0.8523 - 0.4154\ i,\ L_{31} = -0.1707,\ L_{32} = 0.9481 + 0.4621\ i.$

*a—Fe (b.c.c. lattice)*

*Orientation* 2: $p_{1,2} = \mp 0.6247 + 0.8658\ i,\ p_3 = 0.6399\ i,$
$A_{11} = \bar{A}_{12} = -0.4288 - 0.4912\ i,\ A_{21} = -\bar{A}_{22} = -0.1574 - 0.6781\ i,\ A_{33} = 1.3402\ i.$

*Orientation* 3: $p_1 = 0.4670\ i,\ p_{2,3} = \pm\ 0.4361 + 1.4099\ i,$
$A_{11} = -0.1590,\ \ A_{12} = -0.9473 + 0.3885\ \ i,\ \ A_{13} = 0.6487 + 0.8335\ i,$
$A_{21} = -0.7591\ i,\ A_{22} = 0.0452 - 0.3915\ i,\ A_{23} = -0.1603 + 0.3736\ i,$
$A_{31} = 0.3765\ i,\ A_{32} = 0.6510 - 0.9509\ i,\ A_{33} = 0.0952 + 1.1850\ i,$
$L_{13} = -1.1133 + 1.0384\ i,\ L_{23} = -0.8951 - 0.5127\ i,\ L_{31} = -0.1759,\ L_{32} = 0.8411 + 0.4818\, i.$

*Orientation* 4: $p_{1,2} = \mp 0.3839 + 0.6398\ i,\ p_3 = 1.9092\ i,$
$A_{11} = \bar{A}_{12} = -0.2852 - 0.3172\ i,\ A_{21} = -\bar{A}_{22} = -0.0935 - 0.6706\ i,$
$A_{13} = -1.6054\ i,\ A_{23} = 0.4348,\ A_{31} = \bar{A}_{32} = -0.2204 - 0.7287\ i,\ A_{33} = 1.5824\ i,$
$L_{13} = 2.2972,\ L_{23} = 1.2032\ i.\ \ L_{31} = -\bar{L}_{32} = 0.4605 - 0.1393\ i.$

*NaCl (rock-salt lattice)*

*Orientation* 1: $p_{1,2} = \pm 0.4606 + 0.8876\ i,\ p_3 = i,$
$A_{11} = -\bar{A}_{12} = -2.7412 + 2.2565\ i,\ A_{21} = -\bar{A}_{22} = 0.7401 - 3.4725\ i,\ A_{33} = 7.8247\ i.$

*Orientation* 2: $p_1 = 0.6279\ i,\ p_2 = 1.5020\ i,\ p_3 = 1.1947\ i,$
$A_{11} = -1.6162,\ A_{12} = -6.2085,\ A_{21} = -3.8983\ i,\ A_{22} = -2.4274\ i,\ \ A_{33} = 6.5497\ i.$

*Orientation* 4: $p_{1,2} = \mp 0.2426 + 0.8130\ i,\ p_3 = 1.4023\ i,$
$A_{11} = \bar{A}_{12} = -2.5044 - 1.2567\ i,\ A_{21} = -\bar{A}_{22} = -0.4140 - 3.7922\, i,$
$A_{13} = 8.7490\ i,\ A_{23} = -4.3871,\ A_{31} = \bar{A}_{32} = 1.1424 + 2.7747\ i,\ A_{33} = 6.2046\, i,$
$L_{13} = -2.2078,\ L_{23} = -1.5744\ i,\ \ L_{31} = -\bar{L}_{32} = -0.4472 + 0.1841\ i.$

*Zn (h.c.p. lattice)*[1]

*Orientations* 5 *and* 6: $p_{1,2} = \pm 0.6714 + 1.0747\ i,\ p_3 = 1.2809\ i,$
$A_{11} = -\bar{A}_{12} = -1.2890 + 1.2085\, i,\ A_{21} = -\bar{A}_{22} = 0.4334 - 2.1967\, i,\ A_{33} = 2.0127\, i.$

---

[1] There is no difference, as regards the elastic constants, between the orientations 5 and 6 of the $x_k$-axes. Indeed, they differ only by a rotation around **c**, which does not change the second-order elastic constants of hexagonal crystals, for they have transverse isotropy (cf. Sect. 5).

## 10.5. Green's functions for the elastic state of generalized plane strain

There exists a closed connection between straight dislocations and line forces [1]. Indeed, let us resume the case of the straight dislocation considered in Sect. 10.3, but assuming that the tractions acting per unit length of the cylindrical surface $\rho = r_0$ have a non-vanishing resultant force **P**, which is independent of $x_3$.

Requiring again that the stresses and the elastic rotation be continuous across the negative $x_1$-axis and vanish at infinity, we can still write the elastic state in the general form

$$u_k = 2\,\mathrm{Re} \sum_{\alpha=1}^{3} A_{k\alpha} f_\alpha(z_\alpha) + u_k^0, \qquad k = 1, 2, 3, \tag{10.82}$$

$$T_{k1} = -2\,\mathrm{Re} \sum_{\alpha=1}^{3} p_\alpha L_{k\alpha} f'_\alpha(z_\alpha), \qquad T_{k2} = 2\,\mathrm{Re} \sum_{\alpha=1}^{3} L_{k\alpha} f'_\alpha(z_\alpha), \tag{10.83}$$

$$f_\alpha(z_\alpha) = \frac{D_\alpha}{2\pi i} \ln z_\alpha + \sum_{m=1}^{\infty} \frac{g_{\alpha m}}{z_\alpha^m}, \qquad \alpha = 1, 2, 3, \tag{10.84}$$

where $z_\alpha = x_1 + p_\alpha x_2$, whereas $D_\alpha$, $g_{\alpha m}$ are undetermined complex constants. Next, imposing the condition

$$r_0 \int_{-\pi}^{\pi} t_k^*(\theta)\, \mathrm{d}\theta = P_k,$$

and taking into account (10.24), we find that now (10.41) must be replaced by

$$2\,\mathrm{Re} \sum_{\alpha=1}^{3} L_{k\alpha} D_\alpha = P_k, \qquad k = 1, 2, 3, \tag{10.85}$$

while the jump conditions (10.44)

$$2\,\mathrm{Re} \sum_{\alpha=1}^{3} A_{k\alpha} D_\alpha = -\,b_k, \qquad k = 1, 2, 3, \tag{10.86}$$

provide three more real algebraic equations for determining the three complex constants $D_\alpha$. The system (10.85), (10.86) can be solved as before, by using the orthogonality relations (10.15), to give

$$D_\alpha = \frac{A_{k\alpha} P_k - L_{k\alpha} b_k}{2 A_{m\alpha} L_{m\alpha}}, \qquad \alpha = 1, 2, 3, \tag{10.87}$$

where the summation over repeated Latin indices is implied, as usual. This result shows that the logarithmic terms of the complex potentials $f_\alpha(z_\alpha)$ are completely determined by the Burgers

[1] This connection was first noticed by Stroh [324] and further developed in a six-dimensional form by Malén and Lothe [225], and Malén [226].

vector **b** and the resultant force **P** of the tractions acting on $\Sigma_0$, whereas the coefficients $g_{\alpha m}$ occurring in (10.84) depend on the detailed distribution of the tractions on $\Sigma_0$, as shown in Sect. 10.3. Clearly, equations (10.87) establish an algebraic equivalence between a straight dislocation and a line force.

Let us consider now the problem of defining the Green's tensor function of the generalized plane strain. We have seen in Sect. 6.4 that the three-dimensional Green's tensor function $\mathbf{G(x)}$ can be interpreted in terms of the displacement fields produced by concentrated forces in an infinite elastic medium. Namely, a unit concentrated force $\mathbf{P} = \mathbf{e}_s$, acting at the origin and directed along the $x_s$-axis, produces the displacement field $u_k^{(s)}(\mathbf{x}) = G_{ks}(\mathbf{x})$ in the infinite elastic medium. Similarly, by putting $b_k = 0$ and $P_k = \delta_{ks}$ in (10.87) we obtain the particular value $D_\alpha^{(s)}$ of $D_\alpha$ corresponding to the case where the tractions acting on $\Sigma_0$ have the resultant force $\mathbf{P} = \mathbf{e}_s$, namely

$$D_\alpha^{(s)} = \frac{A_{s\alpha}}{2A_{m\alpha}L_{m\alpha}}, \qquad \alpha = 1, 2, 3. \tag{10.88}$$

However, this particularization does not determine uniquely the displacement field (10.82) since the coefficients $g_{\alpha m}$ are still arbitrary. Consequently, by analogy with the three-dimensional case, the *Green's tensor function* $G_{ks}(x_1, x_2)$ *of the generalized plane strain* is defined by taking $u_k^0 = 0$ in (10.82) and retaining only the principal singularity of the solution, i.e. the leading logarithmic term of the expression (10.84) of $f_\alpha(z_\alpha)$, with $D_\alpha$ replaced by (10.88). Thus

$$G_{ks}(\xi) \equiv u_k^{(s)}(\xi) = \mathrm{Re} \sum_{\alpha=1}^{3} \frac{A_{k\alpha}D_\alpha^{(s)}}{\pi i} \ln z_\alpha = \frac{1}{\pi} \mathrm{Im} \sum_{\alpha=1}^{3} A_{k\alpha}D_\alpha^{(s)} \ln z_\alpha, \tag{10.89}$$

where

$$\xi = x_1\mathbf{e}_1 + x_2\mathbf{e}_2, \qquad z_\alpha = x_1 + p_\alpha x_2, \tag{10.90}$$

and the corresponding stresses result from (10.83) as

$$T_{k1}^{(s)}(\xi) = -\frac{1}{\pi} \mathrm{Im} \sum_{\alpha=1}^{3} \frac{p_\alpha L_{k\alpha}D_\alpha^{(s)}}{z_\alpha}, \qquad T_{k2}^{(s)}(\xi) = \frac{1}{\pi} \mathrm{Im} \sum_{\alpha=1}^{3} \frac{L_{k\alpha}D_\alpha^{(s)}}{z_\alpha}. \tag{10.91}$$

From the reasoning above it is apparent that Green's tensor function $\mathbf{G}(\xi)$ has been selected by imposing a certain singularity of the elastic state for $\rho \to 0$, where

$$\rho = \|\xi\| = \sqrt{x_1^2 + x_2^2}. \tag{10.92}$$

In fact, the neglected terms correspond to self-equilibrated traction distributions on $\Sigma_0$, or to elastic states that possess self-equilibrated singularities, such as force multipoles, at the origin. This leads us to the following equivalent definition of $\mathbf{G}(\xi)$.

We call *fundamental singular solution* or *Green's tensor function of the generalized plane strain*, the second-order tensor field $\mathbf{G}(\xi)$ with the following properties:

(i) For any point of the $x_1x_2$-plane with position vector $\xi \neq \mathbf{0}$ and for each $s = 1, 2, 3$, the displacement field $u_k^{(s)}(\xi) = G_{ks}(\xi)$ defines a (regular) elastic state of generalized plane strain, corresponding to zero body forces. In particular, by (10.2),

$$T_{k1,1}^{(s)} + T_{k2,2}^{(s)} = 0, \qquad k = 1, 2, 3, \tag{10.93}$$

where

$$T_{k\alpha}^{(s)}(\xi) = c_{k\alpha m1}\, G_{ms,1}(\xi) + c_{k\alpha m2}\, G_{ms,2}(\xi), \qquad k = 1, 2, 3; \quad \alpha = 1, 2, \tag{10.94}$$

are stress components corresponding to the displacement field $u_k^{(s)}(\xi)$.

(ii) $\mathbf{G}(\xi)/\ln\rho = O(1)$ and $\mathbf{T}^{(s)}(\xi) = O(\rho^{-1})$ as $\rho \to 0$ and also as $\rho \to \infty$.

(iii) For all $\eta > 0$ and $s = 1, 2, 3,$

$$\oint_{\Gamma_\eta} \mathbf{T}^{(s)}\,\mathbf{n}\,\mathrm{d}l = \mathbf{e}_s, \qquad s = 1, 2, 3,$$

or, in component form,

$$\oint_{\Gamma_\eta} (T_{k1}^{(s)} n_1 + T_{k2}^{(s)} n_2)\,\mathrm{d}l = \delta_{ks}, \qquad k,s = 1, 2, 3, \tag{10.95}$$

where $\Gamma_\eta$ is the circle with radius $\eta$ centred at the origin in the $x_1x_2$-plane, and $\mathbf{n} = n_1\mathbf{e}_1 + n_2\mathbf{e}_2$ is the inward unit normal to $\Gamma_\eta$.

It is easily seen that the singular elastic state (10.89), (10.91) satisfies (i)–(iii). Conversely, it may be shown that these properties uniquely determine the singular elastic state. Indeed, (i) implies that each $\underset{\alpha}{f}'(z_\alpha)$ is analytic and single-valued outside any circle $\Gamma_\eta$ centred at the origin, and hence can be developed in a Laurent series. Consequently, integrating with respect to $z_\alpha$, for each $\alpha = 1, 2, 3$, we may write

$$f_\alpha(z_\alpha) = \frac{D_\alpha}{2\pi i} \ln z_\alpha + \sum_{m=-\infty}^{\infty} g_{\alpha m} z_\alpha^{-m}.$$

Next, (ii) eliminates all positive *and* negative powers of $z_\alpha$ from this expression. Finally, (iii) and the condition that the displacement be single-valued, determine the coefficient $D_\alpha$ of the remaining logarithmic term in the form (10.88), thus leading to the expression (10.89) of $\mathbf{G}(\xi)$.

To derive the differential equation satisfied by $\mathbf{G}(\xi)$ in the sense of the theory of distributions we use a reasoning similar to that employed in Sect. 6.4. For conciseness, we assume in the remaining part of this subsection that Greek subscripts take only the values 1 and 2, and extend the summation convention to repeated Greek subscripts. Let $\varphi(\xi)$ denote an arbitrary function of class $C^\infty$ and of compact support on the $x_1x_2$-plane. According to the definition of the derivatives of a distribution, we have

$$(c_{i\alpha k\beta} G_{ks,\alpha\beta}(\xi), \varphi(\xi)) = (c_{i\alpha k\beta} G_{ks,\alpha}(\xi), \varphi_{,\beta}(\xi)) = (c_{i\alpha k\beta} G_{ks}(\xi), \varphi_{,\alpha\beta}(\xi)). \tag{10.96}$$

Denote by $\Delta_\rho$ the exterior domain bounded by the circle $\Gamma_\rho$ of radius $\rho$ and the centre at the origin in the $x_1x_2$-plane. Integrating by parts twice, taking into account that $\varphi$ vanishes together with all its derivatives for sufficiently large values of $\rho$, and considering that, by virtue of (10.93) and (10.94),

$$c_{k\alpha m\beta} G_{ms,\alpha\beta}(\xi) = 0 \qquad \text{for any } \xi \neq \mathbf{0} \text{ and any } k = 1,2,3,$$

we successively obtain

$$\int_{\Delta_\rho} c_{k\alpha m\beta} G_{ms}(\xi)\, \varphi_{,\alpha\beta}(\xi)\,\mathrm{d}s = \int_{\Delta_\rho} (c_{k\alpha m\beta} G_{ms}(\xi)\, \varphi_{,\alpha}(\xi))_{,\beta}\,\mathrm{d}s - \int_{\Delta_\rho} c_{k\alpha m\beta} G_{ms,\beta}(\xi) \varphi_{,\alpha}(\xi)\,\mathrm{d}s =$$

$$= \int_{\Gamma_\rho} c_{k\alpha m\beta} G_{ms}(\xi)\, \varphi_{,\alpha}(\xi)\, n_\beta\,\mathrm{d}l - \int_{\Delta_\rho} (c_{k\alpha m\beta} G_{ms,\beta}(\xi)\, \varphi(\xi))_{,\alpha}\,\mathrm{d}s + \int_{\Delta_\rho} c_{k\alpha m\beta} G_{ms,\alpha\beta}(\xi)\, \varphi(\xi)\,\mathrm{d}s =$$

$$= \int_{\Gamma_\rho} c_{k\alpha m\beta} G_{ms}(\xi)\, \varphi_{,\alpha}(\xi)\, n_\beta\,\mathrm{d}l - \int_{\Gamma_\rho} c_{k\alpha m\beta} G_{ms,\beta}(\xi)\, \varphi(\xi)\, n_\alpha\,\mathrm{d}l.$$

Next, making use of the mean theorem of the integral calculus and considering (ii) and (iii), we find that

$$\lim_{\rho\to 0}\int_{\Gamma_\rho} c_{k\alpha m\beta} G_{ms}(\xi)\, \varphi_{,\alpha}(\xi)\, n_\beta \, dl = 0,$$

$$\lim_{\rho\to 0}\int_{\Gamma_\rho} c_{k\alpha m\beta} G_{ms,\beta}(\xi)\, \varphi(\xi)\, n_\alpha \, dl = \lim_{\rho\to 0}\int_{\Gamma_\rho} T^{(s)}_{k\alpha}\, n_\alpha\, \varphi(\xi)\, dl = \delta_{ks}\varphi(0),$$

and hence

$$\lim_{\rho\to 0}\int_{\Delta_\rho} c_{k\alpha m\beta} G_{ms}(\xi)\, \varphi_{,\alpha\beta}(\xi)\, ds = -\,\delta_{ks}\varphi(0).$$

Combining this result with (10.96), we deduce that the regular functionals associated to $c_{k\alpha m\beta}G_{ms,\alpha\beta}(\xi)$ on the regions $\Delta_\rho$ tend to $-\,\delta_{ip}\delta(\xi)$ as $\rho \to 0$. Therefore, the components of the distribution associated to $\mathbf{G}(\xi)$ satisfy the equations

$$c_{k\alpha m\beta} G_{ms,\alpha\beta}(\xi) + \delta_{ks}\delta(\xi) = 0, \qquad k, s = 1, 2, 3. \tag{10.97}$$

Assume now that the elastic medium is subjected to the action of a body force $\mathbf{f}(\xi)$, which is independent of $x_3$ and of class $C^1$ on the $x_1x_2$-plane, and which satisfies the condition [1]

$$\mathbf{f}(\xi)\ln\rho = O(\rho^2) \quad \text{as} \quad \rho \to \infty \tag{10.98}$$

Making use of the properties of the convolution and taking into account (10.97), we may write

$$f_k(\xi) = \delta_{ks}\delta(\xi - \xi') * f_s(\xi') = -\,c_{k\alpha m\beta}G_{ms,\alpha\beta}(\xi - \xi') * f_s(\xi')$$

$$= -\,c_{k\alpha m\beta}\{G_{ms}(\xi - \xi') * f_s(\xi')\}_{,\alpha\beta},$$

where the derivatives are taken with respect to $x_\alpha$ and $x_\beta$. We conclude that

$$u_m(\xi) = G_{ms}(\xi - \xi') * f_s(\xi'), \qquad m = 1,2,3, \tag{10.99}$$

is a particular solution of the equilibrium equations of the generalized plane strain

$$c_{k\alpha m\beta} u_{m,\alpha\beta} + f_k = 0, \qquad k = 1,2,3.$$

Finally, by taking into account the way in which the distribution $\mathbf{G}(\xi)$ has been generated, as well as the continuity of the convolution, (10.99) may be rewritten as

$$u_m(\xi) = \int_E G_{ms}(\xi - \xi')\, f_s(\xi')\, ds', \tag{10.100}$$

---

[1] This condition is satisfied *a fortiori* if $\mathbf{f}(\xi) = O(\rho^3)$ as $\rho \to \infty$.

where $E$ denotes the $x_1x_2$-plane, the convergence of the improper integral in the right-hand side being granted by the conditions (ii) and (10.98).

Using a procedure similar to that presented in Sect. 6.3 for the three-dimensional case, it is also possible to extend the concept of Green's tensor function to the generalized plane strain of cylindrical bodies with free boundaries at *finite* distance. Such Green's functions have been constructed by Sinclair and Hirth [472] for an anisotropic infinite elastic body containing a planar crack, and then subsequently used to study the interaction between the crack and some rod shaped inclusions or coherent precipitates parallel to the crack front.

## 10.6. Somigliana dislocation in an anisotropic elastic medium

In this subsection we treat, following [478], a Somigliana dislocation that produces a state of generalized plane strain in an infinite anisotropic elastic medium with an infinite circular cylindrical hole of radius $r_0$. The results obtained will be applied to the simulation of crystal dislocations by non-linear elasticity (see Sect. 14).

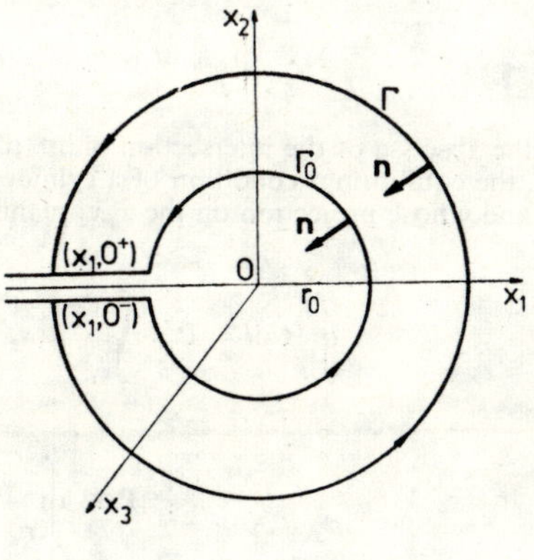


Fig. 10.7. Cut along the half-plane $x_2 = 0$, $x_1 \leqslant -r_0$, used to generate a Somigliana dislocation in an infinite elastic medium with a circular cylindrical hole.

Let us take the axis of the hole as $x_3$-axis of a Cartesian system of co-ordinates and the half-plane $x_2 = 0$, $x_1 \leqslant -r_0$ as cut $S$ for producing the Somigliana dislocation (Fig. 10.7). Having in mind further applications, we first consider a more general case where not only the displacement vector $\mathbf{u}$ is allowed to be discontinuous across the cut $S$, but also the tractions applied on the faces of the cut are not equilibrated, being statically equivalent to a distribution of surface forces on $S$. Specifically, we assume that (7.19) and (7.20) are replaced by the conditions

$$\mathbf{u}(x_1, 0^+) - \mathbf{u}(x_1, 0^-) = \mathbf{g}(x_1), \qquad \mathbf{t}(x_1, 0^+) + \mathbf{t}(x_1, 0^-) = \mathbf{h}(x_1)$$

for $x_1 \in (-\infty, -r_0]$, where $\mathbf{g}$ and $\mathbf{h}$ are analytic vector-valued functions, such that $\mathbf{g} = O(1)$ and $\mathbf{h} = O(x_1^{-2})$ as $|x_1| \to \infty$. These conditions assure that the jump of the displacement vector across the cut and the resultant force of the tractions acting on the cut faces per unit length of the $x_3$-axis are both bounded. Developing the functions $\mathbf{g}(x_1)$ and $\mathbf{h}(x_1)$ in power series for $x_1 \in (-\infty, -r_0]$, we have

$$u_k(x_1, 0^+) - u_k(x_1, 0^-) = \sum_{m=0}^{\infty} \frac{g_{km}}{x_1^m}, \tag{10.101}$$

$$t_k(x_1, 0^+) + t_k(x_1, 0^-) = \sum_{m=2}^{\infty} \frac{h_{km}}{x_1^m}, \tag{10.102}$$

for $k = 1, 2, 3$, where $g_{km}$ and $h_{km}$ are known real constants.

We look for the solution of the field equations of the generalized plane strain satisfying the boundary condition (10.22) at infinity, and either of the boundary conditions (10.20), (10.27) on $\Gamma_0$. We begin by rewriting (10.102) in terms of the stress functions $\Phi_k$ introduced at the end of Sect. 10.1. Let $\Gamma$ be a smooth closed curve in the $x_1x_2$-plane and encircling $\Gamma_0$ anticlockwise (Fig. 10.7). Then, by virtue of (10.16), we have

$$t_k = T_{k1}n_1 + T_{k2}n_2 = -T_{k1}\frac{dx_2}{dl} + T_{k2}\frac{dx_1}{dl} = \frac{d\Phi_k}{dl}, \tag{10.103}$$

where $\mathbf{n}$ is the inward unit normal to $\Gamma$, and $l$ is the curvilinear abscissa on $\Gamma$. Integrating this relation once around $\Gamma$ yields

$$\oint_\Gamma t_k \, dl = \Phi_k(x_1, 0^+) - \Phi_k(x_1, 0^-), \tag{10.104}$$

where $x_1$ is the abscissa of the intersection point of $\Gamma$ with the negative $x_1$-axis.

Next, the equilibrium condition of a cylinder of unit length whose generators are parallel to the $x_3$-axis, and whose projection on the $x_1x_2$-plane is the region bounded by $\Gamma_0$, $\Gamma$, and the cut, gives

$$\oint_\Gamma \mathbf{t} \, dl = \mathbf{P} + \int_{x_1}^{-r_0} [\mathbf{t}(x_1, 0^+) + \mathbf{t}(x_1, 0^-)] dx_1, \tag{10.105}$$

where

$$\mathbf{P} \equiv \oint_{\Gamma_0} \mathbf{t}^* \, dl \tag{10.106}$$

is the resultant force of the tractions acting on the elastic medium per unit length of the cylindrical surface $\rho = r_0$.

By substituting (10.104) and (10.102) into (10.105) and performing the integration with respect to $x_1$, we find

$$\Phi_k(x_1, 0^+) - \Phi_k(x_1, 0^-) = \sum_{m=0}^{\infty} \frac{q_{km}}{x_1^m}, \tag{10.107}$$

with the notation

$$\left.\begin{aligned} q_{k0} &= P_k + \sum_{m=1}^{\infty} \frac{(-1)^{m+1} h_{k,m+1}}{m r_0^m}, \\ q_{km} &= \frac{h_{k,m+1}}{m} \quad \text{for } m \geqslant 1. \end{aligned}\right\} \tag{10.108}$$

The solution may be found by superposing two elastic states:

$$u_k(x_1, x_2) = \tilde{u}_k(x_1, x_2) + \hat{u}_k(x_1, x_2) + \overset{0}{u}_k, \tag{10.109}$$

$$\Phi_k(x_1, x_2) = \tilde{\Phi}_k(x_1, x_2) + \hat{\Phi}_k(x_1, x_2), \tag{10.110}$$

such that $\tilde{u}_k$ and $\tilde{\Phi}_k$ satisfy the jump conditions (10.101), (10.107) on the cut and the boundary condition (10.22) at infinity, whereas $\hat{u}_k$ and $\hat{\Phi}_k$ are continuous across the cut, vanish at infinity, and satisfy the boundary condition on $\Gamma_0$ corrected by the contribution of $\tilde{u}_k$ and $\tilde{\Phi}_k$.

In view of the results in Sect. 10.1, we take

$$\left.\begin{aligned} \tilde{u}_k(x_1, x_2) &= 2\,\mathrm{Re} \sum_{\alpha=1}^{3} A_{k\alpha} f_\alpha(z_\alpha), \\ \tilde{\Phi}_k(x_1, x_2) &= 2\,\mathrm{Re} \sum_{\alpha=1}^{3} L_{k\alpha} f_\alpha(z_\alpha), \end{aligned}\right\} \tag{10.111}$$

$k = 1, 2, 3,$ where

$$f_\alpha(z_\alpha) = \frac{1}{2\pi i} \sum_{m=0}^{\infty} \frac{b_{\alpha m}}{z_\alpha^m} \ln z_\alpha, \qquad \alpha = 1, 2, 3, \tag{10.112}$$

and $b_{\alpha m}$ are some undetermined complex constants. Introducing (10.111) into (10.101) and (10.107), and equating like powers of $x_1$, we find that, for each $m = 0, 1, 2, \ldots$, the three complex constants $b_{\alpha m}$, $\alpha = 1, 2, 3$, must satisfy the system of six real algebraic equations

$$\left.\begin{aligned} 2\,\mathrm{Re} \sum_{\alpha=1}^{3} A_{k\alpha} b_{\alpha m} &= g_{km}, \qquad k = 1, 2, 3, \\ 2\,\mathrm{Re} \sum_{\alpha=1}^{3} L_{k\alpha} b_{\alpha m} &= q_{km}, \qquad k = 1, 2, 3. \end{aligned}\right\} \tag{10.113}$$

Multiplying $(10.113)_1$ by $A_{k\beta}$, $(10.113)_2$ by $L_{k\beta}$, summing up for $k = 1, 2, 3$, adding the two relations thus obtained, and taking into account the orthogonality relations (10.15), we find[1]

$$b_{\alpha m} = \frac{A_{k\alpha} q_{km} + L_{k\alpha} g_{km}}{2 A_{s\alpha} L_{s\alpha}}, \tag{10.114}$$

for $\alpha = 1, 2, 3$; $m = 0, 1, 2, \ldots$, the summation being performed over $k$, $s = 1, 2, 3$.

---

[1] Clearly, the case of a straight dislocation of Burgers vector $b_k$ combined with a line force $P_k$, both of which lying along the $x_3$-axis, may be refound, under neglection of the boundary conditions on $\Gamma_0$, by putting $g_{ko} = -b_k$, $g_{km} = 0$ for $m \geqslant 1$; $h_{km} = 0$ for $m \geqslant 2$; $q_{ko} = P_k$, $q_{km} = 0$ for $m \geqslant 1$. Then, it results from (10.114) that $b_{\alpha m} = 0$ for $m = 1, 2, \ldots$, and $b_{\alpha o} = D_\alpha$, where $D_\alpha$ is given by (10.87).

We pass now to the boundary condition on $\Gamma_0$. Since $\hat{u}_k$ and $\hat{\Phi}_k$ must be continuous across the cut and vanish at infinity, we take

$$\left.\begin{aligned}\hat{u}_k(x_1, x_2) &= 2\,\mathrm{Re}\sum_{\alpha=1}^{3} A_{k\alpha}\varphi_\alpha(\zeta_\alpha),\\ \hat{\Phi}_k(x_1, x_2) &= 2\,\mathrm{Re}\sum_{\alpha=1}^{3} L_{k\alpha}\varphi_\alpha(\zeta_\alpha),\end{aligned}\right\} \tag{10.115}$$

$k = 1, 2, 3$, where

$$\varphi_\alpha(\zeta_\alpha) = \sum_{m=1}^{\infty} \frac{a_{\alpha m}}{\zeta_\alpha^m}, \quad \alpha = 1, 2, 3, \tag{10.116}$$

$$\zeta_\alpha = \frac{z_\alpha}{1 - ip_\alpha}\left[1 + \sqrt{1 - \frac{r_0^2(1 + p_\alpha^2)}{z_\alpha^2}}\right], \tag{10.117}$$

and $a_{\alpha m}$ are some undetermined complex constants.

Let us consider first the traction boundary condition (10.20) on $\Gamma_0$. Denoting by $\tilde{\mathbf{t}}^*(\theta)$ the traction corresponding to the displacement field $\tilde{\mathbf{u}}$ on $\Gamma_0$, and repeating the reasoning that has led to (10.24), we find

$$\tilde{t}_k^*(\theta) = \frac{2}{r_0}\,\mathrm{Re}\sum_{\alpha=1}^{3} L_{k\alpha}\frac{\mathrm{d}}{\mathrm{d}\theta} f_\alpha(r_0\cos\theta + p_\alpha r_0 \sin\theta), \qquad k = 1,2,3, \tag{10.118}$$

where $f_\alpha(z_\alpha)$, $\alpha = 1, 2, 3$, are given by (10.112). Next, from (10.20) it follows that the modified boundary condition on $\Gamma_0$ may be written as

$$\hat{T}_{k1} n_1 + \hat{T}_{k2} n_2 = \hat{t}_k^*(\theta) \quad \text{on } \Gamma_0, \quad k = 1, 2, 3, \tag{10.119}$$

where

$$\hat{t}_k^*(\theta) \equiv t_k^*(\theta) - \tilde{t}_k^*(\theta). \tag{10.120}$$

Since the form of this condition is the same as that considered in Sect. 10.2, we may derive at once the expression of the unknown coefficients $a_{\alpha m}$ by using (10.42). It then results that

$$a_{\alpha m} = \frac{i}{m} L_{\alpha k}^{-1} \bar{f}_{km} r_0^{m+1}, \qquad \alpha = 1, 2, 3; \quad m = 1, 2, \ldots, \tag{10.121}$$

where[1]

$$f_{km} = \frac{1}{2\pi}\int_{-\pi}^{\pi} \hat{t}_k^*(\theta)\, \mathrm{e}^{-im\theta}\, \mathrm{d}\theta, \quad k = 1, 2, 3; \quad m = 1,2, \ldots \tag{10.122}$$

[1] By (10.120), we have $\int_{-\pi}^{\pi} \hat{t}_k^*(\theta)\mathrm{d}\theta = 0$, and hence the Fourier coefficient $f_{k0}$ is zero.

are the Fourier coefficients of $\hat{t}_k^*(\theta)$. Thus, the solution of the problem formulated at the beginning of this subsection is completely determined by (10.109–112), (10.114–117), and (10.121). The constants $u_k^0$, corresponding to a rigid translation of the elastic medium, may be further determined by imposing the value of the displacement field at an arbitrary point of the medium.

Next, we consider the *displacement boundary condition* (10.27) on $\Gamma_0$. Denoting by

$$\tilde{u}_k^*(\theta) \equiv \tilde{u}_k(r_0 \cos\theta, r_0 \sin\theta) \qquad (10.123)$$

the value of $\tilde{u}_k(x_1, x_2)$ on $\Gamma_0$, we find that $\hat{u}_k(x_1, x_2)$ must satisfy the modified boundary condition

$$\hat{u}_k = \hat{u}_k^*(\theta) \quad \text{on } \Gamma_0, \quad k = 1, 2, 3, \qquad (10.124)$$

where

$$\hat{u}_k^*(\theta) = u_k^*(\theta) - \tilde{u}_k^*(\theta). \qquad (10.125)$$

Since the form of (10.124) coincides with (10.27), and that of the representation $(10.115)_1$, (10.116 with (10.43), (10.37), except that now $D_\alpha = 0$, we may directly infer the values of the unknown parameters $u_k^0$ and $a_{\alpha m}$ by using (10.51) and (10.53). We thus obtain

$$u_k^0 = d_{k0}, \qquad k = 1, 2, 3, \qquad (10.126)$$

$$a_{\alpha m} = r_0^m A_{\alpha k}^{-1} d_{km}, \qquad \alpha = 1, 2, 3; \qquad m = 1, 2, \ldots, \qquad (10.127)$$

where

$$d_{km} = \frac{1}{2\pi} \int_{-\pi}^{\pi} \hat{u}_k^*(\theta)\, e^{-im\theta}\, d\theta, \qquad k = 1, 2, 3; \quad m = 0, 1, 2, \ldots \qquad (10.128)$$

are the Fourier coefficients of $\hat{u}_k^*(\theta)$. The solution of the problem is again completely determined by (10.109–112), (10.114–117), but with $u_k^0$ and $a_{\alpha m}$ given by (10.126) and (10.127), respectively.

It is easily proved that the solutions of both boundary-value problems considered above are unique. Indeed, the displacement vector and the stress functions corresponding to the difference of any two solutions, must be continuous across the cut. Then, by Bézier's theorem (Sect. 6.2), the solution is unique when the displacements are given on $\Gamma_0$, and is uniquely determined to within an infinitesimal translation, when the tractions are prescribed on $\Gamma_0$.

Finally, if we are interested only in the solution corresponding to the Somigliana dislocation, we must simply put $h_{km}=0$ for any $k = 1, 2, 3$; $m = 2, 3, \ldots$ Then, from (10.108) it follows that

$$q_{k0} = P_k, \quad q_{km} = 0 \quad \text{for } m \geq 1, \qquad (10.129)$$

and (10.114) gives

$$\left.\begin{aligned} b_{\alpha 0} &= \frac{A_{k\alpha} P_k + L_{k\alpha} g_{k0}}{2 A_{s\alpha} L_{s\alpha}}, \\ b_{\alpha m} &= \frac{L_{k\alpha} g_{km}}{2 A_{s\alpha} L_{s\alpha}} \quad \text{for } m \geq 1. \end{aligned}\right\} \qquad (10.130)$$

## 10.7. Influence of the boundaries on the anisotropic elastic field of straight dislocations

By using Lekhnitsky's representation (10.6), (10.7), it should be possible to determine the stress field of a straight dislocation whose line coincides with the axis of an anisotropic circular elastic cylinder. However, there is only one solution of this type available so far, due to Eshelby [112], which concerns the screw dislocation lying in the axis of a circular cylinder of finite length, for a particular case of material symmetry. The solving of the problem for general anisotropy and mixed dislocations would require the solution of an infinite set of linear algebraic equations, having as unknowns the coefficients of the Laurent expansions of the functions $f_\alpha(z_\alpha)$, $\alpha = 1, 2, 3$, and expressing the boundary conditions on the core surface and the outer surface of the cylinder.

The elastic field of a straight dislocation lying in an infinite plate has been determined in the particular case where the dislocation line is parallel to the faces of the plate and to a two-fold axis of material symmetry. Thus, Spence [316] and Chou [70] have determined the elastic field of a screw and edge dislocation, respectively, for the case where the normal to the plate faces is also a two-fold axis of material symmetry, while Siems, Delavignette, and Amelincks [296] have considered an edge dislocation near the basal plane of a hexagonal crystal. Finally, Lothe [220] has derived an elegant formula giving the force exerted by the free boundary of an infinite elastic half-space on a dislocation of arbitrary inclination with respect to the boundary.

A problem of particular interest is the interaction of dislocations with phase and grain boundaries. This situation is generally modelled by a dislocation lying near the plane interface between two different anisotropic elastic half-spaces. The solution is given by either using Stroh's formalism or Fourier-transform techniques (Pastur, Fel'dman, A. M. Kosevich [264], Gemperlova and Saxl [131], Gemperlova [132], Tucker and Crocker [359], Tucker [360]), but the results are rather cumbersome and expressed in a form which is rather inconvenient for numerical applications. More explicit solutions have been obtained by Chou [408] and Pande and Chou [457] for the case where the adjacent grains possess rhombic symmetry with respect to the plane of the interface. Kurihara [205] has used complex-variable techniques to determine the elastic field of an edge dislocation in an anisotropic half-space coated by a thin layer of anisotropic material. Numerical calculations done for an edge dislocation whose glide plane is perpendicular to the interface have shown that the dislocation may have a stable equilibrium position near the interface for certain combinations of elastic constants.

Interfacial dislocations in anisotropic two-phase media have been given a special attention owing to their role in the mechanical behaviour of polycrystalline and composite materials. Chou and Pande [409] have calculated the elastic field of interfacial screw dislocations again for the case where both half-spaces possess rhombic symmetry with respect to the plane of the interface, while Chou, Pande, and Yang [410] have solved the same problem for edge dislocations.

A review of the problems involving interfacial dislocations for general anisotropy has been made by Nakahara and Willis [453], who corrected a previous approach by Brækhus and Lothe [38], and gave a general formulation by using Lekhnitsky's representation of the elastic field for both anisotropic half-spaces, as follows [1]. Let us consider two anisotropic half-spaces welded along the interface $x_2 = 0$, and assume that an infinite straight dislocation of Burgers vector **b** lies along the $x_3$-axis at the interface. The quantities pertaining to the half-spaces $x_2 > 0$ and $x_2 < 0$ will be denoted by the superscripts (1) and (2), respectively. Ignoring the core boundary conditions, the elastic field is given (cf. Sect. 10.3) by

$$\left.\begin{aligned} u_k^{(s)}(x_1, x_2) &= \frac{1}{\pi} \operatorname{Im} \sum_{\alpha=1}^{3} A_{k\alpha}^{(s)} D_\alpha^{(s)} \ln z_\alpha^{(s)} + u_k^{0(s)}, \\ T_{1k}^{(s)}(x_1, x_2) &= -\frac{1}{\pi} \operatorname{Im} \sum_{\alpha=1}^{3} \frac{p_\alpha^{(s)} L_{k\alpha}^{(s)} D_\alpha^{(s)}}{z_\alpha^{(s)}}, \\ T_{2k}^{(s)}(x_1, x_2) &= \frac{1}{\pi} \operatorname{Im} \sum_{\alpha=1}^{3} \frac{L_{k\alpha}^{(s)} D_\alpha^{(s)}}{z_\alpha^{(s)}}, \\ \Phi_k^{(s)}(x_1, x_2) &= \frac{1}{\pi} \operatorname{Im} \sum_{\alpha=1}^{3} L_{k\alpha}^{(s)} D_\alpha^{(s)} \ln z_\alpha^{(s)}, \quad z_\alpha^{(s)} = x_1 + p_\alpha^{(s)} x_2, \end{aligned}\right\} \tag{10.131}$$

where $s$ takes the values 1 and 2. We choose as before as cut the negative $x_1$-axis to define single-valued functions $\ln z_\alpha^{(s)}$ and a single-valued displacement field. Then the physical requirements to be fulfilled are: the continuity of the displacement vector across the positive $x_1$-axis,

$$u_k^{(1)}(x_1, 0^+) - u_k^{(2)}(x_1, 0^-) = 0 \quad \text{for } x_1 > 0, \tag{10.132}$$

the prescribed jump of the displacement vector across the negative $x_1$-axis,

$$u_k^{(1)}(x_1, 0^+) - u_k^{(2)}(x_1, 0^-) = -b_k \quad \text{for } x_1 < 0, \tag{10.133}$$

the continuity of the stress components $T_{2k}$ across the boundary

$$T_{2k}^{(1)}(x_1, 0^+) - T_{2k}^{(2)}(x_1, 0^-) = 0 \quad \text{for } x_1 \neq 0, \tag{10.134}$$

and finally the condition of zero resultant force of the tractions acting on any cylindrical surface $\Sigma$, surrounding the dislocation line, which leads, in view of (10.104), to

$$\Phi_k^{(1)}(x_1, 0^+) - \Phi_k^{(2)}(x_1, 0^-) = 0 \quad \text{for } x_1 \neq 0. \tag{10.135}$$

[1] Cf. also Barnett and Lothe [398], Dupeux and Bonnet [413].

Next, introducing (10.131) into (10.132—35) gives a set of twelve real equations to determine the six complex unknowns $D_\alpha^{(1)}, D_\alpha^{(2)}$, $\alpha = 1, 2, 3$:

$$\left.\begin{aligned}
&\operatorname{Im} \sum_{\alpha=1}^{3} (A_{k\alpha}^{(1)} D_\alpha^{(1)} - A_{k\alpha}^{(2)} D_\alpha^{(2)}) = 0,\\
&\operatorname{Re} \sum_{\alpha=1}^{3} (A_{k\alpha}^{(1)} D_\alpha^{(1)} + A_{k\alpha}^{(2)} D_\alpha^{(2)}) = -b_k,\\
&\operatorname{Im} \sum_{\alpha=1}^{3} (L_{k\alpha}^{(1)} D_\alpha^{(1)} - L_{k\alpha}^{(2)} D_\alpha^{(2)}) = 0,\\
&\operatorname{Re} \sum_{\alpha=1}^{3} (L_{k\alpha}^{(1)} D_\alpha^{(1)} + L_{k\alpha}^{(2)} D_\alpha^{(2)}) = 0,
\end{aligned}\right\} \tag{10.136}$$

where $k = 1, 2, 3$, as well as the supplementary restrictions $u_k^{0(1)} = u_k^{0(2)}$, $k = 1, 2, 3$, i.e. the equality of the rigid translations adopted for the two half-spaces. As shown by Dupeux and Bonnet [413], equations $(10.136)_4$ imply also the vanishing of the resultant couple of the tractions acting on $\Sigma$, just as for an infinite straight dislocation in a homogeneous medium. The energy factor $K$ of the dislocation can be calculated by (10.60), where $L_{k\alpha}$ and $A_{k\alpha}$ may be replaced by the values corresponding to either of the half-spaces.

There exists so far no explicit solution of system (10.136) similar to that obtained by Stroh for the homogeneous medium. Nevertheless, Dupeux and Bonnet [413] elaborated a computer program for building up this system from geometrical and physical data, to solve it, and to calculate the energy factor of the dislocation.

## 11. Dislocation loops in anisotropic media

### 11.1. The method of Lothe, Brown, Indenbom, and Orlov

We have seen in Sect. 9.1 that a dislocation loop of line $L$ and true Burgers vector $\mathbf{b}$ produces the displacement field

$$u_p(\mathbf{x}) = -\int_{S+} b_i c_{ijkl} G_{kp,l}(\mathbf{x} - \mathbf{x}')\, \mathrm{d}s_j', \tag{11.1}$$

whose gradient may be expressed by one of the equivalent formulae

$$H_{pr}(\mathbf{x}) = -\int_{S}^{+} b_i c_{ijkl} G_{kp,lr}(\mathbf{x}-\mathbf{x}')\,\mathrm{d}s'_j = \oint_L \epsilon_{njr} b_i c_{ijkl} G_{kp,l}(\mathbf{x}-\mathbf{x}')\,\mathrm{d}x'_n, \tag{11.2}$$

where $S$ is a smooth and two-sided surface bounded by $L$, $\mathbf{n}$ is the unit normal to $S$ in the sense given by the right-hand rule with respect to the positive sense chosen on $L$, and $S^+$ denotes the face of $S$ into which points $\mathbf{n}$. Furthermore, the stress field of the dislocation loop is given by

$$T_{km}(\mathbf{x}) = c_{kmpr} H_{pr}(\mathbf{x}). \tag{11.3}$$

Equations (11.1—3) are valid for general anisotropy, but the integrals involved cannot be calculated analytically except for infinite straight dislocations. However, starting from results obtained by Lothe [219] on dislocation bends in anisotropic media, Brown [43] succeeded to show how the in-plane stress field of a planar dislocation loop could be determined directly from the stress field of a straight dislocation and the derivatives of this field with respect to variables describing the direction of the straight dislocation. In view of the importance of this result we reproduce below Brown's proof, by slightly modifying his argument.

**Theorem.** *Let $L$ be a planar, piecewise smooth dislocation loop lying in an infinite anisotropic elastic medium, $M$ a current point on $L$, and $P$ an arbitrary point situated in the plane of the loop, $P \notin L$. Arbitrarily choose a positive sense on $L$ and denote by $\alpha$ and $\theta$ the angles measured in the same sense from a fixed reference direction in the plane of the loop to the tangent at $M$ to $L$ and to the vector $\overrightarrow{MP}$, respectively. Then the stress field of the dislocation loop at $P$ is given by*

$$\mathbf{T}(P) = \frac{1}{2}\oint_L \frac{1}{R^2}\left[\boldsymbol{\sigma}(\theta) + \frac{\mathrm{d}^2\boldsymbol{\sigma}(\theta)}{\mathrm{d}\theta^2}\right]\sin(\theta-\alpha)\,\mathrm{d}l, \tag{11.4}$$

*where $R = \|\overrightarrow{MP}\|$, and $\boldsymbol{\sigma}(\theta)$ is the stress field of an infinite, straight dislocation line $L_\theta$ with the same Burgers vector as $L$ and directed parallel to $\overrightarrow{MP}$, evaluated a unit distance away from $L_\theta$ in the direction $\mathbf{n} \times \overrightarrow{MP}$, where $\mathbf{n}$ is the unit normal to the plane in the sense given by the right-hand rule around $L$.*

*Proof.* Let us choose a Cartesian frame $Ox_1x_2x_3$ with the $x_3$-axis directed along $\mathbf{n}$ and the $x_1$-axis along the fixed reference direction (Fig. 11.1a). Denote by $(x_1, x_2, 0)$ and $(x'_1, x'_2, 0)$ the co-ordinates of $P$ and $M$, respectively. It then follows from $(11.2)_1$ and (11.3) that the stress components of the dislocation loop at $P$ are given by

$$T_{km}(P) = -\int_S c_{kmpr} b_i c_{i3hl} G_{hp,lr}(\mathbf{x} - \mathbf{x}')|_{x_3 = x_3' = 0} \mathrm{d}x_1' \mathrm{d}x_2', \tag{11.5}$$

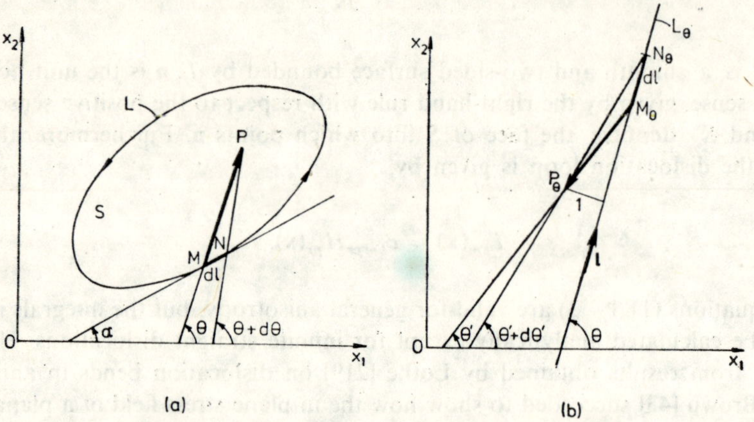


Fig. 11.1. On Brown's theorem. (a) Dislocation loop lying in the $x_1x_2$-plane. (b) An infinite straight dislocation making an angle $\theta$ with the $x_1$-axis.

where $S$ is the region bounded by $L$. Since none of the subscripts in (11.5) are relevant to the argument that follows we may simply rewrite this relation as

$$T(x_1, x_2) = \int_S f(x_1 - x_1', x_2 - x_2') \mathrm{d}x_1' \, \mathrm{d}x_2', \tag{11.6}$$

where $T(x_1, x_2)$ denotes any component of the stress tensor $\mathbf{T}(P)$. On the other hand, taking into account that Green's tensor function $\mathbf{G}$ satisfies the identity (cf. Sect. 6.4)

$$\mathbf{G}(\lambda \mathbf{x}) = |\lambda|^{-1} \mathbf{G}(\mathbf{x}) \tag{11.7}$$

for any $\mathbf{x} \neq \mathbf{0}$ and for any non-zero real number $\lambda$, it is easily seen from (11.5) that the function $f$ satisfies the identity

$$f(\lambda X_1, \lambda X_2) = |\lambda|^{-3} f(X_1, X_2), \tag{11.8}$$

for any $R = \sqrt{X_1^2 + X_2^2} \neq 0$, where $X_1 = x_1 - x_1'$, $X_2 = x_2 - x_2'$.

Putting $\lambda = R^{-1}$ into (11.8) and taking into consideration that $X_1 = R \cos \theta$, $X_2 = R \sin\theta$ (Fig. 11.1a), it results

$$f(X_1, X_2) = R^{-3} \Theta(\theta), \tag{11.9}$$

where $\Theta(\theta) = f(\cos\theta, \sin\theta)$. For further use we also note that, by virtue of (11.8),

$$\Theta(\theta + \pi) = f(-\cos\theta, -\sin\theta) = f(\cos\theta, \sin\theta) = \Theta(\theta). \tag{11.10}$$

Next, by making use of Euler's theorem on homogeneous functions, we deduce from (11.8) that

$$-3f(X_1, X_2) = X_1 \frac{\partial f(X_1, X_2)}{\partial X_1} + X_2 \frac{\partial f(X_1, X_2)}{\partial X_2},$$

and hence

$$-f(X_1, X_2) = \frac{\partial}{\partial X_1}[X_1 f(X_1, X_2)] + \frac{\partial}{\partial X_2}[X_2 f(X_1, X_2)],$$

wherefrom it follows that

$$f(x_1 - x_1', x_2 - x_2') = \frac{\partial}{\partial x_1'}[(x_1 - x_1')f(x_1 - x_1', x_2 - x_2')] +$$

$$+ \frac{\partial}{\partial x_2'}[(x_2 - x_2')f(x_1 - x_1', x_2 - x_2')].$$

Introducing this result into (11.6) and transforming the surface integral into a line integral by Green's theorem, we have

$$T(x_1, x_2) = \oint_L \{-(x_2 - x_2')f\mathrm{d}x_1' + (x_1 - x_1')f\mathrm{d}x_2'\}. \tag{11.11}$$

On the other hand, inspection of Fig. 11.1a shows that

$$\mathrm{d}x_1' = \mathrm{d}l\cos\alpha, \quad \mathrm{d}x_2' = \mathrm{d}l\sin\alpha, \quad x_1 - x_1' = R\cos\theta, \quad x_2 - x_2' = R\sin\theta,$$

and hence (11.11) becomes, considering also (11.9),

$$T(x_1, x_2) = -\oint_L \frac{\Theta(\theta)\sin(\theta - \alpha)}{R^2}\,\mathrm{d}l. \tag{11.12}$$

In order to determine $\Theta(\theta)$, we consider an infinite straight dislocation line $L_\theta$ directed parallel to $\overrightarrow{MP}$ and having the same Burgers vector as $L$ (Fig. 1.11b). Let us denote by $\boldsymbol{\sigma}(\theta)$ the stress field of $L_\theta$ evaluated at a point $P_\theta$ situated a unit distance away from $L_\theta$ in the direction of $\mathbf{n} \times \overrightarrow{MP}$, and by $\sigma(\theta)$ the same component of $\boldsymbol{\sigma}(\theta)$ as $T(x_1, x_2)$ for $\mathbf{T}(P)$. Then, we may calculate $\sigma(\theta)$ by using the same formula (11.12), provided that $\alpha$ is replaced by $\theta$, and $\theta$ by the angle $\theta'$ between $Ox_1$ and the vector $\overrightarrow{M_\theta P_\theta}$, where $M_\theta$ is a current point on $L_\theta$. Thus

$$\sigma(\theta) = -\int_{L_\theta} \frac{\Theta(\theta') \sin(\theta' - \theta)\, dl'}{\|\overrightarrow{M_\theta P_\theta}\|^2} \tag{11.13}$$

But now

$$\|\overrightarrow{M_\theta P_\theta}\| = \operatorname{cosec}(\theta' - \theta), \qquad dl' = \operatorname{cosec}^2(\theta' - \theta) d\theta',$$

and hence (11.13) reduces to

$$\sigma(\theta) = -\int_\theta^{\theta+\pi} \Theta(\theta') \sin(\theta' - \theta) d\theta'. \tag{11.14}$$

Differentiating this equation with respect to $\theta$ yields

$$\frac{d\sigma(\theta)}{d\theta} = \int_\theta^{\theta+\pi} \Theta(\theta') \cos(\theta' - \theta) d\theta',$$

and a second differentiation gives

$$\frac{d^2\sigma(\theta)}{d\theta^2} = \int_\theta^{\theta+\pi} \Theta(\theta') \sin(\theta' - \theta) d\theta' - \Theta(\theta + \pi) - \Theta(\theta),$$

wherefrom, in view of (11.14) and (11.10), it follows that

$$\Theta(\theta) = -\frac{1}{2}\left[\sigma(\theta) + \frac{d^2\sigma(\theta)}{d\theta^2}\right]. \tag{11.15}$$

Finally, by substituting this result into (11.12), we find

$$T(x_1, x_2) = \frac{1}{2}\oint_L \frac{1}{R^2}\left[\sigma(\theta) + \frac{d^2\sigma(\theta)}{d\theta^2}\right] \sin(\theta - \alpha)\, dl,$$

and the theorem is proved, since $T(x_1, x_2)$ and $\sigma(\theta)$ are corresponding, but otherwise arbitrary components of $\mathbf{T}(P)$ and $\boldsymbol{\sigma}(\theta)$. From the above proof and $(11.2)_2$ it is apparent that similar relations hold for the strain field and the displacement gradient of a planar dislocation loop.

Since the tensor $\boldsymbol{\sigma}(\theta)$ is known in the general anisotropic case and for an arbitrary orientation of the dislocation line, formulae of the type (11.4) allow the computation of the elastic field of planar and, as we shall see below, even of non-planar dislocation loops. This has stimulated the occurrence of a series of papers devoted to the numerical calculation of $\boldsymbol{\sigma}(\theta)$ and of its angular derivatives by either starting from the analysis given in Sect. 10.3 (Malén [224], Malén and Lothe [225]) or using an integral formalism that avoids the solution of the sextic equation (10.11) and is more adequate for the purposes of numerical computation (Asaro, Hirth, Barnett, and Lothe [9]).

When considering complicated dislocation configurations, it is frequently more convenient to decompose them into finite dislocation segments, the stress field of each segment being calculated with the aid of (11.4), and then to sum up the individual contributions of all segments. With the notation in Fig. 11.2 we have for the dislocation segment $AB$:

$$R = \|\mathbf{x} - \mathbf{x}'\| = p \operatorname{cosec}(\theta - \alpha), \qquad dl = p \operatorname{cosec}^2(\theta - \alpha)\, d\theta,$$

where $p$ denotes the distance from the field point $P$ to the segment $AB$. Substituting these expressions into (11.4) and integrating by parts twice the term containing the second derivative of $\boldsymbol{\sigma}(\theta)$, we obtain (Brown [43], Asaro and Hirth [8]):

$$\mathbf{T}(P) = \frac{1}{2p}\left[-\boldsymbol{\sigma}(\theta)\cos(\theta - \alpha) + \frac{d\boldsymbol{\sigma}(\theta)}{d\theta}\sin(\theta - \alpha)\right]\Bigg|_{\theta_A}^{\theta_B}. \tag{11.16}$$

As the $x_1x_2$-plane can be arbitrarily rotated around $AB$ it is obvious that (11.16) completely determines the stress field of the dislocation segment. In view of the wide applicability of this formula, Asaro and Barnett [10] have elaborated a nu-

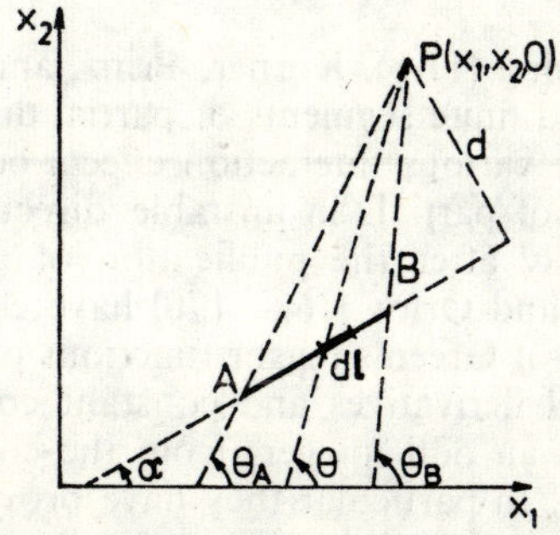


Fig. 11.2. The geometry of a straight dislocation segment $AB$.

merical method which permits the direct calculation of the functions $\boldsymbol{\sigma}(\theta)$ and $d\boldsymbol{\sigma}(\theta)/d\theta$, without solving the sextic equation (10.11). The attractive feature of formulae (11.4) and (11.16) is that it is possible to calculate the values of $\boldsymbol{\sigma}(\theta)$ and of its angular derivatives over a sufficiently large range of orientations and to store the results in some convenient form for later use (see also Sect. 12.3).

Clearly, the elastic field of any planar or non-planar dislocation loop may be approximated by that of a corresponding polygonal loop, i.e. of a union of finite dislocation segments, but is not obvious *a priori* that going from a segment in one plane to another in a different plane does not introduce some "termination" errors, especially when angular derivatives do intervene in the calculations. However, Bacon, Barnett, and Scattergood [396] succeeded to demonstrate that (11.16) actually gives the three-dimensional contribution of a dislocation segment, by using the following argument (see Fig. 11.3).

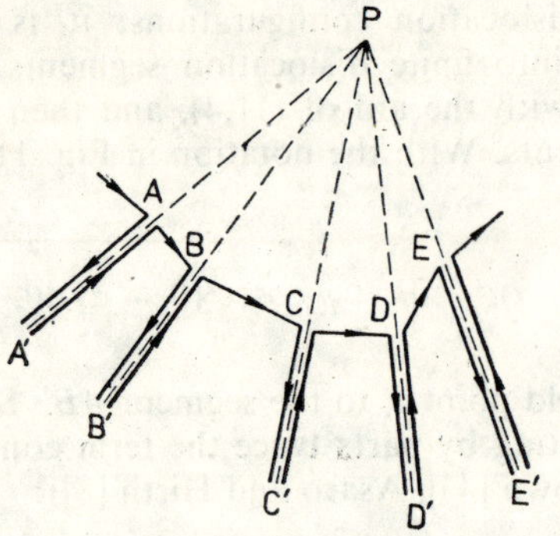


Fig. 11.3. Schematic construction of a polygonal dislocation from bi-angular dislocations.

An arbitrary non-planar polygonal dislocation *ABCDE* ... is replaced by a union of by-angular dislocations *A'ABB'*, *B'BCC'*, ..., such that all (infinite) rays *AA'*, *BB'*, ... pass through the current field point *P*. Clearly, the stress field of each by-angular dislocation can be computed by formulae (11.4) and (11.16) without termination errors. On the other hand, when summing these stress fields, the rays *AA'* and *A'A*, *BB'* and *B'B*, etc. bring no net contribution, since they have equal Burgers vectors and opposite directions. Consequently, the total field is given simply by (11.16) successively applying (11.16) to the dislocation segments *AB*, *BC*, *CD*, ...

By using (11.6), Korner, Prinz, and Kirchner [439] have constructed isostress lines around finite segments of partial dislocations in silver crystals, and used them to visualize various interaction effects between segments of partials, as well as the zigzagging of partials in unstable directions.

Shortly after the publication of the results of Lothe and Brown in 1967, Indenbom and Orlov [168—170] have elaborated a very ingenious procedure for the calculation of Green's tensor functions of linear homogeneous differential operators with partial derivatives and constant coefficients for $n$-dimensional infinite spaces (where $n$ is an odd integer) from the corresponding Green's tensor functions of a hyperplane. In particular they have proved the following generalization of Brown's theorem.

**Theorem** (Indenbom and Orlov [168]). *The stress field* $\mathbf{T}(\mathbf{x})$ *of a piecewise smooth dislocation loop L is given by*

$$\mathbf{T}(\mathbf{x}) = \frac{1}{2} \oint_L \tau_k \tau_m \frac{\partial^2 \boldsymbol{\sigma}(\boldsymbol{\tau}, \mathbf{x} - \mathbf{x}')}{\partial x_k \partial x_m} \mathrm{d}l' \tag{11.17}$$

*where* $\boldsymbol{\tau}$ *is the unit vector tangent at* $\mathbf{x}'$ *to L, and* $\boldsymbol{\sigma}(\mathbf{x}, \boldsymbol{\tau})$ *denotes the stress field at* $\mathbf{x}$ *of an infinite, straight dislocation with the same Burgers vector as L, passing through the origin, and parallel to* $\boldsymbol{\tau}$.

*Proof.* First, note that $(11.2)_2$ may be rewritten as

$$H_{pr}(\mathbf{x}) = \oint_L \epsilon_{njr}\, b_i\, T_{ij}^{(p)}(\mathbf{x}-\mathbf{x}')\, \tau_n\, \mathrm{d}l' \tag{11.18}$$

where $\boldsymbol{\tau}$ is the unit vector tangent at $\mathbf{x}'$ to $L$ and

$$T_{ij}^{(p)}(\mathbf{x}) = c_{ijkl} G_{kp,l}(\mathbf{x}), \quad T_{ij,j}^{(p)}(\mathbf{x}) = 0 \text{ for } \mathbf{x} \neq \mathbf{0}. \tag{11.19}$$

Moreover, it may be shown that (11.18) is equivalent to

$$H_{pr}(\mathbf{x}) = -\oint_L \epsilon_{njq}(x_q - x_q')\, b_i\, T_{ij,r}^{(p)}(\mathbf{x}-\mathbf{x}')\, \tau_n\, \mathrm{d}l'. \tag{11.20}$$

Indeed, by making use of Stokes' formula (1.54),

$$\oint_L u_n \mathrm{d}x_n' = \int_{S^+} \epsilon_{lmn} u_{n,m'}\, \mathrm{d}s_l',$$

and considering (1.11) and $(11.19)_2$, the difference between the right-hand sides of (11.20) and (11.18) may be transformed into the surface integral

$$\int_{S^+} b_i\, [3T_{il,r}^{(p)}(\mathbf{x}-\mathbf{x}') + (\mathbf{x}_m - \mathbf{x}_m')\, T_{il,rm}^{(p)}(\mathbf{x}-\mathbf{x}')]\, \mathrm{d}s_l'.$$

On the other hand, $T_{il,r}^{(p)}(\mathbf{x}-\mathbf{x}')$ is a homogeneous function of order $-3$ in the components of $\mathbf{x}-\mathbf{x}'$, i.e.

$$T_{il,r}^{(p)}(\lambda\mathbf{x}) = |\lambda|^{-3} T_{il,r}^{(p)}(\mathbf{x}) \quad \text{for any } \lambda \neq 0, \ \mathbf{x} \neq \mathbf{0}, \tag{11.21}$$

and hence, by Euler's theorem, the integrand of the surface integral vanishes.

Let us apply now the relation (11.20) to an infinite straight dislocation passing through the origin, parallel to $\boldsymbol{\tau}$, and having the same Burgers vector as $L$. Denoting the corresponding displacement gradient by $\boldsymbol{\beta}(\mathbf{x}, \boldsymbol{\tau})$ and putting $\mathbf{x}' = l'\boldsymbol{\tau}$, we obtain

$$\beta_{pr}(\mathbf{x}, \boldsymbol{\tau}) = -\epsilon_{njq} x_q b_i \tau_n \int_{-\infty}^{\infty} T_{ij,r}^{(p)}(\mathbf{x} - \boldsymbol{\tau} l')\, \mathrm{d}l'. \tag{11.22}$$

It can be readily proved, by considering (11.21), that $\boldsymbol{\beta}(\mathbf{x}, \boldsymbol{\tau})$ is a homogeneous function of order $-1$ in the components of $\mathbf{x}$ and of order 0 in the components of $\boldsymbol{\tau}$, i.e.

$$\boldsymbol{\beta}(\lambda\mathbf{x}, \mu\boldsymbol{\tau}) = |\lambda|^{-1} \boldsymbol{\beta}(\mathbf{x}, \boldsymbol{\tau}) \quad \text{for any } \lambda, \mu \neq 0, \quad \mathbf{x}, \boldsymbol{\tau} \neq \mathbf{0}. \tag{11.23}$$

Next, decomposing the integration interval in (11.22) into the union $(-\infty, 0) \cup$ $\cup\, (0, \infty)$, operating in each integral the change of integration variable $l' = 1/s$, and taking into account (11.21), we find

$$\beta_{pr}(\mathbf{x}, \boldsymbol{\tau}) = -\epsilon_{njq}\, x_q\, b_i \tau_n \int_{-\infty}^{\infty} |s|\, T^{(p)}_{ij,r}(\boldsymbol{\tau} - \mathbf{x}s)\, \mathrm{d}s. \tag{11.24}$$

Now let us apply on both sides of this equation the operator $x_k x_m \partial^2/\partial\tau_k \partial\tau_m$. Since

$$x_k \frac{\partial}{\partial\tau_k}(\epsilon_{njq} x_q \tau_n) = \epsilon_{njq} x_n x_q = 0, \qquad x_k \frac{\partial}{\partial\tau_k} T^{(p)}_{ij,r}(\boldsymbol{\tau} - \mathbf{x}s) = -\frac{\mathrm{d}}{\mathrm{d}s} T^{(p)}_{ij,r}(\boldsymbol{\tau} - \mathbf{x}s),$$

we obtain

$$x_k x_m \frac{\partial^2 \beta_{pr}(\mathbf{x}, \boldsymbol{\tau})}{\partial\tau_k \partial\tau_m} = -\epsilon_{njq} x_q b_i \tau_n \int_{-\infty}^{\infty} |s| \frac{\mathrm{d}^2 T^{(p)}_{ij,r}(\boldsymbol{\tau} - \mathbf{x}s)}{\mathrm{d}s^2}\, \mathrm{d}s.$$

Dividing again the integration interval into $(-\infty, 0) \cup (0, \infty)$, integrating by parts twice each integral, and taking into consideration that

$$\lim_{s \to \pm\infty} s \frac{\mathrm{d}T^{(p)}_{ij,r}(\boldsymbol{\tau} - \mathbf{x}s)}{\mathrm{d}s} = 0, \qquad \lim_{s \to \pm\infty} T^{(p)}_{ij,r}(\boldsymbol{\tau} - \mathbf{x}s) = 0,$$

we get

$$x_k x_m \frac{\partial^2 \beta_{pr}(\mathbf{x}, \boldsymbol{\tau})}{\partial\tau_k \partial\tau_m} = -2\epsilon_{njq} x_q b_i \tau_n T^{(p)}_{ij,r}(\boldsymbol{\tau}).$$

Next, we replace $\mathbf{x}$ by $\boldsymbol{\tau}$, and $\boldsymbol{\tau}$ by $\mathbf{x} - \mathbf{x}' \neq \mathbf{0}$, the latter substitution being possible on account of (11.23). Then

$$\epsilon_{njq}(x_q - x'_q)\, b_i \tau_n T^{(p)}_{ij,r}(\mathbf{x} - \mathbf{x}') = \frac{1}{2} \tau_k \tau_m \frac{\partial^2 \beta_{pr}(\boldsymbol{\tau}, \mathbf{x} - \mathbf{x}')}{\partial x_k\, \partial x_m}.$$

Finally, introducing this result into (11.22) yields

$$\mathbf{H}(\mathbf{x}) = -\frac{1}{2} \oint_L \tau_k \tau_m \frac{\partial^2 \boldsymbol{\beta}(\boldsymbol{\tau}, \mathbf{x} - \mathbf{x}')}{\partial x_k \partial x_m}, \tag{11.25}$$

and the theorem is proved, since (11.25) and (11.3) imply (11.17).

Indenbom and Orlov [168] have also proved that (11.17) generalizes Brown's result (11.4), and have applied (11.25) and (11.23) to show that the displacement gradient of a dislocation ray with initial point at the origin and directed parallel to the unit vector $\boldsymbol{\tau}$ is

$$\mathbf{H}(\mathbf{x}) = \frac{1}{2}\left[\boldsymbol{\beta}(\mathbf{x}, \boldsymbol{\tau}) - \tau_n \frac{\partial \boldsymbol{\beta}(\boldsymbol{\tau}, \mathbf{x})}{\partial x_n}\right]. \tag{11.26}$$

Formulae of the type (11.26) can be also used to obtain the field of polygonal dislocations and to study dislocation interactions (cf. also Indenbom and Dubnova [432], Orlov and Indenbom [456]).

Asaro, Hirth, Barnett, and Lothe [9] have derived formulae which allow the numerical computation of the directional derivatives involved in (11.25). However, when handling complex dislocation configurations, it seems preferable to use the two-dimensional formalism based on Brown's theorem, as explained above.

In the next section, we shall come back to the application of Lothe-Brown-Indenbom-Orlov geometrical techniques in connection with the calculation of self-energies and interaction energies of dislocations.

## 11.2. Willis' method

Willis [383] has obtained a direct evaluation of the integrals occurring in (11.1) and $(11.2)_2$ by using the expression of Green's tensor function obtained by Fourier transformation of the equilibrium equations (Sect. 6.4). This method leads in some cases to results which are more explicit than those given by the method of Lothe, Brown, Indenbom, and Orlov. The relationship between the two methods has been investigated by Malén [226] and by Asaro, Hirth, Barnett, and Lothe [9].

With the notation in Sect. 6.4, we deduce from (6.55) that

$$G_{hp}(\mathbf{x} - \mathbf{x}') = \frac{1}{8\pi^3} \operatorname{Re} \int_{\tilde{\mathscr{E}}} \frac{D^*_{hp}(\mathbf{k})}{D(\mathbf{k})} \mathrm{e}^{-i\mathbf{k}\cdot(\mathbf{x}-\mathbf{x}')} \, \mathrm{d}\tilde{v}.$$

Differentiating this relation with respect to $x_l$ and taking into account that $\operatorname{Re}(-iz) = \operatorname{Im} z$, we have

$$G_{hp,l}(\mathbf{x} - \mathbf{x}') = \frac{1}{8\pi^3} \operatorname{Im} \int_{\tilde{\mathscr{E}}} \frac{k_l D^*_{hp}(\mathbf{k})}{D(\mathbf{k})} \mathrm{e}^{-i\mathbf{k}\cdot(\mathbf{x}-\mathbf{x}')} \, \mathrm{d}\tilde{v}$$

and, by substituting this result into (11.1) and $(11.2)_2$, we infer that

$$u_p(\mathbf{x}) = -\frac{1}{8\pi^3} b_t c_{tjhl} \operatorname{Im} \int_{S^+} \mathrm{d}s'_j \int_{\tilde{\mathscr{E}}} \frac{k_l D^*_{hp}(\mathbf{k})}{D(\mathbf{k})} \mathrm{e}^{-i\mathbf{k}\cdot(\mathbf{x}-\mathbf{x}')} \, \mathrm{d}\tilde{v}. \tag{11.27}$$

$$H_{pr}(\mathbf{x}) = \frac{1}{8\pi^3} \epsilon_{qjr} b_t c_{tjhl} \operatorname{Im} \oint_L \mathrm{d}x'_q \int_{\tilde{\mathscr{E}}} \frac{k_l D^*_{hp}(\mathbf{k})}{D(\mathbf{k})} \mathrm{e}^{-i\mathbf{k}\cdot(\mathbf{x}-\mathbf{x}')} \, \mathrm{d}\tilde{v}. \tag{11.28}$$

Willis' procedure consists in evaluating the integrals involved in (11.27) and (11.28) in the real space $\mathscr{E}$ as well as in the phase space $\tilde{\mathscr{E}}$, for various dislocation configurations. We shall briefly review below the results available in the literature for straight dislocations and dislocation loops.

*Infinite straight dislocation.* Consider first an infinite straight dislocation with Burgers vector **b**, passing through the origin, and directed along the unit vector **l**, arbitrarily oriented with respect to the standard crystallographic axes. Let **m** and **n** be two arbitrary orthogonal unit vectors, such that $\mathbf{l} = \mathbf{m} \times \mathbf{n}$ (Fig. 11.4).

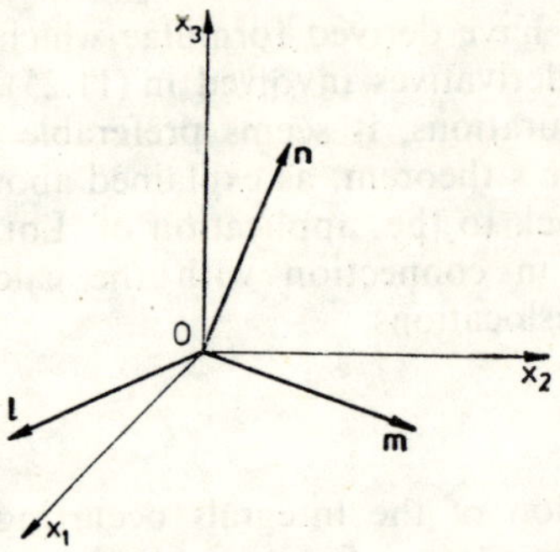


Fig. 11.4. Orthogonal frame associated with a straight dislocation line with unit vector **l**.

Neglecting the influence of the dislocation core, one may choose as cut surface $S$ the infinite strip $\mathbf{x}\cdot\mathbf{n} = 0, -R \leqslant \mathbf{x}\cdot\mathbf{m} \leqslant 0$, which is bounded by the given dislocation line and by a parallel dislocation line distant $R$ from it; $R$ is assumed to be large enough such that the field contribution at **x** be negligible compared to the contribution of the dislocation passing through the origin. Then, by evaluating the integrals in (11.27) and (11.28), Willis [383] obtains

$$u_p(\mathbf{x}) = -\operatorname{Im} \sum_{\lambda=1}^{3} n_j F_{jp}(\mathbf{m}^\lambda) \ln(\mathbf{x}\cdot\mathbf{m}_\lambda / R), \tag{11.29}$$

$$H_{pr}(\mathbf{x}) = \epsilon_{jrs} l_s \operatorname{Im} \sum_{\lambda=1}^{3} F_{jp}(\mathbf{m}^\lambda)/(\mathbf{x}\cdot\mathbf{m}_\lambda). \tag{11.30}$$

Here and in the following we denote for any unit vector **k**:

$$F_{jp}(\mathbf{k}) = \frac{1}{\pi} b_t c_{tjhl} \frac{k_l D^*_{ph}(\mathbf{k})}{n_q \dfrac{\partial D(\mathbf{k})}{\partial k_q}}, \qquad \mathbf{k}^\lambda = \mathbf{k} + \mathbf{n}\omega^\lambda, \tag{11.31}$$

where $\omega^\lambda$, $\lambda = 1, 2, 3$, are the three roots with positive imaginary parts of the sextic equation

$$D(\mathbf{k} + \mathbf{n}\,\omega) = 0. \tag{11.32}$$

Since the orientation of **l** with respect to the co-ordinate axes is arbitrary, it is no more necessary to transform the elastic constants, as in Sect. 10, from the standard crystallographic axes to those associated with the dislocation line. It is easily proved that Willis' solution (11.29) does not depend on the choice of the unit vectors **m** and **n**, as was to be expected. It may be also shown that (11.30) is equivalent to the equation

$$H_{pr}(\mathbf{x}) = -\operatorname{Im}\sum_{\lambda=1}^{3} n_j m_r^{\lambda} F_{jp}(\mathbf{m}^{\lambda})/(\mathbf{x}\cdot\mathbf{m}^{\lambda}),$$

which results directly from (11.29) by partial differentiation with respect to $x_r$.

Clearly, Willis' method requires the solving of a sextic equation of the same type as (10.9). In addition, the solution (11.29) is applicable in this form only when the roots $\omega^{\lambda}$ are simple; the degenerate cases of multiple roots must be considered separately, by using a suitable limiting process [1].

An alternative approach based on the technique elaborated by Barnett [16] for the numerical computation of the derivatives of Green's tensor function (see Sect. 6.4) has been applied to dislocation problems by Barnett and Swanger [15], Asaro and Hirth [8], and Asaro and Barnett [10]. It has the advantage of avoiding the solution of a sextic equation as well as of degenerate cases. Moreover, high accuracy may be obtained after reasonable computation times by using standard Romberg integration schemes (cf. also Meissner [244]).

*Finite straight dislocation segment.* For a finite straight dislocation segment of direction **l**, connecting the points $\boldsymbol{\alpha}$ and $\boldsymbol{\beta}$ and having the Burgers vector **b**, Willis [383] obtains from (11.28) after a somewhat more complicated calculation

$$H_{pr}(\mathbf{x}) = \frac{1}{2p(\mathbf{x})}\,\epsilon_{jrs}\, l_s[f_{jp}(\mathbf{g}(\mathbf{x},\boldsymbol{\beta})) - f_{jp}(\mathbf{g}(\mathbf{x},\boldsymbol{\alpha}))], \tag{11.33}$$

where $p(\mathbf{x})$ is the distance from **x** to the dislocation segment,

$$\left.\begin{aligned} \mathbf{n} &= \frac{(\boldsymbol{\alpha}-\mathbf{x})\times(\boldsymbol{\beta}-\mathbf{x})}{\|(\boldsymbol{\alpha}-\mathbf{x})\times(\boldsymbol{\beta}-\mathbf{x})\|}, \quad \mathbf{g}(\mathbf{x},\boldsymbol{\alpha}) = \frac{\mathbf{n}\times(\boldsymbol{\alpha}-\mathbf{x})}{\|\boldsymbol{\alpha}-\mathbf{x}\|},\\ \mathbf{g}(\mathbf{x},\boldsymbol{\beta}) &= \frac{\mathbf{n}\times(\boldsymbol{\beta}-\mathbf{x})}{\|\boldsymbol{\beta}-\mathbf{x}\|}, \end{aligned}\right\} \tag{11.34}$$

$$f_{jp}(\mathbf{k}) = \operatorname{Im}\sum_{\lambda=1}^{3} F_{jp}(\mathbf{k}^{\lambda}), \tag{11.35}$$

and $\omega^{\lambda}$, $\lambda = 1, 2, 3$, are the three roots with positive imaginary parts of the sextic equation

$$D(\mathbf{g} + \mathbf{n}\omega) = 0,$$

[1] The form assumed by the solution (11.29), (11.30) in the isotropic limit has been given by Meissner [244].

where $\mathbf{g}$ is given by $(11.34)_2$ and $(11.34)_3$ for $\boldsymbol{\alpha}$ and $\boldsymbol{\beta}$, respectively. Inspection of (11.34) reveals that $\mathbf{n}$ is a unit vector perpendicular to the plane passing through the points $\boldsymbol{\alpha}, \boldsymbol{\beta}, \mathbf{x}$ while $\mathbf{g}(\mathbf{x}, \boldsymbol{\alpha})$ and $\mathbf{g}(\mathbf{x}, \boldsymbol{\beta})$ are vectors situated in this plane and perpendicular to $\boldsymbol{\alpha} - \mathbf{x}$, and respectively $\boldsymbol{\beta} - \mathbf{x}$ (Fig. 11.5).

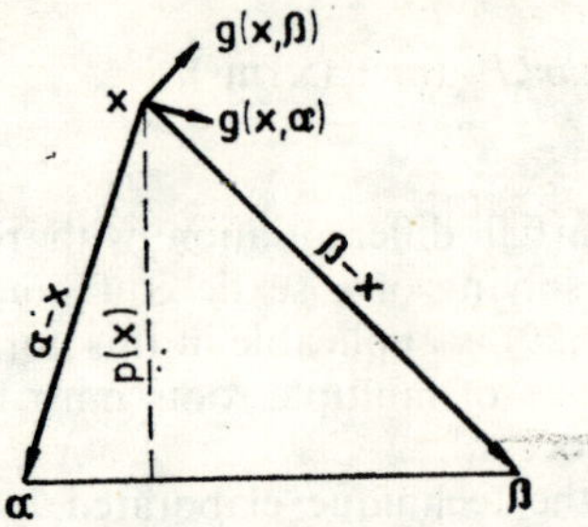


Fig. 11.5. Geometrical variables used in Willis' formula for the displacement gradient of a straight dislocation segment.

It may be seen that, as $\|\boldsymbol{\alpha}\|$ and $\|\boldsymbol{\beta}\|$ tend to infinity, $\mathbf{g}(\mathbf{x}, \boldsymbol{\alpha})$ and $\mathbf{g}(\mathbf{x}, \boldsymbol{\beta})$ tend to $-\mathbf{m}$ and $\mathbf{m}$ respectively, and (11.33) reduces to (11.30), as it should be.

By using a similar approach as Willis', Sekine and Mura [471] have recently determined the displacement field and the displacement gradient of a finite straight dislocation dipole in an unbounded anisotropic elastic medium. The results are expressed in terms of line integrals along a contour on a unit sphere.

*Dislocation loops.* Clearly, the finite dislocation segment considered above is not a real crystal defect, since dislocations cannot end within an otherwise perfect crystal region. In exchange, it provides the elastic field of a "dislocation element", by the integration of which it is possible to obtain the elastic field of a finite dislocation loop. More precisely, let us consider an arbitrary smooth, open or closed, dislocation arc, which is represented parametrically by

$$\mathbf{x}' = \mathbf{x}'(t), \qquad t \in [a, b], \tag{11.36}$$

where $t$ is the arc length. The displacement gradient produced by this dislocation arc may be obtained by putting $\boldsymbol{\alpha} = \mathbf{x}'(t)$, $\boldsymbol{\beta} = \mathbf{x}'(t) + \mathbf{l}(t)\,\Delta t$ in (11.33), performing the appropriate limiting process as $\Delta t \to 0$ and then integrating with respect to $t$. The result reads

$$H_{pr}(\mathbf{x}) = \frac{1}{2}\in_{jrs}\int_a^b \frac{l_s(t)}{p(t)}\frac{\mathrm{d}}{\mathrm{d}t} f_{jp}[\mathbf{g}(\mathbf{x}, \mathbf{x}')]\,\mathrm{d}t, \tag{11.37}$$

where $p(t)$ is the distance of $\mathbf{x}$ to the tangent at $\mathbf{x}'(t)$ to the loop. Integrating by parts, it follows that

$$H_{pr}(\mathbf{x}) = \frac{1}{2}\in_{jrs}\left\{\frac{l_s(b)}{p(b)} f_{jp}[\mathbf{g}(\mathbf{x}, \mathbf{b})] - \frac{l_s(a)}{p(a)} f_{jp}[\mathbf{g}(\mathbf{x}, \mathbf{a})] + \right.$$

$$\left. + \int_a^b f_{jp}[\mathbf{g}(\mathbf{x}, \mathbf{x}')]\,\frac{\mathrm{d}}{\mathrm{d}t}\left(\frac{l_s(t)}{p(t)}\right)\mathrm{d}t\right\}, \tag{11.38}$$

where $\mathbf{a} = \mathbf{x}'(a)$, $\mathbf{b} = \mathbf{x}'(b)$, and $\mathbf{g}(\mathbf{x}, \mathbf{x}') = \mathbf{n} \times (\mathbf{x}' - \mathbf{x})/\|\mathbf{x}' - \mathbf{x}\|$.

Now, the total field produced by a dislocation loop may be found by dividing the loop into smooth arcs and summing up their contributions of the form (11.38). This remark has been the starting point of the calculation done by Bacon, Bullough, and Willis [11] for the self-energy of a *rhombus loop* constrained to slip on a {111} glide prism in f.c.c. metals. By calculating the elastic and the core energy of the loop it has been found that the minimum-energy configuration is close to the {012} orientation, in accordance with experimental data on rhombus-shaped vacancy loops in quenched aluminium.

Another case of curvilinear dislocation studied by Willis [383] is the elliptical loop, defined by the equations

$$\frac{(\mathbf{l}\cdot\mathbf{x}')^2}{a^2}+\frac{(\mathbf{m}\cdot\mathbf{x}')^2}{b^2}=1,\quad \mathbf{n}\cdot\mathbf{x}'=0, \tag{11.39}$$

where $\{\mathbf{l},\mathbf{m},\mathbf{n}\}$ denotes an arbitrary orthonormal frame (Fig. 11.6). In this case, after performing the integrations in (11.6) and (11.7), one obtains

$$u_p(\mathbf{x})=\frac{ab}{2\pi}\,\mathrm{Im}\sum_{\lambda=1}^{3}\oint_\Gamma\frac{n_jF_{jp}(\mathbf{g}^\lambda)(\mathbf{x}\cdot\mathbf{g}^\lambda)\,\mathrm{d}l}{[a^2(\mathbf{g}\cdot\mathbf{l})^2+b^2(\mathbf{g}\cdot\mathbf{m})^2][a^2(\mathbf{g}\cdot\mathbf{l})^2+b^2(\mathbf{g}\cdot\mathbf{m})^2-(\mathbf{g}^\lambda\cdot\mathbf{x})^2]}, \tag{11.40}$$

$$H_{pr}(\mathbf{x})=\frac{ab}{2\pi}\,\mathrm{Im}\sum_{\lambda=1}^{3}\oint_\Gamma\frac{n_jF_{jp}(\mathbf{g}^\lambda)\,g_r^\lambda\,\mathrm{d}l}{[a^2(\mathbf{g}\cdot\mathbf{l})^2+b^2(\mathbf{g}\cdot\mathbf{m})^2-(\mathbf{g}^\lambda\cdot\mathbf{x})^2]^{3/2}}, \tag{11.41}$$

where $\Gamma$ is the unit circle in the plane of the loop and having the same centre as the loop, $\mathbf{g}$ is the unit vector perpendicular to $\mathbf{n}$ and connecting the origin with a current point on $\Gamma$, hence $\|\mathbf{g}\|=1$, $\mathbf{g}\cdot\mathbf{n}=0$.

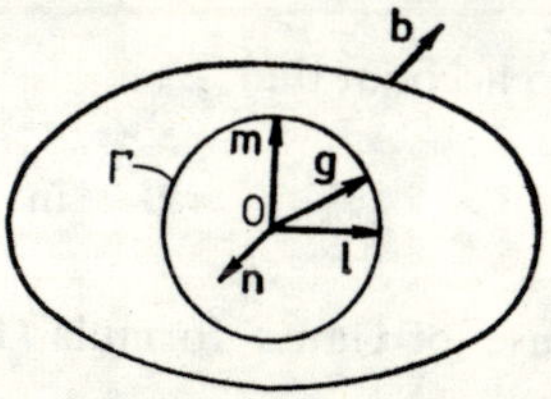

Fig. 11.6. The geometry used to apply Willis' formula to an elliptical dislocation loop.

For a plane circular loop, by setting $a=b=R$ in (11.41) and taking into consideration that $(\mathbf{g}\cdot\mathbf{l})^2+(\mathbf{g}\cdot\mathbf{m})^2=1$, we obtain

$$H_{pr}(\mathbf{x})=\frac{R^2}{2\pi}\,\mathrm{Im}\sum_{\lambda=1}^{3}\oint_\Gamma\frac{n_jF_{jp}(\mathbf{g}^\lambda)g_r^\lambda\,\mathrm{d}l}{[R^2-(\mathbf{g}^\lambda\cdot\mathbf{x})^2]^{3/2}}. \tag{11.42}$$

This formula has been applied by Meissner [244] to calculate the distortions produced by circular dislocation loops in copper and α-uranium, the latter being known as a highly anisotropic material.

### 11.3. Self-energy of a dislocation loop

Consider a dislocation loop $L$ lying in an anisotropic elastic body occupying a region $\mathscr{V}$ of traction-free boundary $\mathscr{S}$. Isolate the dislocation core by a thin tube $\Sigma_0$ of radius $r_0$ and denote by $S$ a cut surface connecting $\Sigma_0$ with $\mathscr{S}$, and by $S_0$ the part of $S$ not enclosed by $\Sigma_0$ (Fig. 11.7).

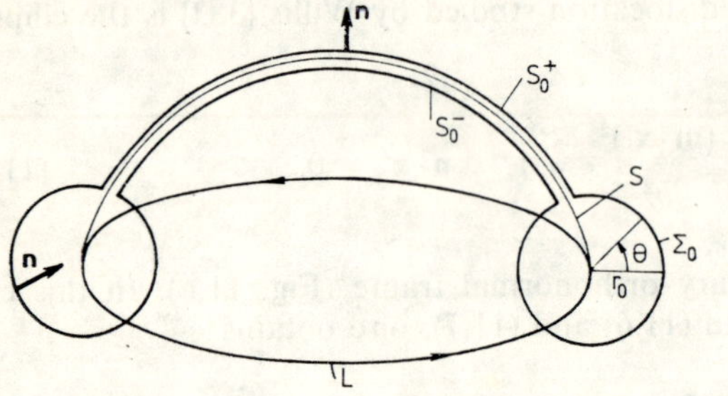


Fig. 11.7. On the definition of the self-energy of a dislocation loop.

The *total self-energy* $\mathscr{W}_t$ of the dislocation loop is defined by

$$\mathscr{W}_t = \mathscr{W}_0 + \mathscr{W},$$

where $\mathscr{W}_0$ is the potential energy of the dislocation core[1], and $\mathscr{W}$ is the strain energy of the part $\mathscr{V}_0$ of the body not enclosed by $\Sigma_0$, i.e.

$$\mathscr{W} = \frac{1}{2}\int_{\mathscr{V}_0} \mathbf{T}\cdot\mathbf{E}\,\mathrm{d}v = \frac{1}{2}\int_{\mathscr{V}_0} T_{km}u_{k,m}\,\mathrm{d}v. \tag{11.43}$$

By taking into account that

$$T_{km,m} = 0 \quad \text{in } \mathscr{V}_0, \qquad T_{km}n_m = 0 \quad \text{on } \mathscr{S},$$

and making use of Gauss' formula (1.52), we obtain from (11.43)

$$\mathscr{W} = \mathscr{W}_{\text{cut}} + \mathscr{W}_{\text{tube}}, \tag{11.44}$$

where

$$\mathscr{W}_{\text{cut}} = -\frac{1}{2}\mathbf{b}\cdot\int_{S_0} \mathbf{T}\,\mathbf{n}\,\mathrm{d}s, \qquad \mathscr{W}_{\text{tube}} = \frac{1}{2}\int_{\Sigma_0} (\mathbf{T}\,\mathbf{n})\cdot\mathbf{u}\,\mathrm{d}s. \tag{11.45}$$

---

[1] This part of the self-energy can be calculated only by a combined atomistic-continuum model (see Sect. 16).

It should be noted that $\mathbf{n}$ denotes in $(11.45)_1$ the unit normal to $S$ in the sense given by the right-hand rule with respect to the positive sense chosen on $L$, while in $(11.45)_2$ it designates the inner unit normal to $\Sigma_0$ (Fig. 11.7).

The expression (11.44) is also valid for a finite dislocation loop in an infinite body, for the stresses decay in this case as $r^{-3}$ as $r = \|\mathbf{x}\| \to \infty$, and hence the surface integral taken on $\mathscr{S}$ vanishes again as $\mathscr{S}$ is continuously deformed to infinity.

By using one of the methods presented above in this section it is possible to numerically evaluate both integrals in (11.45), at least for an arbitrary polygonal loop. Thus, Bacon, Bullough, and Willis [11] have used Willis' solution for the elastic state of a straight dislocation segment in order to calculate the strain energy $\mathscr{W}$ of a rhombus-shaped dislocation loop constrained to glide on a $\{111\}$ prism in aluminium and copper.

A very careful analysis of the self-energy of a *planar* dislocation loop and of the associated self-force has been undertaken by Gavazza and Barnett [419]. Before reviewing their results, however, we will rewrite the expression given in Sect. 10.3 for the strain energy of an infinite straight dislocation in a more invariant form.

Let $L(\mathbf{l})$ be an infinite straight dislocation lying in an infinite anisotropic elastic medium along the unit vector $\mathbf{l}$. Denote by $\mathbf{n}$ the unit normal to the slip plane, and let $\mathbf{m} = \mathbf{n} \times \mathbf{l}$. Choose the cut surface $S$ as the half-plane defined by $\mathbf{x} \cdot \mathbf{n} = 0$, $\mathbf{m} \cdot \mathbf{x} \geqslant 0$, and denote as before by $S_0$ the part of $S$ not enclosed by a thin cylindrical tube $\Sigma_0$ of radius $r_0$ surrounding the dislocation line. Clearly, we may express the results obtained for the straight dislocation in terms of the Cartesian firame $\{\mathbf{l}, \mathbf{m}, \mathbf{n}\}$ by simply noting that the frame used in Sect. 10.3 was $\{\mathbf{e}_1, \mathbf{e}_2, \mathbf{e}_3\}$, where $\mathbf{e}_1 = -\mathbf{m}$, $\mathbf{e}_2 = -\mathbf{n}$, $\mathbf{e}_3 = \mathbf{l}$. Thus, from (10.58) and the subsequent analysis, it follows that the strain energy stored per unit dislocation length between the cylindrical surfaces of radii $r_0$ and $R$ is

$$w(\mathbf{l}) = -\tfrac{1}{2}\mathbf{b} \cdot \int_{S_R} \mathbf{T}\,\mathbf{n}\,\mathrm{d}s, \tag{11.46}$$

where $S_R$ denotes the part of $S$ of unitary length along $L$ and such that $r_0 \leqslant \mathbf{m} \cdot \mathbf{x} \leqslant R$. Next, let us designate by $\boldsymbol{\sigma}(\mathbf{l})$, as in the proof of Brown's theorem, the stress field evaluated a unit distance away from $L(\mathbf{l})$ in the direction of $\mathbf{m}$. Then

$$\mathbf{T}(\mathbf{x} + \lambda\mathbf{m}) = \boldsymbol{\sigma}(\mathbf{l})/\lambda \qquad \text{for any } \lambda \neq 0,\ \mathbf{x} \in L. \tag{11.47}$$

Substituting (11.47) into (11.46) and taking into account that $r_0 \leqslant \lambda \leqslant R$ on $S_R$, we find

$$w(\mathbf{l}) = -\tfrac{1}{2}\mathbf{b} \cdot \int_{r_0}^{R} \boldsymbol{\sigma}(\mathbf{l})\,\mathbf{n}\,\frac{\mathrm{d}\lambda}{\lambda},$$

wherefrom it follows that

$$w(\mathbf{l}) = E(\mathbf{l}) \ln(R/r_0), \tag{11.48}$$

where

$$E(\mathbf{l}) = -\frac{1}{2}\mathbf{b}\cdot\boldsymbol{\sigma}(\mathbf{l})\mathbf{n} \tag{11.49}$$

is the so-called *prelogarithmic factor* of the strain energy of a straight dislocation.

Let us consider now a *planar* dislocation loop $L$ of Burgers vector $\mathbf{b}$ and use the notation in Fig. 11.1. Starting from Brown's formula (11.4), Gavazza and Barnett ([419], App.) have proved the following lemma, which we state without proof.

**Lemma**. *Let* $\mathbf{x}'$ *be a current point on a planar dislocation loop L,* $\mathbf{l}$ *the unit tangent to L at* $\mathbf{x}$, *and* $\mathbf{m}$ *the unit principal normal to L at* $\mathbf{x}'$. *Then the stress field of L at* $\mathbf{x}' + \lambda\mathbf{m}$, $\lambda > 0$, *admits of the following asymptotic representation* [1] *for* $\lambda \to 0$:

$$\mathbf{T}(\mathbf{x}' + \lambda\,\mathbf{m}) = \frac{\boldsymbol{\sigma}(\mathbf{l})}{\lambda} + \frac{1}{2\rho}\left[\boldsymbol{\sigma}(\mathbf{l}) + \frac{\mathrm{d}^2\boldsymbol{\sigma}(\mathbf{l})}{\mathrm{d}\alpha^2}\right]\ln\left(\frac{8\rho}{\lambda}\right) + \mathbf{J}(L, \mathbf{x}'), \tag{11.50}$$

*where* $\alpha$ *is the angle between a fixed direction in the plane of the loop and* $\mathbf{l}$, $\rho$ *is the radius of curvature of L at* $\mathbf{x}'$, *and the tensor field* $\mathbf{J}(L, \mathbf{x}')$ *is bounded as* $\lambda \to 0$.

Inspection of (11.50) reveals that the singularity of the in-plane self-stresses of a dislocation loop $L$ at $\mathbf{x}'$ is that of an infinite straight dislocation $L(\mathbf{l})$ tangent to $L$ at $\mathbf{x}'$ plus a weaker curvature-dependent logarithmic singularity.

Next, Gavazza and Barnett [419] write the variation of the strain energy corresponding to an arbitrary virtual displacement along and normal to $L$, in the form [2]

$$\delta\mathscr{W} = -\oint_L f\,\delta r\,\mathrm{d}l, \tag{11.51}$$

where $f$, $\delta r$ and $\mathrm{d}l$ depend on the current point $\mathbf{x}'$ of $L$. Then, in accordance witt Eshelby [107, 111], the elementary in-plane self-force on and normal to the elemen $\mathrm{d}l$ of $L$ at $\mathbf{x}'$ is defined as the product $f\,\mathrm{d}l$.

Here are the main steps of the argument used by Gavazza and Barnett to derive the expression of $f$. First, a *planar cut* $S_0$ is chosen, passing through $L$ and bounded by its intersection with the tube $\Sigma_0$, say $L_0$. Then, the variation of $\mathscr{W}_{\text{cut}}$ is shown to be given by

$$\delta\mathscr{W}_{\text{cut}} = -\frac{1}{2}\mathbf{b}\cdot\oint_{L_0}\mathbf{T}\mathbf{n}\,\delta r\,\mathrm{d}l_0 - \frac{1}{2}\mathbf{b}\cdot\oint_L \mathbf{T}^0\mathbf{n}\,\delta r\,\mathrm{d}l, \tag{11.52}$$

---

[1] It should be remembered that formulae (11.1–4), and hence also (11.50), hold rigorously only if the influence of the core tractions on the dislocation stresses outside $\Sigma_0$ is neglected.

[2] In the whole analysis the variation $\delta\mathscr{W}_t$ of the potential core energy is not taken into account.

where $\mathbf{T}^0$ is the stress field of a fictitious dislocation loop of line $L_0$ and Burgers vector $\mathbf{b}$. Next, the asymptotic representation (11.50) is used to evaluate the fields $\mathbf{T}$ and $\mathbf{T}^0$ on $L_0$ and $L$, respectively, thus obtaining from (11.52)

$$\delta \mathscr{W}_{\text{cut}} = \frac{1}{2} \oint_L \left\{ \frac{1}{\rho} E(\mathbf{l}) - \frac{1}{\rho} \left[ E(\mathbf{l}) + \frac{\mathrm{d}^2 E(\mathbf{l})}{\mathrm{d}\alpha^2} \right] \ln\left( \frac{8\rho}{r_0} \right) - J(L, \mathbf{x}') \right\} \mathrm{d}l + O(r_0), \tag{11.53}$$

where $J(L, \mathbf{x}') = -\mathbf{b} \cdot \mathbf{J}(L, \mathbf{x}')\, \mathbf{n}$, and $E(\mathbf{l})$ denotes the prelogarithmic factor given by (11.49).

The second step of the proof is the evaluation of the tube contribution $\delta \mathscr{W}_{\text{tube}}$ to $\delta \mathscr{W}$. Here the assumption is being made that the value of $\mathscr{W}_{\text{tube}}$ associated with each elementary segment $\mathrm{d}l$ of $L$ at $\mathbf{x}'$ can be evaluated [1] using only the stresses and displacements of an infinite straight dislocation $L(\mathbf{l})$ tangent to $L$ at $\mathbf{x}'$. By making use of this approximation, $\mathscr{W}_{\text{tube}}$ and its variation are shown to be given by

$$\mathscr{W}_{\text{tube}} = \oint_L F(\mathbf{l})\, \mathrm{d}l, \tag{11.54}$$

$$\delta \mathscr{W}_{\text{tube}} = \oint_L \frac{1}{\rho} \left[ F(\mathbf{l}) + \frac{\mathrm{d}^2 F(\mathbf{l})}{\mathrm{d}\alpha^2} \right] \delta r\, \mathrm{d}l + O(r_0), \tag{11.55}$$

where

$$F(\mathbf{l}) = \frac{1}{2} \int_0^{2\pi} \mathbf{t} \cdot \mathbf{u}\, r_0\, \mathrm{d}\theta \tag{11.56}$$

is precisely the first integral in (10.58) calculated for a straight dislocation directed along $\mathbf{l}$. Finally, by summing (11.53) and (11.55), neglecting terms of the order $O(r_0)$, and comparing with (11.51), one obtains the following result.

**Theorem**. (Gavazza and Barnett [419]). *The component on the principal normal* $\mathbf{m}$ *of the self-force exerted on the element* $\mathrm{d}l$ *of a planar dislocation loop* $L$ *at* $\mathbf{x}'$ *is* $f\mathrm{d}l$, *where*

$$f = \frac{1}{\rho} \left\{ E(\mathbf{l}) - \left[ E(\mathbf{l}) + \frac{\mathrm{d}^2 E(\mathbf{l})}{\mathrm{d}\alpha^2} \right] \ln\left( \frac{8\rho}{r_0} \right) - \left[ F(\mathbf{l}) + \frac{\mathrm{d}^2 F(\mathbf{l})}{\mathrm{d}\alpha^2} \right] \right\} - J(L, \mathbf{x}'). \tag{11.57}$$

Clearly, $E(\mathbf{l})$, $F(\mathbf{l})$ and their angular derivatives may be calculated by using the explicit solutions obtained by Stroh and Willis for the infinite straight dislocation and described in Sects. 10.3 and 11.2, respectively. Alternatively, $E(\mathbf{l})$ and its second

---

[1] Bacon, Bullough, and Willis [11] have used the same approximation and asserted that the error made is of the order $r_0/\Lambda$ as compared to unity, where $\Lambda$ is the length of the dislocation loop.

angular derivative may be calculated by using a numerical method developed by Barnett and Swanger [15], Barnett, Asaro, Gavazza, Bacon, and Scattergood [17], and by Asaro and Hirth [8], which allows the direct evaluation of the in-plane stress components of a planar dislocation loop without solving sextic algebraic equations.

A compilation of numerical values, obtained on these lines for the prelogarithmic factor and its first and second angular derivatives, as well as for the stress vector acting on the slip plane of an infinite straight dislocation, has been given by Bacon and Scattergood [394, 395] for a few slip systems in cubic and hexagonal close-packed crystals. They use a Cartesian frame $\{\mathbf{i}_1, \mathbf{i}_2, \mathbf{i}_3\}$, where $\mathbf{i}_1 = \mathbf{b}/b$, $\mathbf{i}_3 = \mathbf{n}$, the orientation of the dislocation line in the slip plane with normal $\mathbf{n}$ being given by the angle $\theta$ which is chosen such that $\mathbf{b} \times \mathbf{l} = (b \sin \theta)\, \mathbf{n}$, where $\mathbf{l}$ is the unit vector along the dislocation line. The computed quantities are $E(\theta)$, $E'(\theta)$, $E''(\theta)$, and the components $\tau_1(\theta)$, $\tau_2(\theta)$, $\tau_3(\theta)$ of the vector $\boldsymbol{\tau}(\theta) = (1/2)\, \boldsymbol{\sigma}(\theta)\mathbf{n}$. Clearly, in the chosen frame, $E(\theta) = -\tau_1(\theta) b$, while the stress vector acting on the slip plane with unit normal $\mathbf{n}$ at the point $\mathbf{x} + \lambda\mathbf{m}$ with $\mathbf{x} \in L(\mathbf{l})$ is given by $\mathbf{t}(\mathbf{x} + \lambda\mathbf{m}) = 2\boldsymbol{\tau}(\theta)/\lambda$, $\lambda \neq 0$. All quantities are fitted to trigonometric polynomials in $\theta$; it is remarkable that accuracies of better than 0.5% could be generally obtained by using at most 4 to 5 harmonics.

No attempt has been yet made to evaluate the self-energy and the self-force of a non-planar [1] dislocation loop by starting from the stress field given by the formula of Indenbom and Orlov (Sect. 11.1).

# 12. Interaction of single dislocations

## 12.1. Interaction energy between various elastic states

Let us consider a linear elastic body $\mathscr{B}$, which occupies a region $\mathscr{V}$ of boundary $\mathscr{S}$. By a *kinematically admissible state* of $\mathscr{B}$ we mean an admissible state that satisfies the kinematic equations (6.1), the constitutive equations (6.3), and the displacement boundary condition $(6.20)_1$ on the part $\mathscr{S}_1$ of the boundary $\mathscr{S}$. We call *potential energy*[2] $\Phi\{s\}$ corresponding to a kinematically admissible state $s = [\mathbf{u}, \mathbf{E}, \mathbf{T}]$ the difference between the strain energy $\mathscr{W}$ and the work $\mathscr{L}$ done by the body forces $\mathbf{f}$ and by the surface tractions $\mathbf{t}^\circ$ prescribed on the part $\mathscr{S}_2$ of $\mathscr{S}$ ($\mathscr{S}_1 \cup \mathscr{S}_2 = \mathscr{S}$, $\mathscr{S}_1$ and $\mathscr{S}_2$ have no common interior points), i.e.

$$\Phi\{s\} = \mathscr{W} - \mathscr{L} = \frac{1}{2}\int_{\mathscr{V}} \mathbf{T} \cdot \mathbf{E}\, dv - \int_{\mathscr{S}_2} \mathbf{t}^\circ \cdot \mathbf{u}\, ds - \int_{\mathscr{V}} \mathbf{f} \cdot \mathbf{u}\, dv. \tag{12.1}$$

---

[1] Recently, however, Shoeck and Kirchner [468] have proved by using dimensional analysis that the cut contribution $\mathscr{W}_{\text{cut}}$ to the strain energy of a non-planar loop can be expressed as $\mathscr{W}_{\text{cut}} = \left[\oint_L E(\mathbf{l})\, dl\right] \ln(\hat{L}/r_0)$, where $\hat{L}$ is some linear extension of the loop, e.g. the square root of its largest plane projection area.

[2] For *adiabatic* thermoelastic processes, $\mathscr{W}$ coincides with the internal energy $U$, and $\Phi$ is called the *enthalpy*, being usually denoted by $H$. For *isothermal* thermoelastic processes $\mathscr{W}$ coincides with the free energy $F$, and $\Phi$ is called the *free enthalpy*, being usually denoted by $G$ (cf. also Sect. 4).

The importance of this concept for linear elastostatics results from the following extremal property.

**Principle of minimum potential energy.** *Let $\mathfrak{s}$ be a solution of the mixed boundary-value problem of linear elastostatics. Then $\Phi\{\mathfrak{s}\} \leqslant \Phi\{\tilde{\mathfrak{s}}\}$ for any kinematically admissible state $\tilde{\mathfrak{s}}$, and equality holds only if $\tilde{\mathfrak{s}}$ differs from $\mathfrak{s}$ by a rigid displacement.*

*Proof.* By setting $\mathfrak{s}' = \tilde{\mathfrak{s}} - \mathfrak{s}$, we have

$$E'_{kl} = \frac{1}{2}(u'_{k,l} + u'_{l,k}), \qquad T'_{kl} = c_{klmn}E'_{mn} \ \text{ in } \mathscr{V}; \qquad u'_k = 0 \ \text{ on } \mathscr{S}_1. \tag{12.2}$$

On the other hand, definition (12.1) implies that

$$\Phi\{\tilde{\mathfrak{s}}\} - \Phi\{\mathfrak{s}\} = \frac{1}{2}\int_{\mathscr{V}} (\tilde{\mathbf{T}}\cdot\tilde{\mathbf{E}} - \mathbf{T}\cdot\mathbf{E})\,dv - \int_{\mathscr{S}_2} \mathbf{t}^\circ\cdot\mathbf{u}'\,ds - \int_{\mathscr{V}} \mathbf{f}\cdot\mathbf{u}'\,dv. \tag{12.3}$$

Next, in view of $(12.2)_3$, we can replace the integral taken on $\mathscr{S}_2$ by the same integral extended to $\mathscr{S}$, and hence, by Gauss' formula (1.52), and taking into account that $\mathfrak{s}$ satisfies (6.2) and $(6.20)_2$, we obtain

$$\int_{\mathscr{V}} \mathbf{f}\cdot\mathbf{u}'\,dv + \int_{\mathscr{S}} \mathbf{t}^\circ\cdot\mathbf{u}'\,ds = \int_{\mathscr{V}} \{-T_{kl,l}u'_k + (T_{kl}u'_k)_{,l}\}\,dv = \int_{\mathscr{V}} \mathbf{T}\cdot\mathbf{E}\,dv.$$

Substituting this result into (12.4) yields

$$\Phi\{\tilde{\mathfrak{s}}\} - \Phi\{\mathfrak{s}\} = \mathscr{W}(\mathbf{E}') = \int_{\mathscr{V}} W(\mathbf{E}')\,dv.$$

Since the strain energy density $W(\mathbf{E}')$ is positive definite, we conclude that $\Phi\{\mathfrak{s}\} \leqslant \Phi\{\tilde{\mathfrak{s}}\}$ for any $\tilde{\mathfrak{s}}$, and $\Phi\{\mathfrak{s}\} = \Phi\{\tilde{\mathfrak{s}}\}$ only if $\mathbf{E} = \tilde{\mathbf{E}}$, i.e. if $\mathfrak{s}$ and $\tilde{\mathfrak{s}}$ differ by a rigid displacement.

It is worth noting that Kirchhoff's uniqueness theorem for the mixed boundary-value problem follows as a corollary of the principle of minimum potential energy. Indeed, let $\mathfrak{s}$ and $\tilde{\mathfrak{s}}$ be two solutions of the mixed problem. Then, $\Phi\{\mathfrak{s}\} \leqslant \Phi\{\tilde{\mathfrak{s}}\}$, $\Phi\{\tilde{\mathfrak{s}}\} \leqslant \Phi\{\mathfrak{s}\}$, and hence $\mathfrak{s}$ and $\tilde{\mathfrak{s}}$ must be equal to within a rigid displacement.

If the surface tractions $\mathbf{t}^\circ$ are prescribed on the whole boundary of the body, i.e. if $\mathscr{S}_2 = \mathscr{S}$, and if $\mathfrak{s}$ is an *elastic state*, then (12.1) and the theorem of work and energy (6.20) imply that

$$\Phi\{\mathfrak{s}\} = -\mathscr{W} = -\frac{1}{2}\int_{\mathscr{V}} \mathbf{T}\cdot\mathbf{E}\,dv = -\frac{1}{2}\int_{\mathscr{S}} \mathbf{t}^\circ\cdot\mathbf{u}\,ds - \frac{1}{2}\int_{\mathscr{V}} \mathbf{f}\cdot\mathbf{u}\,dv. \tag{12.4}$$

Whenever the potential energy may be expressed as a function of a finite or infinite number of generalized co-ordinates of the system, the partial derivatives of the potential energy with respect to the generalized co-ordinates, corresponding to any kinematically admissible state and taken with opposite signs, may be considered as generalized forces that tend to bring the system to an equilibrium configuration. Such definitions, which are introduced by analogy with similar concepts used in analytical mechanics and thermodynamics, are justified by the fact that the equilibrium state corresponds, according to the above principle of minimum potential energy, to the vanishing of the generalized forces.

By *potential energy of interaction* $\Phi_{\text{int}}\{\mathscr{s}, \mathscr{s}^*\}$ between two kinematically admissible states $\mathscr{s}$ and $\mathscr{s}^*$, we mean the difference between the potential energy of the state $\mathscr{s} + \mathscr{s}^*$ and the sum of the potential energies of the states $\mathscr{s}$ and $\mathscr{s}^*$ taken separately, i.e.

$$\Phi_{\text{int}}\{\mathscr{s}, \mathscr{s}^*\} = \Phi\{\mathscr{s} + \mathscr{s}^*\} - \Phi\{\mathscr{s}\} - \Phi\{\mathscr{s}^*\}. \tag{12.5}$$

For the sake of simplicity, we assume in the following that $\mathscr{S}_2 = \mathscr{S}$ and that $\mathscr{s}$ and $\mathscr{s}^*$ are elastic states corresponding to the external force systems $[\mathbf{f}, \mathbf{t}]$ and $[\mathbf{f}^*, \mathbf{t}^*]$, respectively. It then follows from (12.5) and (12.1) that

$$\Phi_{\text{int}}\{\mathscr{s}, \mathscr{s}^*\} = \frac{1}{2}\int_{\mathscr{V}} (\mathbf{T}\cdot\mathbf{E}^* + \mathbf{T}^*\cdot\mathbf{E})\,\mathrm{d}v - \int_{\mathscr{S}} (\mathbf{t}\cdot\mathbf{u}^* + \mathbf{t}^*\cdot\mathbf{u})\,\mathrm{d}s - \int_{\mathscr{V}} (\mathbf{f}\cdot\mathbf{u}^* + \mathbf{f}^*\cdot\mathbf{u})\,\mathrm{d}v, \tag{12.6}$$

and, by Betti's reciprocal theorem (6.17), we obtain

$$\Phi_{\text{int}}\{\mathscr{s}, \mathscr{s}^*\} = -\int_{\mathscr{V}} \mathbf{T}\cdot\mathbf{E}^*\,\mathrm{d}v = -\int_{\mathscr{S}} \mathbf{t}\cdot\mathbf{u}^*\,\mathrm{d}s - \int_{\mathscr{V}} \mathbf{f}\cdot\mathbf{u}^*\,\mathrm{d}v = -\int_{\mathscr{V}} \mathbf{T}^*\cdot\mathbf{E}\,\mathrm{d}v = -\int_{\mathscr{S}} \mathbf{t}^*\cdot\mathbf{u}\,\mathrm{d}s - \int_{\mathscr{V}} \mathbf{f}^*\cdot\mathbf{u}\,\mathrm{d}v. \tag{12.7}$$

Sometimes it is more convenient to use the potential energy of interaction instead of the total potential energy $\Phi\{\mathscr{s} + \mathscr{s}^*\}$. For instance, if $\mathscr{s}$ and $\mathscr{s}^*$ are singular elastic states produced by two crystal defects, then the potential energy of interaction may assume a finite value, although linear elastostatics predicts infinite values [1] for both $\Phi\{\mathscr{s}\}$ and $\Phi\{\mathscr{s}^*\}$. For this reason, we will adopt (12.6) as a definition of the potential energy of interaction whenever the integrals involved are convergent. In particular, let $\mathscr{s} = [\mathbf{u}, \mathbf{E}, \mathbf{T}]$ and $\mathscr{s}^* = [\mathbf{u}^*, \mathbf{E}^*, \mathbf{T}^*]$ be two singular elastic states produced by the external force systems $[\mathbf{f}, \mathbf{t}, \mathbf{P}]$, and respectively $[\mathbf{f}^*, \mathbf{t}^*, \mathbf{P}^*]$, where $\mathbf{P}$ and $\mathbf{P}^*$ are systems of concentrated loads with disjoint domains $\mathscr{D}$ and $\mathscr{D}^*$, res-

[1] The real finite values of these potential energies may be computed by using an atomistic description of the close neighbourhood of the defects and by taking into account that real crystals have always finite dimensions (see Sects. 16 and 22).

pectively. Then, applying (12.6) to the region obtained from $\mathscr{V}$ by eliminating disjoint balls of radius $\eta$ centred at the points of $\mathscr{D}$, letting $\eta \to 0$, and taking into account (6.68) and (6.69), we find

$$\Phi_{\text{int}}\{\mathfrak{s}, \mathfrak{s}^*\} = \frac{1}{2}\int_{\mathscr{V}} (\mathbf{T}\cdot\mathbf{E}^* + \mathbf{T}^*\cdot\mathbf{E})\,dv - \int_{\mathscr{S}} (\mathbf{t}\cdot\mathbf{u}^* + \mathbf{t}^*\cdot\mathbf{u})\,ds -$$

$$-\int_{\mathscr{V}} (\mathbf{f}\cdot\mathbf{u}^* + \mathbf{f}^*\cdot\mathbf{u})\,dv - \sum_{\mathbf{x}'\in\mathscr{D}} \mathbf{P}(\mathbf{x}')\cdot\mathbf{u}^*(\mathbf{x}') - \sum_{\mathbf{x}'\in\mathscr{D}^*} \mathbf{P}^*(\mathbf{x}')\cdot\mathbf{u}(\mathbf{x}'). \tag{12.8}$$

By the reciprocal theorem for singular elastic states (6.70), this relation may be also rewritten as

$$\Phi_{\text{int}}\{\mathfrak{s}, \mathfrak{s}^*\} = -\int_{\mathscr{V}} \mathbf{T}\cdot\mathbf{E}^*\,dv = -\int_{\mathscr{S}} \mathbf{t}\cdot\mathbf{u}^*\,ds - \int_{\mathscr{V}} \mathbf{f}\cdot\mathbf{u}^*\,dv - \sum_{\mathbf{x}'\in\mathscr{D}} \mathbf{P}(\mathbf{x}')\cdot\mathbf{u}^*(\mathbf{x}') =$$

$$= -\int_{\mathscr{V}} \mathbf{T}^*\cdot\mathbf{E}\,dv = -\int_{\mathscr{S}} \mathbf{t}^*\cdot\mathbf{u}\,ds - \int_{\mathscr{V}} \mathbf{f}^*\cdot\mathbf{u}\,dv - \sum_{\mathbf{x}'\in\mathscr{D}^*} \mathbf{P}^*(\mathbf{x}')\cdot\mathbf{u}(\mathbf{x}'). \tag{12.9}$$

This definition of the potential energy of interaction is still applicable for infinite regions with finite boundaries, provided that the singular elastic states vanish rapidly enough at infinity, e.g. when conditions (6.71) are fulfilled.

If the position of a defect $D$ situated in an infinite elastic medium may be uniquely characterized by the position vector $\mathbf{x}$ of some characteristic point of the defect, then the force exerted by an elastic state $\mathfrak{s}^*$ on the defect is by definition

$$\mathbf{F} = -\operatorname{grad}_{\mathbf{x}} \Phi_{\text{int}}\{D, \mathfrak{s}^*\}. \tag{12.10}$$

We will apply now the above considerations to the interaction between the singular elastic state $\mathfrak{s} = [\mathbf{u}, \mathbf{E}, \mathbf{T}]$ produced by a dislocation of line $L$ and Burgers vector $\mathbf{b}$ and a regular elastic state $\mathfrak{s}^* = [\mathbf{u}^*, \mathbf{E}^*, \mathbf{T}^*]$ produced by the surface tractions $\mathbf{t}^*$. Then

$$T_{kl,l} = 0 \quad \text{in } \mathscr{V}\setminus L, \qquad T_{kl}n_l = 0 \quad \text{on } \mathscr{S}, \tag{12.11}$$

$$T^*_{kl,l} = 0 \quad \text{in } \mathscr{V}, \qquad T^*_{kl}n_l = t^*_k \quad \text{on } \mathscr{S}. \tag{12.12}$$

From the definition (12.6) and considering $(12.11)_2$, we deduce that the interaction energy is given by

$$\Phi_{\text{int}}\{L, \mathfrak{s}^*\} = \frac{1}{2}\int_{\mathscr{V}} (\mathbf{T}\cdot\mathbf{E}^* + \mathbf{T}^*\cdot\mathbf{E})\,dv - \int_{\mathscr{S}} \mathbf{t}^*\cdot\mathbf{u}\,ds.$$

In order to show that the volume integral is convergent we first isolate the dislocation line by the closed surface $\Sigma$ composed of the two faces $S^+$ and $S^-$ of the cut $S$ used to generate the dislocation and of the tube $\Sigma_0$ of radius $r_0$ surrounding the loop $L$ (Fig. 7.5). Then

$$\int_{\mathscr{V}} (\mathbf{T}\cdot\mathbf{E}^* + \mathbf{T}^*\cdot\mathbf{E})\,\mathrm{d}v = \lim_{r_0\to 0}\int_{\mathscr{V}_0} (\mathbf{T}\cdot\mathbf{E}^* + \mathbf{T}^*\cdot\mathbf{E})\mathrm{d}v,$$

where $\mathscr{V}_0$ denotes the region bounded by $\Sigma$ and the outer surface $\mathscr{S}$. Next, taking into account that $\mathbf{T}\cdot\mathbf{E}^* = \mathbf{T}^*\cdot\mathbf{E} = T_{kl}u_{k,l}$, we infer by partial integration and considering (12.12) that

$$\frac{1}{2}\int_{\mathscr{V}_0} (\mathbf{T}\cdot\mathbf{E}^* + \mathbf{T}^*\cdot\mathbf{E})\,\mathrm{d}v = \int_{\mathscr{S}} \mathbf{t}^*\cdot\mathbf{u}\,\mathrm{d}s + \int_{\Sigma} (\mathbf{T}^*\mathbf{n})\cdot\mathbf{u}\,\mathrm{d}s, \tag{12.13}$$

where $\mathbf{n}$ denotes the outer unit normal to $\Sigma$ (with respect to $\mathscr{V}_0$). From the last three relations it follows that

$$\Phi_{\text{int}}\{L, \mathscr{s}^*\} = \lim_{r_0\to 0}\int_{\Sigma} (\mathbf{T}^*\mathbf{n})\cdot\mathbf{u}\,\mathrm{d}s. \tag{12.14}$$

But $\mathbf{T}^*\mathbf{n}$ must be continuous across $S$, while $\mathbf{u}$ satisfies the jump condition (9.1). Thus

$$\Phi_{\text{int}}\{L, \mathscr{s}^*\} = \lim_{r_0\to 0}\left(\int_{\Sigma_0} \mathbf{t}^*\cdot\mathbf{u}\,\mathrm{d}s - \mathbf{b}\cdot\int_{S_0} \mathbf{T}^*\mathbf{n}\,\mathrm{d}s\right),$$

where now $\mathbf{n}$ denotes the unit normal to $S$ in the sense given by the right-hand rule with respect to the positive sense on $L$, while $S_0$ is the part of $S$ not enclosed by $\Sigma_0$. To evaluate the first integral in the right-hand side of (12.14), we decompose it into integrals taken on circular cylindrical surfaces corresponding to the division of $L$ into small straight segments. Then, according to Sect. 10, $\mathbf{u}$ diverges as $\ln r_0$, and hence is $o(r_0^{-1})$, as $r_0 \to 0$. On the other hand, $\mathbf{t}$ is continuous in the vicinity of $L$, and hence, the limiting value of the integral taken on $\Sigma_0$ is zero. Thus, the interaction energy has the finite value

$$\Phi_{\text{int}}\{L, \mathscr{s}^*\} = -\mathbf{b}\cdot\int_S \mathbf{T}^*\mathbf{n}\,\mathrm{d}s, \tag{12.15}$$

where now the integration is extended to the whole cut surface $S$.

In order to determine the force exerted by the stress field $\mathbf{T}$ on an infinitesimal segment $\mathrm{d}\mathbf{l}$ of the dislocation loop, let us assume that this segment undergoes a virtual translation $\delta\mathbf{x}$. The variation of the oriented area element $\mathbf{n}\mathrm{d}s$ is

$$\delta(\mathbf{n}\,\mathrm{d}s) = -\,\mathrm{d}\mathbf{l}\times\delta\mathbf{x}, \qquad \delta(n_m\mathrm{d}s) = -\,\epsilon_{mrs}\mathrm{d}l_r\mathrm{d}x_s, \tag{12.16}$$

and the corresponding variation of the interaction energy is

$$\delta\Phi_{\text{int}}\{L, \mathscr{I}^*\} = b_k T_{km} \in_{mrs} \mathrm{d}l_r \mathrm{d}x_s.$$

By comparing this result with definition (12.10), we see that the force exerted by the stress field $\mathbf{T}$ on the dislocation segment $\mathbf{dl}$ is (Peach and Koehler [265])

$$\mathbf{dF} = -(\mathbf{T}^*\mathbf{b}) \times \mathbf{dl}, \qquad \mathrm{d}F_s = -\in_{mrs} b_k T^*_{km} \mathrm{d}l_r, \tag{12.17}$$

its direction being perpendicular to the dislocation segment. From (12.17), we also conclude that the total force exerted by the stress field $\mathbf{T}$ on the dislocation loop is

$$\mathbf{F} = -\oint_L (\mathbf{T}^*\mathbf{b}) \times \mathbf{dl}. \tag{12.18}$$

Equation (12.13) yields also another interesting result. By making use of Betti's theorem (12.7) for the elastic states $\mathscr{I}$ and $\mathscr{I}^*$, which are regular in $\mathscr{V}_0$, and taking into account (12.11), we successively obtain from (12.13)

$$\lim_{r_0 \to 0} \frac{1}{2} \int_{\mathscr{V}_0} (\mathbf{T} \cdot \mathbf{E}^* + \mathbf{T}^* \cdot \mathbf{E})\, \mathrm{d}v = \lim_{r_0 \to 0} \int_{\Sigma} \mathbf{t} \cdot \mathbf{u}^*\, \mathrm{d}s = \lim_{r_0 \to 0} \int_{\Sigma_0} \mathbf{t} \cdot \mathbf{u}^*\, \mathrm{d}s,$$

the last transformation being permitted because both $\mathbf{t} = \mathbf{Tn}$ and $\mathbf{u}^*$ are continuous across $S$. On the other hand, since $\mathbf{u}^*$ is continuous in the vicinity of $L$, and the resultant force of the tractions acting on $\Sigma_0$ from the dislocation core is zero, we conclude, by making use of the mean theorem of the integral calculus that the above limit vanishes. This result, which is due to Colonnetti [79], may be formulated as follows: *The part of the strain energy that is due to the interaction between an elastic state produced by surface tractions and a state of self-stress is zero.*

## 12.2. Elastic interaction between dislocation loops

It is easily seen that formula (12.15) holds also for the interaction between two non-intersecting loops of lines $L$ and $L^*$, i.e.

$$\Phi_{\text{int}}\{L, L^*\} = -\mathbf{b} \cdot \int_S \mathbf{T}^*\mathbf{n}\, \mathrm{d}s, \tag{12.19}$$

where $\mathbf{T}^*$ is the stress field of the dislocation loop $L^*$, and it is assumed that the cut surface $S$ does not intersect $L^*$. The proof proceeds on the same lines as before, but $L^*$ must be also isolated by a surface $\Sigma^*$ of the same type as $\Sigma$. Then, a reasoning similar to that leading to (12.14) yields

$$\Phi_{\text{int}}\{L, L^*\} = \lim_{r_0 \to 0} \int_{\Sigma} (\mathbf{T}^*\mathbf{n}) \cdot \mathbf{u}\, \mathrm{d}s + \lim_{r_0^* \to 0} \int_{\Sigma^*} (\mathbf{T}^*\mathbf{n}) \cdot \mathbf{u}\, \mathrm{d}s,$$

where $r_0^*$ is the radius of the tube $\Sigma_0^*$ surrounding the loop $L^*$. On the other hand the second limit in the right-hand side vanishes, since **u** is continuous in the vicinity of $L^*$ and the resultant force of the tractions acting on $\Sigma^*$ is zero; hence the last relation reduces indeed to (12.19).

In order to derive the *elastic energy of interaction* between two dislocation loops $L$ and $L^*$ in an *isotropic* medium, we simply introduce (9.20) into (12.19), thus obtaining

$$\Phi_{\text{int}}\{L, L^*\} = -\frac{\mu b_p b_s^*}{4\pi}\int_S n_r\,\mathrm{d}s \oint_{L^*}\left[\frac{1}{2}R_{,mmt}(\epsilon_{rst}\,\mathrm{d}x_p^* + \epsilon_{pst}\,\mathrm{d}x_r^*) + \right.$$

$$\left. + \frac{1}{1-\nu}\epsilon_{nst}(R_{,tpr} - \delta_{pr}R_{,mmt})\,\mathrm{d}x_p^*\right],$$

where $R = \|\mathbf{x} - \mathbf{x}^*\|$. By transforming this equation with the aid of Stokes' formula, we find after some intermediate calculation (Blin [31]) the relation

$$\Phi_{\text{int}}\{L, L^*\} = \frac{\mu}{2\pi}\oint_L \mathrm{d}x_l \oint_{L^*}\left[-\frac{1}{R}\epsilon_{ijk}\epsilon_{kln}b_i b_j^* + \frac{1}{2R}b_l b_n^* + \right.$$

$$\left. + \frac{1}{2(1-\nu)}\epsilon_{ikl}\,\epsilon_{jmn}\,b_k\,b_m^* R_{,ij}\right]\mathrm{d}x_n^*, \tag{12.20}$$

whose symmetry with respect to $L$ and $L^*$ is obvious. Kröner [190] put (12.20) into the more elegant form

$$\Phi_{\text{int}}\{L, L^*\} = b_i b_j^* M_{ij}\{L, L^*\}, \tag{12.21}$$

where

$$M_{ij}\{L, L^*\} = \frac{\mu}{8\pi}\oint_L\oint_{L^*}\epsilon_{ikl}\,\epsilon_{jmn}R_{,km}[\mathrm{d}x_n \mathrm{d}x_l^* +$$

$$+ \delta_{ln}\mathrm{d}x_p \mathrm{d}x_p^* + 2\nu/(1-\nu)\,\mathrm{d}x_l \mathrm{d}x_n^*]. \tag{12.22}$$

The elastic energy of interaction per unit length of two infinite straight dislocations with lines $L$ and $L^*$ parallel to the unit vector **l** and having Burgers vectors **b**, respectively **b***, results from (12.20) as (Nabarro [257]):

$$w(L, L^*) = -\frac{\mu}{2\pi}\left\{\left[(\mathbf{b}\cdot\mathbf{l})(\mathbf{b}^*\cdot\mathbf{l}) + \frac{1}{1-\nu}(\mathbf{b}\times\mathbf{l})\cdot(\mathbf{b}^*\times\mathbf{l})\right]\ln R + \right.$$

$$\left. + \frac{1}{(1-\nu)R^2}[(\mathbf{b}\times\mathbf{l})\cdot\mathbf{R}][(\mathbf{b}^*\times\mathbf{l})\cdot\mathbf{R}]\right\}, \tag{12.23}$$

where $\mathbf{R}$ is any vector perpendicular to $\mathbf{l}$ and connecting $L$ to $L^*$, and $R = \|\mathbf{R}\|$.

We mention one more result due to Kröner [190] and concerning the elastic energy of interaction between two coaxial circular dislocation loops of the same radius $a$, with the same orientation and Burgers vector $\mathbf{b}$ parallel to the axis:

$$\Phi_{\text{int}}\{L, L^*\} = \mu b^2 a k(K - E)/(1 - \nu), \tag{12.24}$$

where $K$ and $E$ are complete elliptic integrals of the first and second kind, respectively, and of modulus $k$ given by the relation $k^2 = 4a^2/(4a^2 + d^2)$, where $d$ is the distance between the planes of the two loops.

The elastic energy of interaction between dislocation loops situated in an *anisotropic* medium may be evaluated by making use of the results presented in the preceding section. Thus, the interaction between two coplanar dislocation loops can be successfully calculated by means of Brown's formula (11.4). Indeed, let us consider two coplanar dislocation loops $L$ and $L^*$ having the same Burgers vector $\mathbf{b}$. Choose the $x_3$-axis of a Cartesian frame parallel to the unit normal $\mathbf{n}$ to the plane of the loops (Fig. 12.1). From (12.19) it results

$$\Phi_{\text{int}}\{L, L^*\} = -\, b_i \int_S T^*_{i3}(x_1, x_2, 0)\, \mathrm{d}s, \tag{12.25}$$

where $S$ is the plane region bounded by $L$, and taken as cut surface. Substituting now (11.4) into (12.25) and taking into account (11.49) we deduce that (Brown [43])

$$\Phi_{\text{int}}\{L, L^*\} = \frac{1}{2}\int_S \mathrm{d}s \oint_{L^*} \frac{1}{R^2}\left[E(\theta) + \frac{\mathrm{d}^2 E(\theta)}{\mathrm{d}\theta^2}\right]\sin(\theta - \alpha)\, \mathrm{d}l^*, \tag{12.26}$$

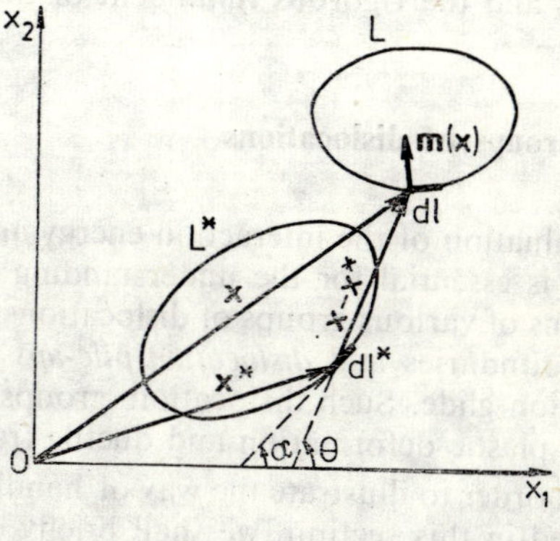


Fig. 12.1. Two dislocation loops lying in the $x_1x_2$-plane.

where $R = \|\mathbf{x} - \mathbf{x}^*\|$, and $E(\theta) = -(1/2)\mathbf{b}\cdot\boldsymbol{\sigma}(\theta)\mathbf{n} = -(1/2)b_i\sigma_{i3}(\theta)$ denotes, as before, the prelogarithmic factor in the expression of the elastic self-energy per unit length of an infinite straight dislocation of Burgers vector $\mathbf{b}$ and directed along $\mathbf{x}-\mathbf{x}^*$.

Moreover, denote by $\mathbf{m}(\mathbf{x})$ the in-plane inward unit normal to $L$ at a current point $\mathbf{x}$ (Fig. 12.1). Then the perpendicular projection $\mathrm{d}F(\mathbf{x})$ on $\mathbf{m}(\mathbf{x})$ of the force $\mathrm{d}\mathbf{F}(\mathbf{x})$ exerted by the dislocation loop $L^*$ on the line element $\mathrm{d}l$ of $L$ at $\mathbf{x}$ is given (Brown [43]) by

$$\mathrm{d}F(\mathbf{x}) = \left\{ \oint_{L^*} \frac{1}{R^2} \left[ E(\theta) + \frac{\mathrm{d}^2 E(\theta)}{\mathrm{d}\theta^2} \right] \sin(\theta - \alpha)\, \mathrm{d}l^* \right\} \mathrm{d}l. \tag{12.27}$$

Indeed, by (12.17) and taking into account that $\mathbf{m} \times \mathrm{d}\mathbf{l} = -\mathbf{n}\,\mathrm{d}l$, we successively have

$$\mathrm{d}F(\mathbf{x}) = \mathbf{m}(\mathbf{x}) \cdot \mathrm{d}\mathbf{F}(\mathbf{x}) = -\mathbf{m}(\mathbf{x}) \cdot [(\mathbf{T}^*\mathbf{b}) \times \mathrm{d}\mathbf{l}] =$$

$$= (\mathbf{T}^*\mathbf{b}) \cdot [\mathbf{m}(\mathbf{x}) \times \mathrm{d}\mathbf{l}] = -(\mathbf{T}^*\mathbf{b}) \cdot \mathbf{n}\,\mathrm{d}l = -\mathbf{b} \cdot (\mathbf{T}^*\mathbf{n})\,\mathrm{d}l,$$

wherefrom (12.27) follows at once by considering (11.4) and the definition of $E(\theta)$.

Since $E(\theta)$ may be explicitly calculated for general anisotropy and an arbitrary orientation of the dislocation line, it is apparent that equations (12.26) and (12.27) allow to solve various problems concerning the interaction between coplanar loops and the stability of plane dislocation configurations by using only straight dislocation data (Lothe [219], Barnett, Asaro, Gavazza, Bacon, and Scattergood [17], Asaro and Hirth [8]).

An approximate analysis has been undertaken by Korner, Svoboda, and Kirchner [437] for the interaction between dislocation segments in regions with free boundaries, and has been subsequently applied by Korner, Karnthaler, and Kirchner [438] to study the trapezoidal splitting of partial dislocations in thin foils of Ag and Cu-10 at% Al; the results obtained seem to be in satisfactory agreement with experimental data. However, as pointed out by these authors, the problem of the interaction of dislocations in *finite* anisotropic bodies involves some still unsolved aspects, e.g. the consideration of end effects for dislocation lines emerging at free surfaces and the rigorous fulfillment of the boundary conditions.

## 12.3. **Groups of dislocations**

The evaluation of the interaction energy and of the interaction forces between dislocations is essential for the understanding and prediction of the equilibrium configurations of various groups of dislocations, e.g. *dislocation walls* building small-angle grain boundaries and *dislocation pile-ups* occurring in front of strong obstacles to dislocation glide. Such dislocation groups are known to play a very important role in both plastic deformation and ductile fracture.

In order to illustrate the way of handling groups of dislocations by the methods presented in this section, we shall briefly consider two basic approaches that allow the determination of the equilibrium configurations of planar dislocation pile-ups.

*The method of orthogonal polynomials.* Assume that $n$ straight dislocations are situated in the plane $x_2 = 0$, have the same Burgers vector $\mathbf{b}$, and are directed along the positive $x_3$-axis. Let $a_1, a_2, \ldots, a_n$ be the values of the co-ordinate $x_1 = x$ corresponding to the equilibrium configuration of the pile-up (Fig. 12.2).

Since now $dl_r = \delta_{r3} dl$, it follows from (12.17) that the in-plane component of the force exerted per unit length of the dislocation $j$ is

$$dF_1/dl = -b_k T^*_{k2}(a_j, 0,0). \tag{12.28}$$

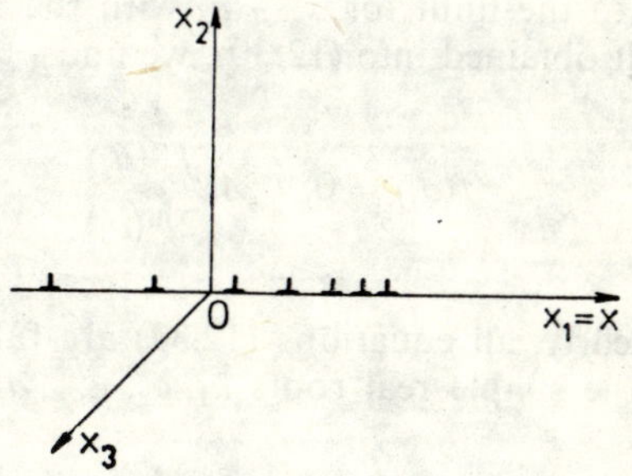


Fig. 12.2. Group of $n$ parallel straight dislocations lying in the $x_1x_3$-plane and having the same Burgers vector.

On the other hand, according to Sect. 10.3, the contribution of the $i$th dislocation to (12.28) is

$$-\frac{1}{\pi} b_k \frac{1}{a_j - a_i} \operatorname{Im} \sum_{\alpha=1}^{3} L_{k\alpha} D_\alpha = \frac{bA}{a_j - a_i}, \tag{12.29}$$

where $A = Kb/(2\pi)$ and $b$ is the magnitude of the Burgers vector. We recall that the energy factor $K$ is given in the general anisotropic case by (10.76), while in the isotropic case it equals $(\mu b)/2\pi$ and $\mu b/[2\pi(1 - \nu)]$ for screw and edge dislocations, respectively.

Next, let $\tau(x)$ be the resolved shear stress (i.e. the component in the glide plane and in the glide direction of the stress field) produced by the external forces and all other crystal defects not belonging to the pile-up. The equilibrium of the pile-up requires the vanishing of the in-plane net force exerted on each dislocation, i.e.

$$\sum_{\substack{i=1 \\ i \neq j}}^{n} \frac{A}{a_j - a_i} + \tau(a_j) = 0, \quad j = 1, 2, \ldots, n. \tag{12.30}$$

We will indicate now the main lines of the method proposed by Eshelby, Frank, and Nabarro [108] for solving this system of equations [1].

Consider the polynomial

$$f(x) = \prod_{i=1}^{n} (x - a_i)$$

---

[1] This method has been applied for the first time by Stieltjes in 1885 to illustrate the possible applications of the orthogonal polynomials to finding out the equilibrium configurations of electric charges.

with roots $a_1, a_2, \ldots, a_n$. It is easily proved that the pile-up, exc ept the $j$th dislocation, produces the resolved shear stress

$$\sum_{\substack{i=1\\i\neq j}}^{n} \frac{A}{x-a_i} = A\left[\frac{f'(x)}{f(x)} - \frac{1}{x-a_j}\right]. \tag{12.31}$$

Passing to the limit for $x \to a_j$ with the aid of l'Hospital's rule and substituting the result obtained into (12.30), we find

$$f(a_j) = 0, \quad A\frac{f''(a_j)}{2f'(a_j)} + \tau(a_j) = 0, \qquad j = 1, \ldots, n. \tag{12.32}$$

Clearly, all equations (12.32) are fulfilled if $f(x)$ is a polynomial of degree $n$ having the simple real roots $a_1, a_2, \ldots, a_n$, and satisfying the differential equation

$$f''(x) + (2/A)\tau(x)f'(x) + q(n, x)f(x) = 0, \tag{12.33}$$

where the function $q(n, x)$ is assumed to be finite for $x = a_j, j = 1, 2, \ldots, n$, but is otherwise arbitrary.

Sometimes it is interesting to consider pile-ups containing, besides the $n$ mobile dislocations, $m - n$ more dislocations that are fixed by various obstacles at points $x = x_\alpha$, $\alpha = n + 1, \ldots, m$. In this case the reduced stress produced by the fixed dislocations must be added to $\tau(x)$, and hence (12.33) becomes

$$f''(x) + 2\left[\frac{\tau(x)}{A} + \sum_{\alpha=n+1}^{m} \frac{1}{x - x_\alpha}\right]f'(x) + q(n, x)\,f(x) = 0. \tag{12.34}$$

Moreover $q(n, x)$ may eventually tend to infinity as $x$ approaches one of the values $x_\alpha$, since it is not necessary that the net reduced stress exerted on fixed dislocations vanishes. Denoting

$$F(x) = f(x) \prod_{\alpha=n+1}^{m} (x - x_\alpha), \tag{12.35}$$

$$Q(n, x) = q(n, x) - 2\frac{\tau(x)}{A}\sum_{\alpha=n+1}^{m}\frac{1}{x - x_\alpha} - \sum_{\substack{\alpha,\beta=n+1\\ \alpha\neq\beta}}^{m} \frac{1}{(x - x_\alpha)(x - x_\beta)}, \tag{12.36}$$

the differential equation (12.34) becomes

$$F''(x) + (2/A)\tau(x)F'(x) + Q(n, x)F(x) = 0, \tag{12.37}$$

i.e. assumes the same form as before. However, the reduced stress generated by all (mobile and fixed) dislocations is now equal to $AF'(x)/F(x)$.

It is interesting that some well-known orthogonal polynomials are solutions of the equation (12.34), having thus a direct application in the theory of dislocation pile-ups. Here are some of the situations that can be easily treated by means of such polynomials.

(i) $n$ dislocations situated in the interval $[-a, a]$, from which $n - 2$ are mobile and two are fixed at points $x = \pm a$; $\tau(x) = 0$. The corresponding solution of (12.34) reads

$$f(x) = P'_{n-1}(x/a), \tag{12.38}$$

where $P'_{n-1}(x)$ is the derivative of Legendre's polynomial of degree $n-1$. The resolved shear stresses acting on the dislocations fixed at points $x = \pm a$ are $\pm n(n-1)A/(4a)$.

(ii) $n$ dislocations situated on the ray $[0, \infty)$, from which $n - 1$ are free and one is fixed at the origin. The external reduced shear stress is assumed to be constant and directed towards the fixed dislocation, i.e. $\tau(x) = -\tau_0$ with $\tau_0 > 0$. The corresponding solution of (12.38) is

$$f(x) = L'_n(2\tau_0 x/A), \tag{12.39}$$

where $L'_n(x)$ is the derivative of Laguerre's polynomial of degree $n$. For large values of $n$, the length of the interval covered by the pile-up on the $x_1$-axis is $L = 2nA/\tau_0$, while the total resolved shear stress acting on the fixed dislocation is $-n\tau_0$.

(iii) $n$ free dislocations under the action of a linearly varying applied shear stress, $\tau(x) = \alpha x$, $\alpha > 0$. Denoting by $[-a, a]$ the interval occupied by the pile-up on the $x_1$-axis in the equilibrium configuration it results that

$$f(x) = H_n(x\sqrt{\alpha/A}), \qquad a = \sqrt{(2n+1)A/\alpha}, \tag{12.40}$$

where $H_n(x)$ is Hermitte's polynomial of degree $n$.

The method of the orthogonal polynomials has been also used to study the elastic field produced by planar dislocation pile-ups in the cases (i) — (iii), as well as in various other situations (fixed dislocations with Burgers vector $n\,\mathbf{b}$, pile-ups of dislocation loops, etc.), by Stroh [322], Chou, Garofalo, and Whitmore [67], Chou and Whitmore [68], Kronmüller and Seeger [195], Chou [69, 72], Head and Thompson [157], Mitchell, Hecker, and Smialek [250], and Smith [307, 308].

Numerical results concerning dislocation pile-ups, obtained by direct solving of the equilibrium equations, have been given by Mitchell [249] and by Hazzledine and Hirsch [153] for straight dislocations, and by Marcinkowski [233] for coaxial circular glide-dislocation arrays.

Seeger and Wobser [293] have numerically investigated a related problem, namely the stable configurations for a pair of parallel straight dislocations gliding on octahedral planes of f.c.c. crystals and in basal planes of h.c.p. crystals. Finally, Kronmüller and Marik [199] have investigated, also by numerical computation, the stability of dislocation pile-ups under a spatially-oscillating shear stress and in the presence of Lomer-Cottrell sessile dislocations, as well as the consequences of the results obtained for work-hardening theories based on long-range stresses.

*The method of singular integral equations.* For large values of $n$, Leibfried [212] proposed to replace the real distribution of the straight dislocations in a pile-up by a continuous distribution with density $D(x)$, defined such that the resultant Burgers vector of the dislocations comprised between $x$ and $x + \mathrm{d}x$ equals $\mathbf{b}\,D(x)$. In this case, the equilibrium equation (12.29) is replaced by the singular integral equation

$$\int_a^{a'} \frac{D(x)\,\mathrm{d}x}{\xi - x} = \frac{\tau(\xi)}{A} \quad \text{for } \xi \in [a, a'], \tag{12.41}$$

where $[a, a']$ is the interval occupied by the dislocation pile-up in the equilibrium configuration. In addition, $D(x)$ has to satisfy some supplementary conditions arising e.g. from the prescription of the total number of the dislocations in the pile-up or of one or both of the values $a$ and $a'$ (fixed dislocations).

For the cases (i) – (iii) considered above the solutions of equation (12.41) with the adequate supplementary conditions are (Leibfried [212]):

(i) $D(x) = (n/\pi)(a^2 - x^2)^{-1/2}$ for $x \in (-a, a)$,

(ii) $D(x) = (2n/\pi l)(L/x - 1)^{1/2}$, $L = \sqrt{2nA/\tau_0}$ for $x \in (0, L)$,

(iii) $D(x) = (\alpha/\pi A)(a^2 - x^2)^{1/2}$, $a = \sqrt{2nA/\alpha}$ for $x \in [-a, a]$,

and $D(x) = 0$ outside the indicated intervals. The graphs of $D(x)$ in these three cases are schematically represented in Fig. 12.3a, b, c, respectively.

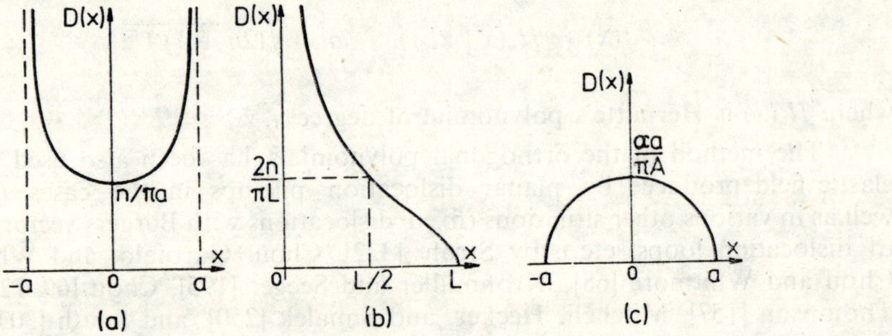


Fig. 12.3. Schematic representation of the surface dislocation density $D(x)$. (a) Case (i), $D(x) = (n/\pi)\ (a^2 - x^2)^{-1/2}$. (b) Case (ii), $D(x) = (2n/\pi L) \times (L/x - 1)^{1/2}$. (c) Case (iii), $D(x) = (\alpha/\pi A)\ (a^2 - x^2)^{1/2}$.

The method of the singular integrals is effective whenever $n$ is large and the elastic field is evaluated at distances sufficiently large from the pile-up. This method has been employed by Kronmüller and Seeger [195], Saxlová-Švábová [281] and Li [214] for calculating the stress field of straight dislocation pile-ups, and by Barnett [13], Chou and Barnett [73], Kuang and Mura [203], Louat [221], and Tucker [361] to investigate the equilibrium configuration of dislocation pile-ups near grain or phase boundaries.

For the study of dislocation walls [1] and of the modelling of crack propagation by coalescence of dislocations at the head of a pile-up, we refer to the book by Hirth and Lothe [162], Sect. 21, where further literature on this topic may be also found.

# 13. Dislocation motion

## 13.1. Dislocation glide and climb

Under suitable energetic conditions, a straight dislocation can move in any direction, except the direction of its own line.

The motion of an *edge dislocation* is said to be a *glide* or a *climb* process, according as the dislocation velocity is contained in the slip plane or is perpendicular to it. The dislocation glide proceeds without material transport, the atomic rearrangement taking place gradually behind the moving dislocation. On the contrary, dislocation climb involves a local change of the crystal density, corresponding to the lengthening or the shortening of the extra atomic half-plane, which may be brought about e.g. by vacancy diffusion away, respectively towards, the dislocation line. Since diffusion requires a considerable specific energy, dislocation climb is generally significant only at sufficiently high temperatures.

In the case of *screw dislocations* the glide plane is not uniquely defined because $\mathbf{b} \| \mathbf{l}$. In other words, any atomic plane passing through $\mathbf{b}$ may serve as slip plane for a screw dislocation. Consequently, the motion of screw dislocations proceeds always by glide, and this explains their higher mobility versus edge dislocations. At sufficiently high temperatures and/or applied stresses screw dislocations can even *cross slip* from one glide plane to another, as long as the intersection line of these planes is parallel to $\mathbf{b}$.

Since dislocations have a very low effective inertia, their speed increases rapidly after overcoming the glide obstacles, until a limiting speed is attained, corresponding to the dynamic equilibrium between the forces exerted on the dislocations by applied tractions and other crystal defects, on one side, and the dragging forces produced by various dissipative mechanisms, on the other side. As the accelerating time is 4 to 5 orders of magnitude smaller than the time of free motion between obstacles, the dislocation motion may be considered mainly as being uniform. Therefore, we shall treat in what follows only uniformly moving dislocations; for accelerating dislocations we refer to Kiusalaas and Mura [180] and Hirth and Lothe [162], Sect. 7.7.

## 13.2. Uniformly moving dislocations in isotropic media

The study of the elastic field of uniformly moving dislocations in isotropic media started some thirty years ago. Thus, Frank [121] and Eshelby [106, 107] have considered uniformly moving edge and screw dislocations, respectively, while Nabarro

[1] In this connection, see also a recent paper by Hirth, Barnett, and Lothe [429], devoted to dislocation arrays at interfaces in bicrystals.

[256] proposed a more general method for studying the motion of a dislocation loop whose shape changes during the motion [1].

Consider first a *screw dislocation* which moves uniformly in an infinite isotropic elastic medium. Choose the $x_3$-axis of a Cartesian system of co-ordinates along the positive direction of the dislocation, and the $x_1$-axis along the dislocation velocity vector. The equations of motion are given by (6.28) with $f_k$ replaced by the inertial term $-\rho_0 a_k$, i.e.

$$(\lambda + \mu)u_{m,mk} + \mu \Delta u_k = \rho_0 \frac{\partial^2 u_k}{\partial t^2}, \qquad k = 1, 2, 3. \tag{13.1}$$

Since the displacement vector must have the form $\mathbf{u}(0, 0, u_3)$, where $u_3$ depends only on $x_1$, $x_2$, and $t$, we see that the first two equations (13.1) are identically satisfied, while the third one reduces to

$$\left(\frac{\partial^2}{\partial x_1^2} + \frac{\partial^2}{\partial x_2^2} - \frac{1}{c_t^2}\frac{\partial^2}{\partial t^2}\right) u_3 = 0, \tag{13.2}$$

where $c_t = (\mu/\rho_0)^{1/2}$ is the speed of the elastic transverse waves.

Making in (13.2) the change of variables of "relativistic" type

$$x_1' = (x_1 - vt)/\gamma_t, \quad x_2' = x_2, \quad x_3' = x_3, \quad t' = (t - vx/c_t^2)/\gamma_t, \tag{13.3}$$

where $v$ is the (constant) dislocation speed and $\gamma_t = (1 - v^2/c_t^2)^{1/2}$, we obtain

$$\left(\frac{\partial^2}{\partial x_1'^2} + \frac{\partial^2}{\partial x_2'^2}\right) u_3 = \frac{1}{c_t^2}\frac{\partial^2 u_3}{\partial t'^2}, \tag{13.4}$$

i.e. the form of (13.2) is preserved in the new variables. On the other hand, since the new frame moves uniformly together with the dislocation, and $v$ is a constant, the right-hand side of (13.4) must vanish. Moreover, the displacement field must satisfy with respect to the moving system of co-ordinates the same "equilibrium" equations and the same jump condition (8.58) as a stationary screw dislocation with respect to a fixed frame.

Hence we conclude that the solution is given by

$$u_3 = -\frac{b\theta'}{2\pi}, \qquad \theta' \in (-\pi, \pi), \tag{13.5}$$

---

[1] In Nabarro's approach the motion and/or extension of a dislocation loop is represented by the sudden creation and annihilation of infinitesimal loops along the primary loop. In particular, the stress tensor of the moving dislocation loop can be calculated by using again (11.3), but replacing G(x) by the time-dependent Green's tensor function G(x, $t$), and performing an extra integration in time over the elementary acts of creation.

where, in view of (8.37) and (13.3),

$$\theta' = \begin{cases} \cot^{-1} \dfrac{x_1 - vt}{\gamma_t x_2} & \text{for } x_2 > 0 \\ 0 & \text{for } x_2 = 0,\ x_1 > vt \\ \cot^{-1} \dfrac{x_1 - vt}{\gamma_t x_2} - \pi & \text{for } x_2 < 0. \end{cases} \tag{13.6}$$

Next, by substituting (13.5) into (6.1) and the result obtained into (6.5), we deduce that the only non-zero components of the stress tensor are

$$T_{13} = \frac{\mu b}{2\pi} \frac{\gamma_t x_2}{(x_1 - vt)^2 + \gamma_t^2 x_2^2}, \qquad T_{23} = -\frac{\mu b}{2\pi} \frac{\gamma_t (x_1 - vt)}{(x_1 - vt)^2 + \gamma_t^2 x_2^2}. \tag{13.7}$$

In order to determine the total energy of the moving dislocation we must add to the strain energy also the kinetic energy, whose density per unit volume is

$$\frac{\rho_0}{2} \left( \frac{\partial u_3}{\partial t} \right)^2 = \frac{\mu}{2c_t^2} \left( \frac{\partial u_3}{\partial t} \right)^2.$$

Then, by taking into account (13.5—7), we deduce, by a calculation similar to that performed in Sect. 8.2, that the total energy per unit dislocation length, stored between two circular cylindrical surfaces of radii $r_0$ and $R$ and having the dislocation line as axis is

$$e = \frac{w}{\gamma_t} = \frac{\mu b^2}{4\pi \gamma_t} \ln \frac{R}{r_0}. \tag{13.8}$$

Since $\gamma_t \to 0$ as $v \to c_t$, the dislocation energy tends to infinity as $v \to c_t$. Hence, within the framework of linear elasticity, the dislocation cannot achieve speeds greater than the limiting value $c_t$. This "relativistic" effect should be probably corrected when the strong distortions in the dislocation core are also taken into account.

By expanding $\gamma_t^{-1}$ in a power series with respect to $(v/c_t)^2$ and retaining (for $v \ll c_t$) only the first term of the expansion, we obtain from (13.8)

$$e = w + \frac{w}{c_t^2} \frac{v^2}{2}. \tag{13.9}$$

Since the second term in the right-hand side of this equation represents the kinetic energy per unit dislocation length, the coefficient

$$m_t = w/c_t^2 \tag{13.10}$$

is sometimes called the *effective mass* of the dislocation per unit length. Moreover, it may be shown that this analogy may be also applied when writing down the equation of motion of the dislocation under the action of applied loads and of various dragging forces.

Let us consider now a uniformly moving *edge dislocation* in an infinite isotropic elastic medium. Choose the $x_3$-axis along the positive direction of the dislocation line, and the $x_1$-axis along the common direction of the Burgers vector and of the dislocation velocity vector (Fig. 13.1). Since the necessary calculation is much more intricate than for the screw dislocation, we confine ourselves to indicate the results concerning the stress field, referring for details to Eshelby [106] and Hirth and Lothe [162], Sect. 7.3. The only non-zero stress components are

$$\left.\begin{aligned} T_{11} &= \frac{bx_2c_t^2}{\pi v^2}\left[\frac{\lambda(1-\gamma_l^2)+2\mu}{\gamma_l r_l^2} - \frac{\mu(1+\gamma_t^2)}{\gamma_t r_t^2}\right], \\ T_{12} &= \frac{\mu b c_t^2(x_1-vt)}{2\pi v^2}\left[\frac{(1+\gamma_t^2)^2}{\gamma_t^3 r_t^2} - \frac{4}{\gamma_l r_l^2}\right], \\ T_{22} &= \frac{bx_2c_t^2}{\pi v^2}\left[\frac{\lambda-\gamma_l^2(\lambda+2\mu)}{\gamma_l r_l^2} + \frac{\mu(1+\gamma_t^2)}{\gamma_t r_t^2}\right], \\ T_{33} &= \nu(T_{11}+T_{22}), \end{aligned}\right\} \quad (13.11)$$

where

$$r_t^2 = (x_1 - vt)^2/\gamma_t^2 + x_2^2, \quad r_l^2 = (x_1 - vt)^2/\gamma_l^2 + x_2^2,$$

$$\gamma_t^2 = 1 - v^2/c_t^2, \quad \gamma_l^2 = 1 - v^2/c_l^2,$$

while $c_t = \sqrt{\mu/\rho_0}$ and $c_l = \sqrt{(\lambda + 2\mu)/\rho_0}$ are the speeds of the transverse and longitudinal elastic plane waves, respectively. It may be proved that the stress state (13.11) reduces as $v \to 0$ to that determined in Sect. 8.1 for the stationary edge dislocation. In addition, it may be shown (Weertman [376]) that the total energy of the dislocation per unit length tends to infinity as $v \to c_t$ and that for $v \ll c_t$ the effective mass per unit length of the edge dislocation is given by

$$m = \frac{w}{c_t^2}\left(1 + \frac{c_t^4}{c_l^4}\right), \quad (13.12)$$

where $w$ is the strain energy (8.54) stored per unit dislocation length.

The above analysis has been recently extended by Moos and Hoover [452] to uniformly moving edge dislocations in an elastic strip of finite width having clamped boundaries.

A problem of particular interest is the interaction of moving dislocations. As shown by Weertman [376], edge dislocations of like sign and gliding in the same plane will attract rather than repel one another provided their speed ranges between the Rayleigh vave speed $c_r$ and $c_t$. This anomalous behaviour may be explained by

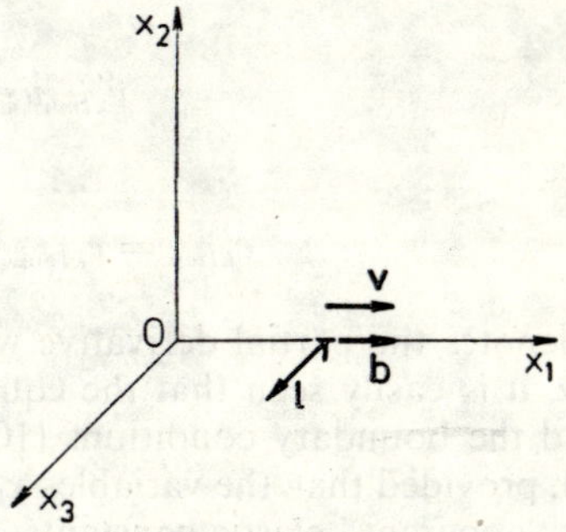


Fig. 13.1. Uniformly moving edge dislocation.

the fact that the kinetic energy of a moving dislocation may surpass its potential energy at sufficiently high speeds ($v > c_r$). Screw dislocations do not display a similar behaviour since their kinetic energy can never be greater than their potential energy.

## 13.3. Uniformly moving dislocations in anisotropic media

The first systematic study of the uniform motion of dislocations in anisotropic media has been undertaken by Sáenz [280]. Later, Bullough and Bilby [45] and Teutonico [346] have investigated the case of the straight dislocations perpendicular to a plane of material symmetry, by extending to the dynamic case the method used by Eshelby, Read, and Shockley [109] for stationary dislocations. The same method has been used by Cotner and Weertman [81], Van Hull and Weertman [367], and Weertman [377, 378] for more complicated situations in which the elastic states produced by edge and screw dislocations do not separate from each other, as well as by Teutonico [347] and Stroh [324], who considered the general anisotropic case and dislocations of arbitrary orientation and character.

Assume that an infinite straight dislocation of Burgers vector **b** moves with the constant velocity **v** in an infinite elastic medium with arbitrary anisotropy. Choosing as before the $x_3$-axis of a Cartesian frame along the positive direction of the dislocation line, and the $x_1$-axis parallel to **v**, the elastic state will be independent of $x_3$ and will depend on $x_1$ and $t$ only through the combination $x_1 - vt$, where $v = \|\mathbf{v}\|$, i.e.

$$\mathbf{u} = \mathbf{u}(x_1 - vt, x_2), \qquad \mathbf{T} = \mathbf{T}(x_1 - vt, x_2). \tag{13.13}$$

Introducing the new independent variables

$$x_1' = x_1 - vt, \quad x_2' = x_2, \quad x_3' = x_3, \tag{13.14}$$

the equations of motion

$$c_{klmn}u_{m,nl} = \rho_0 \frac{\partial^2 u_k}{\partial t^2}$$

become

$$c'_{klmn}u_{m,n'l'} = 0, \tag{13.15}$$

where

$$c'_{klmn} = c_{klmn} - \rho_0 v^2 \delta_{km}\delta_{l1}\delta_{n1}, \tag{13.16}$$

and $(.)_{,l'}$ denotes the partial derivative with respect to $x'_l$, $l = 1, 2$.

Now, it is easily seen that the equations of equilibrium, the jump conditions (10.18), and the boundary conditions (10.19) — (10.22) from the static case remain unchanged, provided that the variables $x_k$ are replaced by $x'_k$ and the elastic constants $c_{klmn}$ by the "apparent" elastic constants $c'_{klmn}$. Nevertheless, the solution of the sextic equation (10.9) requires a special discussion. Since the apparent elastic constants depend on $v$, the nature of the roots of this equation will depend on $v$, too. We have seen in Sect. 10.1 that for $v = 0$ the sextic equation (10.9) admits only complex roots. On the other hand, it may be shown that for sufficiently large values of $v$ all roots are real. Consequently, there exist in general three critical speeds, say $V_1 \geqslant V_2 \geqslant V_3$, such that for $v_\alpha = V_\alpha$, $\alpha = 1, 2, 3$, the pair of complex conjugate roots $p_\alpha$, $\bar{p}_\alpha$ changes from complex to real values. For $v < V_3$ the roots $p_\alpha$ are all complex and the dynamic elastic state will have the same features as the static one, the dislocation motion being described as *subsonic*. For $v > V_3$ at least two roots are real and the motion is accompanied by the generation of waves, being accordingly termed *supersonic* (Stroh [324]).

We confine ourselves to considering only the subsonic case. Then, by neglecting the core boundary conditions, we infer, in view of the results given in Sect. 10.3, that

$$u_k = \frac{1}{\pi} \operatorname{Im} \sum_{\alpha=1}^{3} A'_{k\alpha} D'_\alpha \ln z'_\alpha + u^0_k, \tag{13.17}$$

$$T_{1k} = -\frac{1}{\pi} \operatorname{Im} \sum_{\alpha=1}^{3} \frac{p'_\alpha L'_{k\alpha} D'_\alpha}{z'_\alpha}, \qquad T_{2k} = \frac{1}{\pi} \operatorname{Im} \sum_{\alpha=1}^{3} \frac{L'_{k\alpha} D'_\alpha}{z'_\alpha}, \tag{13,18}$$

where

$$z'_\alpha = x_1 - vt + p'_\alpha x_2, \quad \operatorname{Im} p_\alpha > 0, \quad \alpha = 1, 2, 3, \tag{13.19}$$

while the parameters $p'_\alpha$, $A'_{k\alpha}$, $L'_{k\alpha}$, $D'_\alpha$ can be calculated from $c'_{klmn}$ following the same prescription as that used to derive the parameters $p_\alpha$, $A_{k\alpha}$, $L_{k\alpha}$, $D_\alpha$ from $c_{klmn}$ in the static case.

It may be shown (Stroh [324]) that for $v \to V_3$ the prelogarithmic factor of the total energy tends to infinity. Hence $V_3$ plays in the anisotropic case the role of an upper speed limit, like the speed $c_t$ of the transverse elastic plane waves in the isotropic case. The numerical calculations done by Teutonico [346, 347] have shown that, as in the isotropic case, there exists a speed range, depending on the material symmetry and on the dislocation orientation, for which two like edge dislocations attract rather repel one another. Moreover, in the general anisotropic case, when the elastic states corresponding to screw and edge dislocations do not separate from each other, both screw and edge dislocations can display this anomalous behaviour.

Besides the problems already mentioned, Stroh [324] has studied the waves generated by the supersonic dislocation motion, as well as the uniformly moving dislocations in an anisotropic elastic half-plane. Beltz, Davis, and Málen [24] have extended Brown's formula (11.4) to uniformly moving dislocation loops, while Málen [223] has investigated the stability of moving dislocations.

The motion of single dislocations can be also studied by using the theory of continuous distributions of dislocations, especially by means of Green's tensor functions of elastodynamics (see, e.g. Mura [253], Stenzel [319, 320], and Kosevich [440], Sect. 7, for the linear theory, and Bahr and Schöpf [397] for the non-linear theory). Partial atomic models of moving dislocations have been also considered. Thus, Stenzel [320] has proposed a slightly modified Peierls-Nabarro model to describe the motion of a dislocation under the action of external loads, while Rogula [466] has elaborated a more complex model, called "pseudo-continuum", which includes some of the non-local crystal properties in the continuum description.

CHAPTER III

# NON-LINEAR EFFECTS IN THE ELASTIC FIELD OF SINGLE DISLOCATIONS

We have seen in the preceding chapter that dislocations may be described as line singularities of the elastic field: linear elasticity theory predicts stresses and strains that vary as the inverse first power of the distance from the dislocation line and, therefore, are unbounded as this distance goes to zero. Thus, close to the singularities the strains become very large, and non-linear effects must be taken into account. On the other hand, in regions sufficiently far from dislocations, the stresses and strains are sufficiently small and may be adequately described by the linear theory. For this reason, and also on account of its simplicity, the linear theory of elasticity continues to be successfully applied for simulating crystal defects, e.g. in the study of the long-range stress field of dislocations, the interaction between distant imperfections, and in the calculation of defect energies [1].

The above discussion suggests that one of the main applications of the non-linear theory of elasticity in modelling dislocations could be the study of the dislocation core. However, this requires caution. Indeed, very close to the dislocation line the atomistic nature of the crystal defect is just as important as the deviations from linear elasticity, so that a local continuum theory, even taking into account non-linear effects, cannot give a complete description of the dislocation core. Moreover, the second-order elasticity, which is the only approximation of non-linear constitutive equations for which sufficient experimental data are available at present, proves to be inadequate for the description of the large strains of the order 50 percent or more occurring in the very neighbourhood of the dislocation line, since it does not allow for the potential energy of crystal to be a periodic function of the relative displacement of two neighbouring crystal planes. Therefore, the right solution in applying non-linear elasticity to the study of the dislocation core requires the coupling of the non-linear elastic model with the atomic one, and the use of semidiscrete methods (Sect. 16).

There still exists a different kind of applications of the non-linear theory of elasticity in modelling crystal defects (cf. Seeger [292]). Thus, we may be interested to study a quantity for which the linear theory gives an unrealistic evaluation, e.g. a vanishing result. For instance, linear elasticity predicts a vanishing effect of dislocations on the average strains, and hence on the macroscopic density of crystals;

---

[1] Actually, most of the dislocation energy is stored in the long-range stress field; only about ten percent of the dislocation energy is due to the dislocation core (cf. also Sect. 16.1).

on the contrary, second-order elasticity indicates, in agreement with experimental results, that dislocations produce a positive volume expansion (Sect. 15).

There exist also typical non-linear coupling effects, which simply disappear when applying the superposition principle valid in the linear theory. The best known example is the scattering of elastic waves from dislocations and other imperfections. When the linear elastic field of dislocations and that of the elastic waves are superposed they do not perturb each other. The experimentally observed scattering of elastic waves by strain fields can again be accounted for only within the framework of non-linear elasticity (Sect. 20).

In all the examples above, second-order elastic effects can no longer be ignored, although strains are still not very high. This suggests the solving of the non-linear boundary-value problems by an iteration procedure, in which a linear elastic boundary-value problem is to be solved at each step; generally, the first two steps are sufficient to exhibit the lowest order non-linear effects looked for. That is why we will begin this chapter by expounding an iteration scheme which is adequate for the study of the elastic field of single dislocations (Sect. 14). Similar iteration schemes that are applicable in the case of continuous distributions of dislocations will be given in Sects. 19 and 20.

## 14. Solving of non-linear boundary-value problems by successive approximations

### 14.1. Willis' scheme

The elastic field equations have been linearized in Sects. 2 and 4 under the assumption that the magnitude of the displacement gradient **H** is much smaller than unity, and neglecting second and higher powers of **H**. The solution of non-linear elastic boundary-value problems can be found by an iteration scheme involving the solution of a linear boundary-value problem at each step, the first one being given by the linear theory of elasticity. To develop such a scheme, we again assume that $\|\mathbf{H}\| < 1$, but take also into consideration higher powers of **H**. Specifically, we keep at the $n$'th stage of the iteration all terms up to and including those of $n$'th order in $\|\mathbf{H}\|$.

The first systematic iterative method for the solution of non-linear elastic boundary-value problems has been elaborated in 1930 by Signorini [298], who further developed his ideas in [299, 300]. Later on, Signorini's scheme has been independently generalized by Green and Spratt [146] and by Rivlin and Tapakoglu [279] (see also Truesdell and Noll [358], Sect. 63, and Capriz and Guidugli [55]). Willis [382] has adapted Signorini's scheme to the case of single dislocations and of continuous distributions of dislocations, by using Eulerian co-ordinates. This approach will be preferred in this book, since it avoids the rather complicated discussion implied by the correct definition of the deformation produced by dislocations in terms of Lagrangian co-ordinates[1]. In what follows, Willis' scheme will be extended to include the influence of the core boundary conditions.

Consider an elastic body $\mathscr{B}$, free of surface and body forces, occupying a simply-connected region $\mathscr{V}$ of boundary $\mathscr{S}$ in a deformed configuration ($k$), containing a dislocation of line $L$. Assume that $L$ is either a closed curve in $\mathscr{V}$ or a line ending at $\mathscr{S}$. Denote by $\mathscr{V}_0$ the doubly-connected region obtained by cutting out a thin tube of boundary $\Sigma_0$ around the dislocation line. Let $S$ be a smooth and two-sided barrier rendering $\mathscr{V}_0$ simply-connected. By cutting the body along $S$ and allowing it to relax, it will occupy a natural configuration ($K$). Denote by $\mathbf{X}$ and $\mathbf{x}$ the position vectors of a current material point $X$ in the configurations ($K$) and ($k$), respectively. For the sake of simplicity, we assume that after the deformation from ($k$) to ($K$) the cut faces remain in contact with each other. Then, as shown in Sect. 7.3, the deformation produced by the dislocation may be described by one of the mappings (7.1) or (7.4).

We choose arbitrarily a positive sense on $L$ and define the positive side $S^+$ and the negative side $S^-$ of $S$ according to the convention adopted in Sect. 7.3 and illustrated in Figs. 7.5 and 7.6. Then, the jump of the displacement vector $\mathbf{u}(\mathbf{x})$ across $S$ is given by

$$\mathbf{u}^+(\mathbf{x}) - \mathbf{u}^-(\mathbf{x}) = \mathbf{b}, \tag{14.1}$$

where $\mathbf{x}$ is the position vector of a current point on $S$, $\mathbf{u}^+(\mathbf{x})$ and $\mathbf{u}^-(\mathbf{x})$ denote the limiting values of $\mathbf{u}(\mathbf{x})$ on $S^+$ and $S^-$, respectively, and $\mathbf{b}$ is the true Burgers vector.

The Cartesian components of the finite strain tensor are given by (2.30) as

$$D_{kl} = \tfrac{1}{2}\,(H_{kl} + H_{lk} + H_{pk}H_{pl}). \tag{14.2}$$

When using Eulerian co-ordinates, it is necessary to express $\mathbf{H}$ in terms of $\operatorname{grad}\mathbf{u}$. To this end we introduce (7.2) and (7.5) into the relation $\mathbf{F}\mathbf{F}^{-1} = \mathbf{1}$, thus obtaining

$$(\mathbf{1} + \mathbf{H})(\mathbf{1} - \operatorname{grad}\mathbf{u}) = \mathbf{1},$$

wherefrom it follows that

$$H_{kl} = \mathbf{u}_{k,l} + H_{kr}u_{r,l}, \tag{14.3}$$

where $(\cdot)_{,l}$ denotes, as usual, differentiation with respect to $x_l$.

By taking into account (7.2), the non-linear constitutive equation of elastic materials (4.40) may be written as

$$T_{kl} = j(\delta_{km} + H_{km})\,\frac{\partial W(\mathbf{D})}{\partial D_{mn}}\,(\delta_{ln} + H_{ln}), \tag{14.4}$$

---

[1] In this connection see also Sinclair *et al.* [474], Petrasch [460], and Teodosiu and Soós [479]. Alternative iteration schemes using Lagrangian co-ordinates have been also developed and applied for determining the non-linear elastic field of single straight dislocations (Seeger and Mann [289], Teodosiu [337], vol. 2, Sect. 15, Seeger, Teodosiu, and Petrasch [295], Gairola [418]).

where, by virtue of (7.5) and (4.43),

$$j = \det \mathbf{F}^{-1} = \det [\delta_{pr} - u_{p,r}], \tag{14.5}$$

$$\frac{\partial W(\mathbf{D})}{\partial D_{mn}} = c_{mnpq} D_{pq} + \frac{1}{2} C_{mnpqrs} D_{pq} D_{rs} + \ldots, \tag{14.6}$$

and **c** and **C** are the tensors of the second- and third-order elastic constants, respectively.

In the absence of body forces, the balance of the linear momentum requires that

$$T_{kl,l} = 0. \tag{14.7}$$

Assuming that the external surface $\mathscr{S}$ of the body is free of external loads, we have

$$\mathbf{Tn} = \mathbf{0} \quad \text{on } \mathscr{S}, \tag{14.8}$$

where **n** is the outward unit normal to $\mathscr{S}$. In order to determine the elastic state produced by the dislocation it is necessary to prescribe either the boundary value $\mathbf{u}^*$ of the displacement vector on $\Sigma_0$, or the traction $\mathbf{t}^*$ exerted upon $\Sigma_0$ from the dislocation core[1]. Thus, we must have either

$$\mathbf{u} = \mathbf{u}^* \quad \text{on } \Sigma_0, \tag{14.9}$$

or

$$\mathbf{Tn} = \mathbf{t}^* \quad \text{on } \Sigma_0. \tag{14.10}$$

Since the deformed body is in equilibrium, we require that the resultant force **P** and couple **M** of the tractions acting on $\Sigma_0$ vanish, i.e.

$$\mathbf{P} \equiv \int_{\Sigma_0} \mathbf{t}^* \, \mathrm{d}s = \mathbf{0}, \tag{14.11}$$

$$\mathbf{M} \equiv \int_{\Sigma_0} \mathbf{x} \times \mathbf{t}^* \, \mathrm{d}s = \mathbf{0}. \tag{14.12}$$

When the traction boundary condition (14.10) is satisfied, equations (14.11) and (14.12) are merely restrictions on the given surface traction $\mathbf{t}^*$.

---

[1] The vectors $\mathbf{u}^*$ and $\mathbf{t}^*$ can be evaluated by using semidiscrete methods that combine the elastic and the atomistic models of the dislocation (see Sect. 16). In this case, the equilibrium of the forces acting on $\Sigma_0$ is assured by modifying the positions of the atoms inside the dislocation core at the same time with the elastic state in $\mathscr{V}_0$.

We will solve the traction non-linear boundary-value problem formulated above by an iteration scheme, based on the following hypotheses:

(i) The prescribed traction on $\Sigma_0$ and the true Burgers vector are proportional to a small parameter $\varepsilon$, i.e.

$$\mathbf{t}^* = \varepsilon \mathbf{t}^{*(1)}, \qquad \mathbf{b} = \varepsilon \mathbf{b}^{(1)}. \tag{14.13}$$

This hypothesis is justified by the fact that $\mathbf{t}^*$ vanishes together with $\mathbf{b}$. The numerical choice of $\varepsilon$ is immaterial, since it appears in the final result only through the combinations $\varepsilon \mathbf{t}^{*(1)}$ and $\varepsilon \mathbf{b}^{(1)}$.

(ii) There exists a solution $\mathbf{u}(\mathbf{x})$ of the boundary-value problem that depends analytically on $\varepsilon$ and vanishes for $\varepsilon = 0$, i.e.

$$\mathbf{u} = \varepsilon \mathbf{u}^{(1)} + \varepsilon^2 \mathbf{u}^{(2)} + \dots \tag{14.14}$$

Introducing (14.14) into (14.3), we obtain

$$H_{kl} = \varepsilon u^{(1)}_{k,l} + \varepsilon^2 (u^{(2)}_{k,l} + u^{(1)}_{k,p} u^{(1)}_{p,l}) + \dots \tag{14.15}$$

Next, substituting (14.15) into (14.2) yields

$$D_{kl} = \frac{\varepsilon}{2} (u^{(1)}_{k,l} + u^{(1)}_{l,k}) + \frac{\varepsilon^2}{2} (u^{(2)}_{k,l} + u^{(2)}_{l,k} + u^{(1)}_{k,p} u^{(1)}_{p,l} + u^{(1)}_{p,k} u^{(1)}_{p,l} + u^{(1)}_{p,k} u^{(1)}_{l,p}) + \dots \tag{14.16}$$

Then, introducing (14.14), (14.16) into (14.5), (14.6), respectively, and the results obtained into (14.4), and considering also (14.15), we deduce that

$$T_{kl} = \varepsilon T^{(1)}_{kl} + \varepsilon^2 T^{(2)}_{kl} + \dots, \tag{14.17}$$

where

$$T^{(1)}_{kl} = c_{klmn} u^{(1)}_{m,n}, \qquad T^{(2)}_{kl} = c_{klmn} u^{(2)}_{m,n} + \tau_{kl}, \tag{14.18}$$

$$\tau_{kl} = -u^{(1)}_{m,m} T^{(1)}_{kl} + u^{(1)}_{k,m} T^{(1)}_{ml} + u^{(1)}_{l,m} T^{(1)}_{km} + + c_{klmn} (u^{(1)}_{m,p} u^{(1)}_{p,n} + \tfrac{1}{2} u^{(1)}_{p,m} u^{(1)}_{p,n}) + \tfrac{1}{2} C_{klmnrs} u^{(1)}_{m,n} u^{(1)}_{r,s}. \tag{14.19}$$

Substituting now (14.13), (14.14), (14.17) into (14.1), (14.7), (14.8) and (14.10), and equating like powers of $\varepsilon$, we obtain a sequence of linear traction boundary-value problems, namely, at the *first step*

$$\left.\begin{aligned} & u^{(1)+}_k(\mathbf{x}) - u^{(1)-}_k(\mathbf{x}) = b^{(1)}_k \quad \text{on } S,\\ & T^{(1)}_{kl,l} = 0, \quad T^{(1)}_{kl} = c_{klmn} u^{(1)}_{m,n} \quad \text{in } \mathscr{V}_0 \setminus S,\\ & T^{(1)}_{kl} n_l = \begin{cases} 0 & \text{on } \mathscr{S}\\ t^{*(1)}_k & \text{on } \Sigma_0, \end{cases} \end{aligned}\right\} \tag{14.20}$$

at the *second step*

$$\left.\begin{aligned} u_k^{(2)+}(\mathbf{x}) - u_k^{(2)-}(\mathbf{x}) = 0 \qquad \text{on } S, \\ T_{kl,l}^{(2)} = 0, \qquad T_{kl}^{(2)} = c_{klmn}u_{m,n}^{(2)} + \tau_{kl} \qquad \text{in } \mathscr{V}_0 \setminus S, \\ T_{kl}^{(2)}n_l = 0 \qquad \text{on } \mathscr{S} \cup \Sigma_0, \end{aligned}\right\} \tag{14.21}$$

and so on. In the following we will consider only the first two steps of the iteration, for the subsequent steps involve elastic constants of fourth and higher orders, which, in general, are not available.

The traction boundary-value problems (14.20) and (14.21) can be formulated in terms of the displacement fields $\mathbf{u}^{(1)}$ and $\mathbf{u}^{(2)}$, respectively, as

$$\left.\begin{aligned} & u_k^{(1)+}(\mathbf{x}) - u_k^{(1)-}(\mathbf{x}) = b_k^{(1)} \qquad \text{on } S, \\ & c_{klmn}u_{m,nl}^{(1)} = 0 \quad \text{in } \mathscr{V}_0 \setminus S, \\ & c_{klmn}u_{m,n}^{(1)}n_l = \begin{cases} 0 & \text{on } \mathscr{S} \\ t_k^{*(1)} & \text{on } \Sigma_0 \end{cases} \end{aligned}\right\} \tag{14.22}$$

and

$$\left.\begin{aligned} & u_k^{(2)+}(\mathbf{x}) - u_k^{(2)-}(\mathbf{x}) = 0 \qquad \text{on } S, \\ & c_{klmn}u_{m,nl}^{(2)} + f_k^{(2)} = 0 \quad \text{in } \mathscr{V}_0 \setminus S, \\ & c_{klmn}u_{m,n}^{(2)}n_l = t_k^{*(2)} \quad \text{on } \mathscr{S} \cup \Sigma_0, \end{aligned}\right\} \tag{14.23}$$

where

$$f_k^{(2)} = \tau_{kl,l}, \qquad t_k^{*(2)} = -\tau_{kl}n_l \tag{14.24}$$

play the role of a body force and a traction, respectively. It is easily proved, by using Gauss' theorem, that the resultant force and the resultant couple of the forces (14.24) are zero.

When the traction boundary condition (14.10) on $\Sigma_0$ is replaced by the displacement boundary condition (14.9), we put

$$\mathbf{u}^* = \varepsilon\mathbf{u}^{*(1)}. \tag{14.25}$$

Then, using the same technique as above, we find that the first two steps of the iteration involve the solution of two mixed linear boundary-value problems. Namely, the last equations (14.22) and (14.23) should be replaced by

$$\left.\begin{aligned} & c_{klmn}u_{m,n}^{(1)}n_l = 0 \quad \text{on } \mathscr{S}, \\ & u_k^{(1)} = u_k^{*(1)} \qquad \text{on } \Sigma_0, \end{aligned}\right\} \tag{14.26}$$

and, respectively,

$$\left.\begin{aligned} c_{klmn}u^{(2)}_{m,n}n_l &= t_k \quad \text{on } \mathscr{S}, \\ u^{(2)}_k &= 0 \quad \text{on } \Sigma_0. \end{aligned}\right\} \tag{14.27}$$

Clearly, the boundary-value problems to be solved at the first step correspond to a Volterra dislocation of translational type with prescribed tractions or displacements on the boundary of the dislocation core. Consequently, we shall also require the continuity across $S$ of the infinitesimal strain tensor corresponding to $\mathbf{u}^{(1)}$ together with its partial derivatives of the first two orders, as well as that of the infinitesimal rotation tensor corresponding to $\mathbf{u}^{(1)}$. Since grad $\mathbf{u}^{(1)}$ is continuous across $S$, so is $\boldsymbol{\tau}$, and hence the boundary-value problems corresponding to the second step are classical boundary-value problems of linear elasticity. Assuming as usual that the strain density function $W_2$ is positive definite it follows that any two solutions of the traction boundary-value problems differ by an infinitesimal rigid displacement.

In the original form of Signorini's scheme the arbitrary infinitesimal rotation corresponding to each step of the iteration is determined such that the body forces and surface tractions corresponding to the following step be equilibrated. On the other side, in the case of a single dislocation, the only external forces acting on the part of the body occupying the region $\mathscr{V}_0$ are those applied on the core boundary $\Sigma_0$. Moreover, these forces must be self-equilibrated in the deformed configuration of the body, since the dislocation core itself is in equilibrium. Consequently, when using a semidiscrete method, the infinitesimal rigid rotations occurring at each step must be used as adjustable parameters, together with the tractions on $\Sigma_0$ and the positions of the atoms inside $\Sigma_0$, for minimizing the potential energy of the whole body.

When studying the elastic field of a single dislocation loop lying in an *infinite* elastic medium, we shall require, on physical grounds, that the stress $\mathbf{T}$ vanish and the finite rotation $\mathbf{R}$ approach the unit tensor at infinity, i.e.

$$\lim_{\|\mathbf{x}\|\to\infty} \mathbf{T}(\mathbf{x}) = \mathbf{0}, \qquad \lim_{\|\mathbf{x}\|\to\infty} \mathbf{R}(\mathbf{x}) = \mathbf{1}, \tag{14.28}$$

since the lattice distortion gradually disappears at large distances from the dislocation line.

By using (2.11—13) and (14.14—17), it may be shown [479] that, up to the second step of the iteration, equations (14.28) are equivalent to

$$\lim_{\|\mathbf{x}\|\to\infty} \mathbf{T}^{(n)}(\mathbf{x}) = \mathbf{0}, \qquad \lim_{\|\mathbf{x}\|\to\infty} \boldsymbol{\omega}^{(n)}(\mathbf{x}) = \mathbf{0}, \qquad n = 1, 2, \tag{14.29}$$

where $\boldsymbol{\omega}^{(n)}$ denotes the infinitesimal rotation vector at the $n$'th step. Replacing the boundary conditions on $\mathscr{S}$ by $(14.29)_1$, and the condition (14.12) by $(14.29)_2$, we conclude, by virtue of Bézier's theorem (Sect. 6), that the solutions of the successive linear traction boundary-value problems are unique to within an infinitesimal translation. It is worth noting that, in view of (14.18) and (14.19), equations (14.29) are also equivalent to

$$\lim_{\|\mathbf{x}\|\to\infty} \operatorname{grad} \mathbf{u}^{(n)}(\mathbf{x}) = \mathbf{0}, \quad n = 1, 2. \tag{14.30}$$

Finally, if the displacements instead of the tractions are prescribed on $\Sigma_0$, then, by virtue of Bézier's theorem, each linear boundary-value problem for the infinite medium has at most one solution satisfying (14.30).

## 14.2. Second-order effects in the anisotropic elastic field of an edge dislocation

Second-order effects in the *isotropic* elastic field of an edge dislocation were determined by Pfleiderer, Seeger, and Kröner [270], by disregarding core effects and applying an iteration scheme formulated in Eulerian co-ordinates, which had been previously elaborated by Kröner and Seeger [191].

Seeger, Teodosiu, and Petrasch [295] have determined the second-order effects in the *anisotropic* elastic field of an edge dislocation, under consideration of the core boundary conditions, and using an iteration scheme formulated in Lagrangian co-ordinates. Their results have been completed and partly corrected by Teodosiu and Soós [479], by using Willis' iteration scheme in Eulerian co-ordinates and removing some residual discontinuities occurring across the cut surface in the second step of the iteration, corresponding to a generalized Somigliana dislocation [1]. In the following, we shall summarize the results obtained by these authors.

Consider a straight edge dislocation lying in an infinite anisotropic elastic medium $\mathscr{V}$ and take the axes $x_1$ and $x_3$ of a Cartesian system of co-ordinates along the positive direction of the dislocation line and along the true Burgers vector, respectively. Let us assume that the dislocation lies along a two-fold axis of material symmetry or, equivalently, that any plane which is parallel in the natural state to the $x_1x_2$-plane is a plane of material symmetry.

We apply the non-linear theory of elasticity to the region $\mathscr{V}_0$ situated outside the dislocation core, the latter being considered as an infinite tube bounded by a circular cylindrical surface $\Sigma_0$ of radius $r_0$ and axis $x_3$. Let us denote by $\Gamma_0$ the intersection line of $\Sigma_0$ with the $x_1x_2$-plane and by $\Delta$ the region outside $\Gamma_0$ in this plane (Fig. 10.1). We assume, on physical grounds, that the tractions acting on $\Sigma_0$ from the dislocation core do not depend on $x_3$ and are parallel to the $x_1x_2$-plane, and hence the elastic medium is subjected to a state of plane strain. Then, by taking the half-plane $x_2 = 0$, $x_1 \leqslant -r_0$ as cut surface $S$ rendering $\mathscr{V}_0$ simply-connected, we may define in the region $\mathscr{V}_0 \setminus S$ a single-valued displacement field, whose Cartesian components must have the form [2].

$$u_\alpha = u_\alpha(x_\beta), \qquad u_3 = 0, \tag{14.31}$$

where $x_\beta$ are the co-ordinates in the $x_1x_2$-plane of a current material point $X$ in the deformed configuration. Moreover, since now $\mathbf{b} = b\mathbf{e}_1$, where $b = \varepsilon b^{(1)}$ is the magnitude of the true Burgers vector, the jump conditions $(14.20)_1$ and $(14.21)_1$ corresponding to the first two steps of the iteration assume the simplified form

$$\left.\begin{aligned} u_\alpha^{(1)}(x_1, 0^+) - u_\alpha^{(1)}(x_1, 0^-) &= -\delta_{\alpha 1} b^{(1)}, \\ u_\alpha^{(2)}(x_1, 0^+) - u_\alpha^{(2)}(x_1, 0^-) &= 0 \end{aligned}\right\} \tag{14.32}$$

[1] Similar results have been obtained by Petrasch [460] in the particular case of the orthotropic medium, by using a different reasoning, based on symmetry and continuity conditions.

[2] Throughout this subsection Greek indices range over the values 1, 2; the summation over these indices will be always explicitly indicated by the symbol $\sum_\alpha$.

for $x_1 \in (-\infty, -r_0]$. According to the discussion above, we shall also require that

$$\lim_{\rho \to \infty} u^{(n)}_{\alpha,\beta}(x_1, x_2) = 0, \qquad n = 1, 2, \tag{14.33}$$

where $\rho = \sqrt{x_1^2 + x_2^2}$.

The *first linear boundary-value problem* (14.20), specialized to conditions (14.31—33), has been solved by Teodosiu and Nicolae [338], by using a complex-variable technique. Since this technique is similar to that employed in Sect. 8.1 for the isotropic case, we reproduce here only the main intermediate results that are further used in the second step of the iteration.

By introducing the complex variables

$$z = x_1 + ix_2, \qquad \bar{z} = x_1 - ix_2,$$

the complex displacement

$$U^{(1)} = u_1^{(1)} + iu_2^{(1)}, \tag{14.34}$$

and the complex stresses [1]

$$\Theta^{(1)} = T_{11}^{(1)} + T_{22}^{(1)}, \qquad \Phi^{(1)} = T_{11}^{(1)} - T_{22}^{(1)} + 2iT_{12}^{(1)}, \tag{14.35}$$

we obtain from (14.20) and (14.31—33), like in Sect. 8.1, the jump condition

$$U^{(1)}(x_1, 0^+) - U^{(1)}(x_1, 0^-) = -b^{(1)} \text{ for } x_1 \in (-\infty, -r_0], \tag{14.36}$$

the equilibrium equation

$$\frac{\partial \Theta^{(1)}}{\partial \bar{z}} + \frac{\partial \Phi^{(1)}}{\partial z} = 0, \tag{14.37}$$

the constitutive equations

$$\left.\begin{aligned} 2\frac{\partial U^{(1)}}{\partial \bar{z}} &= A\Phi^{(1)} + B\bar{\Phi}^{(1)} + 2C\Theta^{(1)}, \\ \frac{\partial U^{(1)}}{\partial z} + \frac{\partial \bar{U}^{(1)}}{\partial \bar{z}} &= \bar{C}\Phi^{(1)} + C\bar{\Phi}^{(1)} + 2D\Theta^{(1)}, \end{aligned}\right\} \tag{14.38}$$

and the boundary conditions

$$\tfrac{1}{2}(\Theta^{(1)} + \Phi^{(1)} e^{-2i\theta}) = t_\rho^{*(1)} + it_\theta^{*(1)} \quad \text{for } z = r_0 e^{i\theta}, \quad \theta \in (-\pi, \pi], \tag{14.39}$$

$$\frac{\partial U^{(1)}}{\partial z}, \frac{\partial U^{(1)}}{\partial \bar{z}} \to 0 \text{ as } \rho \to \infty, \tag{14.40}$$

---

[1] Note that the stress component $T_{33} \neq 0$, but it does not intervene in the successive boundary-value problems. Its value can be obtained directly from (14.18) after the calculation of $\mathbf{u}^{(1)}$ and $\mathbf{u}^{(2)}$.

where

$$\frac{\partial}{\partial z} = \frac{1}{2}\left(\frac{\partial}{\partial x_1} - i\,\frac{\partial}{\partial x_2}\right), \quad \frac{\partial}{\partial \bar{z}} = \frac{1}{2}\left(\frac{\partial}{\partial x_1} + i\,\frac{\partial}{\partial x_2}\right), \tag{14.41}$$

$t_\rho^{*(1)}$ and $t_\theta^{*(1)}$ are the radial and the tangential components of $\mathbf{t}^{*(1)}$, respectively, $\theta$ is the polar angle in the $z$-plane, and $A$, $B$, $C$, $D$ are determined by the relations

$$4Ad = c_{66}(c_{11} + c_{22} + 2c_{12}) - (c_{16} + c_{26})^2 + c_{11}c_{22} - c_{12}^2,$$

$$4Bd = c_{66}(c_{11} + c_{22} + 2c_{12}) - (c_{16} + c_{26})^2 - c_{11}c_{22} + c_{12}^2 +$$

$$+ 2i[c_{26}(c_{12} + c_{11}) - c_{16}(c_{12} + c_{22})],$$

$$4Cd = c_{66}(c_{22} - c_{11}) + c_{16}^2 - c_{26}^2 + i[c_{16}(c_{12} - c_{22}) - c_{26}(c_{12} - c_{11})],$$

$$4Dd = c_{66}(c_{11} + c_{22} - 2c_{12}) - (c_{16} - c_{26})^2,$$

$$d = c_{66}(c_{11}c_{22} - c_{12}^2) + 2c_{12}c_{16}c_{26} - c_{11}c_{26}^2 - c_{22}c_{16}^2.$$

Next, the equilibrium condition (14.37) is identically satisfied by setting

$$\Phi^{(1)} = -4\,\frac{\partial^2 F^{(1)}}{\partial \bar{z}^2}, \qquad \Theta^{(1)} = 4\,\frac{\partial^2 F^{(1)}}{\partial z \partial \bar{z}}, \tag{14.42}$$

where $F^{(1)}$ is Airy's stress function. The function $F^{(1)}$ must also satisfy the compatibility equation written in stress components, i.e. the Beltrami-Michell compatibility equations (6.25) specialized to the state of plane strain. We can deduce these equations in a more direct way, by eliminating $U^{(1)}$ between equations (14.38) and taking into account (14.42). After some intermediate calculation, it results

$$\mathscr{L} F^{(1)} = 0, \tag{14.43}$$

where $\mathscr{L}$ is the real differential operator

$$\mathscr{L} \equiv B\,\frac{\partial^4}{\partial z^4} - 4C\,\frac{\partial^4}{\partial z^3 \partial \bar{z}} + 2(A + 2D)\,\frac{\partial^4}{\partial z^2 \partial \bar{z}^2} - 4\bar{C}\,\frac{\partial^4}{\partial z \partial \bar{z}^3} + \bar{B}\,\frac{\partial^4}{\partial \bar{z}^4}. \tag{14.44}$$

This operator can be decomposed into the product

$$\mathscr{L} \equiv \bar{B}\left(\frac{\partial}{\partial \bar{z}} - \gamma_1\,\frac{\partial}{\partial z}\right)\left(\frac{\partial}{\partial \bar{z}} - \gamma_2\,\frac{\partial}{\partial z}\right)\left(\frac{\partial}{\partial \bar{z}} - \gamma_3\,\frac{\partial}{\partial z}\right)\left(\frac{\partial}{\partial \bar{z}} - \gamma_4\,\frac{\partial}{\partial z}\right), \tag{14.45}$$

where $\gamma_k$, $k = 1, \ldots, 4$, are the roots of the algebraic reciprocal equation

$$\bar{B}\gamma^4 - 4\bar{C}\gamma^3 + 2(A + 2D)\gamma^2 - 4C\gamma + B = 0. \tag{14.46}$$

As shown by Green and Zerna [145], when the strain energy function is positive definite, the roots of equation (14.46) are complex, and their moduli cannot equal unity, so that they can be denoted by $\gamma_1, \gamma_2, 1/\bar{\gamma}_1, 1/\bar{\gamma}_2$, where $|\gamma_1| < 1$, $|\gamma_2| < 1$. For further use, we note that

$$\frac{\gamma_1\gamma_2}{\bar{\gamma}_1\bar{\gamma}_2} = \frac{B}{\bar{B}}, \qquad \gamma_1 + \gamma_2 + \frac{1}{\bar{\gamma}_1} + \frac{1}{\bar{\gamma}_2} = \frac{4\bar{C}}{\bar{B}}. \tag{14.47}$$

We assume [1] in the following that $\gamma_1 \neq \gamma_2$. Since the operators $\mathscr{L}_k \equiv \frac{\partial}{\partial \bar{z}} - \gamma_k \frac{\partial}{\partial z}$, $k = 1, \ldots, 4$, are linear independent and have constant coefficients, it follows by Boggio's theorem [401] that the general solution of equation (14.43) equals the sum of the general solutions of the equations $\mathscr{L}_k F^{(1)} = 0$, $k = 1, \ldots, 4$.

By introducing the new complex variables

$$z_1 = z + \gamma_1\bar{z}, \qquad \bar{z}_1 = \bar{z} + \bar{\gamma}_1 z, \tag{14.48}$$

we find that

$$\left(\frac{\partial}{\partial \bar{z}} - \gamma_1 \frac{\partial}{\partial z}\right)\left(\frac{\partial}{\partial \bar{z}} - \frac{1}{\bar{\gamma}_1}\frac{\partial}{\partial z}\right) \equiv -\frac{(1 - \gamma_1\bar{\gamma}_1)^2}{\gamma_1}\frac{\partial^2}{\partial z_1 \partial \bar{z}_1},$$

and hence the general real solution of the equation $\mathscr{L}_1\mathscr{L}_3F^{(1)} = 0$ is

$$F^{(1)}(x_1, x_2) = \Omega_1(z_1) + \overline{\Omega_1(z_1)},$$

where $\Omega_1(z_1)$ is an arbitrary analytic function of $z_1$. Analogously, the general real solution of the equation $\mathscr{L}_2\mathscr{L}_4F^{(1)} = 0$ is

$$F^{(1)}(x_1, x_2) = \Omega_2(z_2) + \overline{\Omega_2(z_2)},$$

where $\Omega_2(z_2)$ is an arbitrary analytic function of $z_2$, and

$$z_2 = z + \gamma_2\bar{z}, \qquad \bar{z}_2 = \bar{z} + \bar{\gamma}_2 z. \tag{14.49}$$

Hence, the general solution of Eq. (14.43) is

$$F^{(1)}(x_1, x_2) = 2\mathrm{Re} \sum_{\alpha} \Omega_\alpha(z_\alpha). \tag{14.50}$$

[1] The case $\gamma_1 = \gamma_2$ may be studied by a method similar to that used in the isotropic case.

Introducing (14.50) into (14.42) yields

$$\Phi^{(1)} = -4 \sum_{\alpha} [\gamma_\alpha^2 \Omega_\alpha''(z_\alpha) + \overline{\Omega_\alpha''(z_\alpha)}], \qquad \Theta^{(1)} = 8\mathrm{Re} \sum_{\alpha} \gamma_\alpha \Omega_\alpha''(z_\alpha). \tag{14.51}$$

Next, by substituting (14.51) into (14.38) and integrating the system of equations obtained, we find [1]

$$U^{(1)}(x_1, x_2) = \sum_{\alpha} [\delta_\alpha \Omega_\alpha'(z_\alpha) + \rho_\alpha \overline{\Omega_\alpha'(z_\alpha)}] + \omega_0^{(1)} iz + u_0^{(1)} + iv_0^{(1)}, \tag{14.52}$$

where $\omega_0^{(1)}$, $u_0^{(1)}$, $v_0^{(1)}$ are arbitrary real constants and

$$\delta_\alpha = -2(A\gamma_\alpha - 2C + B/\gamma_\alpha), \qquad \rho_\alpha = -2(A - 2C\overline{\gamma}_\alpha + B\overline{\gamma}_\alpha^2). \tag{14.53}$$

The expression $\omega_0^{(1)} iz + u_0^{(1)} + iv_0^{(1)}$ corresponds to an infinitesimal rigid displacement and can be determined by prescribing the value of the displacement vector and that of the elastic rotation at an arbitrary point of the medium. By considering (14.52), it may be seen that both conditions (14.40) can be satisfied at the first step by taking

$$\omega_0^{(1)} = 0 \tag{14.54}$$

and requiring that

$$\lim_{\rho \to \infty} \Omega_\alpha''(z_\alpha) = 0, \qquad \alpha = 1, 2. \tag{14.55}$$

Since the strain and rotation components corresponding to the Volterra dislocation occurring at the first step must be continuous across $S$, the functions $\Omega_\alpha''(z_\alpha)$, $\alpha = 1, 2$, must be single-valued in the regions $\Delta_\alpha$ that correspond to $\Delta$ in the $z_\alpha$-planes, and hence can be developed in Laurent series in those regions. Moreover, by virtue of (14.55), these series may contain only negative powers of $z_\alpha$. Consequently, by integrating them with respect to $z_\alpha$, we find

$$\Omega_\alpha'(z_\alpha) = \varkappa_\alpha \ln \frac{z_\alpha}{1 + \gamma_\alpha} + \sum_{k=1}^{\infty} a_{k\alpha}^{(1)} z_\alpha^{-k}, \qquad \alpha = 1, 2, \tag{14.56}$$

where $\varkappa_\alpha$ and $a_k^{(1)}$ ($\alpha = 1, 2$; $k = 1, 2, \ldots$) are arbitrary constants. The denominators $1 + \gamma_\alpha$, $\alpha = 1, 2$, have been introduced as integration constants into the arguments of the logarithmic terms in (14.56) in order to allow the calculation of these terms by means of (8.36) and (8.37) without introducing any cut except the negative $x_1$-axis, which coincides with the discontinuity line of the displacement [2].

[1] The reasoning is very similar to that used to derive the representation (8.30) in the isotropic case.

[2] It may be shown that the ratio $z_\alpha/(1 + \gamma_\alpha)$ coincides with the variable denoted by $z_\alpha$ in Lekhnitsky's representation used in Sect. 10.

The coefficients $\varkappa_\alpha$ will be determined, as usual for single dislocations, by the jump condition (14.36) and by equation (14.11) which expresses the vanishing of the resultant force exerted on $\Gamma_0$ by the dislocation core. By virtue of (10.105), equation (14.11) and the condition that the tractions acting on the cut faces be self-equilibrated at each point of the cut are equivalent with requiring that

$$\oint_\Gamma \mathbf{t}\,\mathrm{d}l = \mathbf{0}, \tag{14.57}$$

where $\Gamma$ is any smooth closed curve encircling $\Gamma_0$ anticlockwise in the $x_1x_2$-plane, and $l$ is the curvilinear abscissa on $\Gamma$. Since in our case $t_3 = 0$, equation (14.57) is in its turn equivalent, to within terms of third order in $\varepsilon$, with the conditions

$$\oint (t_1^{(n)} + it_2^{(n)})\,\mathrm{d}l = 0, \qquad n = 1, 2. \tag{14.58}$$

On the other hand, by making use of (10.103), we successively obtain

$$t_1 + it_2 = T_{11}n_1 + T_{12}n_2 + i(T_{12}n_1 + T_{22}n_2) =$$

$$= -(T_{11} + iT_{12})\frac{\mathrm{d}x_2}{\mathrm{d}l} + (T_{12} + iT_{22})\frac{\mathrm{d}x_1}{\mathrm{d}l} =$$

$$= \frac{i}{2}\left[(T_{11} + T_{22})\frac{\mathrm{d}z}{\mathrm{d}l} - (T_{11} - T_{22} + 2iT_{12})\frac{\mathrm{d}\bar{z}}{\mathrm{d}l}\right]. \tag{14.59}$$

Consequently, considering also (14.35) and (14.42), we have

$$t_1^{(1)} + it_2^{(1)} = \frac{i}{2}\left(\Theta^{(1)}\frac{\mathrm{d}z}{\mathrm{d}l} - \Phi^{(1)}\frac{\mathrm{d}\bar{z}}{\mathrm{d}l}\right) = 2i\frac{\mathrm{d}}{\mathrm{d}l}\left(\frac{\partial F^{(1)}}{\partial \bar{z}}\right)$$

and, introducing this result into (14.58), we deduce for $n = 1$:

$$\frac{\partial F^{(1)}}{\partial \bar{z}}(x_1, 0^+) - \frac{\partial F^{(1)}}{\partial \bar{z}}(x_1, 0^-) = 0 \quad \text{for } x_1 \in (-\infty, -r_0]. \tag{14.60}$$

Next, from (14.48—50) it follows that

$$\frac{\partial F^{(1)}}{\partial \bar{z}}(x_1, x_2) = \sum_\alpha [\gamma_\alpha \Omega'_\alpha(z_\alpha) + \overline{\Omega'_\alpha(z_\alpha)}]. \tag{14.61}$$

Substituting now (14.56) into (14.61) and the result obtained into (14.60) yields

$$\sum_\alpha (\gamma_\alpha \varkappa_\alpha - \bar{\varkappa}_\alpha) = 0. \tag{14.62}$$

On the other hand, introducing (14.52) into the jump condition (14.36), and considering (14.56) we find

$$\sum(\delta_\alpha \varkappa_\alpha - \rho_\alpha \bar{\varkappa}_\alpha) = \frac{ib^{(1)}}{2\pi}. \tag{14.63}$$

Equations (14.62) and (14.63) provide two conditions for the determination of the parameters $\varkappa_1$ and $\varkappa_2$. Following [338], we first simplify (14.63) by taking into consideration (14.53) and (14.62), thus obtaining

$$\sum_\alpha \left( \bar{\gamma}_\alpha^2 \varkappa_\alpha - \frac{\varkappa_\alpha}{\gamma_\alpha} \right) = \frac{ib^{(1)}}{4\pi B}. \tag{14.64}$$

Eliminating $\bar{\varkappa}_1$ and $\bar{\varkappa}_2$ between equations (14.62), (14.64), and their complex conjugates, we get

$$\gamma_2(1-\gamma_1\bar{\gamma}_1)(1-\gamma_1\bar{\gamma}_2)\varkappa_1 + \gamma_1(1-\gamma_2\bar{\gamma}_2)(1-\bar{\gamma}_1\gamma_2)\varkappa_2 = -\frac{ib^{(1)}\gamma_1\gamma_2}{4\pi B},$$

$$(1-\gamma_1\bar{\gamma}_1)(1-\gamma_1\bar{\gamma}_2)\varkappa_1 + (1-\gamma_2\bar{\gamma}_2)(1-\bar{\gamma}_1\gamma_2)\varkappa_2 = -\frac{ib^{(1)}\bar{\gamma}_1\bar{\gamma}_2}{4\pi B}.$$

Making use of (14.47), this system can be further solved to give

$$\varkappa_1 = \frac{ib^{(1)}(1-\gamma_1)\gamma_1\gamma_2}{4\pi B \nu_1 \nu_3 \nu_4}, \quad \varkappa_2 = -\frac{ib^{(1)}(1-\gamma_2)\gamma_1\gamma_2}{4\pi B \nu_2 \nu_3 \nu_4}, \tag{14.65}$$

where

$$\nu_1 = 1 - \gamma_1\bar{\gamma}_1, \quad \nu_2 = 1 - \gamma_2\bar{\gamma}_2, \quad \nu_3 = 1 - \gamma_1\bar{\gamma}_2, \quad \nu_4 = \gamma_1 - \gamma_2. \tag{14.66}$$

In [338], the coefficients $a_{k\alpha}^{(1)}$ $(k = 1, 2, \ldots)$ have been also determined in terms of the complex Fourier coefficients of the function $t_\rho^{*(1)} + it_\theta^{*(1)}$, by using the boundary condition (14.39). However, since the latter are not known *a priori*, it proves more advantageous, when applying semidiscrete methods, to consider as unknowns, besides the positions of the atoms inside $\Sigma_0$, the coefficients $a_{k\alpha}^{(1)}$ themselves, instead of the Fourier coefficients of the tractions acting on $\Sigma_0$. In this case, the solution of the first linear boundary-value problem (14.20) of the iteration scheme is given by (14.51), (14.52), and (14.56), with $\varkappa_\alpha$ determined by (14.65).

We proceed now to solve the *second linear boundary-value problem* (14.21) of the iteration scheme. The only differences from the first problem are the presence of the non-linear term $\tau$ in the expression of $\mathbf{T}^{(2)}$ and the continuity of $\mathbf{u}^{(2)}$ across the cut $S$.

By introducing the complex displacement

$$U^{(2)} = u_1^{(2)} + iu_2^{(2)} \tag{14.67}$$

and the complex stresses

$$\Theta^{(2)} = T_{11}^{(2)} + T_{22}^{(2)}, \qquad \Phi^{(2)} = T_{11}^{(2)} - T_{22}^{(2)} + 2i\, T_{12}^{(2)}, \tag{14.68}$$

we obtain from (14.21) and (14.31—33) the equilibrium equation

$$\frac{\partial \Theta^{(2)}}{\partial \bar{z}} + \frac{\partial \Phi^{(2)}}{\partial z} = 0, \tag{14.69}$$

the constitutive equations

$$\left.\begin{aligned} 2\,\frac{\partial U^{(2)}}{\partial \bar{z}} &= A(\Phi^{(2)} - \Phi_0) + B(\bar{\Phi}^{(2)} - \bar{\Phi}_0) + 2C(\Theta^{(2)} - \Theta_0), \\ \frac{\partial U^{(2)}}{\partial z} + \frac{\partial \bar{U}^{(2)}}{\partial \bar{z}} &= \bar{C}(\Phi^{(2)} - \Phi_0) + C(\bar{\Phi}^{(2)} - \bar{\Phi}_0) + 2D(\Theta^{(2)} - \Theta_0), \end{aligned}\right\} \tag{14.70}$$

and the boundary conditions

$$\Theta^{(2)} + \Phi^{(2)}\, e^{-2i\theta} = 0 \qquad \text{for} \quad z = r_0 e^{i\theta}, \quad \theta \in (-\pi, \pi], \tag{14.71}$$

$$\frac{\partial U^{(2)}}{\partial z}, \quad \frac{\partial U^{(2)}}{\partial \bar{z}} \to 0 \quad \text{as} \quad \rho \to \infty, \tag{14.72}$$

where the functions

$$\Theta_0 = \tau_{11} + \tau_{22}, \qquad \Phi_0 = \tau_{11} - \tau_{22} + 2i\tau_{12} \tag{14.73}$$

depend only on grad $\mathbf{u}^{(1)}$ and hence are known from (14.19), (14.52), and (14.56). By the first equation (14.21) we must also have

$$U^{(2)}(x_1, 0^+) - U^{(2)}(x_1, 0^-) = 0 \quad \text{for } x_1 \in (-\infty, -r_0]. \tag{14.74}$$

The equilibrium condition (14.69) is identically satisfied by putting as before

$$\Phi^{(2)} = -4\,\frac{\partial^2 F^{(2)}}{\partial \bar{z}^2}, \qquad \Theta^{(2)} = 4\,\frac{\partial^2 F^{(2)}}{\partial z\, \partial \bar{z}}, \tag{14.75}$$

where $F^{(2)}$ is Airy's stress function corresponding to the second iteration step. By a reasoning similar to that leading to (14.60) we conclude that (14.58) implies the continuity of $\partial F^{(2)}/\partial \bar{z}$ across the cut $S$, i.e.

$$\frac{\partial F^{(2)}}{\partial \bar{z}}(x_1, 0^+) - \frac{\partial F^{(2)}}{\partial \bar{z}}(x_1, 0^-) = 0 \quad \text{for } x_1 \in (-\infty, -r_0]. \tag{14.76}$$

Clearly, we could derive the compatibility equation to be satisfied by $F^{(2)}$ as above, by eliminating $U^{(2)}$ between equations (14.70). However, the subsequent integration of the equation obtained and of system (14.70) proves to be much more difficult than it was for the first iteration step. That is why it is better to adopt an apparently longer way [295], which, however, greatly simplifies the subsequent procedure.

We recall that the operator $\mathscr{L}$ assumes a particularly simple form when choosing $z_1$ and $\bar{z}_1$ or, alternatively, $z_2$ and $\bar{z}_2$ as independent variables, instead of $z$ and $\bar{z}$. Let us first take $z_2$ and $\bar{z}_2$ as independent variables in (14.70) and (14.75). By using the relations

$$\frac{\partial}{\partial z} = \frac{\partial}{\partial z_2} + \bar{\gamma}_2 \frac{\partial}{\partial \bar{z}_2}, \quad \frac{\partial}{\partial \bar{z}} = \gamma_2 \frac{\partial}{\partial z_2} + \frac{\partial}{\partial \bar{z}_2},$$

we obtain after some algebraic manipulation the system

$$\frac{\partial}{\partial z_2}(\bar{U}^{(2)} - \bar{\gamma}_2 U^{(2)}) = (\rho_2 - \gamma_2\bar{\delta}_2)\frac{\partial^2 F^{(2)}}{\partial z_2^2} + (\bar{\gamma}_2\rho_2 - \bar{\delta}_2)\frac{\partial^2 F^{(2)}}{\partial z_2 \partial \bar{z}_2} +$$

$$+ (4\nu_2)^{-1}[(\bar{\delta}_2 + \bar{\gamma}_2\rho_2)\Theta_0 + \bar{\gamma}_2\bar{\delta}_2\Phi_0 + \rho_2\bar{\Phi}_0], \tag{14.77}$$

$$\frac{\partial U^{(2)}}{\partial z_2} + \frac{\partial \bar{U}^{(2)}}{\partial \bar{z}_2} = 2\mathrm{Re}\left\{\delta_2 \frac{\partial^2 F^{(2)}}{\partial z_2^2} + \left(\frac{\delta_2}{\gamma_2} + \frac{B\nu_2^2}{\gamma_2^2}\right)\frac{\partial^2 F^{(2)}}{\partial z_2 \partial \bar{z}_2} + \right.$$

$$\left. + (2\nu_2)^{-1}[(\bar{\gamma}_2 C + \gamma_2 \bar{C} - 2D)\Theta_0 + (A\bar{\gamma}_2 - 2\bar{C} + B\gamma_2)\Phi_0]\right\}. \tag{14.78}$$

Integrating (14.77) with respect to $z_2$ yields

$$\bar{U}^{(2)} - \bar{\gamma}_2 U^{(2)} = (\rho_2 - \gamma_2\bar{\delta}_2)\frac{\partial F^{(2)}}{\partial z_2} + (\bar{\gamma}_2\rho_2 - \bar{\delta}_2)\frac{\partial F^{(2)}}{\partial \bar{z}_2} +$$

$$+ (4\nu_2)^{-1}\int[(\bar{\delta}_2 + \bar{\gamma}_2\rho_2)\,\Theta_0 + \bar{\gamma}_2\bar{\delta}_2\,\Phi_0 + \rho_2\,\bar{\Phi}_0]\,\mathrm{d}z_2 + \overline{\eta'(z_2)},$$

where $\eta(z_2)$ is an arbitrary analytic function of $z_2$. Eliminating $\bar{U}^{(2)}$ between this equation and its complex conjugate yields

$$\nu_2 U^{(2)} = (\gamma_2\rho_2 - \gamma_2^2\bar{\delta}_2 + \gamma_2\bar{\rho}_2 - \delta_2)\frac{\partial F^{(2)}}{\partial z_2} + (\gamma_2\bar{\gamma}_2\rho_2 - \gamma_2\bar{\delta}_2 + \bar{\rho}_2 - \bar{\gamma}_2\delta_2)\frac{\partial F^{(2)}}{\partial \bar{z}_2} +$$

$$+ (4\nu_2)^{-1}\left\{\gamma_2\int[(\bar{\delta}_2 + \bar{\gamma}_2\rho_2)\Theta_0 + \bar{\gamma}_2\bar{\delta}_2\Phi_0 + \rho_2\bar{\Phi}_0]\,\mathrm{d}z_2 + \right.$$

$$\left. + \int[(\delta_2 + \gamma_2\bar{\rho}_2)\Theta_0 + \gamma_2\delta_2\,\bar{\Phi}_0 + \bar{\rho}_2\Phi_0]\,\mathrm{d}\bar{z}_2\right\} + \eta'(z_2) + \gamma_2\overline{\eta'(z_2)}. \tag{14.79}$$

Finally, introducing (14.79) into (14.78), we find

$$\frac{1}{v_1^2}\left[-\bar{v}_3 v_4\frac{\partial^2 F^{(2)}}{\partial z_2^2}+(v_3\bar{v}_3+v_4\bar{v}_4)\frac{\partial^2 F^{(2)}}{\partial z_2\partial\bar{z}_2}-v_3\bar{v}_4\frac{\partial^2 F^{(2)}}{\partial\bar{z}_2^2}\right]=$$

$$=2\mathrm{Re}\left[k_4\Theta_0+k_5\Phi_0+\frac{\partial h(z_2,\bar{z}_2)}{\partial\bar{z}_2}+\frac{v_2}{k_0}\eta''(z_2)\right], \tag{14.80}$$

where

$$h(z_2,\bar{z}_2)=\int(k_1\Theta_0+k_2\Phi_0+k_3\bar{\Phi}_0)\mathrm{d}z_2, \tag{14.81}$$

$$k_0=\frac{2v_1^2v_2^2B}{\gamma_1\gamma_2},\quad k_1=\frac{\bar{\delta}_2+\bar{\gamma}_2\rho_2}{4k_0},\quad k_2=\frac{\bar{\gamma}_2\bar{\delta}_2}{4k_0},\quad k_3=\frac{\rho_2}{4k_0},$$

$$k_4=\frac{(1+\gamma_2\bar{\gamma}_2)D-C\bar{\gamma}_2-\bar{C}\gamma_2}{k_0},\quad k_5=-\frac{A\bar{\gamma}_2-(1+\gamma_2\bar{\gamma}_2)\bar{C}+\bar{B}\gamma_2}{k_0}.$$

Change now in (14.80) the independent variables $z_2$ and $\bar{z}_2$ by $z_1$ and $\bar{z}_1$. Since

$$z_1=(v_3z_2+v_4\bar{z}_2)/v_2,\qquad \bar{z}_1=(\bar{v}_4z_2+\bar{v}_3\bar{z}_2)/v_2, \tag{14.82}$$

and hence

$$\frac{\partial}{\partial z_2}=\frac{v_3}{v_2}\frac{\partial}{\partial z_1}+\frac{\bar{v}_4}{v_2}\frac{\partial}{\partial\bar{z}_1},\qquad \frac{\partial}{\partial\bar{z}_2}=\frac{v_4}{v_2}\frac{\partial}{\partial z_1}+\frac{\bar{v}_3}{v_2}\frac{\partial}{\partial\bar{z}_1}, \tag{14.83}$$

it follows that

$$\frac{\partial^2F^{(2)}}{\partial z_1\,\partial\bar{z}_1}=2\mathrm{Re}\left[k_4\Theta_0+k_5\Phi_0+\frac{\partial h(z_2,\bar{z}_2)}{\partial\bar{z}_2}+\frac{v_2}{k_0}\eta''(z_2)\right]. \tag{14.84}$$

The general solution of this equation is

$$F^{(2)}(x_1,x_2)=2\mathrm{Re}\sum_\alpha\omega_\alpha(z_\alpha)+F_0(z_1,\bar{z}_1), \tag{14.85}$$

where $\omega_1(z_1)$ is an arbitrary analytic function of $z_1$,

$$\omega_2(z_2)\equiv-\frac{v_1^2v_2}{k_0\bar{v}_3v_4}\eta(z_2)=\frac{\eta(z_2)}{2(\delta_2-\gamma_2\bar{\rho}_2)},$$

and $F_0(z_1,\bar{z}_1)$ is a particular solution of the equation

$$\frac{\partial^2 F_0}{\partial z_1\,\partial\overline{z}_1} = 2k_4\Theta_0 + k_5\Phi_0 + \overline{k_5\Phi_0} + \frac{\partial h(z_2, \overline{z}_2)}{\partial\overline{z}_2} + \overline{\frac{\partial h(z_2, \overline{z}_2)}{\partial z_2}}. \tag{14.86}$$

In order to obtain the expression of $U^{(2)}$, we first note that, by virtue of (14.83),

$$(\gamma_2\rho_2 - \gamma_2^2\delta_2 + \gamma_2\overline{\rho}_2 - \delta_2)\frac{\partial}{\partial z_2} + (\gamma_2\overline{\gamma}_2\rho_2 - \gamma_2\delta_2 + \overline{\rho}_2 - \overline{\gamma}_2\delta_2)\frac{\partial}{\partial\overline{z}_2} =$$

$$= \nu_2\left(\delta_1\frac{\partial}{\partial z_1} + \rho_1\frac{\partial}{\partial\overline{z}_1}\right).$$

Hence, considering also (14.81), equation (14.79) may be rewritten as

$$U^{(2)}(x_1, x_2) = \delta_1\frac{\partial F^{(2)}}{\partial z_1} + \rho_1\frac{\partial F^{(2)}}{\partial\overline{z}_1} + \frac{k_0}{\nu_2^2}[\gamma_2 h(z_2, \overline{z}_2) + \overline{h(z_2, \overline{z}_2)}] +$$

$$+ \frac{1}{\nu_2}[\eta'(z_2) + \gamma_2\overline{\eta'(z_2)}].$$

Next, replacing $F^{(2)}$ by (14.85) yields

$$U^{(2)}(x_1, x_2) = \sum_\alpha [\delta_\alpha\omega_\alpha'(z_\alpha) + \rho_\alpha\overline{\omega_\alpha'(z_\alpha)}] + U_0(x_1, x_2), \tag{14.87}$$

where

$$U_0(x_1, x_2) = \delta_1\frac{\partial F_0}{\partial z_1} + \rho_1\frac{\partial F_0}{\partial\overline{z}_1} + \frac{k_0}{\nu_2^2}[\gamma_2 h(z_2, \overline{z}_2) + \overline{h(z_2, \overline{z}_2)}]. \tag{14.88}$$

Finally, by substituting (14.85) into (14.75), we deduce the complex stresses

$$\left.\begin{aligned} \Phi^{(2)} &= -4\sum_\alpha [\gamma_\alpha^2\,\omega_\alpha''(z_\alpha) + \overline{\omega_\alpha''(z_\alpha)}] - 4\frac{\partial^2 F_0}{\partial\overline{z}^2}, \\ \Theta^{(2)} &= 8\mathrm{Re}\sum_\alpha \gamma_\alpha\omega_\alpha''(z_\alpha) + 4\frac{\partial^2 F_0}{\partial z\partial\overline{z}}. \end{aligned}\right\} \tag{14.89}$$

The complex potentials $\omega_1'(z_1)$ and $\omega_2'(z_2)$ can be determined by using the boundary conditions (14.74), (14.76), provided we are able to calculate two indefinite integrals that are necessary to obtain $h(z_2, \overline{z}_2)$ and $\partial F_0/\partial\overline{z}_1$ from (14.81) and (14.86), respectively. Since the functions $\Theta_0$ and $\Phi_0$ occurring in (14.81) and (14.86) are quadratic in the partial derivatives of $U^{(1)}$, it is easily seen that the amount of algebra necessary to calculate $h(z_2, \overline{z}_2)$ and $\partial F_0/\partial\overline{z}_1$ increases very rapidly with the number of terms taken into account in the expression (14.56) of $\Omega_\alpha'(z_\alpha)$. Therefore, following an idea of Seeger, we content ourselves with determining only those terms which are at most of the order $O(\rho^{-1})$ in the expression of the displacements and $O(\rho^{-2})$ in the expression of the stresses, as $\rho \to \infty$. In this way, we attempt to find out the most significant correction to the classical solution, which neglects boundary conditions on $\Gamma_0$ and second-order effects and retains only the terms of order $O(\rho^{-1})$ in the expression of stresses.

The final result of this calculation reads

$$\frac{\partial F_0}{\partial \bar{z}}(x_1, x_2) = \sum_{\alpha} [\gamma_\alpha \Lambda_\alpha(x_1, x_2) + \overline{\Lambda_\alpha(x_1, x_2)}] \tag{14.90}$$

$$U_0(x_1, x_2) = \Lambda_0(x_1, x_2) + \sum_{\alpha} [\delta_\alpha \Lambda_\alpha(x_1, x_2) + \rho_\alpha \overline{\Lambda_\alpha(x_1, x_2)}], \tag{14.91}$$

where

$$\left.\begin{aligned}
\Lambda_0(x_1, x_2) &= \frac{t_1}{z_1} + \frac{t_2}{\bar{z}_1} + \frac{t_3}{z_2} + \frac{t_4}{\bar{z}_2},\\
\Lambda_1(x_1, x_2) &= \frac{t_5}{\bar{z}_1} + \frac{t_6}{z_2} + \frac{t_7}{\bar{z}_2} + \frac{t_8 \bar{z}_1}{z_1^2} + \frac{1}{z_1}\Bigg(w_1 \ln \frac{z_1}{1 + \bar{\gamma}_1} + \\
&\quad + w_2 \ln \frac{z_2}{1 + \gamma_2} + w_3 \ln \frac{\bar{z}_2}{1 + \bar{\gamma}_2}\Bigg),\\
\Lambda_2(x_1, x_2) &= \frac{t_9 \bar{z}_2}{z_2^2} + \frac{1}{z_2}\left(w_4 \ln \frac{z_1}{1 + \gamma_1} + w_5 \ln \frac{\bar{z}_1}{1 + \bar{\gamma}_1} + w_6 \ln \frac{\bar{z}_2}{1 + \bar{\gamma}_2}\right),
\end{aligned}\right\} \tag{14.92}$$

while $t_1, \ldots, t_9, w_1, \ldots, w_6$ are parameters depending only on the elastic constants of second and third orders and whose explicit expressions are given in the Appendix of [479].

From (14.85), (14.87), (14.90), and (14.91) it follows that

$$\frac{\partial F^{(2)}}{\partial \bar{z}}(x_1, x_2) = \sum_{\alpha} \{\gamma_\alpha[\omega'_\alpha(z_\alpha) + \Lambda_\alpha(x_1, x_2)] + \overline{\omega'_\alpha(z_\alpha)} + \overline{\Lambda_\alpha(x_1, x_2)}\}, \tag{14.93}$$

$$\begin{aligned}
U^{(2)}(x_1, x_2) = \Lambda_0(x_1, x_2) &+ \sum_{\alpha} \{\delta_\alpha [\omega'_\alpha(z_\alpha) + \Lambda_\alpha(x_1, x_2)] + \\
&+ \rho_\alpha[\overline{\omega'_\alpha(z_\alpha)} + \overline{\Lambda_\alpha(x_1, x_2)}]\}.
\end{aligned} \tag{14.94}$$

Inspection of (14.92) shows that $\Lambda_0(x_1, x_2)$ is continuous across $S$, while $\Lambda_1(x_1, x_2)$ and $\Lambda_2(x_1, x_2)$ have the jumps

$$\Lambda_\alpha(x_1, 0^+) - \Lambda_\alpha(x_1, 0^-) = -2\pi i K_\alpha / x_1, \quad \alpha = 1, 2, \tag{14.95}$$

where

$$K_1 = w_1 + w_3 - w_2, \quad K_2 = w_5 + w_6 - w_4. \tag{14.96}$$

In view of the continuity conditions (14.74) and (14.76), it may be shown that the part of the solution (14.93), (14.94) corresponding to the functions $\omega_\alpha'(z_\alpha)$ must represent a generalized Somigliana dislocation of the type considered in Sect. 10.6, i.e. with variable displacement jump across the cut $S$ and a distribution of non-equilibrated tractions acting on the cut faces. As shown in Sect. 10.6, the solution to this problem may be found by setting

$$\omega_\alpha'(z_\alpha) = \tilde{\omega}_\alpha'(z_\alpha) + \hat{\omega}_\alpha'(z_\alpha), \tag{14.97}$$

and requiring that the functions $\tilde{\omega}_\alpha'(z_\alpha)$ satisfy the jump conditions on $S$ and vanish at infinity, while $\hat{\omega}_\alpha'(z_\alpha)$ must be continuous across $S$, vanish at infinity, and fulfil the boundary conditions on $\Gamma_0$, modified by the contribution of $\tilde{\omega}_\alpha'(z_\alpha)$. In our case we may satisfy the jump conditions resulting for $\tilde{\omega}_\alpha'(z_\alpha)$ from (14.74), (14.76), and (14.93—96) by simply taking

$$\tilde{\omega}_\alpha'(z_\alpha) = \frac{K_\alpha}{z_\alpha} \ln \frac{z_\alpha}{1+\gamma_\alpha}, \quad \alpha = 1, 2. \tag{14.98}$$

On the other hand, in agreement with the approximation adopted above, we shall take

$$\hat{\omega}_\alpha'(z_\alpha) = \frac{a_{1\alpha}^{(2)}}{z_\alpha}, \quad \alpha = 1, 2, \tag{14.99}$$

interpreting the coefficients $a_{1\alpha}^{(2)}$ as adjustable parameters.

Summarizing the above considerations we conclude that the non-linear elastic displacement field is given up to terms of order $O(\varepsilon^2)$ and $O(\rho^{-1})$ by the expression

$$U(x_1, x_2) = \sum_\alpha \left\{ \delta_\alpha \left[ \left( \varepsilon \varkappa_\alpha + \frac{\varepsilon^2 K_\alpha}{z_\alpha} \right) \ln \frac{z_\alpha}{1+\gamma_\alpha} + \frac{A_\alpha}{z_\alpha} \right] + \right.$$

$$\left. + \rho_\alpha \left[ \left( \varepsilon \bar{\varkappa}_\alpha + \frac{\varepsilon^2 \bar{K}_\alpha}{\bar{z}_\alpha} \right) \ln \frac{\bar{z}_\alpha}{1+\bar{\gamma}_\alpha} + \frac{\bar{A}_\alpha}{\bar{z}_\alpha} \right] \right\} + \varepsilon^2 U_0(x_1, x_2) + u_0 + i v_0, \tag{14.100}$$

where

$$A_\alpha = \varepsilon a_{1\alpha}^{(1)} + \varepsilon^2 a_{1\alpha}^{(2)}, \quad u_0 = \varepsilon u_0^{(1)}, \qquad v_0 = \varepsilon v_0^{(1)},$$

while $\varkappa_\alpha$, $K_\alpha$, $U_0(x_1, x_2)$ are given by (14.65), (14.96), and (14.91), respectively. Since $\varkappa_1$ and $\varkappa_2$ are proportional to $b^{(1)}$, while $K_1$, $K_2$, and $U_0(x_1, x_2)$ are proportional to the square of $b^{(1)}$, and since $b = \varepsilon b^{(1)}$, the final expression (14.100) of the displacement field does not depend on the choice of the small parameter $\varepsilon$, as was to be expected.

The solution obtained depends linearly on two arbitrary complex constants, $A_1$ and $A_2$. When using a semidiscrete method, these constants should be considered as adjustable parameters in the expression of the total potential energy, together with the positions of the atoms inside the dislocation core, and are to be calculated by minimizing this energy (see also Sect. 16). Finally, $u_0 + iv_0$ gives a rigid translation that can be determined by prescribing the displacement of an arbitrary point of the elastic medium.

### 14.3. Second-order effects in the elastic field of a screw dislocation

Second-order effects in the *isotropic* elastic field of a screw dislocation have been determined by Seeger and Mann [289] as early as 1959, by using an iteration scheme based on Lagrangian co-ordinates, and requiring that the core surface $\Sigma_0$ be traction-free. A slightly more general problem, corresponding to a uniform radial pressure acting on $\Sigma_0$, has been solved by Teodosiu [337], by using a similar formalism. In this subsection we will consider the same problem, but formulated in Eulerian co-ordinates. In Sect. 19, we shall also describe the results obtained in a different way by Willis [382] concerning the second-order effects in the *anisotropic* elastic field of a screw dislocation.

Consider a screw dislocation lying in the axis of an infinite circular isotropic elastic cylinder of radius $R$. Assume that the outer surface of the cylinder is free of external loads.

Let us use a system of cylindrical co-ordinates with the unit vector $\mathbf{e}_z$ directed along the axis of the cylinder in the positive sense of the dislocation line. We apply as above the non-linear elasticity theory outside a circular cylindrical surface of radius $r_0 < R$ and axis $\mathbf{e}_z$. Let $\Gamma_0$ and $\Gamma$ denote the circles of radii $r_0$ and $R$, respectively, situated in an arbitrary cross-section of the cylinder, and let $\Delta$ be the region bounded by $\Gamma_0$ and $\Gamma$ (Fig. 8.1).

Denote by $\rho, \theta, z$ the cylindrical co-ordinates of a current material point in the deformed configuration and assume that the displacement vector $\mathbf{u}(r, \theta, z)$ is of class $C^2$ in the whole elastic cylinder, except the surface $S$: $x_2 = 0, -R \leqslant x_1 \leqslant -r_0$, across which the component $u_z$ experiences the jump

$$u_z(\rho, \pi, z) - u_z(\rho, -\pi, z) = -b, \tag{14.101}$$

where $b$ is the magnitude of the true Burgers vector, while the components $u_r$ and $u_\theta$ are continuous across $S$.

Clearly, the elastic state of the infinite cylinder does not depend on $z$. Moreover, as already mentioned in Sect. 8, in the isotropic case the displacement gradient and the stress tensor do not depend on $\theta$. For the same symmetry reasons the tractions acting on $\Sigma_0$ from the dislocation core must reduce to a pressure [1], say $p$.

---

[1] The value of this pressure could be determined only by a semidiscrete method, combining the atomic and continuum models of the dislocation.

In order to determine the second-order effects by means of the iteration scheme described in Sect. 14.1 we set

$$b = \varepsilon b^{(1)}, \qquad p = \varepsilon p^{(1)}. \tag{14.102}$$

As argued above, this hypothesis is justified by the fact that $p \to 0$ as $b \to 0$.

Since $\mathbf{t}^{*(1)} = p_1 \mathbf{e}_\rho$, $\mathbf{n} = -\mathbf{e}_\rho$ for $\rho = r_0$ and $\mathbf{t}^{*(1)} = \mathbf{0}$, $\mathbf{n} = \mathbf{e}_\rho$ for $\rho = R$, the traction boundary conditions to be satisfied at the *first step of the iteration* are

$$T^{(1)}_{\rho\rho} = \begin{cases} -p^{(1)} & \text{for} \quad \rho = r_0, \\ 0 & \text{for} \quad \rho = R, \end{cases} \tag{14.103}$$

while the jump condition (14.101) yields

$$u^{(1)}_z(\rho, \pi) - u^{(1)}_z(\rho, -\pi) = -b^{(1)}. \tag{14.104}$$

The solution of the first boundary-value problem can be easily deduced from the results obtained in Sect. 8. Indeed, we have seen in Sect. 8.2 that the linear elastic state corresponding to a screw dislocation in an infinite isotropic elastic cylinder is given by

$$u^{(1)}_z = -\frac{b^{(1)}\theta}{2\pi}, \qquad h^{(1)}_{z\theta} = -\frac{b^{(1)}}{2\pi\rho}, \qquad T^{(1)}_{\theta z} = -\frac{\mu b^{(1)}}{2\pi\rho}, \tag{14.105}$$

where $\mathbf{h}^{(1)} = \operatorname{grad} \mathbf{u}^{(1)}$, and the other components of $\mathbf{u}$, $\mathbf{h}$, and $\mathbf{T}$ are zero. Clearly, the elastic state (14.105) satisfies the traction boundary condition $(14.103)_2$. Therefore, we have to superpose on this state a solution of the field equations of linear isotropic elasticity that corresponds to the internal pressure $p^{(1)}$ acting on the surface $\rho = r_0$ and to zero tractions on the surface $\rho = R$. On the other hand, this last solution can be directly derived from the results obtained in Sect. 8.1, by putting $b = 0$, $t^{(1)}_0 = -p^{(1)}$ in (8.43—45), and equating to zero all other coefficients $t^{(1)}_k$ and $t^{(2)}_k$. It then follows that the only non-zero coefficients $a_k$ and $b_k$ are

$$a_0 = \frac{r_0^2 p^{(1)}}{2(R^2 - r_0^2)}, \qquad b_{-2} = \frac{r_0^2 R^2 p^{(1)}}{R^2 - r_0^2}. \tag{14.106}$$

Substituting (14.106) into (8.35), and the result obtained into (8.26) and (8.30) (with $\omega_0 = u_0 = v_0 = 0$), we find

$$U^{(1)} = k[(1 - 2\nu)\rho + R^2/\rho] e^{i\theta}, \tag{14.107}$$

$$T^{(1)}_{\rho\rho} = 2\mu k \left(1 - \frac{R^2}{\rho^2}\right), \qquad T^{(1)}_{\theta\theta} = 2\mu k \left(1 + \frac{R^2}{\rho^2}\right), \qquad T^{(1)}_{zz} = 4\nu\mu k, \tag{14.108}$$

where

$$k = \frac{p^{(1)} r_0^2}{2\mu(R^2 - r_0^2)},$$

and the other displacement and stress components vanish. Next, by (14.34), (1.64)$_2$, and (1.73) we have

$$u_\rho^{(1)} + i u_\theta^{(1)} = U^{(1)} \, \mathrm{e}^{-i\theta},$$

and hence (14.107) implies

$$u_\rho^{(1)} = k[(1-2\nu)\rho + R^2/\rho], \qquad u_\theta^{(1)} = 0, \tag{14.109}$$

while (1.75) yields

$$h_{\rho\rho}^{(1)} = k(1 - 2\nu - R^2/\rho^2), \qquad h_{\theta\theta}^{(1)} = k(1 - 2\nu + R^2/\rho^2), \tag{14.110}$$

the other components of the displacement gradient being zero.

Finally, by superposing the elastic states given by (14.105) and (14.108—110), we find the solution of the first iteration step:

$$\left.\begin{aligned} u_\rho^{(1)} &= k[(1-2\nu)\rho + R^2/\rho], \qquad u_z^{(1)} = -b^{(1)}\theta/(2\pi),\\ h_{\rho\rho}^{(1)} &= k(1 - 2\nu - R^2/\rho^2), \qquad h_{\theta\theta}^{(1)} = k(1 - 2\nu + R^2/\rho^2),\\ h_{z\theta}^{(1)} &= -b^{(1)}/(2\pi\rho),\\ T_{\rho\rho}^{(1)} &= 2\mu k(1 - R^2/\rho^2), \qquad T_{\theta\theta}^{(1)} = 2\mu k(1 + R^2/\rho^2),\\ T_{\theta z}^{(1)} &= -\mu b^{(1)}/(2\pi\rho), \qquad T_{zz}^{(1)} = 4\nu\mu k. \end{aligned}\right\} \tag{14.111}$$

Let us consider now the *second linear boundary-value problem* (14.21) of the iteration scheme. By taking into account (5.26) and (5.33) we deduce that the expression (14.19) of $\tau_{kl}$ becomes in the isotropic case

$$\begin{aligned} \tau_{kl} = \delta_{kl} \left[ \frac{\lambda}{2} h_{mn}^{(1)} h_{mn}^{(1)} + \lambda h_{mn}^{(1)} h_{nm}^{(1)} + \left(\frac{\nu_1}{2} - \lambda\right) (e_{mm}^{(1)})^2 + \nu_2 e_{mn}^{(1)} e_{mn}^{(1)} \right] + \\ + (2\lambda - \mu + \nu_2) e_{mm}^{(1)} e_{kl}^{(1)} - \mu h_{mk}^{(1)} h_{ml}^{(1)} + 4(2\mu + \nu_3) e_{km}^{(1)} e_{ml}^{(1)}, \end{aligned} \tag{14.112}$$

where $e^{(1)}_{km} = (h^{(1)}_{km} + h^{(1)}_{mk})/2$. Next, substituting (14.111) into (14.112), we obtain after some intermediate calculation

$$\left.\begin{aligned}
\tau_{\rho\rho} &= \frac{1}{2}(\lambda + \nu_2)\left(\frac{b^{(1)}}{2\pi\rho}\right)^2 + k^2[f(\rho^2) + g(\rho^2)],\\
\tau_{\theta\theta} &= \frac{1}{2}\left(\lambda + 2\mu + \nu_2 + 2\nu_3\right)\left(\frac{b^{(1)}}{2\pi\rho}\right)^2 + k^2[f(\rho^2) + g(-\rho^2)],\\
\tau_{zz} &= \frac{1}{2}(\lambda + 4\mu + \nu_2 + 2\nu_3)\left(\frac{b^{(1)}}{2\pi\rho}\right)^2 + k^2 f(\rho^2),\\
\tau_{\theta z} = \tau_{z\theta} &= -\frac{kb^{(1)}}{\pi\rho}\left[(\lambda - \mu + \nu_2)(1 - 2\nu) + (2\mu + \nu_3)\left(1 - 2\nu + \frac{R^2}{\rho^2}\right)\right],
\end{aligned}\right\} \tag{14.113}$$

where

$$f(\rho^2) = (-\lambda + 2\nu_1 + 2\nu_2)(1 - 2\nu)^2 + (3\lambda + 2\nu_2)R^4/\rho^4,$$

$$g(\rho^2) = 4(\lambda - \mu + \nu_2)(1 - 2\nu)(1 - 2\nu - R^2/\rho^2) + (7\mu + 4\nu_3)(1 - 2\nu - R^2/\rho^2)^2,$$

while the other components of $\boldsymbol{\tau}$ are zero. As $\boldsymbol{\tau}$ is independent of $\theta$ and $z$, and $\tau_{r\theta} = \tau_{rz} = 0$, we deduce from (14.113) and (1.77) that the only non-zero component of div $\boldsymbol{\tau}$ is

$$(\operatorname{div} \boldsymbol{\tau})_\rho = -a_1\rho^{-3} - a_2\rho^{-5},$$

where

$$a_1 = (\mu + \nu_3)(b^{(1)}/2\pi)^2, \qquad a_2 = 4(3\lambda + 7\mu + 2\nu_2 + 4\nu_3)k^2R^4.$$

It is easily seen now that the boundary-value problem (14.21) is independent of $\theta$ and $z$. Therefore, we shall seek its solution under the form $\mathbf{u}^{(2)} = u^{(2)}_\rho(\rho)\mathbf{e}_\rho$. Hence, by (1.75), the only non-zero components of the displacement gradient $\mathbf{h}^{(2)} = \operatorname{grad} \mathbf{u}^{(2)}$ are

$$h^{(2)}_{\rho\rho} = \frac{du^{(2)}_\rho}{d\rho}, \qquad h^{(2)}_{\theta\theta} = \frac{u^{(2)}_\rho}{\rho}.$$

Introducing this result into $(14.21)_3$ and considering (5.26) yields

$$\left.\begin{aligned} T^{(2)}_{\rho\rho} &= (\lambda+2\mu)\frac{du^{(2)}_\rho}{d\rho} + \lambda\frac{u^{(2)}_\rho}{\rho} + \tau_{\rho\rho}, \\ T^{(2)}_{\theta\theta} &= \lambda\frac{u^{(2)}_\rho}{\rho} + (\lambda+2\mu)\frac{du^{(2)}_\rho}{d\rho} + \tau_{\theta\theta}, \\ T^{(2)}_{zz} &= \lambda\left(\frac{du^{(2)}_\rho}{d\rho} + \frac{u^{(2)}_\rho}{\rho}\right) + \tau_{zz}, \\ T^{(2)}_{\theta z} &= T^{(2)}_{z\theta} = \tau_{\theta z}, \end{aligned}\right\} \tag{14.114}$$

while the other components of $\mathbf{T}^{(2)}$ are zero. Next, by substituting (14.114) into $(14.21)_2$ and taking into account (1.77), we see that the last two equilibrium equations are identically satisfied, whereas the first equation becomes

$$(\lambda+2\mu)\left(\frac{d^2u^{(2)}_\rho}{d\rho^2} + \frac{1}{\rho}\frac{du^{(2)}_\rho}{d\rho} - \frac{u^{(2)}_\rho}{\rho^2}\right) = \frac{a_1}{\rho^3} + \frac{a_2}{\rho^5},$$

which is a differential non-homogeneous equation of Euler type. The general solution of the corresponding homogeneous equation is

$$u^{(2)}_\rho = c_1/\rho + c_2\rho,$$

where $c_1$ and $c_2$ are arbitrary constants. By making use of Lagrange's method, we find the general solution of the non-homogeneous equation

$$u^{(2)}_\rho = \frac{c_1}{\rho} + c_2\rho + \frac{1}{\lambda+2\mu}\left(\frac{a_1}{2}\frac{\ln\rho}{\rho} + \frac{a_2}{8\rho^3}\right). \tag{14.115}$$

Finally, by introducing (14.93) into the first equation (14.114), and the result obtained into the boundary condition $(14.21)_4$, we find

$$-\frac{2\mu c_1}{\rho^2} + 2(\lambda+\mu)c_2 + \frac{a_1}{2\rho^2}\left(1 - \frac{\mu}{\lambda+2\mu}\ln\rho\right) - \frac{\lambda+3\mu}{\lambda+2\mu}\frac{a_2}{4\rho^4} + \tau_{\rho\rho} = 0,$$

a relation that must be satisfied for $r = r_0$ and $r = R$. After determining the constants $c_1$ and $c_2$ from these two conditions we may immediately obtain the displacement, the displacement gradient, and the stress fields by using the formulae

$$\mathbf{u} = \varepsilon\mathbf{u}^{(1)} + \varepsilon^2\mathbf{u}^{(2)}, \qquad \mathbf{H} = \varepsilon\mathbf{H}^{(1)} + \varepsilon^2\mathbf{H}^{(2)}, \qquad \mathbf{T} = \varepsilon\mathbf{T}^{(1)} + \varepsilon^2\mathbf{T}^{(2)},$$

the terms in the right-hand sides being given by (14.111) and (14.113—115). It should be noticed that the parameter $\varepsilon$ does not occur in the final result since it intervenes only through the combinations $\varepsilon b^{(1)} = b$ and $\varepsilon p^{(1)} = p$. It is also worth noting that, in the second-order approximation of non-linear elasticity, the dislocation produces displacements both in the $\rho$ and $z$ directions, an effect entirely absent from the linear approximation.

## 14.4. Determination of second-order elastic effects by means of Green's functions

We end this section by expounding a method elaborated by the author [341] for the calculation of the second-order elastic effects produced by a straight dislocation in an anisotropic medium, by means of Green's functions. This method generalizes the procedure employed by Willis [382] to compute the non-linear elastic field of a screw dislocation, by allowing the consideration of the core conditions, too.

Consider a straight dislocation in an *infinite* anisotropic elastic medium, lying along the $x_3$-axis of a Cartesian system of co-ordinates denoted by $x_1$, $x_2$, $x_3$ (Fig.10.1). We make use throughout of the notation in Sect. 10. Since the elastic state produced by the dislocation depends only on $x_1$ and $x_2$, *the derivatives with respect to $x_3$ vanish identically*, but we will not make use explicitly of this property, in order to avoid complicating notation.

The linear boundary-value problems corresponding to the first two steps of the iteration described at the beginning of this section are given by (14.22), (14.23), and (14.30). They become in our case

$$\left.\begin{aligned}
&u_k^{(1)}(x_1, 0^+) - u_k^{(1)}(x_1, 0^-) = -b_k^{(1)} \quad \text{for } x_1 \in (-\infty, -r_0], \\
&c_{klmn} u_{m,nl}^{(1)} = 0 \quad \text{in } \varDelta, \\
&c_{klmn} u_{m,n}^{(1)} n_l = t_k^{*(1)} \quad \text{on } \varGamma_0, \\
&\lim_{\rho \to \infty} u_{k,l}^{(1)} = 0
\end{aligned}\right\} \tag{14.116}$$

and, respectively,

$$\left.\begin{aligned}
&u_k^{(2)}(x_1, 0^+) - u_k^{(2)}(x_1, 0^-) = 0 \quad \text{for } x_1 \in (-\infty, -r_0], \\
&c_{klmn} u_{m,nl}^{(2)} + f_k^{(2)} = 0 \quad \text{in } \varDelta, \\
&c_{klmn} u_{m,n}^{(2)} n_l = t_k^{*(2)} \quad \text{on } \varGamma_0, \\
&\lim_{\rho \to \infty} u_{k,l}^{(2)} = 0,
\end{aligned}\right\} \tag{14.117}$$

where $\rho = \sqrt{x_1^2 + x_2^2}$, $f_k^{(2)} = \tau_{kl,l}$, $t_k^{*(2)} = -\tau_{kl} n_l$, $n_3 = 0$, $\tau_{kl}$ is given by (14.19), and $(.)_{,l}$ denotes as usual partial differentiation with respect to the co-ordinate $x_l$ in the deformed configuration of the medium.

The boundary-value problem (14.116) has been solved in Sect. 10.2, its solution being given by (10.43), (10.37), and (10.42). Apparently, this solution cannot be applied as well to the boundary-value problem (14.117), due to the presence of the body force $f_k^{(2)}$ in the equilibrium equations. However, this case may be reduced to that of the absence of body forces, by determining a particular solution of $(14.117)_2$ with the aid of Green's tensor function $G_{ks}(x_1, x_2)$ of the generalized plane strain, which was defined in Sect. 10.3. Indeed, since $\mathbf{h}^{(1)} = O(\rho^{-1})$ as $\rho \to \infty$, we have also $\boldsymbol{\tau} = O(\rho^{-2})$, $\mathbf{f}^{(2)} = O(\rho^{-3})$ as $\rho \to \infty$, and hence $\mathbf{f}^{(2)}$ satisfies the condition (10.98). Consequently, by applying (10.100) we infer that a particular solution of $(14.117)_2$ is given by

$$\overline{u}_k(x_1, x_2) = \int_\Delta G_{ks}(x_1 - x_1', \ x_2 - x_2') f_s^{(2)}(x_1', x_2') \, dx_1' \, dx_2', \tag{14.118}$$

where, by virtue of (10.88) and (10.89),

$$G_{ks}(x_1, x_2) = \frac{1}{2\pi} \operatorname{Im} \sum_{\alpha=1}^{3} \frac{A_{k\alpha} A_{s\alpha}}{A_{m\alpha} L_{m\alpha}} \ln(x_1 + p_\alpha x_2). \tag{14.119}$$

From the solution given in Sect. 14.2 for an edge dislocation it is obvious that the particular solution $\overline{\mathbf{u}}$ need not be continuous across the cut $x_2 = 0$, $x_1 \in (-\infty, -r_0]$ and the corresponding tractions acting on the cut faces may not be pointwise equilibrated. Consequently, we have to supplement $\overline{\mathbf{u}}$ by the displacement field of a generalized Somigliana dislocation (cf. Sect. 10.6).

Let us denote by

$$\overline{t}_k = \overline{T}_{kl} n_l = c_{klmn} \overline{u}_{m,n} n_l$$

the stress vector corresponding to the displacement $\overline{\mathbf{u}}$ and assume, like in Sect. 10.6, that both $\overline{\mathbf{u}}(x_1, 0^+) - \overline{\mathbf{u}}(x_1, 0^-)$ and $\mathbf{t}(x_1, 0^+) + \mathbf{t}(x_1, 0^-)$ are analytic vector-valued functions and such that the first is $O(1)$ and the second $O(x_1^{-2})$ as $|x_1| \to \infty$. Developing these functions in power series for $x_1 \in (-\infty, -r_0]$ we have

$$\left.\begin{aligned} \overline{u}_k(x_1, 0^+) - \overline{u}_k(x_1, 0^-) &= -\sum_{m=0}^{\infty} \frac{g_{km}}{x_1^m}, \\ \overline{t}_k(x_1, 0^+) + \overline{t}_k(x_1, 0^-) &= -\sum_{m=2}^{\infty} \frac{h_{km}}{x_1^m} \end{aligned}\right\} \tag{14.120}$$

for $x_1 \in (-\infty, -r_0]$. By using the results in Sect. 10.6 we may find the solution of the boundary-value problem (14.117) in the form

$$u_k^{(2)}(x_1, x_2) = \overline{u}_k(x_1, x_2) + \hat{u}_k(x_1, x_2) + \tilde{u}_k(x_1, x_2) + u_k^0, \tag{14.121}$$

such that both $\hat{\mathbf{u}}$ and $\tilde{\mathbf{u}}$ satisfy the equilibrium conditions in $\Delta$ without body forces, as well as the boundary condition $(14.117)_4$ at infinity, $\hat{\mathbf{u}}$ satisfy the conditions (14.120) with opposite signs, on the cut, while $\tilde{\mathbf{u}}$ be continuous together with its partial derivatives of first two orders across the cut and satisfy the traction boundary condition $(14.117)_3$ on $\Gamma_0$, modified by the contribution of $\bar{\mathbf{u}}$ and $\hat{\mathbf{u}}$. Then, $\tilde{\mathbf{u}}$ will be given by $(10.111)_1$, (10.112), and (10.114), with $q_{km}$ defined by (10.108), where now

$$P_k = \oint_{\Gamma_0} (t_k^{*(2)} - \bar{t}_k)\,\mathrm{d}l.$$

Next, $\hat{\mathbf{u}}$ is given by $(10.115)_1$, (10.116), and (10.121), with $f_{km}$ defined by (10.122), where now

$$\hat{t}_k^*(\theta) = t_k^{*(2)}(\theta) - \bar{t}_k^*(\theta) - \tilde{t}_k^*(\theta)$$

and $\bar{\mathbf{t}}^*(\theta)$, $\tilde{\mathbf{t}}^*(\theta)$ denote the stress vectors corresponding on $\Gamma_0$ to the displacement fields $\bar{\mathbf{u}}$ and $\tilde{\mathbf{u}}$, respectively. Finally, the constants $u_k^0$, corresponding to a rigid translation of the whole medium, may be further determined by imposing the value of the displacement field at an arbitrary point of the medium.

Alternatively, when the traction boundary conditions $(14.116)_3$ and $(14.117)_3$ on $\Gamma_0$ are replaced by the displacement boundary conditions $(14.26)_2$ and $(14.27)_2$, respectively, we may again apply the results obtained at the end of Sect. 10.6. Then, $\bar{\mathbf{u}}$ and $\tilde{\mathbf{u}}$ preserve their expressions, but $\mathbf{u}^0$ is given by (10.126), while the coefficients $a_{\alpha m}$ occurring in the expression of $\hat{\mathbf{u}}$ are determined by (10.127) and (10.128), where now

$$\hat{u}_k^*(\theta) = -\bar{u}_k^*(\theta) - \tilde{u}_k^*(\theta),$$

and $\bar{\mathbf{u}}^*(\theta)$, $\tilde{\mathbf{u}}(\theta)$ denote the values on $\Gamma_0$ of $\bar{\mathbf{u}}$ and $\tilde{\mathbf{u}}$, respectively.

Thus, both the traction and the displacement boundary-value problems occurring at the second step of the iteration have been reduced to the calculation of the integral (14.118), with $\mathbf{G}$ given by (14.119), and $\mathbf{f}^{(2)}$ expressed by (14.19) and $(14.24)_1$ in terms of the solution determined at the first step of the iteration. Such integrals can be calculated, for example, by using the residue theorem, as has been done by Willis [382] to calculate the second-order effects in the elastic field of a screw dislocation lying along a two-fold axis of material symmetry. Unlike Willis, who neglected the boundary conditions on $\Gamma_0$, and eliminated the singularities of the particular solution on a rather intuitive reason, the above approach allows the calculation of the integral (14.118) on $\Delta$, which makes it convergent.

## 15. Influence of single dislocations on crystal density

Experimental evidence proves that dislocations produce a positive volume change of crystals. For isotropic media, this effect has been quantitatively described by Zener [391] in 1942, without making explicit use of second-order elasticity. Zener's formula, based on thermodynamic arguments, was subsequently used by Seeger and Haasen [288] to study the influence of isolated dislocations on crystal density, and has been generalized to cubic crystals by Seeger [287].

In 1960, Zener's formula has been derived from second-order elasticity theory, independently, by Pfleiderer, Seeger, and Kröner [270] for isotropic materials, and by Toupin and Rivlin [354] for isotropic materials and cubic crystals. In this subsection we mainly follow the method used by the latter authors, by employing, however, an Eulerian formulation.

## 15.1. Mean stress theorem and its consequences

Consider a body $\mathscr{B}$ in equilibrium under the action of the surface tractions **t** and body forces **f**. Denote as above by $\mathscr{V}$ the region occupied by $\mathscr{B}$ in the deformed configuration, by $V$ its volume, and by $\mathscr{S}$ the boundary of $\mathscr{V}$. The symmetric tensor

$$\overline{\mathbf{T}}(\mathscr{V}) = \frac{1}{V}\int_{\mathscr{V}} \mathbf{T}\,\mathrm{d}v$$

is called the *mean stress*.

The starting point of our considerations is the so-called *mean stress theorem* (Chree [75]), which is expressed by the relation

$$\overline{\mathbf{T}}(\mathscr{V}) = \frac{1}{V}\left(\int_{\mathscr{S}} \mathbf{x}\,\mathbf{t}\,\mathrm{d}s + \int_{\mathscr{V}} \mathbf{x}\,\mathbf{f}\,\mathrm{d}v\right). \tag{15.1}$$

To prove (15.1), we first note that

$$\int_{\mathscr{V}} T_{km}\,\mathrm{d}v = \int_{\mathscr{V}} [(T_{kp}x_m)_{,p} - x_m T_{kp,p}]\,\mathrm{d}v.$$

By making use now of the integral transformation (1.52) for fixed $k$, m, and taking into account that

$$T_{kp,p} + f_k = 0 \quad \text{in } \mathscr{V}, \qquad T_{kp}n_p = t_k \quad \text{on } \mathscr{S},$$

we obtain

$$\frac{1}{V}\int_{\mathscr{V}} T_{km}\,\mathrm{d}v = \frac{1}{V}\left(\int_{\mathscr{S}} t_k x_m\,\mathrm{d}s + \int_{\mathscr{V}} f_k x_m\,\mathrm{d}v\right), \tag{15.2}$$

a relation which is equivalent to (15.1), since $\mathbf{T} = \mathbf{T}^T$.

Consider now a multiply-connected body in a *state of self-stress* (cf. Sect. 6.2). In this case, we have $\mathbf{t} = \mathbf{f} = \mathbf{0}$, and hence (15.2) reduces to

$$\int_{\mathscr{V}} T_{km}\,\mathrm{d}v = 0, \qquad k,\ m = 1, 2, 3. \tag{15.3}$$

We arrive thus to the following result, which seems to have been formulated for the first time by Albenga [2]: *The mean value of each Cartesian stress component over a body that is in equilibrium in a state of self-stress is zero.* It is interesting to note that this result does not depend on the constitutive equation of the material (except indirectly, through the configuration assumed by the body in the deformed state); in particular it holds for non-linear elasticity, too.

For a homogeneous linearly elastic body it follows at once from (15.3) and (6.4) that

$$\int_{\mathscr{V}} E_{km} \, \mathrm{d}v = 0, \qquad k, m = 1, 2, 3. \tag{15.4}$$

Thus, the *linear* elasticity theory predicts a vanishing value for each Cartesian component of the strain tensor, too; in particular, the mean value of the infinitesimal dilatation of a self-stressed body is also zero. Clearly, this consequence of the mean stress theorem is no longer valid in the non-linear case. However, the general result (15.3) may still be used to compute the (non-zero) mean value of the non-linear volume change.

## 15.2. The volume change produced by single dislocations

In order to apply the above results to the case of single dislocations we shall neglect the effect of the tractions exerted by the dislocation core on the internal boundary of the elastic continuum. A more precise treatment would require the consideration of the contribution of these tractions to the surface integral in the right-hand side of (15.2).

The *mean volume change per unit underformed volume* is defined by

$$\bar{\Theta} = \frac{V - V_0}{V_0}, \tag{15.5}$$

where $V_0$ is the volume of the region occupied by the body $\mathscr{B}$ in the natural state. On the other hand, by (2.26), (2.8), and (7.5) we have

$$\frac{\mathrm{d}V}{\mathrm{d}v} = j = \det \mathbf{F}^{-1} = \det (\mathbf{1} - \mathbf{h}), \tag{15.6}$$

where

$$\mathbf{h}(\mathbf{x}) = \operatorname{grad} \mathbf{u}(\mathbf{x}). \tag{15.7}$$

A direct calculation proves that (15.6) may be rewritten as

$$\frac{\mathrm{d}V}{\mathrm{d}v} = 1 - \mathrm{I}_{\mathbf{h}} + \mathrm{II}_{\mathbf{h}} - \mathrm{III}_{\mathbf{h}}, \tag{15.8}$$

where $I_h$, $II_h$, $III_h$ are the principal invariants of $\mathbf{h}$. Multiplying both sides of (15.8) by $dv$ and integrating over $V$ yields the volume change of $\mathscr{B}$ from the natural state to the deformed one:

$$V - V_0 = \int_{\mathscr{V}} (I_h - II_h + III_h)\, dv. \tag{15.9}$$

Next, we apply the results obtained in Sect. 14.1 concerning the second-order elastic effects of dislocations by writing

$$\mathbf{h} = \varepsilon \mathbf{h}^{(1)} + \varepsilon^2 \mathbf{h}^{(2)} + \dots \tag{15.10}$$

and neglecting terms of third and higher orders in $\varepsilon$. Substituting (15.10) into (15.9) and considering (5.4–6) yields

$$V - V_0 = \int_{\mathscr{V}} (\varepsilon h^{(1)}_{mm} + \varepsilon^2 h^{(2)}_{mm})\, dv + \frac{1}{2}\varepsilon^2 \int_{\mathscr{V}} [h^{(1)}_{mn} h^{(1)}_{nm} - (h^{(1)}_{mm})^2]\, dv. \tag{15.11}$$

The first integral in the right-hand side may be now eliminated by using (15.3). To this end we first introduce (14.17) and (14.18) into (15.3), thus obtaining

$$c_{klmn} \int_{\mathscr{V}} (\varepsilon h^{(1)}_{mn} + \varepsilon^2 h^{(2)}_{mn})\, dv = -\varepsilon^2 \int_{\mathscr{V}} \tau_{kl}\, dv.$$

Next, multiplying both sides of this equation by $s_{ppkl}$, summing up for $k, l = 1, 2, 3$, and taking into account that (4.50) and (4.45) imply

$$s_{ppkl} c_{klmn} = \delta_{pm} \delta_{pn} = \delta_{mn}, \tag{15.12}$$

we obtain

$$\int_{\mathscr{V}} (\varepsilon h^{(1)}_{mm} + \varepsilon^2 h^{(2)}_{mm})\, dv = -\varepsilon^2 \int_{\mathscr{V}} s_{ppkl} \tau_{kl}\, dv. \tag{15.13}$$

Finally, by introducing (15.13) into (15.11) and taking into consideration the expression (14.19) of $\tau_{kl}$, we find

$$V - V_0 = \varepsilon^2 \int_{\mathscr{V}} P_{mnrs} h^{(1)}_{mn} h^{(1)}_{rs}\, dv, \tag{15.14}$$

where

$$P_{mnrs} = \frac{1}{2}(\delta_{mn}\delta_{rs} - \delta_{ms}\delta_{nr} - \delta_{mr}\delta_{ns}) - s_{ppkl}\left(2c_{slmn}\delta_{kr} + \frac{1}{2} C_{klmnrs}\right). \tag{15.15}$$

Formula (15.14) gives the desired expression of the (non-linear) volume change in second-order elasticity theory.

It is important to note that the calculation of the mean dilatation $\overline{\Theta}$ from second-order elasticity involves the elastic constants of third order, but requires only the knowledge of the displacement gradient $\mathbf{h}^{(1)}$ from the first step of the iteration, i.e. from linear elasticity. The parameter $\varepsilon$ occurs in the final result (15.14) only through the combination $\varepsilon\mathbf{h}^{(1)}$, which, as shown in Sect. 14, does not depend on $\varepsilon$. Furthermore, when a Lagrangian description is being used for the iteration scheme (see, e.g. Teodosiu [337], p. 186, Gairola [418], p. 290), formula (15.14) is simply replaced by

$$V - V_0 = \varepsilon^2 \int_{\mathscr{V}_0} P_{mnrs} H^{(1)}_{mn} H^{(1)}_{rs} \, \mathrm{d}V, \tag{15.16}$$

where $\mathscr{V}_0$ is the region occupied by the body in the natural state, $\mathbf{H}^{(1)}$ is the displacement gradient from the first step of the iteration, and $P_{mnrs}$ is given by the same relation (15.15). This is not a surprising result, however, since $\mathrm{d}v = \mathrm{d}V + O(\varepsilon)$ and all terms of third and higher orders in $\varepsilon$ have been neglected in (15.16).

In view of the discussion above, we shall rewrite (15.15) and (15.16) in the unified form

$$V - V_0 = \int_{\mathscr{V}_0} P_{mnrs} H_{mn} H_{rs} \, \mathrm{d}V, \tag{15.17}$$

where $\mathbf{H} = \varepsilon\mathbf{H}^{(1)} = \varepsilon\mathbf{h}^{(1)}$ means the *infinitesimal* displacement gradient, and no distinction is being made between the natural and the deformed configurations when calculating the integral in the right-hand side.

Formula (15.17) assumes a much simpler form in the *isotropic* case. Indeed, from (5.26) and (15.12) it follows that

$$\lambda s_{ppll} \delta_{mn} + 2\mu s_{ppmn} = \delta_{mn},$$

and hence

$$s_{ppll} = \frac{3}{3\lambda + 2\mu}, \qquad s_{ppkl} = \frac{\delta_{kl}}{3\lambda + 2\mu}.$$

Substituting this result into (15.15) yields

$$P_{mnrs} = \tfrac{1}{2}(\delta_{mn}\delta_{rs} - \delta_{ms}\delta_{nr} - \delta_{mr}\delta_{ns}) - (c_{mnrs} + \tfrac{1}{2} C_{ppmnrs})/(3\lambda + 2\mu). \tag{15.18}$$

On the other hand, by (5.33),

$$C_{ppmnrs} = (3\nu_1 + 4\nu_2)\,\delta_{mn}\delta_{rs} + (3\nu_2 + 4\nu_3)(\delta_{mr}\delta_{ns} + \delta_{ms}\delta_{nr}),$$

and hence, considering also (5.26), equation (15.18) becomes

$$P_{mnrs} = -\frac{\lambda - 2\mu + 3\nu_1 + 3\nu_2}{2(3\lambda + 2\mu)}\,\delta_{mn}\delta_{rs} -$$

$$-\frac{1}{2}\left(1 + \frac{4\mu + 3\nu_2 + 4\nu_3}{3\lambda + 2\mu}\right)(\delta_{mr}\delta_{ns} + \delta_{ms}\delta_{nr}). \tag{15.19}$$

Finally, introducing (15.19) into (15.17), we obtain

$$V - V_0 = -\frac{\lambda - 2\mu + 3\nu_1 + 4\nu_2}{2(3\lambda + 2\mu)}\int_{\mathscr{V}_0} (E_{mm})^2\, dv -$$

$$-\left(1 + \frac{4\mu + 3\nu_2 + 4\nu_3}{3\lambda + 2\mu}\right)\int_{\mathscr{V}_0} E_{mn}E_{nm}\, dv, \tag{15.20}$$

where $\mathbf{E} = \frac{1}{2}(\mathbf{H} + \mathbf{H}^T)$ is the infinitesimal strain tensor.

We may further relate the mean value of the dilatation to the mean values of the dilatational and shear parts of the stored energy function. To this end, we first rewrite (6.14) in the form

$$W = \frac{\lambda}{2}(\operatorname{tr}\mathbf{E})^2 + \mu \operatorname{tr}\mathbf{E}^2. \tag{15.21}$$

Introducing the deviator $\mathring{\mathbf{E}}$ of the strain tensor, defined by

$$\mathring{\mathbf{E}} = \mathbf{E} - \tfrac{1}{3}(\operatorname{tr}\mathbf{E})\,\mathbf{1}, \tag{15.22}$$

we have

$$\mathbf{E}^2 = \mathring{\mathbf{E}}^2 + \tfrac{2}{3}(\operatorname{tr}\mathbf{E})\,\mathring{\mathbf{E}} + \tfrac{1}{9}(\operatorname{tr}\mathbf{E})^2\,\mathbf{1},$$

and hence, since $\operatorname{tr}\mathring{\mathbf{E}} = 0$,

$$\operatorname{tr}\mathbf{E}^2 = \operatorname{tr}\mathring{\mathbf{E}}^2 + \tfrac{1}{3}(\operatorname{tr}\mathbf{E})^2. \tag{15.23}$$

Substituting (15.23) into (15.21) yields

$$W = W_d + W_s, \tag{15.24}$$

where

$$W_d = \tfrac{1}{2}K(\operatorname{tr}\mathbf{E})^2, \qquad W_s = \mu \operatorname{tr}\mathring{\mathbf{E}}^2, \tag{15.25}$$

and $K = \lambda + 2\mu/3$ is the bulk modulus. $W_d$ and $W_s$ are called the *dilatational* and, respectively, *the shear part of the stored energy function W*, on account of the following reasoning. Given any homogeneous strain field **E**, it is always possible [1] to find a Cartesian frame $\{0, \mathbf{e}_k\}$ such that

$$\mathbf{E} = \mathbf{E}_d + \mathbf{E}_s, \tag{15.26}$$

where

$$\mathbf{E}_d = e\mathbf{1},$$

$$\mathbf{E}_s = \gamma_1(\mathbf{e}_2\mathbf{e}_3 - \mathbf{e}_3\mathbf{e}_2) + \gamma_2(\mathbf{e}_3\mathbf{e}_1 - \mathbf{e}_1\mathbf{e}_3) + \gamma_3(\mathbf{e}_1\mathbf{e}_2 - \mathbf{e}_2\mathbf{e}_1).$$

In other words, any homogeneous strain field may be decomposed in a uniform dilatation of amount $e$, and three simple shears of amounts $\gamma_1$, $\gamma_2$, and $\gamma_3$, with respect to the direction pairs $(\mathbf{e}_2, \mathbf{e}_3)$, $(\mathbf{e}_3, \mathbf{e}_1)$, and $(\mathbf{e}_1, \mathbf{e}_2)$, respectively. On the other hand, a direct calculation shows that

$$\operatorname{tr} \mathbf{E}_d = \operatorname{tr} \mathbf{E}, \qquad \mathring{\mathbf{E}}_d = \mathbf{0}, \qquad \operatorname{tr} \mathbf{E}_s = 0, \qquad \mathring{\mathbf{E}}_s = \mathring{\mathbf{E}},$$

and hence

$$W_d = \tfrac{1}{2} K (\operatorname{tr} \mathbf{E}_d)^2, \qquad W_s = \mu \operatorname{tr} \mathring{\mathbf{E}}_s^2,$$

these relations justifying the terminology adopted above.

Let us define now the mean values $\overline{W}_d$ and $\overline{W}_s$ of the dilatational and shear parts of the stored energy by the relations

$$\left.\begin{aligned} \overline{W}_d &= \frac{1}{V_0}\int_{\mathscr{V}_0} W_d \, \mathrm{d}V = \frac{K}{2V_0}\int_{\mathscr{V}_0} (\operatorname{tr} \mathbf{E})^2 \, \mathrm{d}V, \\ \overline{W}_s &= \frac{1}{V_0}\int_{\mathscr{V}_0} W_s \, \mathrm{d}V = \frac{\mu}{V_0}\int_{\mathscr{V}_0} (\operatorname{tr} \mathring{\mathbf{E}}^2) \, \mathrm{d}V. \end{aligned}\right\} \tag{15.27}$$

Introducing now (15.23) into (15.20) and taking into account (15.27) we find

$$\overline{\Theta} = -\frac{1}{K}\left(1 + \frac{\nu_1 + 2\nu_2 + 8\nu_3/9}{K}\right)\overline{W}_d - \frac{\lambda + 2\mu + \nu_2 + 4\nu_3/3}{\mu K}\overline{W}_s, \tag{15.28}$$

which is the desired result.

---

[1] See, e.g. Gurtin [150], pp. 35–37.

## 15.3. Derivation of Zener's formula in the isotropic case

The coefficients of $\overline{W}_d$ and $\overline{W}_s$ in equation (15.28) may be given an interesting physical interpretation by considering the *apparent elastic constants* characterizing the response of an elastic material to an infinitesimal deformation superimposed upon a uniform finite dilatation.

We first remark that the constitutive equation of second-order elasticity(5.34) may be rewritten in direct notation as

$$\mathbf{T} = \lambda(\mathrm{tr}\,\mathbf{E})\,\mathbf{1} + 2\mu\mathbf{E} + \left\{\frac{\lambda}{2}\,\mathrm{tr}(\mathbf{H}^T\mathbf{H}) + \left(\frac{\nu_1}{2} - \lambda\right)(\mathrm{tr}\,\mathbf{E})^2 + \nu_2\,\mathrm{tr}\,\mathbf{E}^2\right\}\mathbf{1} +$$

$$+ 2(\lambda - \mu + \nu_2)(\mathrm{tr}\,\mathbf{E})\,\mathbf{E} + \mu\mathbf{H}\mathbf{H}^T + 4(\mu + \nu_3)\,\mathbf{E}^2. \tag{15.29}$$

The *pressure* associated to the stress tensor $\mathbf{T}$ is defined by

$$p = -\tfrac{1}{3}\,\mathrm{tr}\,\mathbf{T}. \tag{15.30}$$

Hence, by (15.29),

$$p = -K\,\mathrm{tr}(\mathbf{E} + \tfrac{1}{2}\mathbf{H}^T\mathbf{H}) + \tfrac{1}{3}(\lambda + 2\mu - 3\nu_1/2 - 2\nu_2)(\mathrm{tr}\,\mathbf{E})^2 -$$

$$- \tfrac{1}{3}(4\mu + 3\nu_2 + 4\nu_3)\,\mathrm{tr}\,\mathbf{E}^2. \tag{15.31}$$

Suppose that the body is subjected to a *uniform finite dilatation* of deformation gradient $\mathbf{F}_0 = \alpha\mathbf{1}$. Then, $\mathbf{H}_0 = \mathbf{E}_0 = (\alpha - 1)\mathbf{1}$ and (15.29) shows that the corresponding stress tensor $\mathbf{T}_0$ reduces to a hydrostatic pressure, namely $\mathbf{T}_0 = -p_0\mathbf{1}$, where, by virtue of (15.31),

$$p_0 = -3(\alpha - 1)[K + (\alpha - 1)(-K/2 + 3\nu_1/2 + 3\nu_2 + 4\nu_3/3)]. \tag{15.32}$$

Let us superimpose now on the uniform finite dilatation a second *uniform infinitesimal dilatation* of amount $\beta$, whose deformation gradient from the first state is $\mathbf{F}_1 = (1 + \beta)\mathbf{1}$. The total deformation gradient is $\mathbf{F} = \mathbf{F}_1\mathbf{F}_0 = (1 + \beta)\,\alpha\mathbf{1}$, and hence $\mathbf{H} = \mathbf{E} = (\alpha + \alpha\beta - 1)\,\mathbf{1}$. Then, the corresponding stress tensor $\mathbf{T}$ reduces again to a hydrostatic pressure, i.e. $\mathbf{T} = -p\,\mathbf{1}$, where $p$ may be obtained directly from (15.32) by simply replacing $\alpha$ with $(\beta + 1)\alpha$. Hence

$$p = -3(\alpha + \alpha\beta - 1)[K + (\alpha + \alpha\beta - 1)(-K/2 + 3\nu_1/2 + 3\nu_2 + 4\nu_3/3)]. \tag{15.33}$$

According to (6.6), the *apparent bulk modulus* can be defined as

$$K^* = -\lim_{\beta\to 0}\frac{p - p_0}{\mathrm{tr}\,\mathbf{E}_1} = -\lim_{\beta\to 0}\frac{p - p_0}{3\beta} =$$

$$= \alpha[K + (\alpha - 1)(-K + 3\nu_1 + 6\nu_2 + 8\nu_3/3)], \tag{15.34}$$

where all constants, including $K$, must be taken with their values in the natural state. Since this state is characterized in our case by $\alpha = 1$ and $p_0 = 0$, we deduce from (15.32) and (15.34) that

$$\left.\frac{\mathrm{d}K^*}{\mathrm{d}p_0}\right|_{p_0=0} = \left.\frac{\dfrac{\mathrm{d}K^*}{\mathrm{d}\alpha}}{\dfrac{\mathrm{d}p_0}{\mathrm{d}\alpha}}\right|_{\alpha=0} = -\frac{\nu_1 + 2\nu_2 + 8\nu_3/9}{K}. \tag{15.35}$$

Clearly, this derivative determines the variation of $K^*$ for moderate values of $p_0$.

Next, we consider an infinitesimal shear of amount $\gamma$, whose deformation gradient is $\mathbf{F}_2 = \mathbf{1} + \gamma \mathbf{e}_1\mathbf{e}_2$, superimposed on the initial uniform dilatation. For the total deformation from the natural state we have

$$\mathbf{F} = \mathbf{F}_2\mathbf{F}_0 = \alpha\mathbf{1} + \alpha\gamma\mathbf{e}_1\mathbf{e}_2, \qquad \mathbf{H} = (\alpha - 1)\,\mathbf{1} + \alpha\gamma\mathbf{e}_1\mathbf{e}_2,$$

$$\mathbf{E} = (\alpha - 1)\,\mathbf{1} + \tfrac{1}{2}\alpha\gamma(\mathbf{e}_1\,\mathbf{e}_2 + \mathbf{e}_2\mathbf{e}_1).$$

Substituting the last two expressions into (15.29) we obtain for the shear stress $T_{12}$ corresponding to the direction pair $(\mathbf{e}_1, \mathbf{e}_2)$ the expression

$$T_{12} = \alpha\gamma[\mu + (\alpha - 1)(3\lambda + 2\mu + 3\nu_2 + 4\nu_3)]. \tag{15.36}$$

The *apparent shear modulus* is defined by

$$\mu^* = \lim_{\gamma\to 0} (T_{12}/\gamma)$$

and hence, by (15.36),

$$\mu^* = \alpha[\mu + (\alpha - 1)(3\lambda + 2\mu + 3\nu_2 + 4\nu_3)]. \tag{15.37}$$

The variation of $\mu^*$ for moderate values of $p_0$ is determined by the derivative

$$\left.\frac{\mathrm{d}\mu^*}{\mathrm{d}p_0}\right|_{p_0=0} = \left.\frac{\dfrac{\mathrm{d}\mu^*}{\mathrm{d}\alpha}}{\dfrac{\mathrm{d}p_0}{\mathrm{d}\alpha}}\right|_{\alpha=1} = -\frac{\lambda + \mu + \nu_2 + 4\nu_3/3}{K}. \tag{15.38}$$

Finally, by substituting (15.35) and (15.38) into (15.28), we derive *Zener's formula* for the mean volume change per unit underformed volume in second-order el asticity

$$\overline{\Theta} = \frac{1}{K}\left(\frac{\mathrm{d}K^*}{\mathrm{d}p} - 1\right)\overline{W}_d + \frac{1}{\mu}\left(\frac{\mathrm{d}\mu^*}{\mathrm{d}p} - \frac{\mu}{K}\right)\overline{W}_s, \tag{15.39}$$

where both moduli $K$ and $\mu$, as well as the derivatives of the apparent moduli, are taken with their values in the natural state. The reasoning leading to (15.39) has been generalized by Toupin and Rivlin [354] to cubic crystals belonging to subsystem 7 (see also Seeger [287]).

Starting from (15.39), Seeger and Haasen [288] have calculated the mean value of the dilatation produced by straight dislocations in isotropic media. In order to derive their result, we make use of equations (8.52) and $(8.60)_2$, which give the strain field of straight dislocations in isotropic media. Then, taking also into account that

$$K = \frac{2\mu(1+\nu)}{3(1-2\nu)},$$

equations (15.27) and (15.39) yield for an *edge dislocation*

$$\overline{W}_d = \frac{Kb^2}{8\pi^2(R^2 - r_0^2)} \left( \frac{1-2\nu}{1-\nu} \right)^2 \ln \frac{R}{r_0},$$

$$\overline{W}_s = \frac{\mu b^2}{6\pi^2(R^2 - r_0^2)} \frac{1-\nu+\nu^2}{(1-\nu)^2} \ln \frac{R}{r_0},$$

$$\overline{\Theta} = \frac{b^2}{4\pi^2(R^2 - r_0^2)} \left[ \frac{1}{2} \left( \frac{1-2\nu}{1-\nu} \right)^2 \left( \frac{\mathrm{d}K^*}{\mathrm{d}p} - 1 \right) + \frac{2}{3} \frac{1-\nu+\nu^2}{(1-\nu)^2} \left( \frac{\mathrm{d}\mu^*}{\mathrm{d}p} - \frac{\mu}{K} \right) \right] \ln \frac{R}{r_0},$$

and for a *screw dislocation*

$$\overline{W}_d = 0, \quad \overline{W}_s = \frac{\mu b^2}{4\pi^2(R^2 - r_0^2)} \ln \frac{R}{r_0},$$

$$\overline{\Theta} = \frac{b^2}{4\pi^2(R^2 - r_0^2)} \left( \frac{\mathrm{d}\mu^*}{\mathrm{d}p} - \frac{\mu}{K} \right) \ln \frac{R}{r_0}.$$

Alternatively, one can correlate the volume expansion per unit dislocation length

$$\mathrm{d}v = \pi(R^2 - r_0^2)\, \overline{\Theta}$$

with the strain energy $w$, stored per unit dislocation length, which is given by (8.54) for an edge dislocation and by (8.62) for a screw dislocation. The result reads

$$\delta v = \frac{1}{3} \left[ \frac{1-\nu-2\nu^2}{1-\nu} \frac{1}{K} \left( \frac{\mathrm{d}K^*}{\mathrm{d}p} - 1 \right) + \frac{1-\nu+\nu^2}{1-\nu} \frac{2}{\mu} \left( \frac{\mathrm{d}\mu^*}{\mathrm{d}p} - \frac{\mu}{K} \right) \right] w$$

for an *edge dislocation*, and

$$\delta v = \frac{1}{\mu}\left(\frac{d\mu^*}{dp} - \frac{\mu}{K}\right)w$$

for a *screw dislocation*. By using the above formulae and the experimental values of the constants for various polycrystalline materials (Cu, Ni, Al, Fe, and NaCl), Seeger and Haasen [288] found in all cases that a dislocation produces a positive expansion, hence a decreasing of the crystal density.

# 16. Study of the core of straight dislocations by fitting the atomic and elastic models

A thorough understanding of the strength of crystalline materials and in particular of metals rests ultimately on knowledge of the atomic arrangements and movements around lattice imperfections (dislocations, point defects, grain boundaries). Indeed, it is these imperfections that make possible the plastic flow and the nucleation and propagation of cracks at applied forces that are several orders of magnitude lower than those necessary to fracture a perfect crystal.

Unfortunately, the highly distorted regions close to crystal defects cannot yet be studied experimentally. On the other hand, the well-developed techniques of continuum mechanics, based on linear or even non-linear elasticity, break down near crystal defects, since they lead to infinite values of the stress and displacement fields which are physically unacceptable. Finally, the fully atomic models based on lattice theory lead to correct predictions of the atomic arrangements around imperfections, but must be restricted to bounded atomic blocks containing a not very large number of atoms, in order to save computing time. Therefore, the only reasonable way seems to be the application of *semidiscrete methods*, which make use of the lattice theory for the close proximity of crystal defects and of the elasticity theory for the remaining of the crystal, each of these theories providing the boundary conditions necessary for the other one.

In the following, we shall confine ourselves to consider the distortions within the core of straight dislocations. After examining the main effects of these distortions on the processes taking place in crystalline materials we shall review the methods used to determine the atomic arrangement around straight dislocations.

## 16.1. Influence of the highly distorted dislocation core on the physical-mechanical behaviour of crystals

To illustrate the high distortions within the dislocation core we reproduce in Fig. 16.1, after Gehlen, Rosenfield and Hahn [127] the atomic arrangement near a $\langle 100 \rangle$ edge dislocation in $\alpha$ iron, as determined by means of a semidiscrete method with 780 atoms in the region treated atomistically. The triangles and the squares represent individual iron atoms on two consecutive (100) planes. The plus sign marks the theoretical position of the dislocation line in the elastic calculation. The unit extensions of the bonds *AB, DE, BE,* and *CD*, calculated in [127] with respect to the perfect lattice [1] were —17.1%, +50.7%, —3.6%, and +9.5%, respectively. It is obvious that for such high strains, not only the linear theory but also the second-order theory of elasticity would yield misleading results.

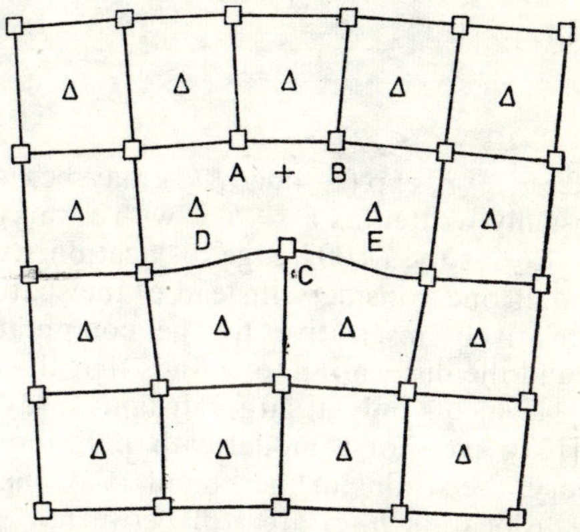


Fig. 16.1. Configuration core of a ⟨100⟩ edge dislocation in α iron (after Gehlen, Rosenfield, and Hahn [127]). The triangles and the squares represent individual iron atoms on two consecutive (100) planes.

Since dislocations interact through their elastic fields with point defects and other dislocations, this interaction is influenced by the actual distortion of the dislocation core. The core contribution to the total interaction energy is the higher, the closer the interacting defects. Thus, the knowledge of the atomic arrangements around dislocations plays an important part in determining such properties as the Peierls stress, the structure and energy of jogs and kinks, the tendency of dislocations to dissociate into partials, the interaction between gliding dislocations and short-range obstacles, the dilatational effect of dislocations, etc. We shall consider below in more detail some of these problems.

*The core energy.* As shown in Sects. 8 and 10, the total energy stored per unit length of a straight dislocation may be written as

$$w_t = \frac{Kb^2}{4\pi} \ln \frac{R}{r_0} + w_0, \tag{16.1}$$

where the first term in the right-hand side is the strain energy of the linear elastic field per unit dislocation length, $R$ is the outer radius of a circular cylinder within which the energy is evaluated (the dislocation line being the axis of that cylinder) $r_0$ is the core radius, and $w_0$ is the core energy per unit dislocation length. Of course, the theory of elasticity cannot provide any information about $r_0$ and $w_0$, which are introduced in (16.1) only to avoid that $w_t \to \infty$ as $r_0 \to 0$. However,

[1] *AB* and *DE* were taken as first-neighbour bonds, *BE* and *CD* as second-neighbour bonds.

by using a semidiscrete method, one can obtain the plot of $w_t$ vs. $\ln R$ and, by comparing it to (16.1), the desired values of $r_0$ and $w_0$. For example, Gehlen *et al.* [127] have found in this way for the $\langle 100 \rangle$ edge dislocation in $\alpha-$Fe values of $r_0$ between 1.25 and 1.65 $b$, and of the core energy per atomic plane between 0.47 and 0.65eV, i.e. about one tenth of the total strain energy. It is also worth mentioning that the comparison of the slope of the $w_t$—vs.—$\ln R$ plot with the factor $Kb^2/4\pi$ in (16.1) has been successfully used to discriminate between different interatomic potentials.

Equation (16.1) may be obviously rewritten as

$$w_t = \frac{Kb^2}{4\pi} \ln \frac{R}{r_i}, \tag{16.2}$$

where $r_i = r_0 \exp(-4\pi w_0/Kb^2)$ may be interpreted as an effective hole radius, and is usually written as $r_i = b/\alpha$, with $\alpha$ varying between 1 and 2 for most metals, e.g. $\alpha = 1.6$ for the $\langle 100 \rangle$ edge dislocation studied by Gehlen *et al.* [127]. On the other hand, if one considers instead of the spatial variation of the strain energy — which is relatively insensitive to the computation method — the atomic configuration around the dislocation, one finds that the linear elastic solution matches the atomic displacements only at larger distances, say 10 atomic spacings. Thus, as pointed out in [127], even for a model with 780 atoms within the region treated by the lattice theory, the strains at the boundary of that region (about 14 atomic spacings from the dislocation line) are still between 1% and 2%.

The evaluation of the parameters $r_0$ and $w_0$ makes possible the application of (16.1) to improving the solution of such problems as: the calculation of the line tension, the determination of the equilibrium angles of dislocation nodes, the study of the tendency of dislocations to zigzag during their glide, and the calculation of the interaction between dislocations and other crystal defects [1].

*Peierls stress.* In the dislocation theory of plastic glide it is generally assumed that each dislocation is influenced by a given stress state only through the so-called *resolved shear stress,* which is the component in the glide direction of the stress vector acting on the glide plane. The resolved shear stress that is necessary to move a dislocation through a crystal in the absence of any other defect is called the *Peierls stress.* The origin of this intrinsic resistance of the lattice is the periodic variation of the misfit energy of the atomic half-planes above and below the glide plane with the position of the dislocation in this plane [2].

When the dislocation density is high the main resistance to glide and the work-hardening are provided by the long-range interaction between dislocations, whereas the contribution of the Peierls stress is negligible. On the contrary, at the beginning of the plastic flow, when the dislocation density is rather low, this contribution may be significant. The Peierls stress determines also the yield stress of the so-called whiskers, which are crystal wires of diameter less than one micron, containing few or only one dislocation running parallel to the axis of the wire. Finally, it is generally

[1] In this connection see also Gehlen, Hirth, Hoagland, and Kanninen [130].

[2] Fore more details, see e.g. Hirth and Lothe [162], Sect. 8.4 and Aczel and Bozan [1], § 8.3.

accepted that the plastic behaviour of b.c.c. metals at low temperature is to a great extent controlled by a Peierls mechanism.

Two main procedures have been used so far to evaluate the Peierls stress by semidiscrete methods:

(i) In the first approach, which has been originally used by Nabarro [255], the dislocation is held in various crystallographically non-equivalent positions, and its core energy $w_t$ is calculated as a function of the abscissa $\xi$ measured in the glide direction. Then, by using some interpolation function to define a smooth dependence $w_t(\xi)$, one calculates the Peierls stress $\tau_P$ by the formula

$$\tau_P = -\frac{1}{b}[w_t'(\xi)]_{\max}.$$

In this way and using a semidiscrete method to determine the various core configurations of the $\frac{1}{2}\langle 111\rangle$ edge dislocation on $\{110\}$ planes in $\alpha$-Fe, Chang and Graham [60] found a Peierls energy barrier of about 0.03 eV per identity distance along the dislocation line and a Peierls stress of 0.0066 $\mu$, in satisfactory agreement with experimental evidence on internal friction. More recently, the same method has been employed by Heinrich, Schellenberger, and Pegel [426], and by Heinrich and Schellenberger [427] for dislocations in b.c.c. crystals. However, such calculations do not take into account the possible changes in the core configuration under an applied stress. As first pointed out by Suzuki [325], this is inconvenient especially for the screw dislocation in b.c.c. lattices, which has a sessile equilibrium configuration, and hence its core must undergo substantial changes before the dislocation can move.

(ii) The second procedure consists in the semidiscrete simulation of a crystallite subject to an applied shear stress. Since the core distortions are strongly non-linear, the superposition principle does not hold, and the presence of the applied stress must be taken into account from the very beginning. This was first done by Kurosawa [445], who calculated the Peierls stress for edge dislocations on $\{110\}$ planes in alkali halides, and more recently by Basinski, Duesbery, and Taylor [19, 20], and by Duesbery, Vitek, and Bowen [411] for screw dislocations in b.c.c. crystals. The calculation proceeds as follows. First, a finite crystallite in the form of a rectangular parallelepiped composed of a certain number of repeat units is chosen. The initial equilibrium core configuration is determined by fixing the lateral boundaries of the crystallite at the positions dictated by the linear anisotropic elasticity for a Volterra dislocation placed in the middle of the crystallite. Along the dislocation line are applied periodic boundary conditions, which make the crystallite effectively infinite in the direction of the dislocation.

The application of an external shear stress is simulated by imposing the corresponding homogeneous strain given by linear elasticity on the initial core configuration, and then allowing the atoms to relax to new equilibrium positions. The calculation always starts with stresses much lower than the Peierls stress; larger stresses are built up by the application of successive small stress increments. Upon choosing a sufficiently large crystallite (this was taken of dimensions $45b \times 45b$ normal to the dislocation line in [411]), the dislocation can move through several atomic spacings without being significantly influenced by the fixed boundaries. The critical stress for which the dislocation starts to move through the lattice is then identified as the Peierls stress. Both positive and negative stresses are considered whenever asymmetries are expected in the dislocation motion: this is for example the case with $\frac{1}{2}\langle 111\rangle$ dislocations moving on $\{112\}$ planes in b.c.c. crystals, for which the Peierls stresses corresponding to the dislocation motion in the twinning and antitwinning sense, respectively, have been evaluated.

In general, two distinct critical shear stresses may exist. The first critical shear stress is the smallest stress for which the core undergoes an irreversible translation. After this first step no further movement is possible until the applied stress is increased to the second critical shear stress, when the dislocation moves freely through the lattice. By using three distinct types of short-range interionic potentials, Duesbery *et al.* [411] have found for the $\frac{1}{2}\langle 111\rangle$ screw dislocation in b.c.c. crystals that the first critical shear stress on $\{110\}$ planes ranges between $0.020\,\mu$ and $0.034\,\mu$ and the second critical shear stress ranges between $0.029\,\mu$ and $0.040\,\mu$. These authors have also demonstrated that the applied stress produces significant internal changes of the core structure before the dislocation can move, and hence the first procedure may lead to a drastic overestimation of the Peierls stress for this dislocation. Using long-range interionic potentials for lithium, Beauchamp, Rabier, and Grilhé [22] arrived essentialy to the same conclusions, except that boundary conditions have an increased influence on the results when employing such potentials. In addition, they showed that the gliding on the $\{112\}$ planes is strongly asymmetric, being easier in the twinning sense than in the antitwinning sense.

Yamaguchi and Vitek [386, 488] thourougly studied the Peierls stresses and the effects of an applied stress on the core structure of non-screw $\frac{1}{2}\langle 111\rangle$ dislocations in b.c.c. crystals by using three different central interatomic forces. They showed that the cores of non-screw dislocations lying on $\{110\}$ planes do not undergo any drastic changes under an external stress except a certain widening prior to the net movement of the dislocation. As a consequence, both procedures described above lead to similar values of the Peierls stress. On the contrary, for non-screw dislocations lying on $\{112\}$ planes, the first method appreciably overestimates the magnitude of the Peierls stress and does not show any difference between the stresses needed for the movement of the dislocations in the twinning and antitwinning senses. It was argued that this difference arises from the fact that the cores of dislocations lying on $\{110\}$ planes are confined to a single atomic layer, while the cores of dislocations lying on $\{112\}$ planes spread across three layers [488].

*The interaction between gliding dislocations and obstacles.* A gliding dislocation may encounter two types of obstacles: *extended obstacles*, e.g. dislocation pile-ups or inclusions, and *local obstacles*, e.g. point defects or dislocations threading the glide plane.

Since extended obstacles produce a long-range stress field, the contribution of the dislocation core to the interaction with such obstacles is relatively small (less than 10%). In exchange, the interaction between dislocations and short-range stress fields of local obstacles is strongly influenced by the high strains in the dislocation core (Hirth and Nix [163]).

Indeed, the total force exerted by a gliding dislocation segment may be written as (see, e.g. Teodosiu *et al.* [343—345]):

$$f(\xi) = f_i(\xi) - \tau^* bl,$$

where $f_i(\xi)$ is the force exerted by the local obstacle on the dislocation segment, $\xi$ is some abscissa measured in the glide direction, $\tau^*$ is the difference between the reduced stress produced by the applied forces and the reduced stress produced by extended obstacles, and $l$ is the length of the dislocation segment. Clearly, the variation and the maximum value of $f_i(\xi)$ strongly depend on the atomic arrangement around the dislocation segment and on the change of this arrangement during the overcoming of the obstacles. Consequently, the determination of the real core configuration is of great importance for a correct evaluation of the interaction between gliding dislocations and local obstacles, and hence for a quantitative microstructural analysis of the viscoplastic behaviour of metals.

*Variation of the crystal density*. As shown in Sect. 15.2, the volume dilatation produced by dislocations is a typical non-linear effect. Indeed, the linear theory of elasticity predicts zero mean volume dilatation for any source of self-stress, hence also for dislocations.

On the other hand, although the second-order elasticity theory predicts a non-vanishing mean volume dilatation that decreases as $R^{-2}$ as $R \to \infty$, this effect is substantially influenced by the boundary conditions on the dislocation core, which must be taken into account by a combined atomistic and elastic calculation (Gehlen *et al.* [130], Sinclair *et al.* [474]). The preliminary results obtained by Granzer *et al.* [143], with the aid of a semidiscrete method and using a non-linear elastic model outside the region treated atomistically, have led to almost exactly one-atom-volume-per-plane overall volume dilatation, in agreement with other theoretical and experimental results.

The considerations above emphasize the significant influence exerted by the core distortion on various processes taking place in crystals. In the following we will focus our attention on the semidiscrete methods used to simulate crystal dislocations and to determine their core configurations.

## 16.2. The semidiscrete method with rigid boundary

In order to simulate a single dislocation by a *semidiscrete method* it is customary to divide the crystal into two regions. Region I, which is the next neighbourhood of the dislocation line, is treated as a *discrete lattice*. In this highly distorted region the continuum theory does not apply; atom positions are considered individually, with some interatomic potential being assumed to give the potential energy in terms of the atom positions. Region II, the remainder of the crystal, is considered as an *elastic continuum*; clearly, any displacement field satisfying the equilibrium equations of the elasticity theory provides also the equilibrium atom positions when evaluated at the discrete lattice points. However, the only atoms which need to be explicitly considered in region II are those whose positions are required for the calculation of the forces exerted on the atoms of region I.

On coupling the atomistic and continuum models of the dislocation, suitable boundary conditions must be introduced on the separation surface $\Sigma_0$ between

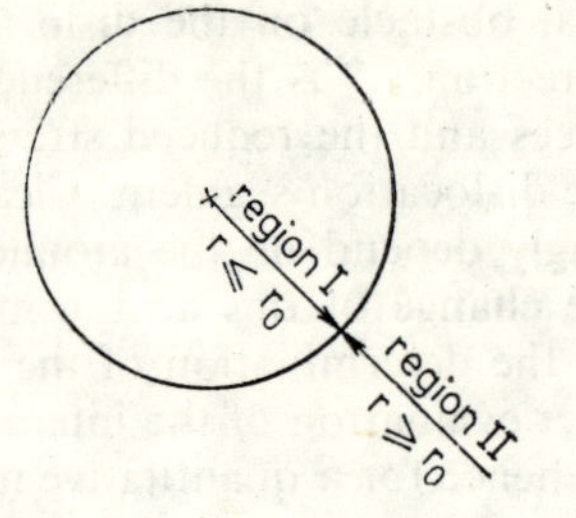


Fig. 16.2. Concentric arrangement of regions for the rigid-boundary method and for Sinclair's flexible-boundary method.

regions I and II. For a straight dislocation, the internal boundary $\Sigma_0$ is generally chosen as a circular cylindrical surface of radius $r_0$ having the dislocation line as axis (Fig. 16.2), or as the lateral boundary of a rectangular parallelepiped having the dislocation line in the middle. In both cases, region I is limited to a repeat distance in the direction of the dislocation line, with periodic boundary conditions being introduced on the surfaces normal to the dislocation line in order to simulate an infinitely deep crystallite.

Until recently, the most frequently used method for simulating crystal dislocations has been the *semidiscrete method with rigid boundary*, in short: the *rigid-boundary method*. This method consists in the calculation of the relaxed atom positions in region I, while the atoms situated on $\Sigma_0$ and in region II are kept fixed in their positions given by the linear elastic displacement field as obtained by neglecting the core boundary conditions [1].

The rigid-boundary method has been first applied in 1955 by Huntington, Dickey, and Thompson [166] to simulate an edge dislocation in NaCl, the linear isotropic elastic displacement field (8.50) being used for region II. A similar calculation has been done by Englert and Tompa [103] in 1961 for a two-dimensional edge dislocation in argon, with the aid of an IBM 650 computer.

The development of high-speed computers coupled with recent advances in elaborating more realistic interatomic potentials has greatly improved the simulation of crystal defects by semidiscrete methods. The first truly *three-dimensional* atomistic treatment (i.e. taking into account the interaction of the atoms situated in a plane perpendicular to the dislocation line with neighbouring atomic planes within the repeat distance) has been worked out for edge and screw dislocations by Doyama and Cotterill [82, 89] in the middle sixties. Later on, the rigid-boundary method has been used to simulate various types of straight dislocations in both f.c.c. and b.c.c. crystals by Chang and Graham [60], Cotterill and Doyama [83, 91],

[1] For a straight dislocation lying in an anisotropic elastic medium this displacement field is given by (10.56). It is sometimes called the *Volterra solution*, although Volterra [373] has derived the linear *isotropic* elastic solution for the case when $\Sigma_0$ is *traction-free*.

Chang [61], Bullough and Perrin [46, 47], Granzer, Wagner, and Eisenblätter [141], Suzuki [325], Vitek [368, 369], Gehlen, Rosenfield, and Hahn [127], Gehlen [128], Vitek, Perrin, and Bowen [370], Diener, Heinrich, and Schellenberger [87], Rabier and Grilhé [273], Puls and Norgett [462], most of these authors employing the *anisotropic* elasticity theory for a more exact evaluation of the atomic displacements in region II. At the same time, region I has been gradually increased from 20 atomic rows in the pioneering work of Huntington *et al.* [166] to 2565 atomic rows in the work of Duesbery, Vitek, and Bowen [411]. A systematic investigation of the effect of an *applied stress* on the core configuration and on the dislocation mobility has been undertaken by Basinski, Duesbery, and Taylor [19, 20], Gehlen [129], Yamaguchi and Vitek [386, 488], Duesbery *et al.* [411], Beauchamp, Rabier, and Grilhé [22]. Finally, the rigid-boundary method has been used to study the *interaction between dislocations and point defects* (Perrin, Englert, and Bullough [267]), the *kinks* and *jogs* and the *dissociation of a full jog into two half-jogs* in NaCl (Eisenblätter [102]), the *dissociation of an edge dislocation into two partials* in NaCl (Belzner and Granzer [399]), as well as the *structure of a grain-boundary dislocation* in Al (Vitek, Sutton, Smith, and Pond [484]).

In spite of the progress achieved in the description of the core configurations by means of the semidiscrete method with rigid boundary, it is apparent that this method has some essential drawbacks. The most important of them is that the elastic solution used for region II does not take into account the boundary conditions on $\Sigma_0$, and hence cannot be improved during the calculation of the relaxed atom positions in the dislocations core.

Moreover, when adopting a linear elastic solution for region II, the rigid boundary introduces some artificial constraints upon region I. For example, the mean volume dilatation of region I must vanish (cf. Sect. 15.1), at variance with theoretical considerations and experimental evidence [1].

In fact, the core configuration calculated by the rigid-boundary method consists of a narrow dilated region around the dislocation line, which is compensated by an artificially compressed shell separating the inner core region from the elastic continuum, such that the mean volume dilatation of the crystal bounded by $\Sigma_0$ be zero (Granzer *et al.* [143]). Furthermore, the atom positions given by the linear elasticity theory, which is based on a *harmonic* interaction potential, abruptly change across $\Sigma_0$ into atom positions computed on the basis of *anharmonic* interatomic potentials, thus making impossible a smooth passage from the dislocation core to its surroundings (see also Petrasch and Belzner [459]).

In addition, calculations done with an increasing number of atoms in region I show a rather slow convergence even for the atom positions in the very centre of the dislocation core, which should be less sensitive to boundary conditions imposed on $\Sigma_0$ (Gehlen *et al.* [127]). That is why, several methods have been elaborated in the last decade, which allow the boundary $\Sigma_0$ to be "flexible", i.e. to relax together with the atoms in region I. Before examining these methods, we will briefly review

[1] As repeatedly emphasized by Seeger (see, e.g. [470]), this unrealistic effect is likely to be avoided by using the non-linear elasticity theory for region II *and* taking into account the boundary conditions on $\Sigma_0$.

the main procedures used to obtain the atom positions in region I: the static relaxation and the dynamic relaxation.

The *static relaxation* permits the determination by successive approximations of the atom equilibrium positions.

Let $\mathbf{X}$ denote the vector of the generalized co-ordinates of a system, for instance the co-ordinates of the atoms in region I with respect to a Cartesian frame. Assume that the potential energy $E$ of the system is known as a function of $\mathbf{X}$, i.e. $E = E(\mathbf{X})$. The generalized force $\mathbf{F}$ associated to the vector $\mathbf{X}$ is defined by

$$\mathbf{F} = -\frac{\partial E(\mathbf{X})}{\partial \mathbf{X}} = \mathbf{F}(\mathbf{X}). \tag{16.3}$$

If $\mathbf{X}=\mathbf{X}^{(e)}$ is an equilibrium configuration of the system which corresponds to a minimum of the potential energy $E$, then, obviously, $\mathbf{F}(\mathbf{X}^{(e)})=\mathbf{0}$. On the other hand, when the system is not in equilibrium, but is closed to the equilibrium configuration $\mathbf{X} = \mathbf{X}^{(e)}$, we may write in a first approximation

$$\mathbf{F} = \mathbf{M}(\mathbf{X} - \mathbf{X}^{(e)}), \tag{16.4}$$

where

$$\mathbf{M} = \left.\frac{\partial \mathbf{F}}{\partial \mathbf{X}}\right|_{\mathbf{X}=\mathbf{X}^{(e)}}. \tag{16.5}$$

Clearly, Eq. (16.4) may be rewritten as

$$\mathbf{X}^{(e)} = \mathbf{X} - \mathbf{M}^{-1}\mathbf{F}. \tag{16.6}$$

Since $\mathbf{M}$ depends on $\mathbf{X}^{(e)}$, the determination of $\mathbf{X}^{(e)}$ from (16.6) is generally done by successive approximations.

Let $\mathbf{X} = \mathbf{X}_0$ be an initial configuration of the system. From (16.5) it results $\mathbf{M}_0 = (\partial\mathbf{F}/\partial\mathbf{X})_{\mathbf{X}=\mathbf{X}_0}$, and (16.3) yields $\mathbf{F}_0=\mathbf{F}(\mathbf{X}_0)$. Introducing this result into (16.6) gives the first approximation, $\mathbf{X}_1 = \mathbf{X}_0 - \mathbf{M}_0^{-1}\mathbf{F}_0$. If required, this iteration step can be repeated until the magnitude of $\mathbf{F}$ becomes sufficiently small. Of course, the (numerical) inversion of the matrices $\mathbf{M}_0, \mathbf{M}_1, \ldots$ is very time-consuming. Therefore, this method is adequate only when merely small adjustments of the atomic configuration are required. In such cases, although the linear relation (16.4) does not hold exactly and (16.6) gives only a first approximation of the equilibrium configuration, it is usually possible to obtain convergence by repeating the application of (16.6) with $\mathbf{M}$ replaced by $\mathbf{M}_0$, even though its components change slightly (Sinclair [302]).

Another variant of the static relaxation, which proves to be advantageous when large adjustments of the atomic configuration are required, consists in relaxing each atom, independently, to a provisional equilibrium position, dictated by the vanishing of the resultant force exerted on this one atom by its neighbours. Then, this step is repeated until the whole atomic array is covered, and the relaxation of the whole array is repeated until the changes in coordinates of all atoms between successive iterations and/or residual forces are within preset limits (see, e.g. Chang and Graham [60]).

*The dynamic relaxation.* This method has been initially used mainly in connection with radiation damage (see Gibson *et al.* [138] and especially Larsen [208], where the so-called GRAPE programs developed for the numerical application of the method are described in great detail). Later on, the dynamic relaxation has been also successfully applied for determining the core configurations (Bullough and Perrin [46, 47], Gehlen, Rosenfield, and Hahn [127], Gehlen [128]).

The method of dynamic relaxation consists in the numerical integration of the classical equations of motion of the atoms in region I, which are considered as a system of material points. The atoms situated on $\Sigma_0$ and in region I are kept fixed in their positions given by the linear elastic solution obtained by neglecting the core boundary conditions, just like in the rigid-boundary method.

For a system of $N$ atoms of mass $m$ in region I, one writes $N$ vectorial equations of motion at time $t$:

$$m\dot{\mathbf{v}}_i = \mathbf{F}_i, \qquad i = 1, 2, \ldots, N, \tag{16.7}$$

where $\mathbf{v}_i = \dot{\mathbf{x}}_i$ is the velocity of the atom $i$, while $\mathbf{x}_i$ denotes its position vector; $\mathbf{F}_i$ is the force exerted on this atom, which may be derived from the central-force pair potential $V$ by the relation

$$\mathbf{F}_i = -\sum_{\substack{j=1 \\ j\neq i}}^{N} \frac{\partial V(r_{ij})}{\partial \mathbf{x}_i}, \tag{16.8}$$

where $r_{ij} = \|\mathbf{x}_i - \mathbf{x}_j\|$. In order to numerically integrate system (16.7), the functions $\mathbf{x}_i(t)$ are developed in Mac-Laurin series around the current time $t$, thus obtaining

$$\mathbf{x}_i(t+\Delta t) = \mathbf{x}_i(t) + \dot{\mathbf{x}}_i(t)\,\Delta t + \tfrac{1}{2}\ddot{\mathbf{x}}_i(t)(\Delta t)^2 + \ldots,$$

wherefrom it follows, considering also (16.7), that

$$\mathbf{x}_i(t+\Delta t) = \mathbf{x}_i(t) + \mathbf{v}_i(t)\,\Delta t + \frac{1}{2m}\mathbf{F}_i(t)(\Delta t)^2 + \ldots, \tag{16.9}$$

$$\mathbf{v}_i(t+\Delta t) = \mathbf{v}_i(t) + \frac{1}{m}\mathbf{F}_i(t)\,\Delta t + \ldots \tag{16.10}$$

Usually, in order to reduce the number of the terms involved, equations (16.9) and (16.10) are replaced by the simplified relations

$$\mathbf{x}_i(t+\Delta t) = \mathbf{x}_i(t) + \mathbf{v}_i\,(t + \Delta t/2), \tag{16.11}$$

$$\mathbf{v}_i(t+\Delta t) = \mathbf{v}_i(t - \Delta t/2) + \frac{\Delta t}{m}\mathbf{F}_i. \tag{16.12}$$

As initial conditions, it is assumed that all atoms have been at rest, i.e. $\mathbf{v}_i(t - \Delta t/2) = \mathbf{0}$, and that the atoms have occupied the positions predicted by the linear elastic solution extrapolated to region I. At each iteration step, the forces $\mathbf{F}_i$ are calculated by (16.8) and the position vectors and velocities of all atoms are determined by means of the six scalar equations (16.11) and (16.12). The iteration proceeds until the potential energy reaches a minimum (for details, see Gehlen *et al.* [127]).

The choice of the time interval is very important. In the initial stages of the iteration, when the forces $\mathbf{F}_i$ are large, the time interval $\Delta t$ must be sufficiently small, say $10^{-4}$ sec., in order to avoid large speeds producing unrealistic displacements. After some relaxation, however, $\Delta t$ may be substantially increased, in order to save computation time. In order to reach equilibrium in the shortest possible time, the crystal is "quenched", i.e. all velocities are put to zero, every time the potential energy reaches a minimum. Calculations are then resumed and the kinetic energy can only increase again, while the potential energy must decrease towards the equilibrium.

One of the advantages of the dynamic relaxation is the possibility to avoid metastable positions, by suitably increasing $\Delta t$ such that the atoms be made to move beyond potential barriers in one time interval. In addition, the computation time required is lower than the time required by static relaxation. In exchange, dynamic relaxation has the disadvantage of preserving all symmetry present at the beginning of the relaxation. Therefore, when less symmetric configurations have to be investigated, the symmetry must be destroyed before relaxation is started, by adequately choosing the initial configuration.

### 16.3. Semidiscrete methods with flexible boundary

Due to the above mentioned shortcomings of the rigid-boundary method, several semidiscrete methods have been developed in the last decade, which allow the relaxation of $\Sigma_0$ and the modification of the elastic displacement field used for region II together with the atom positions in region I. These methods, which are called *flexible-boundary methods*, permit a considerable reduction of the number of atoms in region I along with the improvement of the elastic solution used for region II, and release some of the strong constraints imposed by a rigid boundary. At the same time, by appreciably reducing the number of the atoms which need to be treated independently for a given level of accuracy, the flexible-boundary methods make possible the consideration of more complex situations, such as the interaction of dislocations with other crystal defects.

*Sinclair's method (Flex-S).* As seen in Sect. 10, the displacement field produced by a straight dislocation in an infinite elastic medium with general anisotropy may be expressed as

$$\mathbf{u}(\mathbf{x}) = \mathbf{u}_V(\mathbf{x}) + 2\mathrm{Re} \sum_{\alpha=1}^{3} \mathbf{A}_\alpha \sum_{m=1}^{\infty} a_{\alpha m} z_\alpha^{-m}, \tag{16.13}$$

where [1]

$$\mathbf{u}_V(\mathbf{x}) = \frac{1}{\pi} \operatorname{Im} \sum_{\alpha=1}^{3} \mathbf{A}_\alpha D_\alpha \ln \frac{z_\alpha}{r_0} + \mathbf{u}^0 \tag{16.14}$$

is the Volterra solution, which gives the principal part of the displacemet field at large distances from the dislocation line, $a_{\alpha m}$ are coefficients that depend on the initially unknown boundary conditions on $\Sigma_0$, whereas the vectors $\mathbf{A}_\alpha = A_{k\alpha}\mathbf{e}_k$ and the scalars $D_\alpha$ and $p_\alpha$, $\alpha = 1, 2, 3$, are completely determined by the elastic constants, the direction of the dislocation line, and the Burgers vector.

Sinclair [302] approximates the series (16.13) by the finite sum

$$\mathbf{u}(\mathbf{x}) = \mathbf{u}_V(\mathbf{x}) + 2\operatorname{Re} \sum_{\alpha=1}^{3} \sum_{m=1}^{n} a_{\alpha m} \mathbf{U}_{\alpha m}(\mathbf{x}), \tag{16.15}$$

where

$$\mathbf{U}_{\alpha m}(\mathbf{x}) = \mathbf{A}_\alpha (x_1 + p_\alpha x_2)^{-m}.$$

The basic idea of Sinclair's method is to search a minimum of the potential energy considered as a function of the position vectors $\mathbf{x}_i$, $i = 1, \ldots, N$, of the atoms in region I, and of the $3n$ coefficients $a_{\alpha m}$ occurring in the expression (16.15) of the displacement field adopted for region II. Hence, by using the notation introduced in the preceding subsection,

$$E = E(\mathbf{x}_i, a_{\alpha m}), \tag{16.16}$$

where $i = 1, \ldots, N$; $\alpha = 1, 2, 3$; $m = 1, \ldots, n$.

At first, let us ignore the division of the crystal into two regions. Then (16.16) may be rewritten as

$$E = E(\mathbf{x}_i), \quad i = 1, \ldots, \infty, \tag{16.17}$$

the total potential energy $E$ resulting now by summing up the interaction potentials of all atoms. When only two-body (central) interaction potentials are being used, the dependence (16.17) becomes

$$E = \sum_{\substack{l,m=1 \\ l \neq m}}^{\infty} V(r_{lm}), \tag{16.18}$$

where $r_{lm} = \|\mathbf{x}_l - \mathbf{x}_m\|$. The force $\mathbf{F}_i$ acting on the atom $i$ is given by

$$\mathbf{F}_i = -\frac{\partial E}{\partial \mathbf{x}_i} = -\sum_{\substack{j=1 \\ j \neq i}}^{\infty} \frac{\partial V(r_{ij})}{\partial \mathbf{x}_i}. \tag{16.19}$$

---

[1] The introduction of $r_0$ in the argument of the logarithm in (16.13) can be always done by modifying accordingly the rigid-body displacement $\mathbf{u}^0$ (for a detailed discussion of the effects of this choice within a *non-linear* analysis, see Sinclair *et al.* [474], Sect. III C).

By taking into account that

$$\frac{\partial r_{ij}}{\partial \mathbf{x}_i} = \frac{\mathbf{x}_i - \mathbf{x}_j}{r_{ij}} = -\frac{\partial r_{ij}}{\partial \mathbf{x}_j} \quad \text{(no sum)}, \tag{16.20}$$

the last relation may be rewritten as

$$\mathbf{F}_i = -\sum_{\substack{j=1\\ j\neq i}}^{\infty} V'(r_{ij})\frac{\mathbf{x}_i - \mathbf{x}_j}{r_{ij}}. \tag{16.21}$$

As seen in Sect. 16.2, in order to perform the relaxation of the atoms in region I we need to know not only the expression of the forces, but also that of their derivatives with respect to the generalized co-ordinates of the system. Since

$$\frac{\partial}{\partial \mathbf{x}}(a\mathbf{v}) = \mathbf{v}\frac{\partial a}{\partial \mathbf{x}} + a\frac{\partial \mathbf{v}}{\partial \mathbf{x}},$$

where $a(\mathbf{x})$ is a scalar field and $\mathbf{v}(\mathbf{x})$ a vector field, both of which being supposed of class $C^1$ but otherwise arbitrary, it follows from (16.21) for $i, l = 1, \ldots, N$, $i \neq l$, that

$$\frac{\partial \mathbf{F}_i}{\partial \mathbf{x}_l} = -(\mathbf{x}_i - \mathbf{x}_l)\frac{\partial}{\partial \mathbf{x}_l}\left[\frac{V'(r_{il})}{r_{il}}\right] + \left[\frac{V'(r_{il})}{r_{il}}\right]\mathbf{1} \quad \text{(no sum)},$$

or

$$\frac{\partial \mathbf{F}_i}{\partial \mathbf{x}_l} = \frac{(\mathbf{x}_i - \mathbf{x}_l)(\mathbf{x}_i - \mathbf{x}_l)}{r_{il}^2}\left[V''(r_{il}) - \frac{V'(r_{il})}{r_{il}}\right] + \frac{V'(r_{il})}{r_{il}}\mathbf{1} \quad \text{(no sum)}. \tag{16.22}$$

The case $i = l$ may be reduced to the preceding one by noting that (16.19) and (16.20) yield for any *fixed* $l$

$$\frac{\partial \mathbf{F}_l}{\partial \mathbf{x}_l} = \sum_{\substack{j=1\\ j\neq l}}^{\infty} \frac{\partial^2 V(r_{lj})}{\partial \mathbf{x}_j \partial \mathbf{x}_l} = -\sum_{\substack{j=1\\ j\neq l}}^{\infty} \frac{\partial \mathbf{F}_l}{\partial \mathbf{x}_j}. \tag{16.23}$$

When the division of the crystal into two regions is taken into account, and hence (16.17) is replaced by (16.16), formula (16.21) still holds, but the summation extends to a finite number of atoms (the atoms in region I and those in region II interacting with atoms in region I). On the other hand, (16.22) and (16.23) give now only the part of the tensor **M** corresponding to the *atom-atom interaction*, with the same observation about the summation over $j$. In order to obtain the other components of **M**, let us consider the generalized forces

$$F_{\alpha m} = -\frac{\partial E}{\partial a_{\alpha m}} \tag{16.24}$$

associated to the adjustable parameters $a_{\alpha m}$ of the elastic displacement field. Since (16.16), represents merely another form of (16.16), we deduce from (16.24), considering also (16.19), that

$$F_{\alpha m} = -\sum_{i=1}^{\infty} \frac{\partial E}{\partial \mathbf{x}_i} \cdot \frac{\partial \mathbf{x}_i}{\partial a_{\alpha m}} = \sum_{i=1}^{\infty} \mathbf{F}_i \cdot \frac{\partial \mathbf{x}_i}{\partial a_{\alpha m}} .$$

But $\partial \mathbf{x}_i / \partial a_{\alpha m} = \mathbf{0}$ for region I, and (16.15) yields $\partial \mathbf{x}_i / \partial a_{\alpha m} = \mathbf{U}_{\alpha m}(\mathbf{x}_i)$ for region II. Consequently

$$F_{\alpha m} = \sum_{\mathbf{x}_i \in \mathrm{II}} \mathbf{F}_i \cdot \mathbf{U}_{\alpha m}(\mathbf{x}_i). \tag{16.25}$$

Moreover, the sum in the right-hand side extends again to the atoms in region II situated near $\Sigma_0$, since the atoms situated further away are always in elastic equilibrium ($\mathbf{F}_i = \mathbf{0}$).

The remaining components of **M** can now be easily derived with the aid of the basic formula (16.22). Thus, we have for the *atom-field interaction:*

$$\frac{\partial F_{\alpha m}}{\partial \mathbf{x}_i} = \frac{\partial \mathbf{F}_i}{\partial a_{\alpha m}} = \sum_{\mathbf{x}_s \in \mathrm{II}} \frac{\partial \mathbf{F}_i}{\partial \mathbf{x}_s} \mathbf{U}_{\alpha m}(\mathbf{x}_s), \tag{16.26}$$

and for the *field-field interaction:*

$$\frac{\partial F_{\alpha m}}{\partial a_{\beta l}} = \frac{\partial F_{\beta l}}{\partial a_{\alpha m}} = \sum_{\mathbf{x}_s \in \mathrm{II}} \frac{\partial F_{\alpha m}}{\partial \mathbf{x}_s} \cdot \mathbf{U}_{\beta l}(\mathbf{x}_s). \tag{16.27}$$

In Sinclair's method the numerical calculation proceeds by static relaxation, taking the components of $\mathbf{x}_i$, $i = 1, \ldots, N$, and the adjustable parameters $a_{\alpha m}$, $\alpha = 1, 2, 3$; $m = 1, \ldots, n$, as generalized co-ordinates of the system. Thus, (16.3) must be replaced by (16.21) and (16.25), while (16.5) must be replaced by (16.22), (16.23), (16.26), and (16.27).

By applying the above procedure to the $\langle 100 \rangle$ edge dislocation in $\alpha$ iron, Sinclair [302] has shown that the introduction of a flexible boundary between regions I and II may lead to a substantial computer-time saving. In particular, the results concerning the bond lengths obtained by Gehlen, Rosenfield, and Hahn [127] on a rigid-boundary model with 780 atoms could be recovered by Sinclair on a flexible-boundary model with only 100 atoms in region I, which obviously means a considerable reduction of the required computation. It is worth noting that Sinclair used both variants of the static relaxation explained in Sect. 16.2, namely the individual atom relaxation at the beginning of the iteration, when large adjustments of the configuration are required, and the simultaneous relaxation of the whole atomic array for the final steps of the iteration. This obviously represents a compromise, since the first variant is more practical but requires very many force calculations, whereas the second one requires few force calculations but expensive matrix manipulations.

Later on, Sinclair [473] successfully applied his method to crack modelling problems.

*The flexible-boundary method with overlapping regions (Flex-I).* This method has been proposed in 1972 independently by Teodosiu and Nicolae [338] and by Gehlen, Hirth, Hoagland, and Kanninen [130]. The basic idea of the method is to make regions I and II overlap. For a straight dislocation we can take for instance

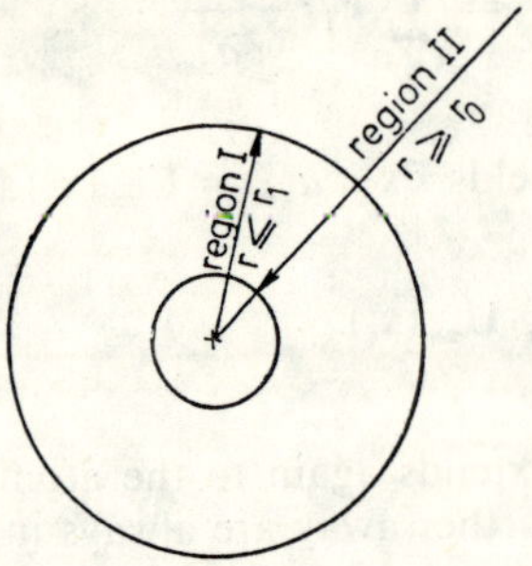


Fig. 16.3. Concentric arrangement of regions for the flexible-boundary method with overlapping regions (Flex-I).

the interior of a circular cylinder of radius $r_1$ as region I and the exterior of a circular cylinder of radius $r_0 < r_1$ as region II, both cylinders having the dislocation line as axis (Fig. 16.3).

Denote by $\Sigma_0$ and $\Sigma_1$ the circular cylindrical surfaces of radius $r_0$, respectively $r_1$. One can now use the following scheme of successive approximations. First, the relaxation is performed for region I, by keeping fixed the atoms situated on and outside $\Sigma_1$ in their positions given by the Volterra solution. The next step is to calculate the linear elastic solution for region II, by taking into account this time the displacement or traction boundary conditions on $\Sigma_0$ as derived from the first step, by using some interpolation procedure. Then, one performs again the relaxation of the atoms in region I, but keeping fixed the atoms situated on and outside $\Sigma_1$ in their positions resulted from the second step, and so on.

The linear elastic solutions satisfying prescribed boundary conditions on $\Sigma_0$ have been derived by Teodosiu and Nicolae [338] for an edge dislocation lying along a two-fold axis of material symmetry, and by Teodosiu, Nicolae, and Paven [342] for an arbitrary straight dislocation lying in elastic medium with general anisotropy. They are given in Sect. 10, namely by (10.43), (10.37), (10.42), when the tractions are prescribed on $\Sigma_0$ and by (10.43), (10.45) when the displacements are prescribed on $\Sigma_0$.

The method of the overlapping regions has been applied by Gehlen *et al.* [130], again to the $\langle 100 \rangle$ edge dislocation in $\alpha$ iron, by using, however, the linear elastic solution for a dislocation lying in an *isotropic* medium. An interesting remark made by these authors is that, when applied to a perfect lattice, the overlap method would yield small but non-zero atomic displacements. The explanation lies in the non-locality of the interaction potential, which prevents the forces exerted on region II atoms from being replaced by a perfectly equivalent distribution of surface tractions acting on $\Sigma_0$. In order to remove this secondary effect, the tractions acting on $\Sigma_0$ must be calculated as differences between the actuall interatomic forces and the forces acting across $\Sigma_0$ in the perfect lattice.

Another significant result obtained by Gehlen *et al.* [130] concerns the possibility of simulating the non-linear effects of the dislocation core by a pair of orthogonal line-force dipoles of different intensities located on the dilatational side of the dislocation field, i.e. opposite to the supplementary atomic half-plane of the edge dislocation (Fig. 16.4). Such a force distribution is equivalent, as regards the elastic far-field, with an elliptical cylindrical inclusion forced into a circular cylindrical hole. Clearly, this simulation holds only at sufficiently large distances $\rho$ from the

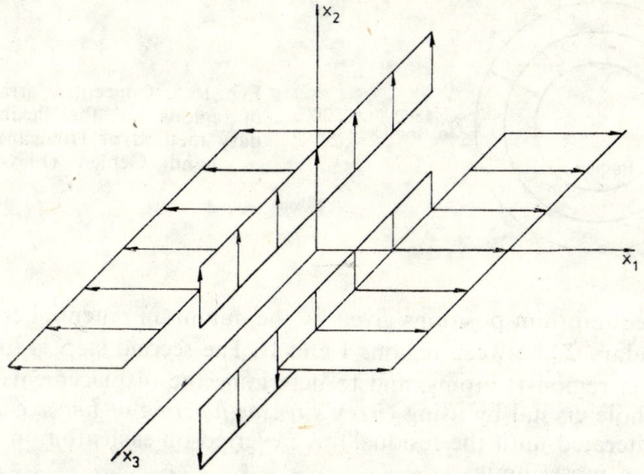


Fig. 16.4. Pair of orthogonal line-force dipoles of different intensities located opposite to the extra atomic half-plane of an edge dislocation, used by Gehlen, Rosenfield, and Hahn [127] to simulate the dilatational effect of the dislocation core.

dislocation line, where terms of order $O(\rho^{-2})$ or higher in the displacement field can be neglected with respect to the terms of order $O(\rho^{-1})$ corresponding to the line-force dipoles. Nevertheless, it enables to approximate the entire long-range elastic field in an analytical form by only two linear elastic fields.

*The method of Hoagland, Hirth, and Gehlen (Flex-II).* This flexible-boundary method has been proposed by Hoagland [430] and developed by Hoagland, Hirth, and Gehlen [431]. Recently, Sinclair, Gehlen, Hoagland, and Hirth [474] introduced several refinem ents of the method and extended it to allow the computation of the mean volume dilatation of a dislocated crystal. In what follows we give a brief description of the method; for details, the reader is referred to the original papers cited above.

In Flex-II three regions are explicitly considered around a straight dislocation. For illustration, these regions are represented in a concentric arrangement in Fig. 16.5, although their shape need not be circular. Like in the other flexible-boundary methods, the atoms in region I are relaxed individually, while the atoms in regions II, III, and in the remainder of the crystal are displaced collectively, according to linear elasticity theory. However, in Flex-II, the atoms of all three regions are supposed to interact via the same interatomic potential. Region II contains all atoms on which a force may be exerted by at least one region-I atom, while region

III is that part of the remainder of the crystal whose atoms interact with region-II atoms. Clearly, the thickness of both regions II and III should equal the maximum range of the interatomic force law.

The initial positions of the atoms throughout the crystal are usually chosen according to the Volterra solution. The first step is to relax the atoms of region I

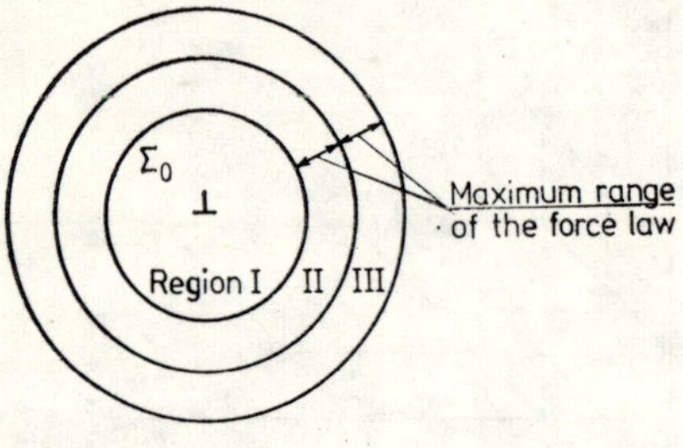


Fig. 16.5. Concentric arrangement of regions for the flexible-boundary method of Hoagland, Hirth, and Gehlen (Flex-II).

towards their equilibrium positions given by the minimum potential energy, keeping fixed the boundary $\Sigma_0$ between regions I and II. The second step is to calculate the forces acting on region-II atoms, and to determine the displacements produced by them in the whole crystal by using *Green's tensor function of linear elasticity*. These two steps are iterated until the residual force exerted on each atom in regions I and II is less than a preset limit.

Since the crystal outside region I is supposed to behave linearly elastic, no unequilibrated forces will develop in region III or beyond. Indeed, all atoms of these regions interact only with atoms that are displaced according to the equilibrium equations of linear elasticity. Moreover, the atom positions beyond region III need not be stored during the computation, for they can be calculated at the end of the iteration, by summing up the effects of the unequilibrated forces exerted on region-II atoms at all iteration steps.

As Green's tensor function of the generalized plane strain diverges logarithmically at the origin (cf. Sect. 10.5), it cannot be used to calculate the displacements produced by each force acting on a region-II atom in the close proximity of its application point. Therefore, in the evolved form of the Flex-II procedure, the continuum Green's function is locally replaced by the lattice Green's function; this latter is calculated by relaxing a small block of perfect lattice with the central row of atoms acted by a line force and the boundary atoms kept fixed at their positions given by the linear elastic Green's function.

Comparative studies (Puls and Woo [461], Sinclair *et al.* [474]) of the Flex-II method with other flexible-boundary methods have shown that, in terms of computational efficiency for a given accuracy, Flex-II is superior. Consequently, the size of region I can be considerably reduced for the same level of accuracy [1]. In exchange, the diminishing of the number of atoms in region I requires caution, since it implies

---

[1] For instance, in the computation of Woo and Puls [487] region I has been reduced to a rectangle $4.5b \times 6.5b$, while the boundary between regions II and III has been located only $7b$ away from the dislocation line.

the treatment by linear elasticity of regions relatively close to the dislocation line, where the strains are still of the order of a few percentages. Another apparent disadvantage of the Flex-II method is that it does not provide directly an improved analytic form of the linear elastic field at sufficiently large distances from the dislocation line. However, such a form can be deduced, e.g. by supplementing the Volterra solution by the linear elastic solutions corresponding to the multipolar moments (with respect to the dislocation line) of the line forces acting on region-II atoms accumulated throughout the iteration (Sinclair *et al.* [474], Sect. D).

The Flex-II method has been successfully used by Woo and Puls [487] to calculate the core configuration of $a/2\langle 110\rangle\{1\bar{1}0\}$ edge dislocations in MgO, in conjunction with Sangster's central-force breathing shell model, which allows taking into account many-body effects in the interaction between ions. These authors have also found that the edge dislocation causes a volume expansion of about 0.9 at. vol/plane, in quantitative agreement with the result obtained by Granzer *et al.* [143] for NaCl. Finally, Woo and Puls have shown that the displacement field can be satisfactorily approximated beyond an average distance of $5b$ from the dislocation line by superimposing on the Volterra solution the displacement field produced by two orthogonal line double forces, one of strength 12.3 eV/Å$^2$, acting parallel to the $x_1$-axis and centred on the $x_2$-axis, 0.211 Å above the slip plane, and the other of strength 8.8 eV/Å$^2$, acting parallel to the $x_2$-axis, and centred on the $x_2$-axis, 5.27 Å below the slip plane. It is worth noting that the initial attempt to use orthogonal line double forces located at the same distance from the dislocation line, i.e. an elliptical dilatation centre of the type employed by Gehlen *et al.* [130], has led in this case to a much worse fit of the atomic displacements.

Sinclair, Gehlen, Hoagland, and Hirth [474] have recently used the Flex-II method for the calculation of the overall dilatation of a finite body due to a dislocation. In order to account for non-linear effects beyond region II, they have first calculated the quasi-body forces $f_k = \tau_{kl,l}$, where $\tau_{kl}$ is determined by (14.18) and (14.19) and represents the non-linear contribution of the linear displacement gradient to second-order elastic constitutive equations. Then, the supplementary dilatation produced by the moments of this force distribution have been added to the linear elastic dilatation. Clearly, this approach involves several approximations with respect to a fully non-linear analysis based on second-order elasticity (cf. Sect. 14). Even so, the analysis of Sinclair *et al.* has emphasized once more the importance of the *non-linear* elastic contribution to the overall dilatation produced by dislocations and has shown that, at least for the $\langle 100\rangle$ edge dislocation in $\alpha$ iron and the crystal array investigated, this contribution is quantitatively comparable with that of the *linear* elastic displacements corresponding to the relaxation of the boundary between regions I and II.

Sinclair *et al.* [474] have also systematically studied the alternative choosing of Eulerian or Lagrangian co-ordinates for the description of the dislocated crystal, and, in the latter case, also the effect of using one of four different cuts for defining a single-valued displacement field [1]. Their results show that the relaxed core confi-

[1] In this connection, see also Teodosiu and Soós [479].

gurations are almost identical in all cases. Similarly, the change in area of a circuit encircling the dislocation line, calculated *with respect to a perfect-lattice area* containing the same number of atoms, is to a large extent independent of the path taken to create the dislocation and of the choice of the co-ordinate system. On the other hand, the change in area during relaxation, sometimes improperly used as a measure of the volume of dislocation formation per unit dislocation length, obviously depends on the initial configuration. Therefore, the discrepancy between the evaluations of this quantity, reported by various authors, merely requires interpretation, and does not correspond to any intrinsic property of the dislocation.

*Semidiscrete methods based on non-linear elasticity.* All flexible-boundary methods presented so far are based on the use of linear elasticity theory beyond region I. A fundamentally different approach, which has been proposed by Seeger in 1968, consists in improving the elastic solution by using second-order elasticity *and* satisfying the boundary conditions on $\Sigma_0$ (Fig. 16.2)[1]. A brief outline of the method has been given by Teodosiu and Nicolae [338]. Subsequently, the method has been applied by Granzer *et al.* [143], and by Petrasch and Belzner [459] for the simulation of the $\langle 110 \rangle$ edge dislocation in NaCl.

The principle of the method is straightforward. The stress vector (or, alternatively, the displacement vector) on $\Sigma_0$ is taken as a Fourier series of the polar angle with initially undetermined coefficients. Since higher harmonics lead to terms in the elastic solution that vanish rapidly with increasing distance from the dislocation core, it is in general sufficient to consider only the first two or three harmonics. The non-linear elastic solution is found by an iterative procedure involving the solution of a linear elastic boundary-value problem at each step, as has been shown in Sect. 14. Then, the total potential energy is minimized as a function of the adjustable parameters occurring in the boundary conditions and of the positions vectors of the atoms located in region I.

For instance, if we attempt to find out the most significant correction to the Volterra solution, then we should retain terms up to the order $O(\rho^{-1})$ in the expression of the displacement field and $O(\rho^{-2})$ in that of the stress field as $\rho \to \infty$. Then, as shown in Sect. 14.2, the displacement field produced by an edge dislocation lying along a two-fold symmetry axis in an anisotropic elastic medium is given within the framework of second-order elasticity theory by Eq. (14.100), which may be rewritten as

$$U(x_1, x_2) = u_N(x_1, x_2) + \sum_\alpha \left( \frac{\delta_\alpha A_\alpha}{z_\alpha} + \frac{\rho_\alpha \bar{A}_\alpha}{\bar{z}_\alpha} \right), \tag{16.28}$$

where

$$u_N(x_1, x_2) = \sum_\alpha \left[ \delta_\alpha \left( \varepsilon\varkappa_\alpha + \frac{\varepsilon^2 K_\alpha}{z_\alpha} \right) \ln \frac{z_\alpha}{1+\gamma_\alpha} + \right.$$
$$\left. + \rho_\alpha \left( \varepsilon\bar{\varkappa}_\alpha + \frac{\varepsilon^2 \bar{K}_\alpha}{\bar{z}_\alpha} \right) \ln \frac{\bar{z}_\alpha}{1+\bar{\gamma}_\alpha} \right] + \varepsilon^2 U_0(x_1, x_2) + u_0 + iv_0, \tag{16.29}$$
$$z_\alpha = z + \gamma_\alpha \bar{z}, \quad z = x_1 + ix_2, \quad \alpha = 1, 2.$$

[1] A different attempt to including second-order elastic effects in region II without considering, however, the boundary conditions on $\Sigma_0$, has been developed by Bullough and Sinclair [404], by using Willis' solution for the screw dislocation (cf. Sect. 19.1).

By comparing (16.28) with (16.13), it may be seen that Sinclair's method (Flex-S) can be applied exactly as before, the only difference being that now $u_N(x_1, x_2)$ includes the non-linear *known* contribution given by $U_0(x_1, x_2)$ and the terms with coefficients $K_\alpha$ and $\bar{K}_\alpha$. In particular (16.16) becomes now

$$E = E(x_{1i}, x_{2i}, A_1, A_2), \qquad i = 1, 2, \ldots, N,$$

and hence the potential energy of the system depends on the $2N$ co-ordinates of the atoms located in region I and on the two adjustable complex parameters $A_1$, $A_2$.

Clearly, the non-linear elastic solution can be also used in conjunction with the flexible-boundary method with overlapping regions. For instance this can been done for an edge dislocation by starting from (16.28) and (16.29) and using the following iteration steps (cf. also Fig. 16.3):

(i) Elastic computation of the initial atom positions in the whole array according to (16.29).

(ii) Static relaxation of the atoms in region I with atoms on and outside $\Sigma_1$ being kept fixed.

(iii) Determination of the adjustable complex parameters $A_1$, $A_2$ from the Fourier analysis of the atomic displacements on $\Sigma_0$.

(iv) Elastic recomputation of the atom positions in the whole array, this time according to (16.28).

(v) Repetition of steps (ii) — (iv) until the changes in the atom positions and in the values of the adjustable parameters between successive iterations lie within preset limits.

A procedure of this type has been employed by Petrasch and Belzner [459] to determine the core configuration of edge dislocations in sodium chloride and silver chloride. In fact, they supplemented the non-linear elastic solution (16.28) by terms in $z_\alpha^{-2}$ and $\bar{z}_\alpha^{-2}$ with adjustable coefficients, which certainly leads to a better approximation of higher-order effects. Moreover, the use of three-body interatomic potentials, in addition to the two-body (central) interaction potentials corresponding to the Born-Mayer repulsion and to the Van der Waals attraction, allowed a much better fit to the elastic constants, which was beneficial, especially for AgCl. The results obtained after 13 cycles for NaCl and 9 cycles for AgCl show a significant improvement against the rigid-boundary method as regards the continuity across $\Sigma_0$ of the residual forces exerted on the ions at the end of the iteration. Almost equal values of the core energies have been obtained for both NaCl and AgCl (about 0.97 eV per plane for 3b-cores).

The major disadvantage of the flexible-boundary methods based on non-linear elasticity is the need for an analytical solution of the non-linear boundary-value problem for region II. On the other side, after determining the adjustable parameters occurring in the solution, just this particularity becomes one of the main advantages of the method, since the analytical expression of the elastic far-field includes both the non-linear elastic and the core effects. In addition, these methods allow reducing region I to a relatively small number of atoms, by taking fully into account, however, the non-linear effects arising from the high strains at short and moderate distances from the dislocation line.

## 16.4. Lattice models of straight dislocations

The first models attempting to give a discrete description of the dislocation core, which are due to Peierls [266], Nabarro [255], and van der Merwe [366], have preceded the use of semidiscrete methods. Actually, they were *partially discrete models* which introduced interaction forces only between the atoms situated immediately above and below the glide plane (the Peierls-Nabarro model), or between the atoms situated on both sides of the extra atomic half-plane of an edge dislocation (the van der Merwe model), the remaining of the crystal being considered as a linear elastic continuum. Although such models cannot provide an accurate representation of the dislocation behaviour in real crystalline solids, they have the merit of providing simple analytical formulae describing some characteristic properties of the dislocation core. In particular, the Peierls-Nabarro model permitted for the first time to obtain realistic values of the order 0.001 $\mu$ of the yield stress.

We shall now very briefly review some of the *lattice models of dislocations*, although they do not belong to the very topic of this book [1]. The first fully atomic model has been developed by Maradudin [231] for the $\langle 110 \rangle$ screw dislocation in alkali halides, and has been subsequently used with slight modifications by Celli [57] for the calculation of the stored strain energy of screw dislocations in diamond structures. These calculations are based on the harmonic lattice theory, in which the interatomic forces are assumed to linearly depend on the relative displacements of the atoms [2]. Moreover, they take into account merely the nearest neighbour interactions and allow only atomic displacements parallel to the dislocation line, i.e. the screw dislocation is generated by a rigid-body translation of each atomic row along its length. These two simplifying assumptions have been subsequently given up by Boyer and Hardy [36], who applied Maradudin's model to screw dislocations in aluminium, potassium, and $\alpha$ iron, still neglecting, however, anharmonic effects. In spite of the various further improvements of the fully atomic models, achieved especially by using the lattice Green's function (Bullough and Tewary [53], Tewary [348], [480]), these purely harmonic models cannot accurately describe either the core configuration or the atomic interactions across the cut surface used to generate the dislocation, since large relative displacements occur in both these regions [3]. Actually, fully atomic models are inferior from this point of view to semidiscrete models. Indeed, the latter allow the correct simulation of both the harmonic and the anharmonic response of the lattice, provided the interatomic potential is carefully constructed (cf. also Heinisch and Sines [424]).

---

[1] For a detailed discussion of the partial and fully atomic models of dislocations we refer to Bullough [50] and Bullough and Tewary [53].

[2] The force constants occurring in these linear laws can be determined by fitting the theoretical phonon dispersion curves to the experimental data obtained by neutron scattering and the long-wave approximation of the force law to the measured elastic constants.

[3] Hölzler and Siems [165] have partly overcome this difficulty, by introducing a sinusoidal force law of the Peierls-Nabarro type in order to describe the interaction between the atom rows parallel to the line of a screw dislocation. However, the generalization of their hypothesis to arbitrary dislocations is by no means straightforward.

CHAPTER IV

# CONTINUOUS DISTRIBUTIONS OF DISLOCATIONS

## 17. Elastostatics of continuous distributions of dislocations

As shown in Sect. 17, each single dislocation may be considered as a state of self-stress in the classical elasticity theory. Sometimes, however, we are more interested to know the mean values of the strains and stresses produced in a crystal by a large number of dislocations. It is then convenient to consider the limiting case of a *continuous distribution of dislocations*, for which the number of dislocations tends to infinity, while the Burgers vector of each tends to zero, in such a way that the product remains finite in any bounded region. In performing this limiting process it is natural to assume that the product of the number of dislocations and of their individual core volumes tends to zero.

In order that the theory of elasticity be applicable to continuous distributions of dislocations it is necessary that the mean strain produced by dislocations be macroscopically continuous, and this may happen only when each macroscopic volume element contains a large number of dislocations. A rough evaluation shows that this is really the case for many situations of practical interest. Indeed, a cube with the side of 1 mm may usually be considered as sufficiently small with respect to the size of the body and the characteristic wave lengths of its elastic state, and thus may be taken as macroscopic volume element. On the other hand, the total length of the dislocation lines amounts, even in good annealed metals, to $10^3$-$10^4$ mm/mm$^3$ and increases during deformation to $10^8$-$10^{10}$ mm/mm$^3$. This shows that the real dislocation density is generally sufficiently high to assure a continuous variation of the mean value of the deformation produced by dislocations from one volume element to another.

It should be mentioned that the theory of continuous distributions of dislocations may be also used to describe other continuously distributed non-mechanical sources of self-stress, e.g. inhomogeneous thermal or magnetic fields [1], which are sometimes called *quasi-dislocations*.

Furthermore, by using distributions associated with lines or surfaces, it is also possible to use the concepts and methods of the theory of continuous distributions of dislocations to study surface distributions of dislocations or even single dislocations (see, e.g. Kröner [190], Kunin [204], Teodosiu [335]).

[1] See Rieder [465], Kröner [192], Anthony [5].

In what follows we shall briefly present the elastostatics of the continuous distributions of dislocations. Consider an elastic body $\mathscr{B}$ which occupies a region $\mathscr{V}$ in a self-stressed state produced by a continuous distribution of dislocations; let $(k)$ be the configuration of the body in this state. In this case there exists no *global natural configuration*, i.e. a stress-free configuration of the whole body. Let $N(X)$ denote a material neighbourhood of a material point $X$. We call *local natural configuration* $(\varkappa)$ of $N(X)$ the (real or ideal) configuration $N(X)$ would assume

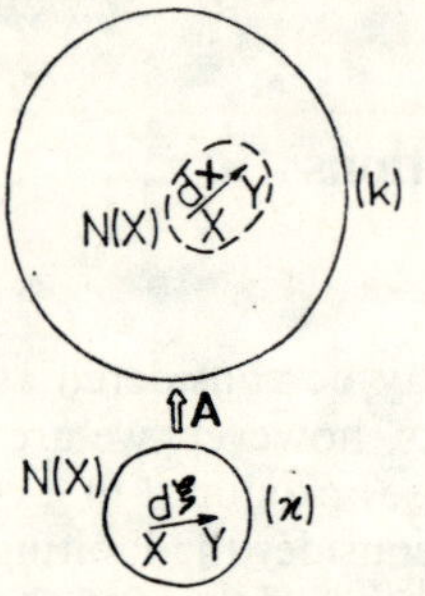


Fig. 17.1. On the definition of the elastic distortion **A**.

if it were cut out of the body and released from the constraints exerted by the rest of the body, the positions of all other crystal defects in $N(X)$ being kept fixed during this process. The last condition prevents the occurrence of any supplementary inelastic deformation during the cutting and releasing of $N(X)$.

Let now $Y$ be another material point of $N(X)$ and denote by $d\mathbf{x}$, $d\boldsymbol{\xi}$ the position vectors of $Y$ with respect to $X$ in the configurations $(k)$ and $(\varkappa)$, respectively (Fig. 17.1). The tensor $\mathbf{A}$ defined by

$$d\mathbf{x} = \mathbf{A}\, d\boldsymbol{\xi} \tag{17.1}$$

is called after Kröner [193] the *elastic distortion*. We assume that, for sufficiently small neighbourhoods $N(X)$, the so-defined value of $\mathbf{A}$ does not depend on the choice of the neighbourhood $N(X)$ and of the particle $Y \in N(X)$. Consequently, by repeating the same procedure for all particles of the body, we may define (and eventually measure) the field $\mathbf{A}(\mathbf{x})$ of the elastic distortion, where $\mathbf{x}$ is the position of $X$ in $(k)$. Moreover, we assume that $\mathbf{A}(\mathbf{x})$ is one-to-one and of class $C^1$ in $\mathscr{V}$. Hence, there exists a continuously differentiable tensor field $\mathbf{A}^{-1}$ such that

$$d\boldsymbol{\xi} = \mathbf{A}^{-1}\, d\mathbf{x}. \tag{17.2}$$

Next, assume that the body is torn into small volume elements which are released individually. Then, the local natural configurations thus obtained would generally not fit together, unless some suitable constraints are exerted on each volume element. The self-stresses existing in the body in the configuration $(k)$ are just the constraints which are necessary to re-establish the continuity of the body. Consequently, unlike $\mathbf{F}$, the distortion $\mathbf{A}$ is no longer the gradient of a vector field, since, otherwise, the elastic strains and stresses ought to vanish in the absence of external loads.

Let us extend now to continuous distributions of dislocations the definition of the Burgers vector that was given in Sect. 7 for single dislocations. Consider a smooth surface $s$ in $\mathscr{V}$, bounded by the closed line $c$, and containing the point $\mathbf{x}$. By analogy with the reasoning leading to (7.7), we define the *true Burgers vector* $\mathbf{b}$ of the dislocations piercing through $s$, as the sum of the infinitesimal vectors $\mathrm{d}\boldsymbol{\xi}$ that result by the cutting and releasing operations from the infinitesimal vectors $\mathrm{d}\mathbf{x}$ taken along $c$, i.e.

$$\mathbf{b} = \oint_c \mathrm{d}\boldsymbol{\xi} = \oint_c \mathbf{A}^{-1}\,\mathrm{d}\boldsymbol{\xi}. \tag{17.3}$$

By virtue of (1.55) this definition may be transformed into

$$\mathbf{b} = \int_s \boldsymbol{\alpha}\mathbf{n}\,\mathrm{d}s \tag{17.4}$$

where [1]

$$\boldsymbol{\alpha} = -\operatorname{curl}\mathbf{A}^{-1}, \qquad \alpha_{km} = -\,\epsilon_{rsm} A^{-1}_{kr,s}, \tag{17.5}$$

and $\mathbf{n}$ is the unit normal to $s$, the positive sense on $c$ being chosen clockwise when sighting down along $\mathbf{n}$.

The tensor $\boldsymbol{\alpha}$ is called the *true dislocation density*. This name is justified by the fact that $\boldsymbol{\alpha} = \mathbf{0}$ implies that $\mathbf{A}^{-1}$ and $\mathbf{A}$ are gradients of some vector fields, and hence the vanishing of the self-stresses. In addition, the definition (17.5) can be also applied to a finite number of single dislocations, provided that the positive sense on each dislocation line $L$ piercing through $s$ is chosen such that $\mathbf{n}\cdot\mathbf{l}$ be positive, where $\mathbf{l}$ denotes the unit tangent vector to $L$ at the intersection point with $s$. In particular, one obtains for a single dislocation (Kunin [204])

$$\boldsymbol{\alpha}(\mathbf{x}) = \mathbf{b}\,\mathbf{l}(\mathbf{x}_L)\delta(L), \qquad \alpha_{km}(\mathbf{x}) = b_k l_m(\mathbf{x}_L)\delta(L), \tag{17.6}$$

where $\mathbf{x}_L$ is the position vector of a current point on $L$, while $\delta(L)$ is the delta-function associated to the line $L$ and characterized by the property

$$(\delta(L), \varphi) = \int_L \varphi(\mathbf{x})\,\mathrm{d}l \tag{17.7}$$

for any function $\varphi$ of class $C^\infty$ and of bounded support on $\mathscr{E}$.

[1] We denote as usual by $(.)_{,s}$ the partial differentiation with respect to the co-ordinate $x_s$ of the current material point $X$ in the deformed configuration $(k)$.

If $\boldsymbol{\alpha}$ is a continuous function of $\mathbf{x}$ and $s$ is infinitesimal, we derive from (17.4) that

$$\mathrm{d}\mathbf{b} = \boldsymbol{\alpha}\mathbf{n}\,\mathrm{d}s, \tag{17.8}$$

where $\mathrm{d}\mathbf{b}$ is the *infinitesimal true Burgers vector* of the dislocations threading $\mathrm{d}s$. Clearly, $\mathbf{A}$ plays for continuous distributions of dislocations the same role as $\mathbf{F}$ for single dislocations. This remark enables us to use (7.12) in order to define the *infinitesimal local Burgers vector* $\mathrm{d}\mathbf{b}^*$ and *the local dislocation density* $\boldsymbol{\alpha}^*$ at $\mathbf{x}$ by the relations

$$\mathrm{d}\mathbf{b}^* = \mathbf{A}\,\mathrm{d}\mathbf{b}, \qquad \boldsymbol{\alpha}^* = \mathbf{A}\boldsymbol{\alpha} = -\mathbf{A}\,\mathrm{curl}\,\mathbf{A}^{-1}. \tag{17.9}$$

Along with $\boldsymbol{\alpha}$ and $\boldsymbol{\alpha}^*$ we shall also make use of another dislocation density, say $\tilde{\boldsymbol{\alpha}}$, which was first considered by Noll [260] and defined by the relation

$$\boldsymbol{\alpha} = j\tilde{\boldsymbol{\alpha}}\mathbf{A}^T, \qquad \alpha_{km} = j\tilde{\alpha}_{ks}A_{ms}, \tag{17.10}$$

where $j = \det \mathbf{A}^{-1}$. In order to understand the significance of Noll's dislocation density, let us consider an oriented material surface element $\mathbf{n}\,\mathrm{d}s$ through the material point $X$ in the configuration $(k)$. By releasing a neighbourhood $N(X)$ of $X$ containing the surface element, the magnitude and the orientation of the vector $\mathbf{n}\,\mathrm{d}s$ will change to $\tilde{\mathbf{n}}\,\mathrm{d}\tilde{s}$, say. Then, by (2.25), we have

$$\tilde{\mathbf{n}}\,\mathrm{d}\tilde{s} = j\,\mathbf{A}^T\mathbf{n}\,\mathrm{d}s,$$

and hence (17.8) and (17.10) yield

$$\mathrm{d}\mathbf{b} = \tilde{\boldsymbol{\alpha}}\,\tilde{\mathbf{n}}\,\mathrm{d}s. \tag{17.11}$$

Since $\mathrm{d}\mathbf{b}$ and $\tilde{\mathbf{n}}\,\mathrm{d}\tilde{s}$ do not depend on the elastic distortion $\mathbf{A}$, $\tilde{\boldsymbol{\alpha}}$ does not either, in contradistinction to the true and local dislocation densities, which vary under superimposed elastic deformations. This property will be used in Sect. 20.

Clearly, each local natural configuration $(\varkappa)$ is defined so far to within a rigid rotation. We remove this indeterminacy by requiring the mean lattice orientation, i.e. the preferred crystallographic directions sufficiently far from crystal defects, be the same for all natural configurations. Then, the elastic constitutive equations will be the same for all particles, provided that $\mathbf{F}$ is replaced by $\mathbf{A}$ in (4.40).

The fundamental problem of the elastostatics of continuous distributions of dislocations is the *determination of the stresses produced by a given dislocation density*. The kinematic equations of this problem are (17.5) and the definition $(2.15)_2$ of the finite strain tensor, which becomes now

$$\mathbf{D} = \tfrac{1}{2}(\mathbf{A}^T\mathbf{A} - \mathbf{1}), \qquad D_{mn} = \tfrac{1}{2}(A_{rm}A_{rn} - \delta_{mn}). \tag{17.12}$$

They must be supplemented by the constitutive equations

$$T_{kl} = j\, A_{kr} A_{ls} \frac{\partial W(\mathbf{D})}{\partial D_{rs}} \tag{17.13}$$

$$= jA_{kr}A_{ls}(c_{rsmn}D_{mn} + \tfrac{1}{2} C_{rsmnpq}D_{mn}D_{pq} + \ldots),$$

the equilibrium equations

$$T_{kl,l} = 0 \quad \text{in } \mathscr{V}, \tag{17.14}$$

and the traction boundary conditions

$$T_{kl}n_l = 0 \quad \text{on } \mathscr{S}, \tag{17.15}$$

where $\mathbf{n}$ is the outward unit normal to the boundary $\mathscr{S}$ of $\mathscr{V}$. In addition, when the dislocation density is given, it must fulfil the identity

$$\operatorname{div} \boldsymbol{\alpha} = \mathbf{0}, \qquad \alpha_{km,m} = 0, \tag{17.16}$$

which is a consequence of (17.5) and $(1.50)_2$.

In order to linearize the above equations we put [1]

$$\mathbf{A}^{-1} = \mathbf{1} - \boldsymbol{\beta}, \quad A^{-1}_{km} = \delta_{km} - \beta_{km}, \tag{17.17}$$

assume that $\|\boldsymbol{\beta}\| \ll 1$, and neglect all terms of second and higher orders in $\|\boldsymbol{\beta}\|$. Then, (17.5), (17.9), (17.10), (17.12), and (17.13) yield $\boldsymbol{\alpha} = \boldsymbol{\alpha}^* = \tilde{\boldsymbol{\alpha}}$ and

$$\boldsymbol{\alpha} = \operatorname{curl} \boldsymbol{\beta}, \quad \alpha_{lj} = \epsilon_{nmj}\beta_{ln,m}, \tag{17.18}$$

$$\mathbf{E} = \tfrac{1}{2}(\boldsymbol{\beta} + \boldsymbol{\beta}^T), \quad E_{km} = \tfrac{1}{2}(\beta_{km} + \beta_{mk}), \tag{17.19}$$

$$T_{km} = c_{kmrs}E_{rs}, \tag{17.20}$$

while the equilibrium equations (17.14) and the traction boundary conditions (17.15) remain unchanged.

The following *uniqueness theorem* (Teodosiu [337], vol. 2) holds for the linear boundary-value problem.

*If the strain energy function is positive definite, and if the dislocation density* $\boldsymbol{\alpha}$ *is a given continuous function whose support is contained in the simply-connected and bounded region* $\mathscr{V}$ *occupied by the elastic body, then the solution of the boundary-*

[1] The tensor $\boldsymbol{\beta}$ is called the *infinitesimal elastic distorsion*. In the case of a single dislocation, $\mathbf{A}^{-1}$ reduces to $\mathbf{F}^{-1} = \operatorname{grad} \boldsymbol{\chi}^{-1}(\mathbf{x})$, and $\boldsymbol{\beta}$ to $\mathbf{h} = \operatorname{grad} \mathbf{u}(\mathbf{x})$.

*value problem* (17.18—20), (17.14), *and* (17.15), *is determined to within an infinitesimal rigid displacement.*

*Proof.* Let $\boldsymbol{\beta}'$ and $\boldsymbol{\beta}''$ be two solutions of the boundary-value problem, and denote by $\mathbf{E}'$ and $\mathbf{E}''$ the corresponding infinitesimal strain tensors and by $\mathbf{T}'$ and $\mathbf{T}''$ the corresponding Cauchy stress tensors. Let

$$\boldsymbol{\beta} = \boldsymbol{\beta}' - \boldsymbol{\beta}'', \qquad \mathbf{E} = \mathbf{E}' - \mathbf{E}'', \qquad \mathbf{T} = \mathbf{T}' - \mathbf{T}''. \tag{17.21}$$

From (17.17) it follows that curl $\boldsymbol{\beta} = \mathbf{0}$, and hence, by $(1.49)_2$, we have $\boldsymbol{\beta} = \text{grad}\,\mathbf{u}$, where $\mathbf{u}$ is an arbitrary vector field. On the other hand, (17.19), (17.20), (17.14), and (17.15) imply that the difference-solution (17.21) satisfies (6.22) and $(6.23)_2$. Consequently, we may apply Kirchhoff's uniqueness theorem and deduce that $\mathbf{u}$ is an infinitesimal rigid displacement, and hence $\mathbf{E} = \mathbf{T} = \mathbf{0}$, $\mathbf{E}' = \mathbf{E}''$, $\mathbf{T}' = \mathbf{T}''$ in $\mathscr{V}$, which completes the proof.

The first step towards constructing a theory of continuous distributions of dislocations has been taken by Moriguti [252], who calculated the self-stresses produced by an incompatible elastic deformation; however, his paper being written in Japanese, it remained unknown for almost two decades. In 1953, Nye [261] introduced the concept of dislocation density which gave a particular momentum to the research on dislocation theory. Thus, starting from 1954, Bilby [27], Eshelby [111], and Kröner [189] rediscovered and generalized Moriguti's results. Almost in the same period, Kondo [184], followed independently by Bilby, Bullough, and Smith [28], established the relation between the kinematics of continuous distributions of dislocations and the geometry of non-Euclidean spaces with affine connexion. Namely, they showed that, when the geometry of the local natural configurations is described in terms of the co-ordinates of the material points in the configuration $(k)$, one obtains an affine connexion with vanishing curvature tensor and whose torsion tensor equals the dislocation density. This idea has been further developed by Kondo [185], Bilby [29], and especially by Kröner [192] and Kröner and Seeger [191], who incorporated the geometric theory into a non-linear elastic theory of continuous distributions of dislocations, and developed a method of successive approximations for the solving of non-linear boundary-value problems. Subsequently, Günther [148] constructed a non-linear dynamic theory of dislocations, and Teodosiu [332—334] extended the non-linear theory of continuous distributions of dislocations to materials of grade two, by assuming that the strain energy function depends not only on $\mathbf{A}$ but also on Grad $\mathbf{A}$.

The plan of this book does not allow us to consider the connection between the geometry of the deformation produced by continuous distributions of dislocations and that of affine or anholonomic spaces. We will adopt, therefore, an approach that is closer to the modern developments in continuum mechanics (Truesdell and Noll [358]) and can be easily extended to the dynamic case (Teodosiu [335], Teodosiu and Seeger [336]). We shall devote the remaining part of this chapter to the solving of the linear and non-linear boundary-value problems formulated at the beginning of this section. For details concerning the theory of continuous distributions of dislocations we refer to the monograph by Kröner [190] and to the review articles of Eshelby [111], Bilby [29], de Wit [385], Bullough [50], Sects. 4, 5, 13, Landau and Lifshits [207], Chap. 4, Rieder [276—277].

In spite of its genuine elegance, the theory of continuous distributions of dislocations has relatively few applications in modelling crystal defects. Indeed, although single dislocations may be described within this theory with the aid of delta-functions, this approach leads essentially to the same results as the direct approach presented in Chapters II and III. On the other hand, in the attempt to develop a macroscopic theory of the elasto-plasticity on the basis of the laws governing the production, interaction, and motion of dislocations, the dislocation density tensor proves to be inadequate for characterizing the dislocation arrangement and its evolution. The explanation lies in the fact that dislocations are generated during deformation as closed loops, along which the Burgers vector is constant. When such a loop is intersected by an arbitrary plane, it occurs as a pair of dislocations of opposite signs, and hence gives a vanishing contribution to the dislocation density tensor. This has led to the introduction of other dislocation measures, such as the total length per unit volume of the dislocations belonging to each glide system (Zarka [387—389], Teodosiu [335, 344], Kröner and Teodosiu [194]), the density of dislocation loops (Kröner [442]), or some statistical measures of the dislocation arrangement (Kröner [443]).

However, as pointed out at the beginning of this section, the theory of continuous distributions of dislocations possesses its own field of application containing especially the two-dimensional dislocation arrangements such as dislocation pile-ups and low-angle grain boundaries. It addition, it allows to trace a fruitful analogy of heuristic value to the theory of relativity and the theory of non-mechanical stresses (Kröner [441], Anthony [5]).

## 18. Determination of the stresses produced by continuous distributions of dislocations

We will consider in this section various methods for solving the linear boundary-value problem defined by (17.18—20), (17.14), and (17.15). By applying the operator $\in_{ikl}\,\partial/\partial x_k$ on both sides of (17.18) and taking the symmetric part of the relation obtained, we find

$$\operatorname{inc}\mathbf{E} = \boldsymbol{\eta}, \qquad -\in_{ikl}\in_{jmn}E_{ln,km} = \eta_{ij}, \tag{18.1}$$

where $\boldsymbol{\eta}$ is the so-called *incompatibility tensor* and is defined by

$$\eta_{ij} = \tfrac{1}{2}(\in_{ikl}\alpha_{lj,k} + \in_{jkl}\alpha_{li;k}). \tag{18.2}$$

It is easily seen now that our boundary-value problem almost coincides with the traction boundary-value problem of classical elasticity theory (Sect. 6.2), the only difference being the replacement of the compatibility equations (6.10) by the non-homogeneous equations (18.1). In what follows we shall analyse several ways of approaching this modified problem.

### 18.1. Eshelby's method

Eshelby [111] starts from the remark that, by virtue of the tensor identity $(1.51)_1$, any class $C^2$ solution of (18.1) may be written as

$$\mathbf{E} = \operatorname{sym} \operatorname{grad} \mathbf{u} + \tilde{\mathbf{E}}, \tag{18.3}$$

where $\tilde{\mathbf{E}}$ is a particular solution of the equation and $\mathbf{u}$ is an arbitrary vector field of class $C^3$. On the other hand, if on the boundary $\mathscr{S}$ of the body we have $\boldsymbol{\eta} = \mathbf{0}$, or at least $\boldsymbol{\eta}\mathbf{u} = \mathbf{0}$, then a particular solution $\tilde{\mathbf{E}}$ of Eq. (18.1) can be obtained by analogy with that derived by Eddington ([101, p. 128) in the general relativity theory, namely,

$$\tilde{\mathbf{E}}(\mathbf{x}) = \frac{1}{4\pi} \int_{\mathscr{V}} \frac{\boldsymbol{\eta}(\mathbf{x}') - [\operatorname{tr}\boldsymbol{\eta}(\mathbf{x}')]\mathbf{1}}{\|\mathbf{x} - \mathbf{x}'\|} \, dv'. \tag{18.4}$$

In the general case where $\boldsymbol{\eta}\mathbf{u} \neq \mathbf{0}$ on $\mathscr{S}$, one can still derive a particular solution $\tilde{\mathbf{E}}$ of (18.1) by means of the formula (18.4), provided that $\boldsymbol{\eta}$ is replaced by another tensor field $\tilde{\boldsymbol{\eta}}$ of class $C^3$ in the whole space $\mathscr{E}$, which is obtained by extending $\boldsymbol{\eta}$ through $\mathscr{S}$ and vanishes rapidly enough at infinity for the integral in (18.4) to be convergent. It may shown that the field $\tilde{\mathbf{E}}(\mathbf{x})$ thus obtained satisfies (18.1) in $\bar{\mathscr{V}} = \mathscr{V} \cup \mathscr{S}$ on condition that the integration in (18.4) be equally extended to the whole space $\mathscr{E}$.

Next, by substituting (18.3) into the other field equations and boundary conditions, one obtains a traction boundary-value problem of classical elasticity theory, in which $\mathbf{u}$ plays the role of a displacement field, while the body forces in $\mathscr{V}$ and the surface tractions on $\mathscr{S}$ are given, respectively, by

$$f_k = c_{klmn}\tilde{E}_{mn,l}, \quad t_k = -c_{klmn}\tilde{E}_{mn}n_l.$$

### 18.2. Kröner's method

As shown by Kröner [187] the problem of determining the self-stresses produced by dislocations can be successfully dealt with by means of Beltrami's solution (6.26), provided suitable supplementary conditions are imposed on the stress functions, We limit ourselves to the isotropic case[1], for which the constitutive equations (17.20). are replaced by (6.5) or (6.8). Taking into account the identity (1.10) we may rewrite (18.1) as (cf. also Sect. 2.7):

$$\Delta E_{kl} - E_{lm,km} - E_{km,lm} + (E_{mp,mp} - \Delta E_{mm})\delta_{kl} + E_{mm,kl} = \eta_{kl}.$$

---

[1] Kröner [188, 190] has also considered the possibility of extending this method to the anisotropic case, for which, however, the efficiency of the procedure is greatly diminished.

Substituting (6.8) into this relation yields

$$\Delta T_{kl} + \frac{1}{1+\nu}(T_{mm,kl} - \delta_{kl}\Delta T_{mm}) = 2\mu\eta_{kl}. \tag{18.5}$$

The equilibrium equations (17.14) may be identically satisfied by using Beltrami's solution (6.26) in terms of the stress function tensor $\boldsymbol{\chi}$. However, Kröner [187] ingeniously replaces the tensor $\boldsymbol{\chi}$ by another stress function tensor $\boldsymbol{\chi}'$ defined by

$$\chi_{kl} = 2\mu\left(\chi'_{kl} + \frac{\nu}{1-\nu}\chi'_{mm}\delta_{kl}\right) \tag{18.6}$$

and subjected to the supplementary conditions [1]

$$\chi'_{km,m} = 0. \tag{18.7}$$

The easiest way to derive the relation between $\mathbf{T}$ and $\boldsymbol{\chi}'$ is to use the obvious analogy between (18.1), (17.14), (6.8) on one hand, and, respectively, (6.26), (18.7) (18.6) on the other hand. Indeed, the last group of equations may be obtained from the first one by replacing $\mathbf{E}, \boldsymbol{\eta}, \mathbf{T}, \nu, 2\mu$ with $\boldsymbol{\chi}, \mathbf{T}, \boldsymbol{\chi}', -\nu, (2\mu)^{-1}$, respectively. By virtue of this analogy we directly derive from (18.5):

$$\Delta\chi'_{kl} + \frac{1}{1-\nu}(\chi'_{mm,kl} - \delta_{kl}\Delta\chi'_{mm}) = \frac{1}{2\mu}T_{kl}. \tag{18.8}$$

Finally, by introducing (18.8) into (18.5), we obtain

$$\Delta\Delta\chi'_{kl} = \eta_{kl}. \tag{18.9}$$

Thus, the new stress functions $\chi'_{kl}$ must satisfy the *uncoupled* equations (18.9) and the supplementary conditions (18.7). On the other hand, from the theory of bipotential equations it is known that in the case of an *infinite* medium the solution of (18.9) that vanishes at infinity is given by

$$\boldsymbol{\chi}' = -\frac{1}{8\pi}\int_{\mathscr{E}} \boldsymbol{\eta}(\mathbf{x}')\,\|\mathbf{x} - \mathbf{x}'\|\,\mathrm{d}v'. \tag{18.10}$$

By (17.16) and (18.2) this solution satisfies the supplementary conditions (18.7), too. Consequently, the corresponding self-stress results from (18.8) and (18.10).

In the case of a *finite* body, as shown by Kröner [190], the formula (18.10) still gives a particular solution of the field equations provided that $\boldsymbol{\eta}$ be replaced by

[1] Kröner [187] proved that these conditions are admissible, i.e. they do not restrict the generality of the possible stress states.

the tensor $\tilde{\boldsymbol{\eta}}$ satisfying the same conditions as in Eshelby's method. Finally, the solution of the boundary-value problem may be obtained by adding to the particular solution (18.10) a solution of the homogeneous equations

$$\Delta\Delta\chi'_{kl} = 0, \qquad \chi'_{km,m} = 0,$$

such that their sum satisfy the boundary conditions (17.15) on $\mathscr{S}$. Clearly, this comes to the solving of a classical boundary-value problem of linear elasticity, since now $\boldsymbol{\eta} = \mathbf{0}$, i.e. the strains are compatible.

Resuming the case of the *infinite* medium and substituting (18.2) into (18.10), we obtain

$$\chi'_{ij}(\mathbf{x}) = -\frac{1}{16\pi}\int_{\mathscr{E}} [\epsilon_{iks}\alpha_{sj,k'}(\mathbf{x}') + \epsilon_{jks}\alpha_{si,k'}(\mathbf{x}')]R\, \mathrm{d}v',$$

where $(\,.\,)_{,k'} = \partial(\,.\,)/\partial x'_k$ and $R = \|\mathbf{x} - \mathbf{x}'\|$. Assuming that $\boldsymbol{\alpha}$ is of bounded support, integrating by parts, and taking into account that $R_{,k'} = -R_{,k}$, the last relation becomes

$$\chi'_{ij}(\mathbf{x}) = -\frac{1}{16\pi}\int_{\mathscr{E}} [\epsilon_{iks}\alpha_{sj}(\mathbf{x}') + \epsilon_{jks}\alpha_{si}(\mathbf{x}')]R_{,k}\, \mathrm{d}v'. \tag{18.11}$$

In particular, for a *single* dislocation, introducing (17.6) into (18.11) and considering (17.7), we find

$$\chi'_{ij}(\mathbf{x}) = -\frac{b_s}{16\pi}(\epsilon_{iks}\psi_{j,k} + \epsilon_{jks}\psi_{i,k}), \tag{18.12}$$

where

$$\psi_j(\mathbf{x}) = \int_L R\, \mathrm{d}x'_j \tag{18.13}$$

and $\mathrm{d}x'_j = l_j(\mathbf{x}')\mathrm{d}l'$. Finally, it may be shown (de Wit [385]) that substituting (18.12) into (18.8) leads to the formula of Peach and Koehler for the self-stresses produced by a single dislocation loop in an infinite medium (cf. Sect. 9.3).

By employing the above method, Kröner [190] succeeded to undertake a systematic study of the self-stresses produced by dislocations, of the elastic interaction of dislocations, and of the stress concentration near the dislocation core.

### 18.3. Mura's method

We shall present now a last method for the calculation of the self-stresses produced by dislocations, which makes use of Green's tensor function. Having in mind further applications of this method to the determination of second-order effects we insert into (17.14) a term corresponding to the body forces, i.e.

$$T_{kl,l} + f_k = 0. \tag{18.14}$$

Introducing $(17.19)_2$ into (17.20) and the result obtained into (18.14), we find

$$c_{klmn}\beta_{mn,l} + f_k = 0. \tag{18.15}$$

We consider the case of an *infinite* medium and suppose that

$$\mathbf{f} = O(r^{-3}), \quad \boldsymbol{\beta} = O(r^{-2}) \quad \text{as} \quad r = \|\mathbf{x}\| \to \infty. \tag{18.16}$$

By virtue of (1.11), it follows from (17.18) that

$$\beta_{mr,n} - \beta_{mn,r} = \in_{rnl}\alpha_{ml}. \tag{18.17}$$

Differentiating (18.15) with respect to $x_r$ and taking into account (18.17) we obtain

$$c_{klmn}\beta_{mr,nl} + f_{k,r} - c_{klmn}\in_{rnt}\alpha_{mt,l} = 0. \tag{18.18}$$

On the other hand, it was shown in Sect. 6.4 that Green's tensor function of an infinite elastic medium satisfies the equation

$$c_{klmn}G_{mp,nl}(\mathbf{x}-\mathbf{x}') + \delta_{kp}\delta(\mathbf{x}-\mathbf{x}') = 0.$$

Multiplying this equation in the sense of the convolution product by

$$f_{p,r'}(\mathbf{x}') - c_{plqn}\in_{rnt}\alpha_{qt,l'}(\mathbf{x}')$$

and comparing with (18.18), we infer that

$$\beta_{mr}(\mathbf{x}) = \int_{\mathscr{E}} [f_{p,r'}(\mathbf{x}') + c_{plqn}\in_{nrt}\alpha_{qt,l'}(\mathbf{x}')]G_{mp}(\mathbf{x}-\mathbf{x}')\mathrm{d}v'. \tag{18.19}$$

Finally, integrating by parts and taking into account that the surface integral vanishes since $\mathbf{G}(\mathbf{x}-\mathbf{x}') = O(R^{-1})$ as $R = \|\mathbf{x}-\mathbf{x}'\| \to \infty$, we find

$$\beta_{mr}(\mathbf{x}) = \int_{\mathscr{E}} [f_p(\mathbf{x}')G_{mp,r}(\mathbf{x}-\mathbf{x}') + c_{plqn}\in_{nrt}\alpha_{qt}(\mathbf{x}')G_{mp,l}(\mathbf{x}-\mathbf{x}')]\,\mathrm{d}v'. \tag{18.20}$$

This solution has been first obtained in a different way for the dynamic case and for $\mathbf{f} = \mathbf{0}$ by Mura [253]. The above proof is due to Willis [382], App. 1. For the extension of formula (18.20) to the case of bounded elastic bodies we refer to the papers by Vaisman and Kunin [364] and by Simmons and Bullough [301].

## 19. Second-order elastic effects

### 19.1. Solving of non-linear boundary-value problems by successive approximations

In this section we shall deal with the determination of second-order effects in the elastic field of continuous distributions of dislocations. We begin by constructing an algorithm for solving the non-linear boundary-value problem (17.5), (17.12—17), which is similar to that used for single dislocations in Sect. 14.1 and is based on the following hypotheses:

(i) The true dislocation density is given, satisfies the identity (17.16), and is proportional to a small parameter $\varepsilon$, i.e.[1]

$$\boldsymbol{\alpha}(\mathbf{x}) = \varepsilon\boldsymbol{\alpha}^{(1)}(\mathbf{x}). \tag{19.1}$$

(ii) The body is free of surface tractions and body forces.

(iii) There exists a solution $\boldsymbol{\beta}(\mathbf{x})$ of the boundary-value problem that depends analytically on $\varepsilon$ and vanishes for $\varepsilon = 0$, i.e.

$$\boldsymbol{\beta} = \varepsilon\boldsymbol{\beta}^{(1)} + \varepsilon^2\boldsymbol{\beta}^{(2)} + \dots \tag{19.2}$$

By introducing (19.2) into (17.17), we obtain

$$A_{km}^{-1} = \delta_{km} - \varepsilon\beta_{km}^{(1)} - \varepsilon^2\beta_{km}^{(2)} - \dots, \tag{19.3}$$

and from $\mathbf{A}\,\mathbf{A}^{-1} = \mathbf{1}$ it results that

$$A_{km} = \delta_{km} + \varepsilon\beta_{km}^{(1)} + \varepsilon^2(\beta_{km}^{(2)} + \beta_{kp}^{(1)}\beta_{pm}^{(1)}) + \dots \tag{19.4}$$

Next, by substituting (19.4) into (17.12), we find

$$D_{mn} = \varepsilon E_{mn}^{(1)} + \varepsilon^2[E_{mn}^{(2)} + \tfrac{1}{2}(\beta_{mp}^{(1)}\beta_{pn}^{(1)} + \beta_{pm}^{(1)}\beta_{pn}^{(1)} + \beta_{pm}^{(1)}\beta_{np}^{(1)}) + \dots \tag{19.5}$$

where

$$E_{mn}^{(1)} = \tfrac{1}{2}(\beta_{mn}^{(1)} + \beta_{nm}^{(1)}), \qquad E_{mn}^{(2)} = \tfrac{1}{2}(\beta_{mn}^{(2)} + \beta_{nm}^{(2)}).$$

By (19.3), we also have

$$j = \det \mathbf{A}^{-1} = 1 - \varepsilon\beta_{mm}^{(1)} + \dots \tag{19.6}$$

Introducing now (19.4—6) into (17.13) yields

$$T_{kl} = \varepsilon T_{kl}^{(1)} + \varepsilon^2 T_{kl}^{(2)} + \dots, \tag{19.7}$$

---

[1] The choice of the parameter $\varepsilon$ is again immaterial, since the final results comprise only the combination $\varepsilon\boldsymbol{\alpha}^{(1)} = \boldsymbol{\alpha}$.

where

$$\left.\begin{aligned} T_{kl}^{(1)} &= c_{klmn}\beta_{mn}^{(1)}, \qquad T_{kl}^{(2)} = c_{klmn}\beta_{mn}^{(2)} + \tau_{kl}, \\ \tau_{kl} &= -\beta_{mm}^{(1)}T_{kl}^{(1)} + \beta_{km}^{(1)}T_{ml}^{(1)} + \beta_{lm}^{(1)}T_{km}^{(1)} + \\ &+ c_{klmn}(\beta_{mp}^{(1)}\beta_{pn}^{(1)} + \tfrac{1}{2}\beta_{pm}^{(1)}\beta_{pn}^{(1)}) + \tfrac{1}{2}C_{klmnrs}\beta_{mn}^{(1)}\beta_{rs}^{(1)}. \end{aligned}\right\} \qquad (19.8)$$

Cleary, (19.4—8) reduce to (14.15—19) when $\boldsymbol{\beta} = \operatorname{grad} \mathbf{u}$ as in the case of single dislocations.

Substituting now (19.1), (19.3) into (17.5) and (19.7), (19.8) into (17.14), (17.15), and equating like powers of $\varepsilon$, we obtain a sequence of linear boundary-value problems, namely, at the *first step* of the interation

$$\left.\begin{aligned} &\epsilon_{nmj}\beta_{ln,m}^{(1)} = \alpha_{lj}^{(1)}, \\ T_{kl,l}^{(1)} = 0, \qquad &T_{kl}^{(1)} = c_{klmn}\beta_{mn}^{(1)} \quad \text{in } \mathscr{V}, \\ &T_{kl}^{(1)}n_l = 0 \text{ on } \mathscr{S}, \end{aligned}\right\} \qquad (19.9)$$

at the *second step*

$$\left.\begin{aligned} &\epsilon_{nmj}\beta_{ln,m}^{(2)} = 0, \\ T_{kl,l}^{(2)} = 0, \qquad &T_{kl}^{(2)} = c_{klmn}\beta_{mn}^{(2)} + \tau_{kl} \quad \text{in } \mathscr{V}, \\ &T_{kl}^{(2)}n_l = 0 \text{ on } \mathscr{S}, \end{aligned}\right\} \qquad (19.10)$$

and so on. The term $\tau_{kl}$ occurring at the second step is obviously known, by $(19.8)_3$, from the first step of the iteration. It is also worth noting that the first step requires the solution of a boundary-value problem of the type considered in the preceding section, while the subsequent steps involve only traction boundary-value problems of classical elasticity theory, i.e. with $\boldsymbol{\alpha} = \mathbf{0}$. Consequently, the determination of the second-order effects in the elastic field of continuous distributions of dislocations reduces to the solving of two known boundary-value problems, namely (19.9) and (19.10).

*The method of Kröner and Seeger.* Kröner and Seeger [191] elaborated an algorithm similar to the above, adapted for the determination of the stresses produced by dislocations in an infinite isotropic elastic medium with the aid of stress functions. The solution of the boundary-value problem (19.9) is found like in Sect. 18.2 by putting

$$T_{kl}^{(1)} = 2\mu\left[\Delta\chi_{kl}^{\prime(1)} + \frac{1}{1-\nu}(\chi_{mm,kl}^{\prime(1)} - \delta_{kl}\Delta\chi_{mm}^{\prime(1)})\right], \qquad (19.11)$$

the stress functions $\chi_{kl}^{\prime(1)}$ being determined by the equations

$$\Delta\Delta\chi_{kl}^{\prime(1)} = \eta_{kl}^{(1)}, \qquad \chi_{kl,l}^{\prime(1)} = 0, \qquad (19.12)$$

where

$$\eta_{kl}^{(1)} = \tfrac{1}{2}(\in_{krs}\alpha_{sl,r}^{(1)} + \in_{lrs}\alpha_{sk,r}^{(1)}). \tag{19.13}$$

The boundary-value problem (19.10) differs from (19.9) only in that equations $(19.10)_1$ are homogeneous and the constitutive equations $(19.10)_3$ contain the non-linear term $\tau_{kl}$, which is known, however, from the first step of the iteration. From $(19.10)_1$ it follows, as in Sect. 18.2, that

$$-\in_{ikl}\in_{jmn}E_{ln,km}^{(2)} = 0. \tag{19.14}$$

In the *isotropic* case, by solving $(19.10)_3$ with respect to $\mathbf{E}^{(2)}$ and substituting the result obtained into (19.14), we find

$$\Delta T_{kl}^{(2)} + \frac{1}{1+\nu}(T_{mm,kl}^{(2)} - \delta_{kl}\Delta T_{mm}^{(2)}) = P_{kl}, \tag{19.15}$$

where

$$P_{kl} = \tau_{lm,km} + \tau_{km,lm} - \frac{1}{1+\nu}(\tau_{mm,kl} - \delta_{kl}\tau_{mm}). \tag{19.16}$$

Next, by putting

$$T_{kl}^{(2)} = 2\mu\left[\Delta\chi_{kl}^{\prime(2)} + \frac{1}{1-\nu}(\chi_{mm,kl}^{\prime(2)} - \delta_{kl}\Delta\chi_{mm}^{\prime(2)})\right], \tag{19.17}$$

with the supplementary conditions

$$\chi_{kl,l}^{\prime(2)} = 0, \tag{19.18}$$

we deduce that the stress functions $\chi_{kl}^{\prime(2)}$ must satisfy the equations

$$\Delta\Delta\chi_{kl}^{\prime(2)} = P_{kl}. \tag{19.19}$$

Consequently, the determination of linear and second-order effects in the elastic field of a continuous distribution of dislocations reduces to the solving of two identical systems of equations, namely (19.12) at the first step and (19.18), (19.19) at the second step of the iteration. The comments in Sect. 18.2 concerning the solution of the field equations in the case of infinite or bounded elastic bodies are obviously valid for both steps of the iteration.

Actually, the iteration scheme used by Kröner and Seeger [191] was a bit more complicated since they assumed as given the local rather than the true dislocation density; this brought about the occurrence of some supplementary terms in (19.19), originating from the non-linear geometric relation between the two dislocation densities. The method of Kröner and Seeger was subsequently used by Pfleiderer, Seeger, and Kröner [270] for the calculation of second-order effects in the elastic field of a continuous distribution of parallel edge or screw dislocations, and in particular in the elastic field of a single straight dislocation in an infinite isotropic medium.

*Willis' method.* Willis [382] employed Mura's method presented in Sect. 18.3 for solving the boundary-value problems (19.9) and (19.10) in the case of an *infinite* elastic medium. By making use of (18.20) one may easily derive the elastic distortions corresponding to the first two steps of the iteration, namely

$$\beta^{(1)}_{mr}(\mathbf{x}) = \int_{\mathscr{E}} c_{plqn} \in_{nrt} \alpha^{(1)}_{qt}(\mathbf{x}') G_{mp,l}(\mathbf{x} - \mathbf{x}') \mathrm{d}v', \tag{19.20}$$

$$\beta^{(2)}_{mr}(\mathbf{x}) = \int_{\mathscr{E}} \tau_{pq,q'}(\mathbf{x}') G_{mp,r}(\mathbf{x} - \mathbf{x}') \mathrm{d}v', \tag{19.21}$$

provided that Green's tensor function $\mathbf{G}$ of the infinite elastic medium is known.

By means of (19.20) and (19.21), Willis [382] determined the second-order effects in the elastic field of a screw dislocation lying along a two-fold axis of material symmetry. Since Willis neglected the boundary conditions on the dislocation core, he considered the single dislocation as the limiting case of a continuous distribution of dislocations and retained only the finite part of the integrals (19.21), in order to derive a physically realistic solution.

Willis' method was subsequently extended by Teodosiu and Seeger [336] to the case of uniformly moving dislocation densities and of infinitesimal motions superimposed upon a strain produced by dislocations (see Sect. 20).

## 19.2. Influence of the continuous distributions of dislocations on crystal density

In order to calculate the mean volume change produced by a continuous distribution of dislocations, we may again use the mean stress theorem, as in Sect. 15.2. In fact, since the whole reasoning leading to (15.14) involved only the displacement gradient $\mathbf{h} = \operatorname{grad} \mathbf{u}$ and not the displacement field itself, it still holds for the incompatible strains produced by a continuous distribution of dislocations. Therefore, we may derive the non-linear volume change by simply replacing $\mathbf{h}^{(1)}$ with $\boldsymbol{\beta}^{(1)}$ in the final result (15.14), thus obtaining

$$V - V_0 = \varepsilon^2 \int_{\mathscr{V}} P_{mnrs} \beta^{(1)}_{mn} \beta^{(1)}_{rs} \mathrm{d}v, \tag{19.22}$$

where $P_{mnrs}$ is given by (15.15).

Equations (19.22) and (15.15) show that the calculation of the mean dilatation $\bar{\Theta} = (V - V_0)/V_0$ produced by a continuous distribution of dislocations involves the elastic constants of third order, but requires only the knowledge of the elastic distortion $\boldsymbol{\beta}^{(1)}$ from the first step of the iteration, i.e. from linear elasticity. The parameter $\varepsilon$ occurs in the final result only through the combination $\varepsilon\boldsymbol{\beta}^{(1)}$ which does not depend on $\varepsilon$. Therefore, we may also rewrite (19.22) under the form

$$V - V_0 = \int_{\mathscr{V}} P_{mnrs}\beta_{mn}\beta_{rs}\,\mathrm{d}v, \tag{19.23}$$

where $\boldsymbol{\beta}$ means the *infinitesimal* elastic distortion produced by the continuous distribution of dislocations.

## 20. Infinitesimal motion superimposed upon a finite elastic distortion produced by dislocations

In this chapter, we have considered so far only the *elastostatics* of continuous distributions of dislocations, since their motion produces in general an inelastic deformation of the body, which lies beyond the scope of this book [1]. There exists, however, a category of *dynamic* dislocation problems that can be still treated within the theory of elasticity, namely the propagation and scattering of elastic waves in media with initial strains produced by dislocations. Such problems occur in a few important fields of solid state physics (Seeger [292], Seeger and Brand [294]). Thus, *physical acoustics* studies plane waves whose wave lengths are long enough, and hence can be studied by a continuum approach. The elastic interaction of such waves with dislocations and point defects is used, for instance, to explain and evaluate the internal friction in solids. In the *theory of heat conduction* the thermal motion of the crystalline lattice at low temperatures is modelled by a superposition of elastic waves. The elastic interaction between these waves and dislocations, which is called phonon-dislocation interaction, is of great consequence for the understanding of the thermal conduction in solids and for the evaluation of the phonon drag on moving dislocations.

In most of these applications the amplitude of the elastic waves is sufficiently small to permit their treatment as an infinitesimal motion superimposed upon an initial strain produced by dislocations. In this case it is necessary to involve the non-linear theory of elasticity in order to point out the expected scattering effects. On the other hand, the deformations produced by dislocations are small enough to allow considering only the second-order elastic effects.

We begin by presenting in a slightly different form the results obtained by Teodosiu and Seeger [336] concerning the infinitesimal motions superimposed upon a finite strain produced by a continuous distribution of dislocations in an anisotropic elastic medium. At the end of this section we shall outline the method used to study the influence of dislocations on the low-temperature thermal conductivity and shall briefly review some of the results available in the literature.

---

[1] The connexion between the *kinematics* of the elastoplastic deformation and the motion of continuous distributions of dislocations has been studied by Kosevich [182, 183] in the linear case and by Günther [148] and Teodosiu [335] in the non-linear case. For a substantiation of the theory of the elastoviscoplasticity of crystalline materials on the basis of dislocation *dynamics* we refer to the papers by Zarka [387–389], Teodosiu [335, 344], Teodosiu and Sidoroff [345], and Perzyna [268, 269].

## 20.1. Infinitesimal elastic waves superimposed upon an elastic distortion produced by dislocations

Let us consider an infinitesimal displacement field

$$\mathbf{u} = \mathbf{u}(\mathbf{x}, t) \tag{20.1}$$

superimposed upon an elastic distortion produced by a continuous distribution of dislocations in an anisotropic elastic medium. For the sake of simplicity we suppose that the medium is infinite and free of body forces and that the stress tensor produced by dislocations vanishes at infinity. Such a problem appears for example, when studying the propagation of plane waves in an initially stressed medium.

Fig. 20.1. Infinitesimal displacement field superimposed upon an elastic distortion produced by dislocations.

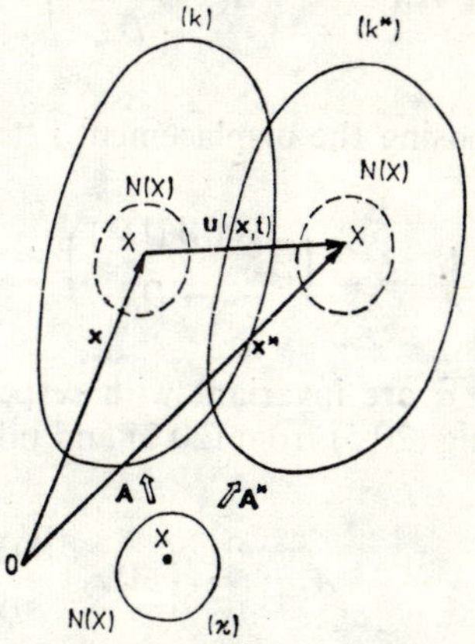


We denote by $(k)$ and $(k^*)$ the current configuration of the body at time $t$ before and after superimposing the displacement (20.1) and label by a star all quantities associated with configuration $(k^*)$ (Fig. 20.1). Let

$$\mathbf{h} = \operatorname{grad} \mathbf{u}, \qquad h_{km} = \frac{\partial u_k(\mathbf{x}, t)}{\partial x_m} \tag{20.2}$$

be the gradient of the superimposed displacement field with respect to the positions $\mathbf{x}$ of the material particles in the configuration $(k)$. Then, we have

$$\mathbf{A}^* = (\mathbf{1} + \mathbf{h})\mathbf{A}. \tag{20.3}$$

In order to deduce the modification of the equations of motion when passing from $(k)$ to $(k^*)$ it proves useful to employ a somewhat different form, which points out the terms of these equations that are invariant with respect to a superimposed elastic motion. Starting from (17.5) and (17.10), it may be shown ([336], App. 1) that

$$(jA_{ls})_{,l} = j\epsilon_{smn}\tilde{\alpha}_{mn}, \tag{20.4}$$

where $\tilde{\boldsymbol{\alpha}}$ is Noll's dislocation density, and $j = \rho/\tilde{\rho}$ is the ratio between the mass density $\rho$ in the current configuration $(k)$ and the mass density $\tilde{\rho}$ in the local natural configuration $(\varkappa)$. On the other hand, in view of $(17.13)_1$, equations $(3.22)_2$ may be written in the dynamic case as

$$\frac{\partial}{\partial x_l}\left(jA_{kr}A_{ls}\frac{\partial W(\mathbf{D})}{\partial D_{rs}}\right) = \rho\ddot{x}_k, \qquad k = 1, 2, 3. \tag{20.5}$$

Next, by making use of (20.4), we deduce from (20.5) that

$$A_{ls}\frac{\partial}{\partial x_l}\left(A_{kr}\frac{\partial W(\mathbf{D})}{\partial D_{rs}}\right) + A_{kr}\frac{\partial W(\mathbf{D})}{\partial D_{rs}}\in_{smn}\tilde{\alpha}_{mn} = \tilde{\rho}\ddot{x}_k. \tag{20.6}$$

By superimposing the displacement field (20.1) this equation becomes

$$A^*_{ls}\frac{\partial}{\partial x^*_l}\left(A^*_{kr}\frac{\partial W(\mathbf{D}^*)}{\partial D^*_{rs}}\right) + A^*_{kr}\frac{\partial W(\mathbf{D}^*)}{\partial D^*_{rs}}\in_{smn}\tilde{\alpha}_{mn} = \tilde{\rho}\ddot{x}^*_k, \tag{20.7}$$

since $\tilde{\rho}$ and $\tilde{\boldsymbol{\alpha}}$ are invariant with respect to a superimposed elastic motion. Next, by subtracting (20.5) from (20.7) and taking into account that $A^*_{ls}\,\partial/\partial x^*_l = A_{ls}\,\partial/\partial x_l$, we find

$$A_{ls}\frac{\partial}{\partial x_l}\left(A^*_{kr}\frac{\partial W(\mathbf{D}^*)}{\partial D^*_{rs}} - A_{kr}\frac{\partial W(\mathbf{D})}{\partial D_{rs}}\right) +$$

$$+\left(A^*_{kr}\frac{\partial W(\mathbf{D}^*)}{\partial D^*_{rs}} - A_{kr}\frac{\partial W(\mathbf{D})}{\partial D_{rs}}\right)\in_{smn}\tilde{\alpha}_{mn} = \tilde{\rho}\ddot{u}_k.$$

Finally, by multiplying both sides of this relation by $j = \rho/\tilde{\rho}$, and considering again (20.4), we obtain

$$\frac{\partial}{\partial x_l}\left[j\,A_{ls}\left(A^*_{kr}\frac{\partial W(\mathbf{D}^*)}{\partial D^*_{rs}} - A_{kr}\frac{\partial W(\mathbf{D})}{\partial D_{rs}}\right)\right] = \rho\ddot{u}_k, \quad k = 1, 2, 3. \tag{20.8}$$

It is worth noting that although the equations (20.8) depend on $\mathbf{A}$ they have the same form irrespective of $\mathbf{A}$ being produced by stationary or moving dislocations. Let us assume now that $\|\text{grad}\,\mathbf{u}\| \ll 1$ and that there exists a solution $\mathbf{u}(\mathbf{x}, t)$ of (20.8) which depends analytically on a small parameter $\nu$, i.e.

$$\mathbf{u} = \nu\mathbf{u}^{(1)} + \nu^2\mathbf{u}^{(2)} + \dots$$

By putting

$$\mathbf{h}^{(1)} = \text{grad}\,\mathbf{u}^{(1)}, \qquad \mathbf{h}^{(2)} = \text{grad}\,\mathbf{u}^{(2)}, \dots$$

expanding $\partial W(\mathbf{D}^*)/\partial \mathbf{D}^*$ in power series of $\mathbf{D}^* - \mathbf{D}$, and making use of (17.12) and (17.13), we obtain after a straightforward but tedious calculation [1]

$$\nu[(B_{ijkl}(\mathbf{A})h_{kl}^{(1)})_{,j} - \rho\, \ddot{u}_i^{(1)}] +$$
$$+ \nu^2\{[(B_{ijkl}(\mathbf{A})h_{kl}^{(2)} + \psi_{ij}(\mathbf{A}, \mathbf{h}^{(1)})]_{,j} - \rho \ddot{u}_i^{(2)}\} + \ldots = 0, \quad i = 1, 2, 3, \tag{20.9}$$

where

$$\psi_{ij}(\mathbf{A}, \mathbf{h}^{(1)}) = \frac{1}{2} G_{ijkl}(\mathbf{A}) h_{ik}^{(1)} h_{il}^{(1)} + G_{tjkl}(\mathbf{A}) h_{it}^{(1)} h_{kl}^{(1)} + \frac{1}{2} F_{ijklmn}(\mathbf{A}) h_{kl}^{(1)} h_{mn}^{(1)},$$

$$B_{ijkl}(\mathbf{A}) = j\delta_{ik} A_{jq} A_{ls} \frac{\partial W(\mathbf{D})}{\partial D_{qs}} + G_{ijkl}(\mathbf{A}),$$

$$G_{ijkl}(\mathbf{A}) = jA_{ip} A_{jq} A_{kr} A_{ls} \frac{\partial^2 W(\mathbf{D})}{\partial Dpq\, \partial D_{rs}},$$

$$F_{ijklmn}(\mathbf{A}) = jA_{ip} A_{jq} A_{kr} A_{ls} A_{mt} A_{nu} \frac{\partial^3 W(\mathbf{D})}{\partial D_{pq} \partial D_{rs} \partial D_{tu}}.$$

Equations (20.9) have been derived without making any restrictive assumption on the magnitude of the initial dislocation strain. Let us assume now that the elastic distortion $\mathbf{A}$ produced by dislocations may be obtained by one of the methods presented in Sect. 19.1, under the form of the power series (19.4), i.e.

$$\mathbf{A} = \mathbf{1} + \varepsilon \boldsymbol{\beta}^{(1)} + \varepsilon^2(\boldsymbol{\beta}^{(2)} + \boldsymbol{\beta}^{(1)}\boldsymbol{\beta}^{(1)}) + \ldots, \tag{20.10}$$

where $\varepsilon$ is another small parameter, characterizing the magnitude of the dislocation strain.

According to the relative magnitude of the small parameters $\nu$ and $\varepsilon$ we differentiate the following three cases.

(i) $\varepsilon \ll \nu < 1$. In this case $\nu\varepsilon \ll \nu^2$ and hence, by taking $\mathbf{A} = \mathbf{1}$, $j = 1$, $\mathbf{D} = \mathbf{0}$, the corresponding error in (20.9) will be much smaller than $\nu^2$. Furthermore, by (4.41) and (4.44), we have $B_{ijkl}(\mathbf{1}) = G_{ijkl}(\mathbf{1}) = c_{ijkl}$, $F_{ijklmn}(\mathbf{1}) = C_{ijklmn}$, while $\psi_{ij}(\mathbf{1}, \mathbf{h}^{(1)}) = \sigma_{ij}(\mathbf{h}^{(1)})$, where

$$\sigma_{ij}(\mathbf{h}^{(1)}) = c_{sjkl} h_{is}^{(1)} h_{kl}^{(1)} + \frac{1}{2} c_{ijkl} h_{sk}^{(1)} h_{sl}^{(1)} + \frac{1}{2} c_{ijklmn} h_{kl}^{(1)} h_{mn}^{(1)}.$$

Consequently, equations (20.9) are equivalent up to an error of the order $\nu^2$ with the systems

$$c_{ijkl} h_{kl,j}^{(1)} = \rho \ddot{u}_i^{(1)}, \quad c_{ijkl} h_{kl,j}^{(2)} + \tau_{ij,j}(\mathbf{h}^{(1)}) = \rho \ddot{u}_i^{(2)}, \quad i = 1, 2, 3, \tag{20.11}$$

---

[1] For details see [336], p. 896.

which determine, respectively, $\mathbf{u}^{(1)}$ and $\mathbf{u}^{(2)}$. When $\mathbf{u}$ is a plane wave, these systems describe the lowest-order anharmonic effects due to the non-linearity of the constitutive equations. Under the hypothesis (i) the influence of the strains produced by dislocations is negligible up to the order $O(\nu^2)$.

(ii) $\varepsilon \approx \nu \ll 1$. In this case we have $\nu\varepsilon \approx \nu^2$, and hence the terms containing $\nu\varepsilon$ in the first line of (20.9) must be included in the terms of order $O(\nu^2)$. Accordingly, by equating to zero the coefficients of $\nu$ and $\nu^2$ in (20.9), it results that

$$c_{ijkl}h^{(1)}_{kl,j} = \rho\ddot{u}^{(1)}_i, \qquad i = 1, 2, 3. \tag{20.12}$$

$$[c_{ijkl}h^{(2)}_{kl} + \sigma_{ij}(\mathbf{h}^{(1)}) + (\varepsilon/\nu)\varphi_{ijkl}(\boldsymbol{\beta}^{(1)})h^{(1)}_{kl}]_{,j} = \rho\ddot{u}^{(2)}_i, \qquad i = 1,2,3, \tag{20.13}$$

where

$$\varphi_{ijkl}(\boldsymbol{\beta}^{(1)}) = \delta_{ik}c_{jlpr}\beta^{(1)}_{pr} - c_{ijkl}\beta^{(1)}_{mm} + c_{iqkl}\beta^{(1)}_{jq} + c_{ijks}\beta^{(1)}_{ls} + \\ + c_{pjkl}\beta^{(1)}_{ip} + c_{ijrl}\beta^{(1)}_{kr} + C_{ijkltu}\beta^{(1)}_{tu}. \tag{20.14}$$

The systems (20.12) and (20.13) allow the determination of $\mathbf{u}^{(1)}$ and $\mathbf{u}^{(2)}$. When $\mathbf{u}$ is a plane wave, system (20.13) describes the anharmonicity due to both the strains produced by dislocations and the non-linearity of the constitutive equations.

(iii) $\nu \ll \varepsilon \ll 1$. In this case $\nu^2 \ll \nu\varepsilon$ and hence, by retaining in (20.9) only terms of the orders $O(\nu)$ and $O(\nu\varepsilon)$, we may describe the lowest anharmonic effects due to the strains produced by dislocations, which now predominate over those arising from the non-linearity of the constitutive equation. We then obtain

$$\{[c_{ijkl} + \varepsilon\varphi_{ijkl}(\boldsymbol{\beta}^{(1)})]h^{(1)}_{kl}\}_{,j} = \rho\ddot{u}_i, \qquad i = 1, 2, 3. \tag{20.15}$$

Finally, it is important to note that terms accounting for the anharmonicity produced by dislocations in (20.14) and (20.15) are completely determined by the linear approximation $\boldsymbol{\beta}^{(1)}$ to the elastic distortion produced by dislocations.

## 20.2. The influence of dislocations on the low-temperature lattice conductivity

As already mentioned at the beginning of this section, the thermal motion in solids at low temperatures may be described by a superposition of elastic waves (phonons). The interaction between a dislocation and phonons arises mainly [1] through the large distortion field of the dislocation, which has a dissipative effect, called *phonon scattering*, on the travelling elastic waves, thus limiting the thermal conductivity. The same process is important in evaluating the phonon drag on moving dislocations, and hence the low-temperature mobility of dislocations (Seeger and Engelke [469], Gruner [423]).

At temperatures well below the Debye temperature, phonons have large wave-lengths and are, therefore, more effectively scattered by dislocations, which possess a far-reaching distortion field, than by point defects, which distort the lattice only in a small spatial region (cf. Chap. V).

The influence of the phonon scattering by the lattice defects on thermal conductivity has been first investigated by Klemens [435], who calculated the scatter-

[1] An incident phonon may also cause a dislocation to oscillate and to emit other phonons (see, e. g. Ninomiya [454]); this fluttering mechanism leads also to phonon scattering.

ing cross sections of various defects and found that each defect has a characteristic wave-length dependence of the scattering cross section and a characteristic temperature dependence of the thermal resistance. For instance, the scattering cross section of a single dislocation is inversely proportional to the phonon wave-length, and its thermal resistance has a quadratic temperature dependence. The attractive feature of this discovery is the possibility to distinguish different types of defects and to study their arrangement by measuring the temperature dependence of thermal conductivity.

Klemen's analysis has been based on *non-linear lattice theory*; however, the anharmonicity of the interatomic forces has been estimated in terms of a single parameter, the Grüneisen constant, by comparing the change in the energy of a crystal produced by a homogeneous dilatation with that produced by thermal expansion. As shown by Bross [403], this rather poor approximation may be considerably improved by using *non-linear elasticity theory*.

The use of continuum mechanics for studying the phonon scattering is fully justified at low temperatures, where the phonon wave-lengths are large in comparison with the lattice parameter. The main advantage of this continuum approach is that third-order elastic constants are known for a variety of materials, whereas it is still difficult to get reliable experimental values for the anharmonic force constants. Moreover, non-linear elasticity theory allows to easily take into account the so-called *three-phonon processes*, which describe the lowest-order scattering of waves from each other.

Bross' method has been successfully used in conjunction with second-order elasticity theory to study the phonon scattering by single screw (Bross, Seeger, and Gruner [39]) and edge dislocations (Bross, Seeger, and Haberkorn [40]), by edge-dislocation dipoles (Gruner and Bross [422]), and by dislocation pile-ups (Gruner [423]) in *infinite isotropic* media. Alshits and Kotowski [392] have shown, on the particular example of prismatic dislocation loops, that the influence of the curvature of dislocation loops becomes significant only at very low temperatures, where the phonon wave-lengths become comparable with the curvature radii of the loops. Finally, Bross, Gruner, and Kirschenmann [41] have described the phonon-phonon interaction (three-phonon processes) in the presence of edge dislocations, by the scattering of two elastic waves propagating in a non-linear elastic medium with dislocations.

Although the temperature dependence of the thermal resistance due to the phonon scattering from stationary dislocations have been verified experimentally in several cases, the calculated values of the thermal conductivity are larger than those found experimentally by a factor of 10 in the case of metals, and by a factor of $10^3$ in the case of ionic crystals.

Recently, Eckhardt and Wasserbäch [414, 415] have considerably enlarged Bross' theory, by considering the *simultaneous* scattering of phonons by stationary dislocations, and by three-phonon processes in a *finite anisotropic* body (crystal plate). The numerical results obtained by these authors for ionic cristals are in much better quantitative agreement with experimental data. Moreover, the theory developed permits to more coherently explain various effects, sometimes considered as anomalous, e.g. the dependence of thermal conductivity on the deformation history.

CHAPTER V

# THE ELASTIC FIELD OF POINT DEFECTS

Unlike dislocations, which are linear defects of the crystal lattice, *point defects* are lattice imperfections having all dimensions of the order of one atomic spacing. The point defect may be a vacant site in the atomic lattice, called a *vacancy*, a foreign atom replacing one atom of the lattice, called a *substitutional atom*, or an atom situated between the normal sites of the lattice, called an *interstitial atom*. An interstitial atom is said to be *intrinsic* or *extrinsic*, according as it is of the same nature or of different nature with the atoms of the host lattice. Sometimes, two or more point defects can build characteristic arrangements which are thermodynamically stable, i.e. their self-energies are smaller than the sum of the self-energies of the individual point defects.

The collective motion of point defects produces viscous effects at a macroscopic scale, which are of great importance for many processes taking place in crystals (see, e.g. Seeger [286], Hirth and Lothe [162], Nowik and Berry [455], chapters 7 and 8). Like dislocations, however, each point defect moves in an almost good crystal, which may be considered to a large extent as an elastic medium. Moreover, the interaction of a point defect with other crystal defects is mostly of elastic nature. That is why we will devote this chapter to elastic models of point defects and to various methods for calculating the elastic interaction of a point defect with dislocations or other point defects.

## 21. Modelling of point defects as spherical inclusions in elastic media

The simplest model of a point defect is given by a spherical rigid or elastic inclusion in an infinite isotropic medium. In both cases the elastic state possesses spherical symmetry with respect to the centre of the inclusion. By using spherical co-ordinates $r, \theta, \varphi$ (Fig. 1.5) and assuming that the displacement vector $\mathbf{u}$ has radial direction and is independent of $\theta$ and $\varphi$, we have

$$\mathbf{u} = u_r(r)\mathbf{e}_r, \qquad \mathbf{u}_\theta = u_\varphi = 0, \tag{21.1}$$

and, by making use of (1.84), we deduce that the infinitesimal strain tensor $\mathbf{E}$ has the physical components

$$E_{rr} = \frac{\mathrm{d}u_r}{\mathrm{d}r}, \qquad E_{\theta\theta} = E_{\varphi\varphi} = \frac{u_r}{r}, \qquad E_{r\theta} = E_{\theta\varphi} = E_{r\varphi} = 0. \tag{21.2}$$

Next, substituting (21.2) into (6.5) gives the physical components of the stress tensor

$$\left.\begin{aligned} T_{rr} &= (\lambda + 2\mu)\frac{\mathrm{d}u_r}{\mathrm{d}r} + 2\lambda\,\frac{u_r}{r}, \\ T_{\theta\theta} &= T_{\varphi\varphi} = \lambda\frac{\mathrm{d}u_r}{\mathrm{d}r} + 2(\lambda + \mu)\,\frac{u_r}{r}, \\ T_{r\theta} &= T_{r\varphi} = T_{\theta\varphi} = 0. \end{aligned}\right\} \tag{21.3}$$

Introducing (21.3) into (6.2), setting $f_k = 0$, and considering (1.85), we see that the last two equilibrium equations are identically satisfied, while the first one reduces to the differential equation of Euler type

$$r^2\,\frac{\mathrm{d}^2u_r}{\mathrm{d}r^2} + 2r\,\frac{\mathrm{d}u_r}{\mathrm{d}r} - 2u_r = 0, \tag{21.4}$$

which admits the general solution

$$u_r(r) = Cr^{-2} + C_1 r, \tag{21.5}$$

where $C$ and $C_1$ are arbitrary constants.

## 21.1. The point defect as rigid inclusion

Let us first simulate a point defect by a rigid ball of radius $r_0$ which is forced into a spherical cavity whose volume is by $\delta v$ smaller than that of the ball and is situated in an *infinite* isotropic medium [1]. On physical grounds, **u** must vanish at infinity, hence $C_1 = 0$, and solution (21.5) reduces to

$$u_r^\infty(r) = Cr^{-2}, \qquad r = \|\mathbf{x}\| = \sqrt{x_1^2 + x_2^2 + x_3^2}. \tag{21.6}$$

Here and in the following the superscript $\infty$ is used to recall that the elastic medium is infinite. Next, from $(21.1)_1$ it results that

$$\mathbf{u}^\infty(r) = Cr^{-2}\mathbf{e}_r = Cr^{-3}\mathbf{x} = -\,C\,\mathrm{grad}(r^{-1}). \tag{21.7}$$

The constant $C$ is a measure of the "strength" of the singularity introduced by the point defect and is completely determined by $\delta v$. Indeed, requiring that $4\pi r_0^2 u_r^\infty(r_0) = \delta v$ gives

$$\delta v = 4\pi C. \tag{21.8}$$

[1] This model has been firstly used by Bitter [30]. The reasoning that follows is mainly due to Eshelby [111].

Introducing (21.6) into (21.2) and (21.3) we find the non-zero strain and stress components:

$$E_{rr} = -2Cr^{-3}, \qquad E_{\theta\theta} = E_{\varphi\varphi} = Cr^{-3}, \qquad \operatorname{tr} \mathbf{E} = 0, \tag{21.9}$$

$$T_{rr} = -4\mu Cr^{-3}, \qquad T_{\theta\theta} = T_{\varphi\varphi} = 2\mu Cr^{-3}, \qquad \operatorname{tr} \mathbf{T} = 0. \tag{21.10}$$

From $(21.9)_3$ and $(21.10)_3$ it is easily seen that the mean pressure and the volume dilatation vanish at any point of the elastic medium ($r > r_0$).

Assume now that the elastic state outside the cavity is extended to the whole space, having the origin as singular point. We then derive from (21.7), considering also (1.47), that

$$\operatorname{div} \mathbf{u}^\infty = -C\Delta(r^{-1}) = 4\pi C\delta(\mathbf{x}),$$

$$\Delta \mathbf{u}^\infty = -C\Delta[\operatorname{grad}(r^{-1})] = -C \operatorname{grad} \Delta(r^{-1}) = 4\pi C \operatorname{grad} \delta(\mathbf{x}),$$

where $\delta(\mathbf{x})$ is Dirac's distribution. By introducing these relations into (6.28) we find that the action of the point defect on the elastic medium is equivalent to that of the body force field

$$\mathbf{f}(\mathbf{x}) = -4\pi C(\lambda + 2\mu) \operatorname{grad} \delta(\mathbf{x}), \tag{21.11}$$

acting on the elastic medium without cavity. Clearly, the field (21.11) represents three mutually orthogonal force dipoles without moment, having equal intensities, and acting at the centre of the point defect (Fig. 21.1). As already mentioned in Sect. 6.4, such a distribution of concentrated loads is called a *spherical dilatation centre*.

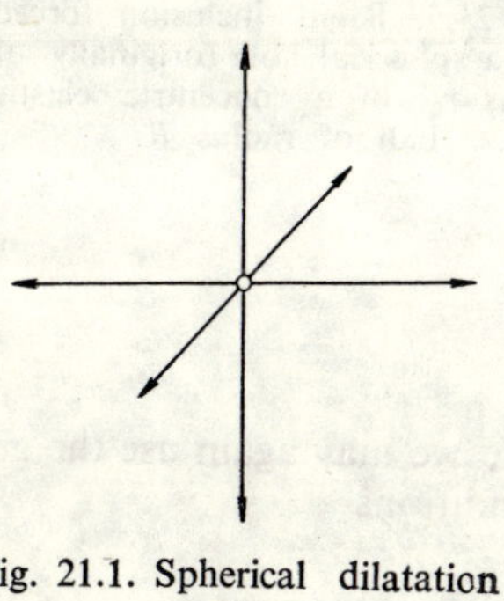

Fig. 21.1. Spherical dilatation centre.

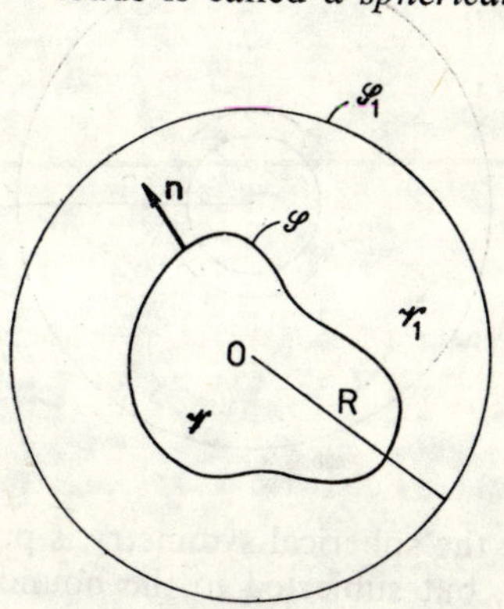


Fig. 21.2. Spherical surface $\mathcal{S}_1$ and arbitrary smooth surface $\mathcal{S}$ around a rigid inclusion in an infinite elastic medium.

The volume change of a part $\mathcal{V}$ of the elastic medium that contains the point defect is

$$\delta V^\infty = \int_{\mathcal{S}} \mathbf{u}^\infty \cdot \mathbf{n} \, ds, \tag{21.12}$$

where $\mathscr{S}$ is the boundary of $\mathscr{V}$, and $\mathbf{n}$ is the outward unit normal to $\mathscr{S}$. It is easily proved that $\delta V^\infty$ is independent of the choice of $\mathscr{V}$. Indeed, let $\mathscr{V}_1$ be a sphere of radius $R$, centred at the origin, and of boundary $\mathscr{S}_1$ (Fig. 21.2). Since $\operatorname{div} \mathbf{u}^\infty = 0$ at any point of the elastic medium outside the inclusion, we deduce from (1.52) that

$$0 = \int_{\mathscr{V}_1 \setminus \mathscr{V}} \operatorname{div} \mathbf{u}^\infty \, dv = \int_{\mathscr{S}_1} \mathbf{u}^\infty \cdot \mathbf{n} ds - \int_{\mathscr{S}} \mathbf{u}^\infty \cdot \mathbf{n} \, ds$$

and hence, by $(21.7)_1$ and (21.12),

$$\delta V^\infty = \int_{\mathscr{S}} \mathbf{u}^\infty \cdot \mathbf{n} \, ds = \int_{\mathscr{S}_1} \mathbf{u}^\infty \cdot \mathbf{n} \, ds = C \int_0^{2\pi} d\varphi \int_0^{\pi} \sin\theta \, d\theta = 4\pi C = \delta v, \tag{21.13}$$

since on $\mathscr{S}_1$ we have: $r = R$, $ds = R^2 \sin\theta \, d\theta \, d\varphi$, and $\mathbf{e}_r \cdot \mathbf{n} = 1$. Equation (21.13) shows that a rigid inclusion increases the volume of each finite region containing it by $4\pi C$, without any elastic dilatation.

When the elastic medium is *finite*, some supplementary elastic displacements occur from the vanishing condition of the tractions on the external boundary of the elastic body. Let us assume for the sake of simplicity that the elastic body is a ball of radius $R$ containing a spherical concentric hole whose volume is by $\delta v$ smaller than that of the rigid inclusion (Fig. 21.3).

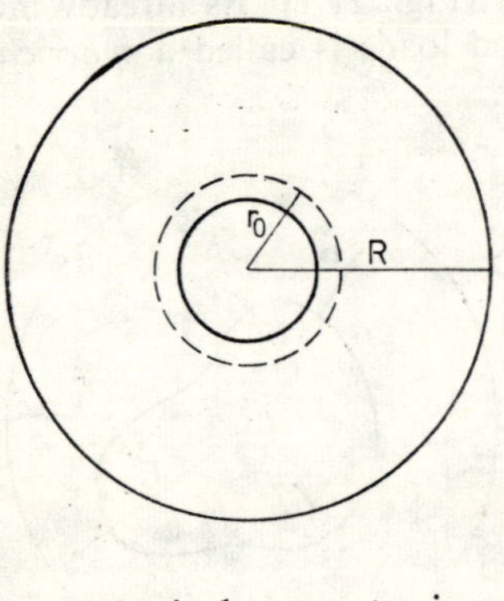


Fig. 21.3. Rigid inclusion forced into a spherical hole (originally of radius $r_0$) of a concentric elastic ball of radius $R$.

Since the spherical symmetry is preserved, we may again use the general solution (21.5), but subjected to the boundary conditions

$$T_{rr}(R) = 0, \qquad 4\pi r_0^2 u_r(r_0) = \delta v, \tag{21.14}$$

which imply

$$C = \frac{\delta v}{4\pi \left(1 + \dfrac{4\mu}{3K} \dfrac{r_0^3}{R^3}\right)}, \qquad C_1 = \frac{4\mu C}{3KR^3}, \tag{21.15}$$

where $K$ is the bulk modulus. Usually $r_0 \ll R$, and thus $(21.15)_1$ reduces again to (21.8). Substituting (21.5) into (21.2), (21.3), and considering $(21.15)_2$, we find the non-zero strain and stress components:

$$E_{rr} = -\frac{2C}{r^3}\left(1 - \frac{2\mu}{3K}\frac{r^3}{R^3}\right),$$

$$E_{\theta\theta} = E_{\varphi\varphi} = \frac{C}{r^3}\left(1 + \frac{4\mu}{3K}\frac{r^3}{R^3}\right), \quad \mathrm{tr}\,\mathbf{E} = \frac{4\mu C}{KR^3}, \tag{21.16}$$

$$T_{rr} = -\frac{4\mu C}{r^3}\left(1 - \frac{r^3}{R^3}\right),$$

$$T_{\theta\theta} = T_{\varphi\varphi} = \frac{2\mu C}{r^3}\left(1 + \frac{2r^3}{R^3}\right), \quad \mathrm{tr}\,\mathbf{T} = \frac{12\mu C}{R^3}. \tag{21.17}$$

Equations $(21.16)_3$ and $(21.17)_3$ show that the boundary condition $(21.14)_1$ leads to the occurrence of a uniform dilatation, and hence to a uniform mean pressure in the isotropic elastic ball.

The volume change produced by the rigid inclusion is

$$\delta V = 4\pi R^2 u_r(R) = 4\pi C\left(1 + \frac{4\mu}{3K}\right), \tag{21.18}$$

or, by taking into account (21.13),

$$\delta V = \left(1 + \frac{4\mu}{3K}\right)\delta V^\infty = \frac{3(1-\nu)}{1+\nu}\,\delta V^\infty. \tag{21.19}$$

When the finite elastic body has a more complicated geometry the boundary-value problem cannot be solved generally in closed form. However, as shown by Eshelby [111], equation (21.18) still holds. To prove this, let us make use of the simulation of the point defect by a dilatation centre, i.e. by the distribution of body forces (21.11). For a finite elastic body occupying a region $\mathscr{V}$ of boundary $\mathscr{S}$ we have (see also Gurtin [150], p. 97)

$$\delta V = \int_{\mathscr{V}} E_{mm}\,\mathrm{d}v = \frac{1}{3K}\int_{\mathscr{V}} T_{mm}\,\mathrm{d}v,$$

wherefrom, in view of (15.2), it follows that

$$\delta V = \frac{1}{3K}\left(\int_{\mathscr{S}} t_k x_k\,\mathrm{d}s + \int_{\mathscr{V}} f_k x_k\,\mathrm{d}v\right). \tag{21.20}$$

In our case $\mathbf{t} = \mathbf{0}$ on $\mathscr{S}$ and, by substituting (21.11) into (21.20), we obtain

$$\delta V = -\frac{4\pi C(\lambda + 2\mu)}{3K} \int_{\mathscr{V}} x_k \frac{\partial \delta(\mathbf{x})}{\partial x_k} \, \mathrm{d}v =$$

$$= \frac{4\pi C(\lambda + 2\mu)}{K} = 4\pi C \left(1 + \frac{4\mu}{3K}\right),$$

i.e. the same result as in the particular case of the elastic ball. Since $u_r(r) = C_1 r$ is a particular solution of the equilibrium equation (21.4), by repeating the reasoning leading to (21.11) and making use of (21.18), it is easily seen that the action of a rigid inclusion on an arbitrary finite elastic body is equivalent with that of the body force

$$\mathbf{f}(\mathbf{x}) = -K(\delta V) \operatorname{grad} \delta(\mathbf{x}). \tag{21.21}$$

By virtue of (12.9), *the potential interaction energy between a point defect D and a regular or singular elastic state* $s^* = [\mathbf{u}^*, \mathbf{E}^*, \mathbf{T}^*]$ is

$$\Phi_{\text{int}}\{D, s^*\} = -\int_{\mathscr{V}} \mathbf{f} \cdot \mathbf{u}^* \, \mathrm{d}v, \tag{21.22}$$

since the surface tractions corresponding to the point defect on $\mathscr{S}$ are zero. Next, by substituting (21.21) into (21.22) we successively have

$$\Phi_{\text{int}}\{D, s^*\} = K(\delta V) \int_{\mathscr{V}} u_k^* \frac{\partial \delta(\mathbf{x})}{\partial x_k} \, \mathrm{d}v = -K(\delta V) \int_{\mathscr{V}} \delta(\mathbf{x}) \frac{\partial u_k^*}{\partial x_k} \, \mathrm{d}v =$$

$$= -K(\delta V) \operatorname{tr} \mathbf{E}^* = -\tfrac{1}{3}(\delta V) \operatorname{tr} \mathbf{T}^* = p^* \, \delta V, \tag{21.23}$$

where $p^*$ is the mean pressure corresponding to $\mathbf{T}^*$ and evaluated at the centre of the point defect.

If $\mathbf{T}^*$ is also generated by a point defect and the elastic medium is infinite, then, by $(21.10)_3$, we have $p^* = 0$, and hence $\Phi_{\text{int}} = 0$. Therefore, the interaction energy of two point defects simulated by small rigid inclusions in an *infinite* isotropic elastic medium is zero [1].

It is interesting to note that the result (21.23) still holds for an anisotropic elastic medium with cubic symmetry, as long as the point defects are simulated by dilatation centres. Indeed, the reasoning leading to (21.23) remains unchanged, since the relation $\operatorname{tr} \mathbf{T} = 3K \operatorname{tr} \mathbf{E}$ is also valid for cubic materials, with $K = (c_{11} + 2c_{12})/3$.

---

[1] We shall see in Sect. 23.2 that a more exact simulation of a point defect by multipolar forces allows to take into account non-local effects, which bring a non-vanishing contribution to $\Phi_{\text{int}}$.

## 21.2. The point defect as elastic inclusion

An improved model of a point defect may be obtained by an *elastic* ball of radius $r_0$ which is forced into a spherical hole, whose volume is by $\delta v'$ smaller than that of the inclusion and which is situated in another elastic ball of radius $R$ (Fig. 21.4).

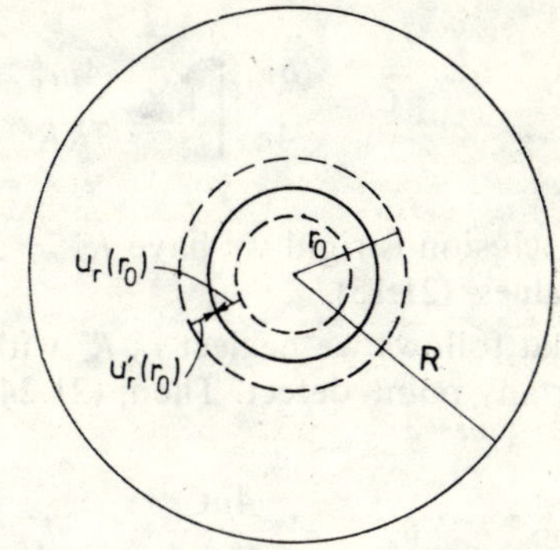


Fig. 21.4. Elastic inclusion (originally of radius $r_0$) forced into a smaller spherical hole of a concentric elastic ball of radius $R$.

Let us denote by $\mathbf{u}'$, $\mathbf{T}'$, $\lambda'$, $\mu'$ and $\mathbf{u}$, $\mathbf{T}$, $\lambda$, $\mu$ the displacement vector, the stress tensor, and Lamé's constants corresponding to the inclusion and the larger ball, respectively. As $\mathbf{u}$ must be finite for $r = 0$, it follows from (21.5) that the only non-zero components of the vectors $\mathbf{u}'$ and $\mathbf{u}$ are

$$u'_r = C'r, \qquad u_r = Cr^{-2} + C_1 r. \tag{21.24}$$

The constants $C'$, $C$ and $C_1$ are to be determined by the condition that the radial stress component $T_{rr}$ be continuous across the surface of the inclusion and vanish on the external surface $r = R$:

$$T'_{rr}(r_0) = T_{rr}(r_0), \qquad T_{rr}(R) = 0, \tag{21.25}$$

as well as by the geometric condition [1]

$$4\pi r_0^2[u_r(r_0) - u'_r(r_0)] = \delta v'. \tag{21.26}$$

Substituting (21.24) into $(21.3)_1$ gives

$$T'_{rr} = 3K'C', \qquad T_{rr} = -4\mu C r^{-3} + 3KC_1, \tag{21.27}$$

[1] We assume as before that $\delta v'$ is small enough to allow the treatment of the problem by linear elasticity theory. Consequently, we make no distinction in (21.25) and (21.27) between the co-ordinates of the material points before and after the deformation. For a treatment of the same problem in the framework of second-order elasticity theory, but with $\lambda = \lambda'$, $\mu = \mu'$, see Seeger and Mann [289].

where $K' = \lambda' + 2\mu'/3$, $K = \lambda + 2\mu/3$. Next, by introducing (21.27) into (21.25), and (21.24) into (21.26), and solving the algebraic system obtained with respect to $C$, $C'$, and $C_1$, we find

$$C_1 = \frac{4\mu C}{3KR^3}, \quad C' = \frac{4\mu C}{3K'}\left(\frac{1}{R^3} - \frac{1}{r_0^3}\right),$$

$$C = \frac{\delta v'}{4\pi}\left[1 + \frac{4\mu r_0^3}{3KR^3} + \frac{4\mu}{3K'}\left(1 - \frac{r_0^3}{R^3}\right)\right]^{-1}.$$

When the inclusion is rigid we have $K' = \infty$, $C' = 0$, and $C$ and $C_1$ reduce to their previous values (21.15).

In what follows we neglect $r_0^3/R^3$ with respect to unity, which is certainly admissible for any point defect. Then, (21.24) and (21.3) yield

$$\left.\begin{aligned} u_r' &= -\frac{4\mu C}{3K' r_0^3} r, \quad u_r = C\left(\frac{1}{r^2} + \frac{4\mu}{3K}\frac{r}{R^3}\right), \\ C &= \frac{\delta v'}{4\pi\left(1 + \dfrac{4\mu}{3K'}\right)}, \end{aligned}\right\} \tag{21.28}$$

$$T'_{rr} = T'_{\varphi\varphi} = T'_{\theta\theta} = -\frac{4\mu C}{r_0^3}, \qquad \operatorname{tr} \mathbf{T}' = -\frac{12\mu C}{r_0^3}. \tag{21.29}$$

$$\left.\begin{aligned} T_{rr} &= -\frac{4\mu C}{r^3}\left(1 - \frac{r^3}{R^3}\right) \\ T_{\theta\theta} &= T_{\varphi\varphi} = \frac{2\mu C}{r^3}\left(1 + \frac{2r^3}{R^3}\right), \quad \operatorname{tr} \mathbf{T} = \frac{12\mu C}{R^3} \end{aligned}\right\} \tag{21.30}$$

while the other components of $\mathbf{u}$ and $\mathbf{T}$ are zero. Equations (21.29) and (21.30) show that the elastic inclusion is subjected to a pure hydrostatic pressure, while the stress state outside the inclusion is characteristic of the presence of a dilatation centre. By comparing (21.30) with (21.17) and $(21.28)_3$ with $(21.15)_1$, it is easily seen that an elastic inclusion with the bulk modulus $K'$ and of strength $C$ exerts the same action on the external elastic ball as a rigid inclusion of strength $C/(1 + 4\mu/3K')$. Finally, we remark that the volume change of the inclusion is

$$\delta v = 4\pi r_0^2 u_r'(r_0) = -\frac{4\mu}{3K'} \cdot \frac{\delta v'}{1 + 4\mu/(3K')},$$

and that of the external ball is

$$\delta V = 4\pi R^2 u_r(R) = \frac{1 + 4\mu/(3K)}{1 + 4\mu/(3K')}\,\delta v'.$$

Actually, the evaluation of the bulk modulus $K'$ for a single point defect, if possible, is always questionable. Therefore, the use of the above model is reasonable only when extrinsic point defects of the same nature agglomerate by diffusion and generate larger inclusions.

## 21.3. The inhomogeneity effect of point defects

The above models of point defects were based on their description as rigid or elastic balls that are forced into holes of a smaller volume. The elastic state generated in this way is called the *size effect* or the *paraelastic effect* of point defects. The last name was introduced by Kröner [192] on the analogy of electrodynamics and was suggested by the simulation of point defects by force dipoles.

Extrinsic point defects may also have an *inhomogeneity effect*, which occurs when they are placed in a stress field different from their own. As shown by Eshelby [111], the change in the elastic state outside the point defect may be ascribed to the induction by the applied stress field of supplementary force dipoles in the extrinsic point defect; for this reason, Kröner [193] called this effect the *diaelastic effect*, again by analogy with the similar situation in electrodynamics.

Usually, the size and the inhomogeneity effects occur simultaneously, one of them being eventually preponderant with respect to the other. In the linear theory of elasticity, however, these effects may be considered separately and then superposed. Since the size effect has been studied above, we will treat in this subsection the pure inhomogeneity effect. Such an ideal situation could be realized if an atom of the lattice were replaced by a foreign atom of the same atomic volume, i.e. without displacing the neighbouring atoms.

Let us consider the point defect as a small elastic ball that replaces a subregion $\mathscr{V}'$ of volume $\Omega$ and boundary $\mathscr{S}'$ of the region $\mathscr{V}$ of boundary $\mathscr{S}$ occupied by the elastic body (Fig. 21.5). Denote by $\mathbf{c}'$ and $\mathbf{c}$ the tensors of the second-order elastic constants in $\mathscr{V}'$ and $\mathscr{V}'' = \mathscr{V} \setminus \mathscr{V}'$, respectively. Let $\mathscr{s} = [\mathbf{u}, \mathbf{E}, \mathbf{T}]$ and $\mathscr{s}' = [\mathbf{u}', \mathbf{E}', \mathbf{T}']$ be the elastic states produced in $\mathscr{V}'$ before and after changing the elastic constants in $\mathscr{V}'$ by an external force system $[\mathbf{t}, \mathbf{f}, \mathbf{P}]$, about which we assume only that it is applied outside $\mathscr{V}'$.

The potential energy of interaction (shortly: the interaction energy) between the inhomogeneity $N$ in $\mathscr{V}'$ and the elastic state $\mathscr{s}$ is by definition the change in the potential energy produced by changing the elastic constants in $\mathscr{V}'$. Hence, we may write

$$\Phi_{\mathrm{int}}\{N, \mathscr{s}\} = \Phi\{\mathscr{s}'\} - \Phi\{\mathscr{s}\} = \Delta\Phi' + \Delta\Phi'', \tag{21.31}$$

where $\Delta\Phi'$ and $\Delta\Phi''$ are the changes in the potential energies of the parts $\mathscr{V}'$, respectively $\mathscr{V}''$, of the body.

By making use of $(12.4)_3$, we first deduce that

$$\Delta\Phi' = -\frac{1}{2}\int_{\mathscr{S}'} (\mathbf{t}' \cdot \mathbf{u}' - \mathbf{t} \cdot \mathbf{u})\, ds,$$

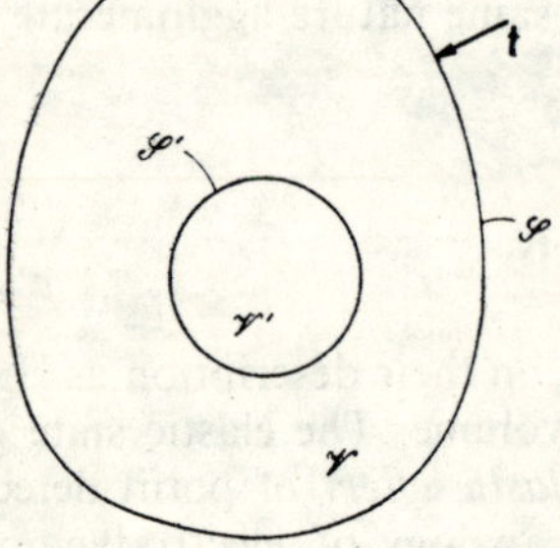


Fig. 21.5. Elastic body containing a small spherical inclusion with different elastic constants.

where $\mathbf{t}'$ and $\mathbf{t}$ are the stress vectors acting on the boundary $\mathscr{S}'$ of $\mathscr{V}'$ before and after changing the elastic constants in $\mathscr{V}'$. On the other hand, the change in the potential energy of $\mathscr{V}''$ results from the change of the surface tractions acting on the subboundary $\mathscr{S}'$ of $\mathscr{V}''$ from $-\mathbf{t}$ to $-\mathbf{t}'$ and in the presence of the external forces acting on $\mathscr{V}''$. It is therefore additively composed of the potential energy of the elastic state produced in $\mathscr{V}''$ by the surface forces $\mathbf{t} - \mathbf{t}'$ acting on $\mathscr{S}$ and by the interaction energy between this state and the state $\mathscr{s}$. Consequently, by $(12.4)_3$ and $(12.9)_2$, we have

$$\Delta\Phi'' = \frac{1}{2}\int_{\mathscr{S}'} (\mathbf{t}' - \mathbf{t}) \cdot (\mathbf{u}' - \mathbf{u})\, ds + \int_{\mathscr{S}} (\mathbf{t}' - \mathbf{t}) \cdot \mathbf{u}\, ds.$$

Substituting the last two relations into (21.31) we obtain

$$\Phi_{\text{int}}\{N, \mathscr{s}\} = \frac{1}{2}\int_{\mathscr{S}'} (\mathbf{t}' \cdot \mathbf{u} - \mathbf{t} \cdot \mathbf{u}')\, ds. \tag{21.32}$$

Finally, by applying (1.53) and taking into account that $\mathscr{s}$ and $\mathscr{s}'$ are regular elastic states in $\mathscr{V}'$ which correspond, respectively, to the surface tractions $\mathbf{t}$ and $\mathbf{t}'$ acting on $\mathscr{S}'$, we successively deduce

$$\Phi_{\text{int}}\{N, \mathscr{s}\} = \frac{1}{2}\int_{\mathscr{S}'} (T'_{kl}\, u_k - T_{kl} u'_k)\, n_l\, ds = \frac{1}{2}\int_{\mathscr{V}'} (T'_{kl} E_{kl} - T_{kl} E'_{kl})\, dv,$$

or

$$\Phi_{\text{int}}\{N, \mathscr{s}\} = \frac{1}{2}\int_{\mathscr{V}'} (c'_{klmn} - c_{klmn}) E'_{kl} E_{mn}\, dv. \tag{21.33}$$

In the *isotropic* case, by substituting (5.26) into (21.33), it results

$$\Phi_{\text{int}}\{N, \mathscr{s}\} = \frac{1}{2}\int_{\mathscr{V}'} [(\lambda' - \lambda)E'_{mm}E_{pp} + 2(\mu' - \mu)E'_{kl}E_{kl}]\mathrm{d}v. \tag{21.34}$$

Next, introducing the deviators $\mathring{\mathbf{E}}, \mathring{\mathbf{E}}'$ of the strain tensors $\mathbf{E}, \mathbf{E}'$, defined by

$$\mathring{E}_{kl} = E_{kl} - \tfrac{1}{3}E_{mm}\delta_{kl}, \qquad \mathring{E}'_{kl} = E'_{kl} - \tfrac{1}{3}E'_{mm}\delta_{kl}, \tag{21.35}$$

equation (21.34) may be rewritten as

$$\Phi_{\text{int}}\{N, \mathscr{s}\} = \frac{1}{2}\int_{\mathscr{V}'} [(K' - K)E'_{mm}E_{pp} + 2(\mu' - \mu)\mathring{E}'_{kl}\,\mathring{E}_{kl}]\,\mathrm{d}v =$$

$$= \frac{1}{2}\int_{\mathscr{V}'} [(K' - K)(\operatorname{tr}\mathbf{E}')(\operatorname{tr}\mathbf{E}) + 2(\mu' - \mu)\,\mathring{\mathbf{E}}'\cdot\mathring{\mathbf{E}}]\,\mathrm{d}v, \tag{21.36}$$

where $K = \lambda + 2\mu/3$, $K' = \lambda' + 2\mu'/3$ are the bulk moduli inside and outside the inclusion, respectively.

In what follows we assume that the strain $\mathbf{E}$ is uniform, i.e. it does not depend on $\mathbf{x}$. Then, it may be shown (Eshelby [110]) that the strain field $\mathbf{E}'$ inside the spherical inclusion is also uniform, being related to $\mathbf{E}$ by

$$\operatorname{tr}\mathbf{E}' = (A\alpha + 1)\operatorname{tr}\mathbf{E}, \qquad \mathring{\mathbf{E}}' = (B\beta + 1)\,\mathring{\mathbf{E}}, \tag{21.37}$$

where

$$\left.\begin{aligned} A &= \frac{K' - K}{(K - K')\alpha - K}, & B &= \frac{\mu' - \mu}{(\mu - \mu')\beta - \mu}, \\ \alpha &= \frac{1}{3}\,\frac{1 + \nu}{1 - \nu}, & \beta &= \frac{2}{15}\,\frac{4 - 5\nu}{1 - \nu}. \end{aligned}\right\} \tag{21.38}$$

Introducing (21.37) into (21.36) and taking into account (6.5—7), we obtain

$$\Phi_{\text{int}}\{N, \mathscr{s}\} = -\frac{\Omega}{2}[KA(\operatorname{tr}\mathbf{E})^2 + 2\mu B\,\mathring{\mathbf{E}}\cdot\mathring{\mathbf{E}}] = -\frac{\Omega}{2}\left[\frac{A}{9K}(\operatorname{tr}\mathbf{T})^2 + \frac{B}{2\mu}\,\mathring{\mathbf{T}}\cdot\mathring{\mathbf{T}}\right], \tag{21.39}$$

where $\mathring{\mathbf{T}}$ is the deviator of the stress tensor $\mathbf{T}$ outside the inclusion.

Usually, the quantities $K' - K$ and $\mu' - \mu$ can be only roughly evaluated. It is therefore preferable to consider the parameters $A$ and $B$ as independent macroscopic constants and to determine them from data on the apparent elastic constants of materials containing a large number of point defects [111]. Indeed, under constant surface tractions producing a uniform strain field $\mathbf{E}$, the strain energy density per unit volume $W$ changes, in view of $(12.4)_1$ and $(21.31)_1$, by $-n\Phi_{\text{int}}\{N, \mathscr{o}\}$, where $n$ is the number of point defects per unit volume and $\Phi_{\text{int}}\{N, \mathscr{o}\}$ is given by (21.39). On the other hand, in the absence of point defects, it follows from (15.24) and (15.25) that

$$W = \tfrac{1}{2}\,[K(\operatorname{tr}\mathbf{E})^2 + 2\mu\mathring{\mathbf{E}}\cdot\mathring{\mathbf{E}}]. \tag{21.40}$$

Consequently, the apparent elastic constants are

$$K_{\text{app}} = K(1 + cA), \qquad \mu_{\text{app}} = \mu(1 + cB), \tag{21.41}$$

where $c = n\Omega$ is the volume concentration of point defects. Clearly, equations (21.41) hold only for small values of $c$, since in deriving them we have neglected the interaction of point defects. However, they permit, at least in principle, the determination of the constants $A$ and $B$ by macroscopic experiments.

We have assumed so far that the strain field $\mathbf{E}$ is uniform and is produced by an external force system. Nevertheless, the final result (21.39) may also be applied to non-uniform strain fields produced by other defects, provided they have a negligible variation on distances comparable with the dimensions of the point defect. On this ground, we shall make use of $(21.39)_1$ for calculating the inhomogeneity interaction between two point defects $D_1$ and $D_2$ in an infinite isotropic elastic medium [1]. Considering the point defects as rigid inclusions and denoting by $\delta v_1$ and $\delta v_2$ the volume changes produced by them, we obtain from (21.9) and (21.8)

$$\operatorname{tr}\mathbf{E}_1 = \operatorname{tr}\mathbf{E}_2 = 0, \qquad \mathring{\mathbf{E}}_1\cdot\mathring{\mathbf{E}}_1 = \frac{3(\delta v_1)^2}{8\pi^2 r^6}, \qquad \mathring{\mathbf{E}}_2\cdot\mathring{\mathbf{E}}_2 = \frac{3(\delta v_2)^2}{8\pi^2 r^6}.$$

Substituting this result into $(21.39)_1$ and taking into account that the integration leading from (21.36) to (21.39) must be performed over the volumes of both point defects, we find

$$\Phi_{\text{int}}(D_1, D_2) = -\frac{3\mu\Omega}{8\pi^2 r^6}\,[B_1(\delta v_1)^2 + B_2(\delta v_2)^2],$$

where now $r = \|\mathbf{x}_2 - \mathbf{x}_1\|$ is the separation distance between point defects. Finally, the force exerted by the defect $D_1$ situated at $\mathbf{x}_1$ on the defect $D_2$ at $\mathbf{x}_2$, originating in the inhomogeneity effect, is

$$\begin{aligned}\mathbf{F} &= -\operatorname{grad}_{\mathbf{x}_2}\Phi_{\text{int}}(D_1, D_2) = \\ &= -\frac{9\mu\Omega}{4\pi^2 r^7}\,[B_1(\delta v_1)^2 + B_2(\delta v_2)^2](\mathbf{x}_2 - \mathbf{x}_1),\end{aligned} \tag{21.42}$$

where again $r = \|\mathbf{x}_2 - \mathbf{x}_1\|$.

[1] We have seen above that the corresponding size effect interaction vanishes.

## 22. Description of point defects by force multipoles

A point defect exerts forces on the neighbouring atoms, which are different from those acting on these atoms in the perfect lattice. Let $\mathbf{P}^\nu$ be the *additional* force exerted by a point defect centred at $\mathbf{x}'$ on the atom situated at $\mathbf{x}' + \mathbf{l}^\nu$. According to (6.32), the force system $\mathbf{P}^\nu$ produces in an *infinite* elastic medium the displacement field

$$u_m(\mathbf{x}) = \sum_{\nu=1}^{N} G_{ms}(\mathbf{x} - \mathbf{x}' - \mathbf{l}^\nu)P_s^\nu, \tag{22.1}$$

where $\mathbf{G}$ is Green's tensor function of the elastic medium, while $N$ is the number of atoms on which extra forces are exerted. Theoretically, $N = \infty$, but, as $\mathbf{P}^\nu$ decays very rapidly when $\|\mathbf{l}^\nu\| \to \infty$, it is usually sufficient to take into account only the forces exerted on the first and second neighbours.

Developing $\mathbf{G}(\mathbf{x} - \mathbf{x}' - \mathbf{l}^\nu)$ in a Taylor series about $\mathbf{x} - \mathbf{x}'$ yields

$$G_{ms}(\mathbf{x} - \mathbf{x}' - \mathbf{l}^\nu) = \sum_{k=0}^{\infty} \frac{1}{k!} G_{ms,q_1' \ldots q_k'}(\mathbf{x} - \mathbf{x}')l_{q_1}^\nu \ldots l_{q_k}^\nu \tag{22.2}$$

$$= G_{ms}(\mathbf{x} - \mathbf{x}') + G_{ms,n'}(\mathbf{x} - \mathbf{x}')l_n^\nu + \frac{1}{2!} G_{ms,n'q'}(\mathbf{x} - \mathbf{x}')l_n^\nu l_q^\nu + \ldots,$$

where $(.)_{,m'} = \partial(.)/\partial x_m'$. Clearly, the expansion (22.2) converges only for sufficiently small values of $\|\mathbf{l}^\nu\|$, i.e. only if the application points of the forces $\mathbf{P}^\nu$ are sufficiently close to the point defect.

By substituting (22.2) into (22.1), it results

$$u_m(\mathbf{x}) = \sum_{k=0}^{\infty} \frac{1}{k!} G_{ms,q_1' \ldots q_k'}(\mathbf{x} - \mathbf{x}')P_{q_1 \ldots q_k}^{(k)} = \tag{22.3}$$

$$= G_{ms}(\mathbf{x} - \mathbf{x}')P_s^{(0)} + G_{ms,n'}(\mathbf{x} - \mathbf{x}')P_{ns}^{(1)} + G_{ms,n'q'}(\mathbf{x} - \mathbf{x}')P_{nqs}^{(2)} + \ldots,$$

where

$$\mathbf{P}^{(0)} = \sum_{\nu=1}^{N} \mathbf{P}^\nu, \qquad P_s^{(0)} = \sum_{\nu=1}^{N} P_s^\nu \tag{22.4}$$

is the *resultant force*, and

$$\mathbf{P}^{(k)} = \sum_{\nu=1}^{N} \underbrace{\mathbf{l}^\nu \mathbf{l}^\nu \ldots \mathbf{l}^\nu}_{k} \mathbf{P}^\nu, \qquad P_{q_1 \ldots q_k s}^{(k)} = \sum_{\nu=1}^{N} l_{q_1}^\nu l_{q_2}^\nu \ldots l_{q_k}^\nu P_s^\nu \tag{22.5}$$

is the *multipolar moment of k'th order*, $k = 1, 2, \ldots$, of the system of additional forces $\mathbf{P}^\nu$ exerted by the point defect on its surroundings. In particular, we call *dipole moment*, *quadrupole moment*, and respectively *octopole moment*, the tensors

$$\mathbf{P}^{(1)} = \sum_{\nu=1}^{N} \mathbf{l}^\nu \mathbf{P}^\nu, \qquad P^{(1)}_{ns} = \sum_{\nu=1}^{N} l^\nu_n P^\nu_s, \tag{22.6}$$

$$\mathbf{P}^{(2)} = \sum_{\nu=1}^{N} \mathbf{l}^\nu \mathbf{l}^\nu \mathbf{P}^\nu, \qquad P^{(2)}_{nqs} = \sum_{\nu=1}^{N} l^\nu_n l^\nu_q P^\nu_s, \tag{22.7}$$

$$\mathbf{P}^{(3)} = \sum_{\nu=1}^{N} \mathbf{l}^\nu \mathbf{l}^\nu \mathbf{l}^\nu \mathbf{P}^\nu, \qquad P^{(3)}_{nqrs} = \sum_{\nu=1}^{N} l^\nu_n l^\nu_q l^\nu_r P^\nu_s. \tag{22.8}$$

Formula (22.3) may be also given a somewhat different interpretation. Namely, we have seen in Sect. 6.5 that Green's tensor function $\mathbf{G}$ satisfies the equation

$$c_{ijmn}(G_{ms,jn}(\mathbf{x} - \mathbf{x}') + \delta_{is}\delta(\mathbf{x} - \mathbf{x}') = 0 \tag{22.9}$$

in the sense of the theory of distributions. Applying on this equation the differential operator

$$\frac{1}{k!} P^{(k)}_{q_1 \ldots q_k s} \frac{\partial^k}{\partial x'_{q_1} \ldots \partial x'_{q_k}}$$

and summing with respect to $s$, we obtain

$$c_{ijmn} \left\{ \frac{1}{k!} P^{(k)}_{q_1 \ldots q_k s} G_{ms, q'_1 \ldots q'_k}(\mathbf{x} - \mathbf{x}') \right\}_{,jn} + \frac{1}{k!} P^{(k)}_{q_1 \ldots q_k s} \delta_{, q'_1 \ldots q'_k}(\mathbf{x} - \mathbf{x}') = 0.$$

Finally, by comparing these relations with the equilibrium equations

$$c_{ijmn} u_{m,jn}(\mathbf{x}) + f_i(\mathbf{x}) = 0, \quad i = 1, 2, 3,$$

we conclude that the body force

$$f_s(\mathbf{x}) = \frac{1}{k!} P^{(k)}_{q_1 \ldots q_k s} \delta_{, q'_1 \ldots q'_k}(\mathbf{x} - \mathbf{x}') \tag{22.10}$$

generates the displacement field

$$u_m(\mathbf{x}) = \frac{1}{k!} G_{ms, q'_1 \ldots q'_k}(\mathbf{x} - \mathbf{x}') P_{q_1 \ldots q_k s}. \tag{22.11}$$

The body force (22.10) is called [1] a *multipolar force of k'th order*, and of *strength*

$$\tilde{P}^{(k)}_{q_1 \ldots q_k s} = \frac{1}{k!} P^{(k)}_{q_1 \ldots q_k s}. \tag{22.12}$$

By making use of this terminology and comparing (22.3) with (22.11), we may say that the action of a point defect on the elastic medium is equivalent to that of a body force field which consists of force dipoles, quadrupoles, octopoles, etc. applied at the centre of the defect, namely

$$f_s(\mathbf{x}) = \sum_{k=0}^{\infty} \tilde{P}^{(k)}_{q_1 \ldots q_k s} \; \delta_{,q_1' \ldots q_k' s}(\mathbf{x} - \mathbf{x}'), \tag{22.13}$$

the strengths of the multipolar forces being completely determined through (22.12) by the multipolar moments associated with the point defect [2].

We shall constantly assume in what follows that the resultant force and couple exerted by a point defect on its surroundings are zero. Then, the equilibrium condition implies

$$\sum_{\nu=1}^{N} \mathbf{P}^{\nu} = \mathbf{0}, \qquad \sum_{\nu=1}^{N} \mathbf{l}^{\nu} \times \mathbf{P}^{\nu} = \mathbf{0}. \tag{22.14}$$

The last relation may be rewritten as

$$\sum_{\nu=1}^{N} \epsilon_{kns}\, l_n^{\nu} P_s^{\nu} = \epsilon_{kns} P^{(1)}_{ns} = 0,$$

wherefrom it follows that the dipole moment $\mathbf{P}^{(1)}$ must be a symmetric tensor. Hence, conditions (22.14) are equivalent with

$$\mathbf{P}^{(0)} = \mathbf{0}, \qquad \mathbf{P}^{(1)} = (\mathbf{P}^{(1)})^T. \tag{22.15}$$

By introducing $(22.15)_1$ into (22.3) and taking into account that

$$G_{ms,n'}(\mathbf{x} - \mathbf{x}') = -G_{ms,n}(\mathbf{x} - \mathbf{x}'),$$

we obtain

$$u_m(\mathbf{x}) = \sum_{k=1}^{\infty} \frac{(-1)^k}{k!} G_{ms,q_1 \ldots q_k}(\mathbf{x} - \mathbf{x}') P^{(k)}_{q_1 \ldots q_k s} = \tag{22.16}$$

$$= -G_{ms,n}(\mathbf{x} - \mathbf{x}') P^{(1)}_{ns} + \frac{1}{2!} G_{ms,nq}(\mathbf{x} - \mathbf{x}') P^{(2)}_{nqs} - \frac{1}{3!} G_{ms,nqr}(\mathbf{x} - \mathbf{x}') P^{(3)}_{nqrs} + \ldots$$

---

[1] This designation generalizes that of "double force" introduced by Love [222], Sect. 132.

[2] Kröner [190, 192, 193] was the first to give a systematic description of point defects with the aid of multipolar moments or as multipolar forces of second and higher orders (cf. also Dehlinger and Kröner [85]).

Equation (22.16) shows that the elastic state produced by a point defect in an infinite elastic medium is completely determined by the multipolar moments $\mathbf{P}^{(k)}$, $k=1, 2, \ldots$, provided Green's tensor function of the medium is known. For an isotropic material this function is (cf. Sect. 6.4)

$$G_{ms}(\mathbf{x}-\mathbf{x}')=\frac{1}{16\pi\mu(1-\nu)}\left[\delta_{ms}(3-4\nu)\frac{1}{r}+\frac{(x_m-x_m')(x_s-x_s')}{r^3}\right], \quad (22.17)$$

where $r=\|\mathbf{x}-\mathbf{x}'\|$. By substituting (22.17) into (22.16), we see that $\mathbf{u}(\mathbf{x})$ is of the order $O(r^{-2})$ as $r\to\infty$, in agreement with the results obtained in modelling point defects by rigid spherical inclusions in an infinite isotropic medium (Sect. 21).

As already mentioned, the elastic field of a point defect is characterized by its multipolar moments. There exist so far three main procedures for evaluating these quantities.

The first method consists in calculating the interaction energy of point defects with other elastic fields and comparing the results obtained with experimental data. Unfortunately, due to the rather limited range of available experimental results, the applicability of this otherwise straightforward method is restricted to point defects and crystal lattices with high symmetry.

A second method is based on the direct calculation of the forces $\mathbf{P}^{\nu}$, exerted by the point defect on its surroundings, by means of the *lattice theory*. Namely, it is assumed that the interatomic forces depend linearly on the separating distances between atoms (harmonic lattice theory), and only the interaction between first and second neighbours is taken into account. The proportionality constants between the interatomic forces and the axial or tangential relative displacements of the atoms are evaluated by using second-order elastic constants and experimental data on phonon or neutron scattering. Then, the radial forces $\mathbf{P}^{\nu}$ exerted by the point defect on its first and second neighbours are calculated by using a relaxation technique to determine the deformed configuration of the lattice around the point defect [1]. This method, which was elaborated by Kanzaki [177], has been refined by Bullough and Hardy [48], who employed it to calculate the displacements produced by a vacancy and the vacancy-vacancy interaction in copper and aluminium. The method has been subsequently applied to study interstitials in copper (Flocken and Hardy [117]) and vacancies in alkali halides (Flocken and Hardy [118, 119]). As shown by Tewary [348], the calculation of the atomic displacements around point defects may be considerably simplified by using the so-called *static Green's tensor function of the lattice* $\mathbf{G}(\mathbf{x}_\alpha-\mathbf{x}_\beta')$ (cf. Sect. 16). The main advantage of using this function is that it may be calculated directly from data on neutron scattering (Tewary and Bullough [349]), without previously determining the interatomic forces.

The discrete method has anyway the advantage that it allows the simultaneous determination of the forces exerted by a point defect on its neighbours *and* of the application points of these forces on the imperfect lattice. The main disadvantage of this method is the use of the *harmonic* lattice theory, which usually proves to be inadequate for the large atomic displacements in the close neighbourhood of the point defect.

---

[1] For a detailed description of this method see Bullough [50].

A third way for the evaluation of the multipolar moments of the defect is the use of a *semidiscrete method* (see, e.g. Seeger and Mann [290], Johnson and Brown [172], Schottky, Seeger, and Schmid [284], Doyama and Cotterill [90], and Johnson [173], [174], where further references on previous literature devoted to this topics may be also found). Like in the case of dislocations (Sect. 16), the crystal is divided into two regions: the region $A$, close to the point defect, in which one applies the lattice theory, and the remaining part of the crystal, say $E$, which is considered as a linearly elastic continuum, the action of the point defect on this region being considered equivalent with a spherical dilatation centre. Then, the variation of the potential energy produced by the introduction of the point defect is minimized as a function of the positions of the atoms in region $A$ and of the strengths of the force dipoles of the dilatation centre. The semidiscrete method has the advantage of allowing the consideration of non-linearities in the lattice theory, by using a suitable interatomic potential for region $A$. However, the correct application of this method requires that the interatomic potential be fitted to the experimental values of second and third-order elastic constants and to data on phonon scattering.

Another version of the semidiscrete method, which allows a smoother passage across the boundary between regions A and E, has been proposed by Seeger, Mann, and v. Jan [291] and used for determining the dipole moments of interstitials in copper. In this approach, the crystal is divided into three regions: an atomic region $A\ (r \leqslant r_0)$, an intermediate region $I\ (r_0 < r < r_1)$, and an elastic region $E(r \geqslant r_1)$ (Fig. 22.1). In region $A$ the atoms are permitted to have individual displacements, and the potential energy is calculated by means of the interatomic potential; in the intermediate region, the atomic displacements are derived from the elastic displacement field but the energy is still calculated with the aid of the interatomic potential; finally, in region $E$, both the displacement field and the energy are calculated by the elasticity theory. Apparently, this version of the method resembles the semidiscrete method of the overlapping regions used for simulating the dislocation

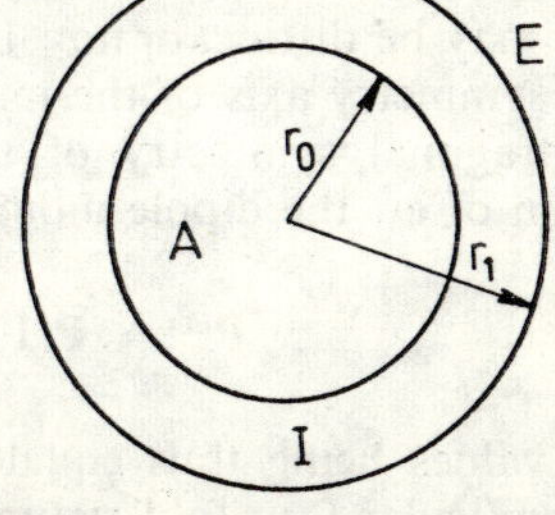


Fig. 22.1. Concentric arrangement of regions for the semidiscrete method of Seeger, Mann, and v. Jan.

core and described in Sect. 16.3. However, the boundary conditions on the surface $r = r_0$ are ignored, and the potential energy is minimized as in Sinclair's method, by taking the dipole moment of the point defect as unique adjustable parameter in the elastic solution.

The multipolar moments (22.5) of a point defect may be reduced to a simpler form by taking into account the *symmetry of the imperfect lattice*, i.e. the symmetry of the lattice containing the point defect. It should be noticed that this *microscopic* or *defect symmetry* is lower than or equal to the macroscopic elastic symmetry of the crystal [1]. Since the symmetry transformations of the imperfect lattice must leave unchanged the place of the point defect, they must form a *point group*. We have seen in Sect. 5 that there exist 32 crystallographic classes, each of which corresponding to a point group; their symbols after Schoenflies [283[, which will be repeatedly used in the following, are shown in Table 5.1. The significance of these symbols may be found in any standard course on crystallography and in many books on general or solid state physics (see, e.g. Landau and Lifshits [206], Sect. 93)[2].

Let $\mathcal{G}_p$ be the point group of an imperfect lattice containing a single point defect. We take the centre of the defect as origin $\omega$ of a Cartesian frame $\{\omega, \mathbf{e}_k\}$. Each transformation belonging to $\mathcal{G}_p$ is a displacement of the crystal that leaves unchanged the point $\omega$ and superposes the imperfect lattice on itself. It is more convenient, however, to consider such a lattice transformation as a change of the orthonormal basis $\{\mathbf{e}_k\}$ into another orthonormal basis $\{\mathbf{e}'_r\}$ which has the same orientation with respect to the preferred crystallographic directions of the imperfect lattice. Then, by denoting $q_{kr} = \mathbf{e}_k \cdot \mathbf{e}'_r$ as in Sect. 1.1, each matrix $[q_{kr}]$ will define a transformation that leaves invariant all multipolar moments of the point defect, for $\mathbf{l}^\nu$ and $\mathbf{P}^\nu$ depend only on the relative positions of the atoms of the imperfect lattice. Hence, by (1.35), we must have

$$P^{(1)}_{rs} = q_{kr} q_{ms} P^{(1)}_{km}, \qquad P^{(2)}_{rst} = q_{kr} q_{ms} q_{nt} P^{(2)}_{kmn}, \ \ldots \tag{22.18}$$

In particular, from (22.18) it follows that whenever the point defect is a *centre of symmetry* of the imperfect lattice, i.e. when $\mathcal{G}_p$ contains the inversion $q_{kr} = -\delta_{kr}$, all multipolar moments of even order $\mathbf{P}^{(2)}, \mathbf{P}^{(4)}, \ldots$ are zero.

The moment dipole $\mathbf{P}^{(1)}$ is of particular interest, since it generally determines the principal singularity of the displacement field produced by the point defect, i.e. the terms which have the slowest decay to 0 as $r \to \infty$. According to the spectral theorem (Sect. 1.1),the component matrix of the symmetric tensor $\mathbf{P}^{(1)}$ may be always reduced to a diagonal form, by choosing the Cartesian axes along the principal directions of the tensor. The three components thus obtained, which are the principal values of $\mathbf{P}^{(1)}$, may be distinct or not. It may be shown, however, that if a three-fold or a four-fold symmetry axis of the imperfect lattice passes through $\omega$ (hexagonal, respectively tetragonal symmetry of the imperfect lattice), then, by taking this axis as direction of $\mathbf{e}_3$, the dipole moment assumes the form

$$\mathbf{P}^{(1)} = \mathbf{P}(\mathbf{1} + \eta \mathbf{e}_3\, \mathbf{e}_3), \tag{22.19}$$

two principal values being thus equal (cf. also (1.37 a)). Furthermore, when two mutually perpendicular four-fold symmetry axes pass through $\omega$, then, by choosing them as co-ordinate axes, the dipole moment reduces to the spherical form

$$\mathbf{P}^{(1)} = P\mathbf{1}, \tag{22.20}$$

---

[1] Defects whose symmetry system is lower than that of the crystal allow the production of anelastic relaxation (cf. Nowik and Berry [455], Sect. 8.2.)

[2] For a detailed investigation of the conditions imposed by the crystal symmetry on tensors of various orders describing material properties see Jagodzinski [171], Chap. II.

i.e. all principal values are equal (cf. also (1.37b)). The same form of $\mathbf{P}^{(1)}$ may be obtained when through $\omega$ there passes a three-fold axis which trisects a trihedron having two-fold symmetry axes as edges. Inspection of Table 5.1 shows that both these situations belong to the case of the cubic symmetry of the imperfect lattice.

The above considerations may be extended to composite defects, made up of more than one foreign, extra, or missing atom. This generalization is particularly straightforward for pairs of like or unlike atoms and for divacancies. Thus, for the like pair the appropriate choice of $\omega$ is obviously the midpoint of the pair, while for the unlike pair any point on the pair axis may be chosen to survey the symmetry of the surrounding lattice.

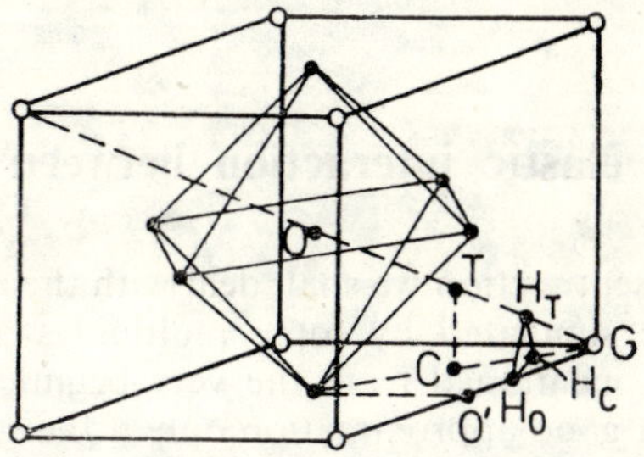


Fig. 22.2. Stable equilibrium positions of intrinsic interstitials in an f.c.c. metal.

In order to illustrate the possible symmetries of point defects [1] we have represented in Fig. 22.2 the stable equilibrium positions of intrinsic interstitials in an f.c.c. metal, which are denoted after Seeger, Mann, and v. Jan [291] by, $O$, $C$, $T$, $H_O$, $H_C$, and $H_T$. By repeating the cubic cell in the three directions parallel to the edges of the cell, it can be easily seen that the f.c.c. lattice may be decomposed into a sequence of regular tetrahedra and octahedra of side $a/\sqrt{2}$, where $a$ is the side of the cube.

The point defect $O$ lies in the centre of the octahedron, i.e. of the cube; that is why it is said to be in an *octahedric position*. The point group of this defect, which coincides with that of a vacancy and of a substitutional atom, is the group $O_h$ of all symmetry transformations of the cube and belongs to the *cubic* system.

The point defect $C$, called a *crowdion*, is situated at the midpoint between two neighbouring atoms of the lattice in $\langle 110 \rangle$-direction. It is also a centre of symmetry of the lattice, through which pass three mutually perpendicular two-fold axes. The point group of the crowdion is $D_{2h}$ and belongs to the *rhombic* system.

The point defect $T$ lies in the centre of the tetrahedron; its position divides the cube diagonal in the ratio 1:3 and is called a *tetrahedric* position. Its point group is the group $T_d$ of all symmetry transformations of the tetrahedron and belongs to the *cubic* system.

If the interstitial $O'$, which occupies a crystallographic site equivalent with that of $O$, is pushed towards the atom situated at the cube corner $G$, this latter will also move in the same direction. Finally, an equilibrium situation may occur, in which the interstitial and the corner atom occupy positions that are symmetrical with respect to $G$, situated on the $\langle 100 \rangle$-direction, and denoted by $H_O$ in Fig.

[1] For a more detailed discussion of various types of point defects in metals we refer to Johnson [173, 174], Kronmüller [197, 198], and especially to Nowik and Berry [455], Sects. 7 and 8.

22.2. This typical configuration, composed of two neighbouring interstitials is called a *dumbbell*. Dumbbells $H_C$ and $H_T$, situated on directions $\langle 110\rangle$, and respectively $\langle 111\rangle$, are defined in a similar way. By inspecting the symmetry of the imperfect lattice with respect to the midpoints of the dummbbells, it may be seen that $H_O$ has the point group $D_{4h}$ of the *tetragonal* system, $H_C$ possesses the same symmetry as $C$, while $H_T$ has the point group $C_{3i}$ of the *hexagonal* system.

According to the above discussion, the moment dipole $\mathbf{P}^{(1)}$ may be reduced to (22.20) for the interstitials $O$ and $T$, and to (22.19) for the dumbbells $H_O$ and $H_T$.

In the case of the crowdion and of the dumbbell $H_C$, the matrix of the components of $\mathbf{P}^{(1)}$ may be also reduced to a diagonal form, but with different diagonal elements [1].

## 23. The elastic interaction between point defects

In the present section we shall deal with the elastic interaction between point defects which are simulated by force multipoles acting in an infinite elastic medium. It should be mentioned from the very beginning that this description of point defects provides a good approximation only if the separation distance between point defects is large enough. Otherwise, a semidiscrete or fully atomic model of the interacting defects must be adopted.

Most of the following considerations in this section are based on a paper by Siems [297].

### 23.1. Interaction energy of two point defects in an infinite elastic medium

The elastic interaction energy between a point defect located at $\mathbf{x}$ and an elastic displacement field $\mathbf{u}$ is given, according to (12.9), by the work done against the forces $\mathbf{P}^\nu$ exerted by the point defect on the neighbouring atoms, i.e.

$$\Phi_{\text{int}} = -\sum_{\nu=1}^{N} \mathbf{P}^\nu \cdot \mathbf{u}(\mathbf{x} + \mathbf{l}^\nu). \tag{23.1}$$

By expanding $\mathbf{u}(\mathbf{x} + \mathbf{l}^\nu)$ in a Taylor series around $\mathbf{x}$, we obtain

$$u_m(\mathbf{x} + \mathbf{l}^\nu) = \sum_{n=0}^{\infty} \frac{1}{n!} u_{m,j_1\ldots j_n}(\mathbf{x}) \mathbf{l}^\nu_{j_1\ldots j_n} =$$

$$= u_m(\mathbf{x}) + u_{m,i}(\mathbf{x})\, l_i^\nu + \frac{1}{2!} u_{m,ij}(\mathbf{x})\, l_i^\nu l_j^\nu + \frac{1}{3!} u_{m,ijk} l_i^\nu l_j^\nu l_k^\nu + \ldots \tag{23.2}$$

[1] By using the semidiscrete method described above, Seeger Mann, and v. Jan [291] succeeded to prove that the dumbbell $H_O$ represents for Cu the position of absolute minimum self-energy with respect to all other interstitial configurations.

Substituting this expansion into (23.1) and considering (22.4), (22.5), and (22.15), we find

$$\Phi_{\text{int}} = -\sum_{n=1}^{\infty} \frac{1}{n!} u_{m,j_1\ldots j_n}(\mathbf{x})\, P^{(n)}_{j_1\ldots j_n m} = \tag{23.3}$$

$$= -\left\{ P^{(1)}_{im} u_{m,i}(\mathbf{x}) + \frac{1}{2!} P^{(2)}_{ijm} u_{m,ij}(\mathbf{x}) + \frac{1}{3!} P^{(3)}_{ijkm} u_{m,ijk}(\mathbf{x}) + \ldots \right\}.$$

In a *homogeneous* strain field we have $u_{m,i}(\mathbf{x}) = \text{const.}$ and (23.3) reduces to

$$\Phi_{\text{int}} = -P^{(1)}_{im} u_{m,i} = -P^{(1)}_{im} E_{im}, \tag{23.4}$$

i.e. only the dipole moments contribute to the interaction energy.

Returning to the general case, we recall that the force exerted on the point defect by the elastic state which generates the displacement field $\mathbf{u}$ is

$$\mathbf{F} = -\operatorname{grad}_{\mathbf{x}} \Phi_{\text{int}}. \tag{23.5}$$

Hence, by taking into account (23.3), we have

$$F_s = -\frac{\partial \Phi_{\text{int}}}{\partial x_s} = \sum_{n=1}^{\infty} \frac{1}{n!} u_{m,j_1\ldots j_n s}(\mathbf{x})\, P^{(n)}_{j_1\ldots j_n m} =$$

$$= P^{(1)}_{im} u_{m,is}(\mathbf{x}) + \frac{1}{2!} P^{(2)}_{ijm} u_{m,ijs}(\mathbf{x}) + \ldots \tag{23.6}$$

We can now easily derive the elastic interaction energy of two point defects situated at points $\mathbf{x}$ and $\mathbf{x}'$ in an infinite elastic medium and having the multipolar moments $\mathbf{P}^{(1)}, \mathbf{P}^{(2)}, \ldots$, and respectively $\tilde{\mathbf{P}}^{(1)}, \tilde{\mathbf{P}}^{(2)}, \ldots$, by substituting the expression (22.16) of the displacement field produced by one of the defects into (23.3). The result reads

$$\Phi_{\text{int}} = -\sum_{n=1}^{\infty} \frac{1}{n!} P^{(n)}_{j_1\ldots j_n m} \sum_{k=1}^{\infty} \frac{(-1)^k}{k!} \tilde{P}^{(k)}_{q_1\ldots q_k s}\, G_{ms,q_1\ldots q_k j_1\ldots j_n}(\mathbf{x}-\mathbf{x}'). \tag{23.7}$$

Since $\mathbf{G}(\mathbf{x}-\mathbf{x}') = O(r^{-1})$ as $r = \|\mathbf{x}-\mathbf{x}'\| \to \infty$, we see that the first three terms of the expansion (23.7) decrease as $r^{-3}$, $r^{-4}$, and $r^{-5}$, respectively, for sufficiently large values of the separation distance $r$ between the point defects.

## 23.2. Point defects with cubic symmetry in an isotropic medium

We will apply now (23.7) for calculating the interaction energy between two point defects with symmetry group $O_h$. Since any such defect is a centre of symmetry of the imperfect lattice, all multipolar moments of even order, and in particular $\mathbf{P}^{(2)}$, vanish.

Owing to the cubic symmetry, only the components of multipolar moments whose indices occur in pairs of equal numbers are non-zero. Moreover, the three co-ordinate axes chosen parallel to the cube edges are crystallographically equivalent, so that

$$P_{11}^{(1)} = P_{22}^{(1)} = P_{33}^{(1)} = P,$$

$$P_{1111}^{(3)} = P_{2222}^{(3)} = P_{3333}^{(3)}, \qquad P_{1122}^{(3)} = P_{1133}^{(3)} = P_{2233}^{(3)}.$$

Therefore, the components of the dipole and octopole moments may be written as

$$P_{ij}^{(2)} = P\delta_{ij}, \quad P_{ijkm}^{(3)} = (\delta_{ij}\delta_{km} + \delta_{ik}\delta_{jm} + \delta_{im}\delta_{jk} - 3\delta_{ijkm})\, P_{1122}^{(3)} + \delta_{ijkm}P_{1111}^{(3)},$$

where the symbol $\delta_{ijkm}$ equals 1 for $i = j = k = m$ and vanishes otherwise. Substituting these expressions into (23.7) and retaining only the first three terms of the expansion, we obtain

$$\Phi_{\text{int}} = P\tilde{P}G_{ms,ms}(\mathbf{x} - \mathbf{x}') + \tfrac{1}{2}(\tilde{P}P_{1122}^{(3)} + P\tilde{P}_{1122}^{(3)})\, \Delta G_{ms,ms}(\mathbf{x} - \mathbf{x}') +$$
$$+ \tfrac{1}{6}\{\tilde{P}(P_{1111}^{(3)} - 3P_{1122}^{(3)}) + P(\tilde{P}_{1111}^{(3)} - 3P_{1122}^{(3)})\}\, G_{ms,m^3s}(\mathbf{x} - \mathbf{x}'). \tag{23.8}$$

Assuming now that the elastic medium is *isotropic*, we deduce from (22.17) that

$$\frac{\partial G_{ms}(\mathbf{x} - \mathbf{x}')}{\partial x_s} = -\frac{1 - 2\nu}{8\pi\mu(1 - \nu)}\, \frac{x_m - x_m'}{r^3}, \qquad \frac{\partial^2 G_{ms}(\mathbf{x} - \mathbf{x}')}{\partial x_m \partial x_s} = 0. \tag{23.9}$$

Consequently, the dipole-dipole interaction, i.e. the first term in the right-hand side of (23.8), vanishes, in agreement with the result obtained in Sect. 21.1 by simulating point defects as rigid spherical inclusions. Next, from $(23.9)_1$ it follows that

$$\frac{\partial^4 G_{ms}(\mathbf{x} - \mathbf{x}')}{\partial x_m^3 \partial x_s} = \frac{21(1 - 2\nu)}{8\pi\mu(1 - \nu)}\, \frac{1}{r^5} \left\{- 3 + 5\, \frac{(x_1' - x_1)^4 + (x_2' - x_2)^4 + (x_3' - x_3)^4}{r^4}\right\}. \tag{23.10}$$

To simplify the writing, we choose the centre of one point defect as origin of the system of co-ordinates and denote the position vector of the other defect with respect to the first one by $\mathbf{x}$. Then, by putting $\mathbf{x} = \mathbf{0}$, $\mathbf{x}' = \mathbf{x}$ and substituting (23.9) and (23.10) into (23.8), we obtain the relation

$$\Phi_{\text{int}} = \frac{7(1 - 2\nu)}{16\pi\mu(1 - \nu)} \left\{\tilde{P}(P_{1111}^{(3)} - 3P_{1122}^{(3)}) + P(\tilde{P}_{1111}^{(3)} - 3\tilde{P}_{1122}^{(3)})\right\} \times$$
$$\times \frac{1}{r^5}\left(3 - 5\, \frac{x_1^4 + x_2^4 + x_3^4}{r^4}\right), \tag{23.11}$$

which shows that the principal singularity of the elastic interaction energy between two point defects with cubic symmetry in an isotropic medium is determined by the dipole-octopole interaction.

Because of the cubic symmetry, all forces $\mathbf{P}^\nu$ have radial direction. Taking into account only the forces exerted by the point defects on their first and second neighbours, we have $\mathbf{P}^\nu = K_1 \mathbf{l}^\nu$, $\nu = 1, \ldots, N_1$ for the $N_1$ nearest neighbours, and $\mathbf{P} = K_2 \mathbf{l}^\nu$, $\nu = 1, \ldots, N_2$ for the $N_2$ next nearest neighbours. Introducing these relations into (22.6) and (22.8), we obtain

$$\mathbf{P}^{(1)} = K_1 \sum_{\nu=1}^{N_1} \mathbf{l}^\nu \mathbf{l}^\nu + K_2 \sum_{\nu=1}^{N_2} \mathbf{l}^\nu \mathbf{l}^\nu, \tag{23.12}$$

$$\mathbf{P}^{(3)} = K_1 \sum_{\nu=1}^{N_1} \mathbf{l}^\nu \mathbf{l}^\nu \mathbf{l}^\nu \mathbf{l}^\nu + K_2 \sum_{\nu=1}^{N_2} \mathbf{l}^\nu \mathbf{l}^\nu \mathbf{l}^\nu \mathbf{l}^\nu. \tag{23.13}$$

For a *vacancy* or a *substitutional atom* in an f.c.c. crystal with lattice parameter $a = 2d$, we have $\mathbf{l}^\nu = d(\mathbf{e}_k \pm \mathbf{e}_l)$, $d(-\mathbf{e}_k \pm \mathbf{e}_l)$, $k, l = 1, 2, 3$, $k < l$, for the $N_1 = 12$ first neighbours and $\mathbf{l}^\nu = \pm 2d\, \mathbf{e}_k$, $k = 1, 2, 3$, for the $N_2 = 6$ second neighbours, where $\mathbf{e}_k$, $k = 1, 2, 3$ are the unit vectors of the co-ordinate axes directed parallel to the cube edges. It then follows from (23.12) and (23.13) that

$$P = P_{11}^{(1)} = 8d^2(K_1 + K_2), \quad P_{1111}^{(3)} = 8d^4(K_1 + 4K_2), \quad P_{1122}^{(3)} = 4d^4 K_1,$$

and (23.11) becomes

$$\Phi_{\text{int}} = -\frac{28(1-2\nu)}{\pi\mu(1-\nu)} d^6 K_1 \tilde{K}_1 \left(1 - 7\frac{g + \tilde{g}}{2} - 8g\tilde{g}\right) \frac{1}{r^5} \left(-3 + 5\frac{x_1^4 + x_2^4 + x_3^4}{r^4}\right), \tag{23.14}$$

where $g = K_2/K_1$, $\tilde{g} = \tilde{K}_2/\tilde{K}_1$.

The mean value $\bar{\Phi}_{\text{int}}$ of $\Phi_{\text{int}}$, calculated for all possible orientations of the pair of point defects, which gives the interaction energy between a defect situated at $O$ and a uniform distribution of defects on the surface of a sphere centred at $O$, is zero, as $\overline{(x_k/r)^4} = 1/5$, $k = 1, 2, 3$. The variation of the orientation factor

$$S(x_1, x_2, x_3) = -3 + 5(x_1^4 + x_2^4 + x_3^4)/r^4 \tag{23.15}$$

over the orientation triangle of an f.c.c. lattice is shown in Fig. 23.1, after Siems [297]. The stationary directions of $S$, given by conditions $S_{,k} = 0$, $x_k x_k = \text{const.}$, $k = 1, 2, 3$, are $\langle 100 \rangle$ with $S = 2$, $\langle 111 \rangle$ with $S = -4/3$ and $\langle 110 \rangle$ with $S = -1/2$. For $g \ll 1$, $\tilde{g} \ll 1$ the sign of $\Phi_{\text{int}}$ does not depend on $g$ and $\tilde{g}$. Then, the interaction energy has minima in $\langle 100 \rangle$-directions, maxima in $\langle 111 \rangle$-directions, and saddle points in $\langle 110 \rangle$-directions.

For an *octahedral interstitial* in an f.c.c. lattice, we have $\mathbf{l}^\nu = \pm d\, \mathbf{e}_k$, $k = 1, 2, 3$, for the $N_1 = 6$ first neighbours, and $\mathbf{l}^\nu = d(\pm\mathbf{e}_1 + \mathbf{e}_2 + \mathbf{e}_3)$, $d(\mathbf{e}_1 \pm \mathbf{e}_2 + \mathbf{e}_3)$, $d(\mathbf{e}_1 + \mathbf{e}_2 \pm \mathbf{e}_3)$ for the $N_2 = 8$ second neighbours (see Fig. 22.2). It then folows from (23.12) and (23.13) that

$$P = P^{(1)}_{11} = 2d^2(K_1 + 4K_2), \quad P^{(3)}_{1111} = 2d^4(K_1 + 4K_2), \quad P^{(3)}_{1122} = 8d^4K_2,$$

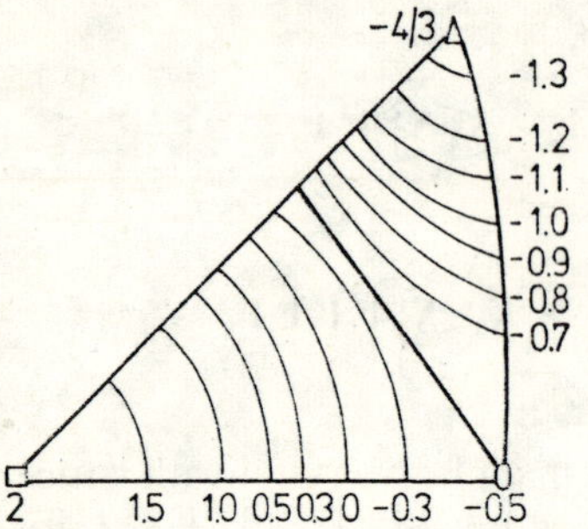


Fig. 23.1. Variation of the orientation factor $S$ over the orientation triangle of an f.c.c. lattice (after Siems [297]).

and (23.11) becomes

$$\Phi_{\text{int}} = \frac{7(1-2\nu)}{2\pi\mu(1-\nu)} d^6 K_1 \tilde{K}_1 \{1 - 2(g + \tilde{g}) - 32 g\tilde{g}\}\, r^{-5} S(x_1, x_2, x_3) \quad (23.16)$$

with $S(x_1, x_2, x_3)$ given by (23.15). Consequently, $\bar{\Phi}_{\text{int}}$ is again zero. However, since $\Phi_{\text{int}}$ has now opposite sign, the interaction energy has minima in $\langle 111\rangle$-directions, maxima in $\langle 100\rangle$-directions, and saddle points in $\langle 110\rangle$-directions.

The extrema of the interaction energy are of particular interest, for they determine the stable and metastable equilibrium positions of point defects, and hence the diffusion of vacancies and interstitials and the reorientation of defect pairs, phenomena playing a decisive role in mechanical and magnetic relaxation.

Siems' results have also a special theoretical value, as they show that simulating point defects by force multipoles acting in an elastic continuum allows to correctly determine the order $O(r^{-5})$ of the principal singularity in the interaction energy, in accordance with results obtained by the lattice theory (Hardy and Bullough [152]).

## 23.3. Point defects with tetragonal symmetry in an isotropic medium

As shown in Sect. 22, the stable dumbbell $H_O$ in f.c.c. lattices has tetragonal symmetry. The same type of symmetry is encountered at the C-atoms occupying one of the octahedric positions in the b.c.c. lattice of $\alpha$-Fe (Fig. 23.2).

In the case of tetragonal symmetry the dipole-dipole interaction is no longer zero. On the other hand, since this interaction is preponderant for sufficiently large separation distances between point defects, we shall neglect in what follows higher-order interactions.

By choosing the principal directions of the dipole moment $\mathbf{P}^{(1)}$ as co-ordinate axes [1], it assumes the reduced form (22.19). Clearly, the interaction energy depends on whether the principal value of the dipole moment which differs from the other two is the same diagonal element for both point defects or not. We are thus led to consider the following two cases.

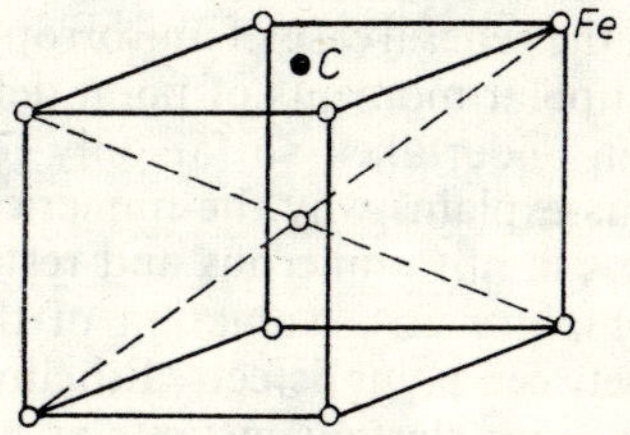


Fig. 23.2. Carbon interstitial in an octahedric position of the b.c.c. lattice of $\alpha$ iron.

*Case 1*:

$$P_{ij}^{(1)} = P(\delta_{ij} + \eta_1 \delta_{1i}\delta_{1j}), \quad \tilde{P}_{ij}^{(1)} = \tilde{P}(\delta_{ij} + \tilde{\eta}_1 \delta_{1i}\delta_{1j}). \tag{23.17}$$

Assuming again that one of the defects is situated in the origin and the other at $\mathbf{x}$, introducing (23.17) into (23.7), and taking into account $(23.9)_2$, we obtain the dipole-dipole interaction energy

$$\Phi_{\text{int}} = P\tilde{P}\{(\eta_1 + \tilde{\eta}_1) G_{1s,1s} + \eta_1 \tilde{\eta}_1 G_{11,11}\},$$

which, in view of (22.17), may be rewritten as

$$\Phi_{\text{int}} = -\frac{P\tilde{P}}{16\pi\mu(1-\nu)} \frac{1}{r^3} \left\{ 2(\eta_1 + \tilde{\eta}_1)(1-2\nu)\left(1 - \frac{3x_1^2}{r^2}\right) + \right.$$

$$\left. + \eta_1 \tilde{\eta}_1 \left[ 1 - 4\nu + 6(1+2\nu)\frac{x_1^2}{r^2} - 15\frac{x_1^4}{r^4} \right] \right\}. \tag{23.18}$$

*Case 2*:

$$P_{ij}^{(1)} = P(\delta_{ij} + \eta_1 \delta_{1i}\delta_{1j}), \quad \tilde{P}_{ij}^{(1)} = \tilde{P}(\delta_{ij} + \tilde{\eta}_2 \delta_{i2}\delta_{j2}). \tag{23.19}$$

Proceeding as above, we find

$$\Phi_{\text{int}} = -\frac{P\tilde{P}}{16\pi\mu(1-\nu)} \frac{1}{r^3} \left\{ 2(1-2\nu)\left[ \eta_1 \left(1 - 3\frac{x_1^2}{r^2}\right) + \tilde{\eta}_2 \left(1 - 3\frac{x_2^2}{r^2}\right) \right] + \right.$$

$$\left. + \eta_1 \tilde{\eta}_2 \left[ -1 + \frac{3(x_1^2 + x_2^2)}{r^2} - \frac{15 x_1^2 x_2^2}{r^4} \right] \right\}. \tag{23.20}$$

As $\overline{(x_1/R)^2} = \overline{(x_2/R)^2} = 1/3$ and $\overline{(x_1 x_2/R^2)^2} = 1/15$, we again obtain $\bar{\Phi}_{\text{int}} = 0$ in both cases 1 and 2.

[1] For the dumbbell $H_o$ in f.c.c. lattices and for the C-atom in $\alpha$-Fe the principal directions are parallel to the cube edges.

### 23.4. Point defects in anisotropic media

The calculation of the elastic interaction energy between two point defects by means of (23.7) requires the knowledge of Green's tensor function of the elastic medium and of the multipolar moments of both defects. The first condition may be presently considered as fulfilled by the available methods for tabulation of Green's tensor function in the general case of anisotropy (cf. Sect. 6.4). In exchange, the determination of the multipolar moments of point defects by one of the methods described in Sect. 22 has been undertaken so far only for a reduced number of crystals with cubic lattice. This explains why the numerical results concerning the interaction of point defects are still not numerous and restricted to cubic crystals.

To obtain a first evaluation of the influence of anisotropy on the elastic interaction between point defects, Eshelby [110, 111] proposed to additively decompose the second-order elastic constants as $c_{KM} = c^0_{KM} + c'_{KM}$, such that $c^0_{KM}$ satisfy the isotropy relations [1]. Unfortunately, such a decomposition is not unique, and there exists no way to ensure that the condition $\|\mathbf{c}'\| \ll \|\mathbf{c}\|$ be fulfilled in the general case. For materials with cubic symmetry, Eshelby's result [110] reads

$$\Phi_{\text{int}} = \frac{3P\tilde{P}c_{44}}{4\pi(c^0_{11})^2}\frac{H}{r^3}S(x_1, x_2, x_3) + O(H^2), \tag{23.21}$$

where $H = 1 - (c_{11} - c_{12})/c_{44}$ is the *anisotropy factor*, and $S$ is the orientation factor (23.15). The constant $c^0_{11}$ obviously depends on the choice of $c^0_{KM}$. Barnett [14] has retaken the problem of expanding $\Phi_{\text{int}}$ in power series of $H$, determined the higher-order corections, and studied the effect of choosing different initial approximations for Lamé's constants.

Formulae of the type (23.21) have a limited application, since the first anisotropic correction can be considered as satisfactory only for few weakly anisotropic cubic crystals, such as aluminium and diamond. However, (23.21) has the merit of showing that the vanishing of the dipole-dipole interaction is typical for isotropic materials, whereas for anisotropic materials the principal singularity in this interaction is of the order $O(r^{-3})$ for large values of $r$.

## 24. The elastic interaction between dislocations and point defects

### 24.1. Various types of interaction between dislocations and point defects

The interaction between dislocations and point defects governs the kinetics of migration of point defects to dislocations, and hence the segregation of point defects from supersaturated solid solutions to dislocations, a process which plays an impor-

---

[1] Leibfried [213] used a similar method to determine the first order anisotropic corrections in the elastic field of straight dislocations; he also suggested to calculate $c^0_{KM}$ as mean values of the anisotropic elastic constants $c_{KM}$ over all possible orientations of the crystal.

tant role in radiation damage, internal friction [1], dislocation motion during plastic deformation, electric conduction, etc. In this subsection we shall briefly indicate the various interactions that can exist between a point defect and a dislocation [2].

*The linear elastic interaction* is undoubtedly the most important interaction between a point defect and a dislocation and will be analyzed in more detail in the next subsection. When the point defect is simulated by an *elastic inclusion*, we may decompose this interaction into a *first-order size interaction*, generated by the difference in size between the inclusion and the hole into which it is being forced, and an *inhomogeneity interaction*, which is due to the difference between the elastic constants of the inclusion and those of the host lattice. Alternatively, when the point defect is simulated by *force multipoles* acting upon an elastic continuum, both types of linear elastic interaction are simultaneously taken into account.

*The non-linear elastic interaction* arises from the second-order effects in the elastic field of the dislocation and of the point defect. This type of interaction becomes significant especially when the linear elastic interaction vanishes, as in the case of the first-order size interaction between an interstitial and a screw dislocation in an infinite isotropic elastic medium.

By using the solution obtained by Pfleiderer, Seeger, and Kröner [270] for the non-linear elastic field of a straight dislocation in an isotropic medium, Bullough and Newman [49] have shown that the second-order elastic interaction between a straight (screw or edge) dislocation and a point defect situated at a distance $r$ from the dislocation line decreases as $r^{-2}$ when $r \to \infty$.

*The interaction between the dislocation core and the point defect*, which is also a non-linear effect, can be satisfactorily described only by using a semidiscrete method that uses the lattice theory for the neighbourhood of the dislocation line, and the theory of elasticity for the remaining part of the crystal. Such a calculation has been performed by Perrin, Englert, and Bullough [267], who evaluated the interaction between interstitials and various edge dislocations in Cu and used the results obtained for explaining the growth in irradiated materials.

*Electrical interaction.* Due to the preferential attraction of electrons by the tensile zones around edge dislocations, it is to be expected that the supplementary electric charges thus created will interact with solute atoms, if the latter also carry an effective charge. This electrical interaction is usually negligible in metals, compared to the size-effect interaction, because of screening effects (Friedel [124]), but may become preponderant in polar crystals (Bullough [50]).

*Chemical interaction (of Suzuki type)* occurs owing to the local change in the chemical potential in the stacking fault region of dislocations in f.c.c. crystals, which makes the solid solubility of impurity atoms be different in the fault region from the surrounding crystal. Although of short range, this interaction may cause diffusion flow of impurity atoms into the faults.

---

[1] For a thorough analysis of the influence exerted by the interaction between dislocations and point defects on the internal friction in metals see Nowik and Berry [455] and Aczel and Bozan [1].

[2] In this connection see also Bullough and Newman [49], Bullough [50], and Hirth and Lothe [162], Part III.

*Anisotropy interaction (Snoeck ordering)* arises from the supplementary anisotropy induced by the strain field around dislocations. This anisotropy introduces a discrimination between sites and/or directions that were energetically equivalent in the perfect lattice, thus leading to a certain ordering, correlated with the strain state in the vicinity of dislocation lines. This ordering involves single atomic jumps and/or reorientations of defect pairs, but does not directly contribute to the migration of point defects to dislocations.

## 24.2. The linear elastic interaction between dislocations and point defects

*The first-order size interaction.* The potential energy of interaction between a dislocation and a point defect in an infinite isotropic elastic medium can be calculated with the aid of (21.23), by simply replacing $\delta V$ with $\delta v$, and $\mathbf{E}^*$ with the strain field $\mathbf{E}$ of the dislocation.

For an *edge dislocation* with Burgers vector $\mathbf{b}(b, 0, 0)$ we have from (8.52)

$$\operatorname{tr} \mathbf{E} = \frac{b(1-2\nu)}{2\pi(1-\nu)} \frac{\sin\theta}{r}, \tag{24.1}$$

and hence

$$\Phi_{\text{int}} = -\frac{\mu b(1+\nu)\,\delta v}{3\pi(1-\nu)} \frac{\sin\theta}{r}, \tag{24.2}$$

where $r, \theta$ are cylindrical co-ordinates of the point defect, the positive direction of the dislocation line being chosen as $z$-axis.

For a *screw dislocation*, it follows from (8.60) that $\operatorname{tr} \mathbf{E} = 0$, and hence the first-order size effect vanishes. It should be noticed, however, that this result is no longer valid when the simplifying hypotheses about the isotropy of the material and the perfect spherical form of the inclusion are given up. In this connection, the result of Boyer and Hardy [37] is particularly illuminating. These authors have used a fully atomic description of the size interaction between a vacancy and a screw dislocation in various b.c.c. metals and deduced that this interaction, admittedly, decreases very rapidly as $r \to \infty$, but is non-zero even for isotropic materials. This shows once again that the continuum theory produces correct results only if the separation distance between defects is sufficiently large.

*The inhomogeneity interaction.* The expression of this interaction energy can be immediately derived from (21.39) by replacing $\mathbf{E}$ with the infinitesimal strain field produced by the dislocation. As already mentioned, the parameters $A$ and $B$ occurring in (21.39) must be considered rather as independent material constants that are to be determined by macroscopic experiments. However, two particular cases, characterizing the extreme situations of a very soft or very hard point defect, may be treated by directly using the results in Sect. 21.3. These are the *vacancy* ($K' = \mu' = 0$), for which (21.38) yields

$$A = \frac{1}{1-\alpha} = \frac{3(1-\nu)}{2(1-2\nu)}, \quad B = \frac{1}{1-\beta} = \frac{15(1-\nu)}{7-5\nu}, \quad \frac{A}{B} = \frac{7-5\nu}{10(1-2\nu)}, \tag{24.3}$$

and the *perfectly rigid inclusion* ($K' = \mu' = \infty$), for which

$$A = -\frac{1}{\alpha} = \frac{3(1-\nu)}{1+\nu}, \quad B = -\frac{1}{\beta} = \frac{15(1-\nu)}{2(4-5\nu)}, \quad \frac{A}{B} = \frac{4-5\nu}{10(1+\nu)}. \tag{24.4}$$

In the case of an *edge dislocation* of Burgers vector $\mathbf{b}(b, 0, 0)$, situated in an infinite isotropic elastic medium, by substituting (8.52) into $(21.39)_2$, we find

$$\Phi_{\text{int}} = -\frac{\mu b^2 \Omega B}{8\pi^2(1-\nu)^2}\frac{1}{r^2}\left\{1 + \frac{2}{3}\left[\frac{A}{B}(1+\nu)(1-2\nu) - (1+2\nu-2\nu^2)\sin^2\theta\right]\right\}, \tag{24.5}$$

where $r, \theta$ are the cylindrical co-ordinates of the point defect, the positive direction of the dislocation line being chosen as above as $z$-axis.

Analogously, for a *screw dislocation* of Burgers vector $\mathbf{b}(0, 0, b)$, it follows from $(8.60)_2$ and $(21.39)_2$ that the interaction energy is

$$\Phi_{\text{int}} = -\frac{\mu b^2 \Omega B}{8\pi^2 r^2}. \tag{24.6}$$

The expressions of the interaction energy corresponding to a vacancy or an infinitely rigid inclusion may be derived by simply introducing (24.3), respectively (24.4), into (24.5) and (24.6); their correct form has been first obtained by Bullough [49]. Inspection of (24.5) and (24.6) reveals that the angular dependence of the inhomogeneity interaction energy is either very weak or entirely absent and that the sign of this interaction depends on the relative hardness of the defect and matrix. Thus, a vacancy is always (for any $\theta$) attracted to a dislocation, whereas a perfectly rigid point defect is always repelled.

As already mentioned above, when the point defect is simulated by *force multipoles* exerted on the neighbouring atoms, then both types of linear elastic interaction are simultaneously taken into account. Such a calculation has been performed by Meissner, Savino, Willis, and Bullough [245], who evaluated the interaction between a ⟨100⟩-dumbbell and a prismatic circular dislocation loop with Burgers vector 1/3 ⟨111⟩, in copper, by using the force multipoles previously calculated by Bullough and Tewary [52] for the dumbbell and the expression of the elastic distortion determined by Willis [383] for the dislocation loop. The result obtained has shown that both the symmetry of the point defect and the anisotropy of the host lattice have a pronounced influence on the interaction energy between the point defect and the dislocation. The most significant distinction from the isotropic calculations is that, in general, anisotropy greatly reduces the drift path length that any interstitial would have to take to be captured by the interstitial loop. Thus, anisotropy increases the efficiency of interstitial capture by an interstitial loop, accelerating the void growth, and hence the development of radiation damage[1].

The same hybrid discrete-continuum method has been applied by Heinisch and Sines [425] to several cases of dislocation-point defect interactions in $\alpha$ iron, potassium, and some hypothetical b.c.c. metals. These authors have also thoroughly compared the results obtained by the hybrid method with those given by anisotropic elasticity and by lattice statics. In particular, they found that the hybrid method retains the versatility of the continuum approach and also includes the essential atomistic features of the dislocation-point defect interactions at separation distances greater than a few atomic spacings. In comparison with lattice statics, the hybrid method has the advantage of requiring less computation time and of being easily adapted to different dislocations in the same crystal structure, by merely changing a few input data.

Clearly, when a point defect lies in the very dislocation core, its interaction with the dislocation can no longer be evaluated by continuum or hybrid methods. In such situations the use of the semidiscrete methods is still the most attractive alternative, especially when accurate interatomic potentials are known. Thus, Perrin, Englert, and Bullough [267] have applied the rigid-boundary method to study the interaction between intrinsic interstitials and glissile edge dislocations or sessile Frank dislocations in copper. More recently, Puls, Woo, and Norgett [463] have elaborated a general method (based on the flexible-boundary method Flex-II) for the evaluation of the point defect-dislocation interaction in cubic ionic crystals and have applied it to calculate the binding energies of both cation and anion vacancies at various positions in the core of an a/2 [110] edge dislocation in MgO. The interaction energy is evaluated in two steps: the first determines the equilibrium configuration of the dislocation core; this forms the input for the subsequent calculation of the point-defect formation energy in the dislocated lattice. Then, the difference between the formation energies of the point defect in the dislocation lattice and in the perfect crystal gives the point defect-dislocation interaction energy.

---

[1] The stress-induced point defect-dislocation interaction and its relevance to irradiation creep has been also studied by modelling the point defect as a spherical inclusion (Bullough and Willis [405]), or by making use of the hybrid discrete-continuum method (Savino [467]).

# REFERENCES

For the sake of conciseness, the title is indicated only for books and review articles. Italic numbers in parentheses following the reference indicate the section in which it has been cited; *P* means Preface, while Roman numerals designate the introductions to the corresponding chapters.


1. Aczel, O., Bozan, C., *Dislocations and Internal Damping in Metals* (in Romanian), Facla, Timişoara, 1974. (*16*, *24*)
2. Albenga, G., Atti Acad. Sci. Torino, Cl. Sci. Fis. Mat. Natur.,**54** (1918/19), 864. (*15*)
3. Alers, G. A., Neighbours, J. R., J. Phys. Chem. Solids, **7** (1958), 58. (5)
4. Alers, G. A., Neighbours, J. R., Sato, H., J. Phys. Chem. Solids, **13** (1960), 40. (5)
5. Anthony, K.-H., Arch. Rational Mech. Anal., **37** (1970), 161; **39** (1970) 43; **40** (1971) 50. (*7*, *17*)
6. Armstrong, P. E., Carlson, O. N., Smith, J. F., J. Appl. Phys., **30** (1959), 36. (5)
7. Artman, R. A., Thompson, D. J., J. Appl. Phys., **23** (1952), 470. (5)
8. Asaro, R. J., Hirth, J.. P., J. Phys. F: Metal Phys., **3** (1973), 1659. (*11*, *12*)
9. Asaro, R. J., Hirth, J. P., Barnett, D. M., Lothe, J., phys. stat. sol. (b), **60** (1973), 261. (*11*)
10. Asaro, R. J., Barnett, D. M., J. Phys. F: Metal Phys., **4** (1974), L103. (*11*)
11. Bacon, D. J., Bullough, R., Willis, J. R., Phil. Mag., **22** (1970), 31. (*11*)
12. Bacon, D. J., Groves, P. P., in *Fundamental Aspects of Dislocation Theory*, J. A. Simmons, R. de Wit, and R. Bullough, Eds., Nat. Bur. Stand. Spec. Publ. 317, Washington, vol. 1, p. 35, 1970. (*9*)
13. Barnett, D. M., Acta Met., **15** (1967), 589. (*12*)
14. Barnett, D. M., in *Fundamental Aspects of Dislocation Theory*, J. A. Simmons, R. de Wit, and R. Bullough, Eds., Nat. Bur. Stand. Spec. Publ. 317, Washington, vol. 1. p. 125. (*6*, *23*)
15. Barnett, D. M., Swanger, L. A., phys. stat. sol. (b), **48** (1971), 419. (*6*, *11*)
16. Barnett, D. M., phys. stat. sol. (b), **49** (1972), 741. (*6*, *11*)
17. Barnett, D. M., Asaro, R. J., Gavazza, S. D., Bacon, D. J., Scatergood, R. O., J. Phys. F: Metal Phys., **2** (1972), 854. (*11*, *12*)
18. Barsch, G. R., Shull, H. E., phys. stat. sol. (b), **43** (1971), 637. (*5*)
19. Basinski, Z. S., Duesbery, M. S., Taylor, R., in *Interatomic Potentials and Simulation of Lattice Defects*, Battelle Colloquium, P. C. Gehlen, J. R. Beeler Jr., and R. I. Jafee, Eds., Plenum Press, New York—London, p. 537, 1972. (*16*)
20. Basinski, Z. S., Duesbery, M. S., Taylor, R., Can. J. Phys., **49** (1971), 2160. (*16*)

21. Baštecká, J., Czech. J. Phys., **14** (1964), 430. (*9*)
22. Beauchamp, P., Rabier, J., Grilhé, J., J. Physique, **34** (1973), 923. (*16*)
23. Beltrami, E., Atti Accad. Lincei Rend. (5), **1** (1892), 141. (*6*)
24. Beltz, R. J., Davis, T. L., Malén, K., phys. stat. sol., **26** (1968), 621. (*13*)
25. Betti, E., Nuovo Cim. (2), **7–8** (1872), 5, 69, 158; **9** (1872) 34; **10** (1873), 58. (*6*)
26. Bézier, P., C. R. Acad. Paris (A) , **265** (1967), 365. (*6*)
27. Bilby, B. A., *Report of 1954 Bristol Conference on Defects in Crystalline Solids*, The Phys. Soc., London, p. 123, 1955. (*17*)
28. Bilby, B. A., Bullough, R., Smith, E., Proc. Roy. Soc. London A **231** (1955), 263. (*17*)
29. Bilby, B. A., *Continuous Distributions of Dislocations*, in *Prog. Solid Mechanics*, I. N. Sneddon and R. Hill, Eds., North Holland, Amsterdam, vol. 1, p. 329–398 (1960). (*17*)
30. Bitter, F., Phys. Rev., **37** (1931), 1527. (*21*)
31. Blin, J., Acta Met., **3** (1955), 199. (*12*)
32. Bogardus, E. H., J. Appl Phys., **36** (1965), 2504. (*5*)
33. Bolef, D. I., J. Appl. Phys., **32** (1961), 100. (*5*)
34. Bolef, D. I., Klerk, J. de, Phys. Rev., **129** (1963), 1063. (*5*)
35. Bolef, D. I., Smith, E. R., Miller, J. G., Phys. Rev. (B), **3** (1971) 4100. (*5*)
36. Boyer, L. L., Hardy, J. R., Phil. Mag., **24** (1971), 647. (*16*)
37. Boyer, L. L., Hardy, J. R., Phil. Mag., **26** (1972), 225. (*24*)
38. Braækhus, J., Lothe, J., phys. stat. sol. (b), **43** (1971), 651. (*10*)
39. Bross, H., Seeger, A., Gruner, P., Ann. Phys., **11** (1963), 230. (*20*)
40. Bross, H., Seeger, A., Haberkorn, R., phys. stat. sol., **3** (1963), 1126. (*20*)
41. Bross, H., Gruner, P., Kirshenmann, P., Z. Naturforsch. (a), **20** (1965), 1611. (*20*)
42. Bross, H., Z. angew. Math. Phys., **19** (1968), 434. (*6*)
43. Brown, L. M., Phil. Mag., **15** (1967), 363. (*11, 12*)
44. Brugger, K., Phys. Rev. (A), **133** (1964), 1611. (*4*)
45. Bullough, R., Bilby, B. A., Proc. Phys. Soc. (B), **67** (1954), 615. (*13*)
46. Bullough, R., Perrin, R. C., A.E.R.E. Report, 1967. (*16*)
47. Bullough, R., Perrin, R. C., in *Dislocation Dynamics*, A. R. Rosenfield, G. T. Hahn, A. L. Bement Jr., and R. I. Jafee, Eds., McGraw-Hill, New York–San Francisco–Toronto–London–Sydney, p. 175, 1968. (*16*)
48. Bullough, R., Hardy, J. R., Mag., **24** (1971), 647. (*22*)
49. Bullough, R., Newman, R. C., A.E.R.E. Report R–6215, 1969. (*24*)
50. Bullough, R., *Dislocation Theory of Imperfect Crystalline Solids*, Trieste Lectures 1970, I.A.E.A. Vienna 1971, p. 101–218, 1971. (*P, 7, 8, 16, 17, 22, 24*)
51. Bullough, R., Norgett, M. J., Webb, S., J. Phys. F: Metal Phys., **1** (1971), 345. (*6*)
52. Bullough, R., Tewary, V. K., in *Interatomic Potentials and Simulation of Lattice Defects*, Battelle Colloquium, P. C. Gehlen, J. R. Beeler Jr., and R. I. Jafee, Eds., Plenum Press, New York-London, p. 155, 1972. (*24*)
53. Bullough, R., Tewary, V. K., A.E.R.E. Report T.P. 547, 1973. (*16*)
54. Burgers, J. M., Proc. Kon. Nederl. Akad. Wetensch., **42** (1939), 293, 378. (*7, 9*)
55. Capriz, G., Guidhgli, P. P., Arch. Rational Mech. Anal., **57** (1974), 1. (*14*)
56. Carlson, D. E., *Linear Thermoelasticity*, in *Handbuch der Physik*, S. Flügge and C. Truesdell, Eds., Springer, Berlin–Heidelberg–New York, vol. VIa/2, p. 297–345, 1972. (*4*)
57. Celli, V., J. Phys. Chem. Solids, **19** (1961), 100. (*16*)
58. Cesàro, E., Rend. Napoli (3a), **12** (1906), 311. (*2*)

59. Chang, E., Barsch, G. R., J. Phys. Chem. Solids, **34** (1973), 1543. (*5*)
60. Chang, E., Graham, L. T., phys. stat. sol., **18** (1966), 99. (*16*)
61. Chang, R., Phil. Mag., **16** (1967), 1021. (*16*)
62. Chang, Y. A., Neumann, J. P., J. Phys. Chem. Solids, **28** (1967), 2117. (*5*)
63. Chang, Z. P., Phys. Rev. (A), **140** (1965), 1788. (*5*)
64. Chang, Z. P., Barsch, G. R., Miller, D. L., phys. stat. sol., **23** (1967), 577. (*5*)
65. Chang, Z. P., Barsch, G. R., J. Geophysical Res., **74** (1969), 3291. (*5*)
66. Chang, Z. P., Barsch, G. R., J. Phys. Chem. Solids, **32** (1971), 27. (*5*)
67. Chou, Y. T., Garofalo, F., Whitmore, R. W., Acta Met., **8** (1960), 480. (*12*)
68. Chou, Y. T., Whitmore, R. W., J. Appl. Phys., **32** (1961), 1920. (*12*)
69. Chou, Y. T., J. Appl. Phys., **33** (1962), 2747. (*12*)
70. Chou, Y. T., J. Appl. Phys., **34** (1963), 3608. (*10*)
71. Chou, Y. T., phys. stat. sol., **17** (1966), 509. (*8*)
72. Chou, Y. T., phys. stat. sol., **20** (1967), 285. (*12*)
73. Chou, Y. T., Barnett, D. M., phys. stat. sol., **21** (1967), 239. (*12*)
74. Chou, Y. T., Mitchell, T. E., J. Appl. Phys., **38** (1967), 1535. (*10*)
75. Chree, C., Cambridge Phil. Soc. Trans., **15** (1892), 313. (*15*)
76. Cleavelin, C. R., Pederson, D. O., Marshall B. J., Phys. Rev. (B), **5** (1972), 3193. (*5*)
77. Cline, C. F., Dunegan, H. L., Henderson, G. W., J. Appl. Phys., **38** (1967), 1944. (*5*)
78. Coleman, B. D., Noll, W., Arch. Rational Mech. Anal., **15** (1964), 87. (*5*)
79. Colonetti, G., Atti Accad. Naz. Lincei (5), **24** (1915), 404. (*12*)
80. Conniers, G. H., Int. J. Engng. Sci., **5** (1967), 25. (*8*)
81. Cotner, J., Weertman, J., Acta Met., **10** (1962), 515. (*13*)
82. Cotterill, R. M. J., Doyama, M., Phys. Letters, **14** (1965), 79. (*16*)
83. Cotterill, R. M. J., Doyama, M., Phys. Rev., **145** (1966), 465. (*16*)
84. Cottrell, A. H., *Dislocations and Plastic Flow in Crystals*, Oxford University Press, Fair Lawn, N. J., 1953. (*P, 7*)
85. Dehlinger, U., Kröner, E., Z. Metallkunde, **51** (1960), 457. (*22*)
86. Dickinson, J. M., Armstrong, P. E., J. Appl. Phys., **38** (1967), 602. (*5*)
87. Diener, D., Heinrich, R., Schellenberger, W., phys. stat. sol. (b), **44** (1971), 403. (*16*)
88. Dietze, H., Diplomarbeit, Göttingen, 1949. (*8*)
89. Doyama, M., Cotterill, R. M. J., Phys. Letters, **13** (1964), 110. (*16*)
90. Doyama, M., Cotterill, R. M. J., Phys. Rev. (A), **137** (1965), 994. (*22*)
91. Doyama, M., Cotterill, R. M. J., Phys. Rev., **150** (1966), 448. (*16*)
92. Drabble, J. R., Fendley, J., Solid State Comm., **3** (1965), 269. (*5*)
93. Drabble, J. R., Gluyas, M., in *Lattice Dynamics*, R. F. Wallis, Ed., Pergamon Press, Oxford, p. 607, 1965. (*5*)
94. Drabble, J. R., Brammer, A. J., Solid State Comm., **4** (1966), 467. (*5*)
95. Drabble, J. R., Brammer, A. J., Proc. Phys. Soc. (London), **91** (1967), 959. (*5*)
96. Drabble, J. R., Strathen, E. B., Proc. Phys. Soc. (London), **92** (1967), 1090. (*5*)
97. Duncan, T. R., Kuhlmann-Wilsdorf, D., Bull. Am. Phys. Soc., **11** (1966), 46. (*10*)
98. Dundurs, J., Mura, T., J. Mech. Phys. Solids, **12** (1964), 177. (*8*)
99. Dundurs, J., Sedencky, J., J. Appl. Phys., **36** (1965), 3353. (*8*)
100. Eastman, D. E., J. Appl. Phys., **37** (1966), 2312. (*5*)
101. Eddington, A. S., *Mathematical Theory of Relativity*, New York , 1923. (*18*)
102. Eisenblätter, J., phys. stat. sol., **31** (1969), 87. (*16*)
103. Englert, A., Tompa, H., J. Phys. Chem. Solids, **21** (1961), 306. (*16*)
104. Ericksen, J. L., Rivlin, R. S., J. Rational Mech. Anal., **3** (1954), 281. (*5*)

105. Eringen, A. C., *Non-Linear Theory of Continuous Media*, McGraw-Hill, New York, 1962. (*2*)
106. Eshelby, J. D., Proc. Phys. Soc. (A), **62** (1949), 307. (*13*)
107. Eshelby, J. D., Phill. Trans. Roy. Soc. (A), **244** (1951), 87. (*11, 13*)
108. Eshelby, J. D., Frank, F. C., Nabarro, F.R.N., Phil. Mag., **42** (1951), 351. (*12*)
109. Eshelby, J. D., Read, W. T., Shockley, W., Acta Met., **1** (1953), 251. (*6, 10, 13*)
110. Eshelby, J. D., Acta Met., **3** (1955), 487. (*21, 23, 24*)
111. Eshelby, J. D., *The Continuum Theory of Lattice Defects*, in *Solid State Physics*, F. Seitz and D. Turnbull, Eds., Academic Press, New York, vol. 3, p. 79–144, (*P, 17, 18, 21, 23*)
112. Eshelby, J. D., Phil Mag., **3** (1958), 440. (*10*)
113. Eubanks, R. A., Sternberg, E., J. Rational Mech. Anal., **5** (1956), 735. (*6*)
114. Ferris, R. W., Shepard, M. L., Smith, J. F., J. Appl. Phys., **34** (1963), 768. (*5*)
115. Fichera, G., Ann. Scuola Norm. Pisa (3), **4** (1950), 35. (*6*)
116. Fisher, E. S., Manghnani, M. H., Sokolowski, T. J., J. Appl. Phys., **41** (1970), 2991. (*5*)
117. Flocken, J. W., Hardy, J. R., Phys. Rev., **175** (1968), 175. (*22*)
118. Flocken, J. W., Hardy, J. R., Phys. Rev., **177** (1969), 1054. (*22*)
119. Flocken, J. W., Hardy, J. R., in *Fundamental Aspects of Dislocation Theory*, J. A. Simmons, R. de Wit, and R. Bullough, Eds., Nat. Bur. Stand. Spec. Publ. 317, Washington, vol. 1, p. 219, 1970. (*22*)
120. Foreman, A. J. E., Acta Met., **3** (1955), 322. (*10*)
121. Frank, F. C., Proc. Phys. Soc. (A), **62** (1949), 131. (*13*)
122. Frank, F. C., Phil. Mag., **42** (1951), 809. (*7*)
123. Fredholm, I., Acta Math. Stockholm, **23** (1900), 1. (*6*)
124. Friedel, J., *Les Dislocations*, Gauthier-Villars, Paris, 1956. (*P, 24*)
125. Garland, C. W., Dalven, R., Phys. Rev., **111** (1958), 1232. (*5*)
126. Gebbia, M., Ann. di mat. (3a), **10** (1904), 157. (*6*)
127. Gehlen, P.C., Rosenfield, A. R., Hahn, G. T., J. Appl. Phys., **39** (1968), 5246. (*16*)
128. Gehlen, P. C., J. Appl. Phys., **41** (1970), 5165. (*16*)
129. Gehlen, P. C., in *Interatomic Potentials and Simulations of Lattice Defects*, Battelle Colloquium, P. C. Gehlen, J. R. Beeler Jr., and R. I. Jafee, Eds., Plenum Press, New York–London, p. 475, 1972. (*16*)
130. Gehlen, P. C., Hirth, J. P., Hoagland, R. G., Kanninen, M. F., J. Appl. Phys., **43** (1972), 3921. (*16*)
131. Gemperlova, J., Saxl, I., Czech. J. Phys. B, **18** (1968), 1085. (*10*)
132. Gemperlova, J., phys. stat. sol., **30** (1968), 261. (*10*)
133. Gerlich, D., J. Phys. Chem. Solids, **28** (1967), 2575. (*5*)
134. Gerlich, D., Phys. Rev., **168** (1968), 947. (*5*)
135. Gerlich, D., Fisher, E. S., J. Phys. Chem. Solids, **30** (1969), 1197. (*5*)
136. Ghafelehbashi, M., Koliwad, K. M., J. Appl. Phys., **41** (1970) 4010. (*5*)
137. Gelfand, I. M., Shilov, G. E., *Generalized Functions* (in Russian), Fizmatgiz, Moscow, vol. 1, 1958. English Trans.: Academic Press, New York, 1964. (*6*)
138. Gibson, J. B., Goland, A. N., Milgram, M., Vineyard, G. H., Phys. Rev., **120** (1960), 1229. (*16*)
139. Gittus, J. N., *Uranium*, Butterworths, London, 1963. (*5*)
140. Gluyas, M., Brit. J. Appl. Phys., **18** (1967), 913. (*5*)
141. Granzer, F., Wagner, G., Eisenblätter, J., phys. stat. sol., **30** (1968), 587. (*16*)
142. Granzer, F., Acta Met., **18** (1970), 159. (*10*)

143. Granzer, G., Belzner, V., Bücher, M., Petrasch, P., Teodosiu, C., J. Physique, **34** (1973), Colloque C 9, suppl. 11–12, C 9–359. *(16)*
144. Green, A. E., Proc. Roy. Soc. (A), **180** (1942), 173; **184** (1945), 231, 289, 301. *(10)*
145. Green, A. E., Zerna, W., *Theoretical Elasticity*, Oxford at the Clarendon Press, London, 1954. *(10, 14)*
146. Green, A. E., Spratt, E. B., Proc. Roy. Soc London (A), **224** (1954), 347. *(14)*
147. Guinan, M. V., Beshers, D. N., J. Phys. Chem. Solids, **29** (1968), 541. *(5)*
148. Günther, H., *Zur nichtlinearen Kontinuumstheorie bewegter Versetzungen*, Dissertation, Akademie-Verlag, Berlin, 1967. *(17, 20)*
149. Gurtin, M. E., Sternberg, E., Arch. Rational Mech. Anal., **8** (1961), 99. *(6)*
150. Gurtin, M. E., *The Linear Theory of Elasticity*, in *Handbuch der Physik*, S. Flügge and C. Truesdell, Eds., vol. VIa/2, Springer, Berlin–Heidelberg–New York, p. 1–295, 1972. *(1, 2, 6, 8, 15, 21)*
151. Halmos, P. R., *Finite Dimensional Vector Spaces*, D. Van Nostrand Co., Princeton, N. J., 1958. *(1)*
152. Hardy, J. R., Bullough, R., Phil. Mag., **15** (1967) 237. *(23)*
153. Hazzledine, P. M., Hirsch, B. P., Phil. Mag., **15** (1967), 121. *(12)*
154. Head, A. K., Proc . Phys. Soc. (B), **66** (1953), 793. *(8)*
155. Head, A. K., Phil. Mag., **44** (1953), 92. *(8)*
156. Head, A. K., Aust. J. Phys., **13** (1960), 278. *(8)*
157. Head, A. K., Thompson, P. F., Phil. Mag., **7** (1962), 439. *(12)*
158. Head, A. K., phys. stat. sol., **5** (1964), 51; **6** (1964), 461. *(10)*
159. Hearmon, R. F. S., Acta Cryst., **6** (1953), 331. *(5)*
160. Hermann, C., Z. Knistallogr., **68** (1928), 257; **69** (1929), 226, 250, 533; **75** (1930), 159. *(5)*
161. Hiki, Y., Granato, A. V., Phys. Rev., **144** (1966), 411. *(5)*
162. Hirth, J. P., Lothe, J., *Theory of Dislocations*, McGraw-Hill, New York–St. Louis–San Francisco–Toronto–London–Sydney, 1968. *(P, 7, 8, 9, 10, 12, 13, 16, V, 24)*
163. Hirth, J. P., Nix, W. D., phys stat. sol., **35** (1969), 177. *(16)*
164. Ho, P. S., Ruoff, A. L., Phys. Rev., **161** (1967), 864. *(5)*
165. Hölzler, A., Siems, R., in *Fundamental Aspects of Dislocation Theory*, J. A. Simmons, R. de Wit, and R. Bullough, Eds., Nat. Bur. Stand. Spec. Publ. 317, Washington, vol. 1, p. 291, 1970. *(16)*
166. Huntington, H. B., Dickey, J. E., Thompson, R., Phys. Rev. **100** (1955), 1117; **113** (1959), 1696. *(16)*
167. Indenbom, V. L., *Types of Defects. Dislocation Theory. Physics of Crystals with Defects* (in Russian). Tbilisi, vol. 1, 1966. *(P)*
168. Indenbom, V. L., Orlov, S. S., Zh. Eksp. Teor. Fiz. Pisma, **6** (1967), 826 (in Russian); Soviet Phys. JETP Lett. (English transl.), **6** (1967), 274. *(6, 11)*
169. Indenbom, V. L., Orlov, S. S., Kristallografiya, **12** (1967), 971 (in Russian); Sov. Phys. Cryst. (English transl.), **12** (1968), 849. *(6, 11)*
170. Indenbom, V. L., Orlov, S. S., Prikl. Mat. Mekh., **32** (1968), 414 (in Russian); Appl. Math. Mekh. (English transl.), **32** (1968), 414. *(11)*
171. Jagodzinski, H., *Kristallographie*, in *Handbuch der Physik*, S. Flügge, Ed., Springer, Berlin–Göttingen–Heidelberg, vol. VII/1, p. 1–103, 1955. *(22)*
172. Johnson, R. A., Brown, E., Phys. Rev., **127** (1962), 446. *(22)*
173. Johnson, R. A., Phys. Rev. (A), **134** (1964), 1329. *(22)*
174. Johnson, R. A., Phys. Rev., **145** (1966), 423. *(22)*

175. Jones, D. S., *Generalized Functions*, McGraw-Hill, London—New York—Toronto—Sydney, 1966. (*6*)

176. Kammer, E. W., Cardinal, L. C., Vold, C. L., Glicksman, M. E., J. Phys. Chem. Solids, **33** (1972), 1891. (*5*)

177. Kanzaki, H., Phys. Chem. Sol., **2** (1957), 24. (*22*)

178. Kecs, W., Teodorescu, P. P., *Applications of the Theory of Distributions in Mechanics*, Ed. Academiei, Bucharest, and Abacus Press, Tunbridge Wells, Kent, 1974. (*6*)

179. Kirchhoff, G., J. reine u. angew. Math., **56** (1859), 285. (*6*)

180. Kiusalaas, J., Mura, T., Phil. Mag., **9** (1964), 1. (*13*)

181. Koliwad, K. M., Ghate, P. B., Ruoff, A. L., phys. stat. sol., **21** (1967), 507. (*5*)

182. Kosevich, A. M., Zh. Eksp. Teor. Fiz., **42** (1962), 152; **43** (1962), 637 (in Russian); Soviet Phys. JETP (English transl.), **15** (1962), 108; **16** (1963), 455. (*20*)

183. Kosevich, A. M., Usp. Fiz. Nauk, **84** (1964), 579 (in Russian); Soviet Phys. Usp. (English transl.) **7** (1965), 837. (*20*)

184. Kondo, K., Proc. 2nd Jap. Congr. Appl. Math., p. 41, 1952. (*17*)

185. Kondo, K., in *Memoirs of the Unifying Studies of the Basic Problems in Engineering by Means of Geometry*, Gakujutsu Bunken Fukyu-kai, vol. 1, D—I, p. 458, 1955. (*17*)

186. Kröner, E., Z. Phys., **136** (1953), 402. (*6*)

187. Kröner, E., Z. Phys., **139** (1954), 175. (*18*)

188. Kröner, E., Z. Phys., **141** (1955), 386. (*18*)

189. Kröner, E., Z. Phys., **142** (1955), 463. (*17*)

190. Kröner, E., *Kontinuumstheorie der Versetzungen und Eigenspannungen*, Springer, Berlin—Göttingen—Heidelberg, 1958. (*P, 1, 9, 12, 17, 18, 22*)

191. Kröner, E., Seeger, A., Arch. Rational Mech. Anal., **3** (1959), 97. (*14, 17, 19*)

192. Kröner, E., Arch. Rational Mech. Anal., **4** (1959), 273. (*17, 21, 22*)

193. Kröner, E., Phys. kondens. Materie, **2** (1964), 262. (*17, 21, 22*)

194. Kröner, E., Teodosiu, C., in *Problems of Plasticity*, A. Sawczuk, Ed., Noordhoff Int. Publ., Leyden, vol 2, p. 45, 1974. (*17*)

195. Kronmüller, H., Seeger, A., J. Phys. Chem. Solids, **18** (1961), 93. (*12*)

196. Kronmüller, H., in *Moderne Probleme der Metallphysik*, A. Seeger, Ed., Springer, Berlin, vol. 1, p. 126, 1965. (*7*)

197. Kronmüller, H., *Nachwirkung in Ferromagnetika*, Springer, Berlin, 1968. (*22*)

198. Kronmüller, H., in *Vacancies and Interstitials in Metals*, A. Seeger, B. Schumacher, W. Schilling, and J. Diehl, Eds., North-Holland, Amsterdam, p. 667, 1969. (*22*)

199. Kronmüller, H., Marik, J., Phil. Mag., **26** (1972), 523. (*12*)

200. Kroupa, F., Czech. J. Phys. (B), **10** (1960), 284; **12** (1962), 191. (*9*)

201. Kroupa, F., phys. stat. sol., **9** (1965), 27. (*9*)

202. Kroupa, F., in *Theory of Crystal Defects*, B. Gruber, Ed., Academic Press, New York, p. 275, 1966, (*9*)

203. Kuang, J. G., Mura, T., J. Appl. Phys., **39** (1968), 109. (*12*)

204. Kunin, I. A., Prikl. meh. i tehn. fiz., **5** (1965), 76. (*17*)

205. Kurihara, T., Int. J. Engng. Sci., **11** (1973), 891. (*10*)

206. Landau, L., Lifshits, E., *Mécanique quantique*, Mir, Moscow, 1967. (*22*)

207. Landau, L., Lifshits, E., *Théorie de l'élasticité*, Mir, Moscow, 1967. (*17*)

208. Larsen, A., *GRAPE — A Computer Program for Classical Many-Body Problems in Radiation Damage*, Brookhaven Nat. Lab. Assoc. Univ. Inc., 1964. (*16*)

209. Lauricella, G., Ann. Scuola Norm. Pisa, **7** (1895), 1. (*6*)

210. Lekhnitsky, S. G., Prikl. Mat ; Meh., **1** (1973), 1; **2** (1938), 345; **4** (1940), 1; **6** (1942), 3. (*10*)

211. Lekhnitsky, S. G., *Theory of elasticity of an anisotropic body* (in Russian), 2nd ed., Nauka, Moscow, 1977, English trans. of the 1st ed.: Holden Day, San Francisco, 1963. (*10*)
212. Leibfried, G., Z. Phys., **130** (1951), 214. (*12*)
213. Leibfried, G., Z. Phys., **135** (1953), 23. (*6, 9, 23*)
214. Li, J. C. M., in *Fundamental Aspects of Dislocation Theory*, J. A. Simmons, R. de Wit, and R. Bullough, Eds., Nat. Bur. Stand. Spec. Publ. 317, Washington, vol. 1, p. 147, 1970. (*12*)
215. Lie, K. C., Koehler, J. S., Adv. in Phys., **17** (1968), 421. (*6*)
216. Lifshits, I. M., Rozentsveig, L. N., Zh. Eksp. Teor. Fiz., **17** (1947), 783. (*6*)
217. Loje, K. F., Schuele. D. E., J. Phys. Chem. Solids, **31** (1970), 2051. (*5*)
218. Lopatinsky, Ya., V., Mat. Sbornik, **17** (1945), 267. (*10*)
219. Lothe, J., Phil. Mag., **15** (1967), 353. (*11, 12*)
220. Lothe, J., in *Fundamental Aspects of Dislocation Theory*, J. A. Simmons, R. de Wit, and R. Bullough, Eds., Nat. Bur. Stand. Spec. Publ. 317, Washington, vol. 1, p. 11, 1970. (*10*)
221. Louat, N., in *Fundamental Aspects of Dislocation Theory*, J. A. Simmons, R. de Wit, and R. Bullough, Eds., Nat. Bur. Stand. Spec. Publ. 317, Washington, vol. 1, p. 135, 1970. (*12*)
222. Love, A. E. H., *A Treatise on the Mathematical Theory of Elasticity*, 4th ed., Cambridge Univ. Press, Cambridge, 1927. (*2, 6, 7, 22*)
223. Malén, K., in *Fundamental Aspects of Dislocation Theory*, J. A. Simmons, R. de Wit, and R. Bullough, Eds., Nat. Bur. Stand. Spec. Publ. 317, Washington, vol. 1, p. 23, 1970. (*13*)
224. Malén, K., phys. stat. sol., **38** (1970), 259. (*6, 11*)
225. Malén, K., Lothe, J., phys. stat. sol., **39** (1970), 287. (*6, 10, 11*)
226. Malén, K., phys. stat. sol. (b), **44** (1971), 661. (*10, 11*)
227. Malvern, L. E., *Introduction to the Mechanics of a Continuous Medium*, Prentice Hall, Englewood Cliffs, N. J., 1969. (*1, 2, 3*)
228. Manghnani, M. H., Fisher, E. S., Brower, W. S., Jr., J. Phys. Chem. Solids, **33** (1972), 2149. (*5*)
229. Mann, E., Jan, R. v., Seeger, A., phys. stat. sol., **1** (1961), 17. (*6*)
230. Mantea, Şt., Geru, N., Dulămiţă, T., Rădulescu, M., *Physical metallurgy* (in Romanian), Ed. tehnică, Bucharest, 1970. (*10*)
231. Maradudin, A. A., J. Phys. Chem. Solids, **9** (1958), 1. (*16*)
232. Marcinkowski, M. J., Sree Harsha, K. S., J. Appl. Phys., **39** (1968), 1775. (*9*)
233. Marcinkowski, M. J., J. Appl. Phys., **39** (1968), 4552. (*12*)
234. Marshall, B. J., Miller, R. E., J. Appl. Phys., **38** (1967), 4749. (*5*)
235. Marshall, B. J., Cleavelin, C. R., J. Phys. Chem. Solids, **30** (1969), 1905. (*5*)
236. Martinson, R. H., Phys. Rev., **178** (1969), 902. (*5*)
237. Mauguin, Ch., Z. Kristallogr., **76** (1931). (*5*)
238. Maxwell, J. C., Proc. Lond. Math. Soc. (1), **2** (1868), 58. (*6*)
239. Mc Lean, K. O., Smith, C. S., J. Phys. Chem. Solids, **33** (1972) 275. (*5*)
240. Mc Skimin, H. J., J. Appl. Phys., **26** (1955), 406. (*5*)
241. Mc Skimin, H. J., Andreatsch, P., Jr., J. Appl. Phys., **35** (1964), 3312. (*5*)
242. Mc Skimin, H. J., Andreatsch, P., Jr., J. Appl. Phys., **38** (1967), 2610. (*5*)
243. Mc Skimin, H. J., Andreatsch, P., Jr., Glynn, P., J. Appl. Phys., **43** (1972), 985. (*5*)
244. Meissner, N., A.E.R.E. Report, T.P. 556, 1973. (*6, 11*)
245. Meissner, N., Savino, E. J., Willis, J. R., Bullough, R., A.E.R.E. Report, T.P. 543 (1973). (*24*)
246. Michell, J. H., Proc. Lond. Math. Soc., **31** (1900), 100. (*2, 6*)
247. Miller, R. A., Smith, C. S., J. Phys. Chem. Solids, **25** (1964), 1279. (*5*)
248. Miller, R. A., Schuele, D. E., J. Phys. Chem. Solids, **30** (1969), 589. (*6*)

249. Mitchell, T. E., Phil. Mag., **10** (1964), 301. (**12**)
250. Mitchell, T. E., Hecker, S. S., Smialek, R. L., phys. stat. sol., **11** (1965), 585. (*12*)
251. Morera, G., Atti Accad. Lincei Rend. (5), **1** (1892), 137. (*6*)
252. Moriguti, S., Oyo Sugaku Rikigaku, **1** (1947), 29, 87. (*17*)
253. Mura, T., Phil. Mag., **8** (1963), 843. (*9, 13, 18*)
254. Muskhelishvili, N. I., *Some Basic Problems of the Mathematical Theory of Elasticity* (in Russian), Izd. Akad. Nauk SSSR, 5th ed., Moscow, 1966. Engl. transl. of the 4th ed. by J.R.M. Radok, Groningen, Noordhoff, 1963. (*2, 6, 8*)
255. Nabarro, F. R. N., Proc. Phys. Soc., **59** (1947), 256. (*16*)
256. Nabarro, F. R. N., Phil. Mag., **42** (1951), 1224. (*13*)
257. Nabarro, F. R. N., Advan. Phys., **1** (1952), 269. (*12*)
258. Nabarro, F. R. N., *Theory of Crystal Dislocations*, Clarendon Press, Oxford, 1967. (*P, 2, 7*)
259. Neuber, H., Z. Angew. Math. Mech., **14** (1934), 203. (*6*)
260. Noll, W., Arch. Rational Mech. Anal., **27** (1967), 1. (*17*)
261. Nye, J. F., Acta Met., **1** (1953), 153. (*17*)
262. Orowan, E., Z. Phys., **89** (1934), 605. (*7*)
263. Papkovitch, P. F., C. R. Acad. Paris, **195** (1932), 513. (*6*)
264. Pastur, L. A., Fel'dman, E. P., Kosevich, A. M., Kosevich, V. M., Sov. Phys. Solid State, **4** (1963), 1896. (*10*)
265. Peach, M. O., Koehler, J. S., Phys. Rev., **80** (1950), 436. (*9, 12*)
266. Peierls, R. E., Proc. Phys. Soc., **52** (1940), 23. (*16*)
267. Perrin, R. C., Englert, A., Bullough, R., in *Interatomic Potentials and Simulation of Lattice Defects*, Battelle Colloquium, New York—London, p. 509, 1972. (*16, 24*)
268. Perzyna, P., in *Advances in Applied Mechanics*, Academic Press, New York, vol. 11, p. 313, 1971. (*7, 20*)
269. Perzyna, P., *Théorie Physique de la Viscoplasticité*, Acad. Polon. Sci., Centre Sci. Paris, fasc. 104, 1972. (*7, 20*)
270. Pfleiderer, H., Seeger, A., Kröner, E., Z. Naturforschung (a), **15** (1960), 758. (*14, 15, 19, 24*)
271. Polanyi, M., Z. Phys., **89** (1934), 660. (*7*)
272. Potter, W. N., Bartels, R. A., Watson, R. W., J. Phys. Chem. Solids, **32** (1971), 2363. (*5*)
273. Rabier, J., Grilhé, J., J. Phys. Chem. Solids, **34** (1973), 1031. (*16*)
274. Ramji Rao, R., Menon, C. S., J. Appl. Phys., **44** (1973), 3892. (*5*)
275. Read, W. T. Jr., *Dislocations in Crystals*, McGraw-Hill, New York, 1953. (*P, 7*)
276. Rieder, G., Abh. Braunschweig. Wiss. Ges., **14** (1962), 109. (*17*)
277. Rieder, G., Österreich. Ing. Arch., **18** (1964), 173. (*17*)
278. Rowland, W. D., White J. S., J. Phys. F: Metal Phys., **2** (1972), 231. (*5*)
279. Rivlin, R. S., Tapakoglu, C., J. Rational Mech. Anal., **3** (1954), 581. (*14*)
280. Sáenz, A. W., J. Rational Mech. Anal., **2** (1953), 83. (*13*)
281. Saxlova-Švábová, Z. Metallkde., **58** (1967), 326. (*12*)
282. Schaefer, H., Z. Angew. Math. Mech., **33** (1953), 356. (*6*)
283. Schoenflies, A., *Theorie der Kristallstrukturen*, Berlin, 1923. (*5, 22*)
284. Schottky, G., Seeger, A., Schmid, G., phys. stat. sol., **4** (1964), 419; **7** (1964), K25. (*22*)
285. Seeger, A., Schöck, G., Acta Met., **1** (1953), 519. (*10*)
286. Seeger, A., *Theorie der Gitterfehlstellen*, in *Handbuch der Physik*, S. Flügge, Ed., Springer, Berlin—Göttingen—Heidelberg, vol. VII/1, p. 383—665, 1955. (*P, 7, 8, V*)
287. Seeger, A., Suppl. Nuovo Cimento, **7** (1968), 632. (*15*)
288. Seeger, A., Haasen, P., Phil. Mag., **3** (1958), 470. (*15*)

289. Seeger, A., Mann, E., Z. Naturforschung (a), **14** (1959), 154. (*14, 21*)
290. Seeger, A., Mann, E., J. Phys. Chem. Solids, **12** (1960), 326. (*22*)
291. Seeger, A., Mann, E., Jan, R. v., J. Phys. Chem. Solids, **23** (1962), 639. (*22*)
292. Seeger, A., in *Proc. Int. Symp. on Second-Order Effects in Elasticity, Plasticity, and Fluid Dynamics*, Haifa, 1962, Pergamon Press, Oxford – Paris – New York, and Jerusalem Academic Press, Israel, p. 129, 1964. (*III, 20, 21*)
293. Seeger, A., Wobser, G., phys. stat. sol., **17** (1966), 709. (*12*)
294. Seeger, A., Brand, P., in *Small-Angle X-Ray Scattering*, Syracuse University 1965, H. Brumberger, Ed., Gordon and Breach, New York – London – Paris, p. 383, 1967. (*20*)
295. Seeger, A., Teodosiu, C., Petrasch, P., phys. stat. sol. (b), **67** (1975), 207. (*14*)
296. Siems, R., Delavignette, P., Amelincks, S., phys. stat. sol., **2** (1962), 636. (*10*)
297. Siems, R., phys. stat. sol., **30** (1968), 645. (*23*)
298. Signorini, A., in Proc. 3rd Int. Congr. Appl. Mech., Stockholm, vol. 2, p. 80, 1930. (*14*)
299. Signorini, A., in Atti 2° Congr. Un. Mat. Ital., p. 56, 1940. (*14*)
300. Signorini, A., Memoria 2ª Ann. Mat. Pura Appl. (4), **30** (1949), 1. (*14*)
301. Simmons, J. A., Bullough, R., in *Fundamental Aspects of Dislocation Theory*, J. A. Simmons, R. de Wit, and R. Bullough, Eds., Nat. Bur. Stand. Spec. Publ. 317, Washington, vol. 1, p. 89, 1970. (*18*)
302. Sinclair, J. E., J. Appl. Phys., **42** (1971), 5321. (*16*)
303. Slagle, O. D., Mc Kinstry, H. A., J. Appl. Phys., **38** (1967), 437. (*5*)
304. Slagle, O. D., Mc Kinstry, H. A., J. Appl. Phys., **38** (1967), 451. (*5*)
305. Slotwinski, T., Trivisonno, J., J. Phys. Chem. Solids, **30** (1969), 1276. (*5*)
306. Slutsky, L. J., Garland, C. W., Phys. Rev., **107** (1957), 972. (*5*)
307. Smith, E., Phil. Mag., **16** (1967), 1285; **18** (1968), 1067. (*12*)
308. Smith, E., in *Fundamental Aspects of Dislocation Theory*, J. A. Simmons, R. de Wit, and R. Bullough, Eds., Nat. Bur. Stand. Spec. Publ. 317, Washington, vol. 1, p. 151, 1970. (*12*)
309. Smith, G. F., Rivlin, R. S., Trans. Amer. Math. Soc., **88** (1958), 175. (*5*)
310. Smith, G. F., Arch. Rational Mech. Anal., **10** (1962), 108. (*5*)
311. Smith, J. F., Gjevre, J. A., J. Appl. Phys., **31** (1960), 645. (*5*)
312. Smith, P. A., Smith, C. S., J. Phys. Chem. Solids, **26** (1965), 279. (*5*)
313. Sokolnikoff, I. S., *Mathematical Theory of Elasticity*, 2nd ed. McGraw-Hill, New York, 1956. (*6*)
314. Solomon, L., *Elasticité linéaire*, Masson, Paris, 1968. (*8*)
315. Son, R. P., Bartels, R. A., J. Phys. Chem. Solids, **33** (1972), 819. (*5*)
316. Spence, G. B., J. Appl. Phys., **33** (1962), 729. (*10*)
317. Steeds, J. W., *Introduction to Anisotropic Elasticity Theory of Dislocations*, Clarendon Press, Oxford, 1973. (*10*)
318. Steketee, J. A., Can. J. Phys., **36** (1958), 192. (*9*)
319. Stenzel, G., phys. stat. sol., **34** (1969), 365. (*13*)
320. Stenzel, G., phys. stat. sol., **34** (1969), 495. (*13*)
321. Sternberg, E., Eubanks, R. A., J. Rational Mech. Anal., **4** (1955), 135. (*6*)
322. Stroh, A. N., Proc. Roy. Soc. (A), **218** (1953), 391. (*12*)
323. Stroh, A. N., Phil. Mag., **3** (1958), 625. (*10*)
324. Stroh, A. N., J. Math. Phys., **41** (1962), 77. (*6, 10, 13*)
325. Suzuki, H., in *Dislocation Dynamics*, A. R. Rosenfield, G. T. Hahn, A. L. Hahn, A. L. Bement Jr., and R. I. Jafee, Eds., McGraw-Hill, New York – San Francisco – Toronto – London – Sydney, p. 679, 1968. (*16*)
326. Swartz, K. D., J. Acoust. Soc. Am., **41** (1967), 1083. (*5*)

327. Synge, J. L., *The Hypercircle in Mathematical Physics: A Method for the Approximate Solution of Boundary-Value Problems*, Cambridge University Press, Cambridge, 1957. (*6*)

328. Tamate, O., Int. J. Engng. Sci., **5** (1967), 25. (*8*)

329. Tamate, O., Kurihara, T., Int. J. Fracture Mech., **6** (1970), 341. (*8*)

330. Taylor, G. I., Proc. Roy. Soc. (A), **145** (1934), 362. (*7*)

331. Teodosiu, C., Rev. méc. appl., **8** (1963), 639. (*6*)

332. Teodosiu, C., Rev. roum. sci. techn. – méc. appl., **10** (1965), 1461. (*3*, *17*)

333. Teodosiu, C., Bull. Acad. Polon. Sci. – sér. sci. techn., **15** (1967), 95, 103. (3, 17)

334. Teodosiu, C., Rev. roum. sci. techn. – méc. appl., **12** (1967), 961, 1061, 1291. (*3*, *17*)

335. Teodosiu, C., in *Fundamental Aspects of Dislocation Theory*, J. A. Simmons, R. de Wit, and R. Bullough, Eds., Nat. Bur. Stand. Spec. Publ., 317, Washington, vol. 2, p. 837, 1970. (*7*, *17*, *20*)

336. Teodosiu, C., Seeger, A., in *Fundamental Aspects of Dislocation Theory*, J. A. Simmons, R. de Wit, and R. Bullough, Eds., Nat. Bur. Stand. Spec. Publ. 317, Washington, vol. 2, p. 877, 1970. (*17*, *19*, *20*)

337. Teodosiu, C., *Kontinuumsmechanik mit Anwendungen im Bereich der Festkörperphysik*, Vorlesungsmanuskript, vol. 1 and 2, University of Stuttgart, 1971. (*7*, *14*, *15*, *17*)

338. Teodosiu, C., Nicolae, V., Rev. roum. sci. techn. –méc, appl., **17** (1972), 919. (*10*, *14*, *16*)

339. Teodosiu, C., Nicolae, V., Rev. roum. sci. techn. – méc. appl., **25** (1980), 879. (*10*)

340. Teodosiu, C., Nicolae, V., Paven, H., Centre of Solid Mechanics, Report Nr. 33/1974. (*10*)

341. Teodosiu, C., J. Phys. F.: Metal Phys., **4** (1974), 1225. (*14*)

342. Teodosiu, C., Nicolae, V., Paven, H., phys. stat. sol. (a), **27** (1975), 191. (*10*, *16*)

343. Teodosiu, C., Paven, H., Nicolae, V., Marin, M., St. cerc. fiz., **27** (1975), 645. (*16*)

344. Teodosiu, C., *A Physical Theory of the Finite Elastic – Viscoplastic Behaviour of Single Crystals*, Eng. Transactions, **23** (1975), p. 151–184. (*7*, *16*, *17*, *20*)

345. Teodosiu, C., Sidoroff, F., Int. J. Engng. Sci., **14** (1976), 713. (*7*, *16*, *20*)

346. Teutonico, L. J., Phys. Rev., **124** (1961), 1039 ; **125** (1962), 1530. (*13*)

347. Teutonico, L. J., Phys. Rev., **127** (1962), 413. (*13*)

348. Tewary, V. K., A.E.R.E. Report, T.P. 388, 1969. (*16*, *22*)

349. Tewary, V. K., Bullough, R., J. Phys. F: Metal Phys., **1** (1971), 554. (*22*)

350. Thomas, J. F., Jr., Phys. Rev., **175** (1968), 955. (*5*)

351. Thompson, W. (Lord Kelvin), Cambr. Dubl. Math. J., **3** (1848), 87. (*6*)

352. Tikhonov, L. V., Fizika met. i metalloved, **24** (1967), 577. (*9*)

353. Timoshenko, S., Goodier, J. N., *Theory of Elasticity*, 2nd ed., McGraw-Hill, New York, 1951. (*8*)

354. Toupin, R. A., Rivlin, R. S., J. Math. Phys., **1** (1960), 8. (*15*)

355. Toupin, R. A., Bernstein, B., J. Acoust. Soc. Am., **33** (1961), 216. (*5*)

356. Toupin, R. A., Arch. Rational Mech. Anal., **18** (1965), 83. (*8*)

357. Truesdell, C., Toupin, R. A., *The Classical Field Theories*, in *Handbuch der Physik*, S. Flügge, Ed., Springer, Berlin – Göttingen – Heidelberg, vol. III/1, 1960. (*2*)

358. Truesdell, C., Noll, W., *The Non-Linear Field Theories of Mechanics*, in *Handbuch der Physik*, S. Flügge, Ed., Springer, Berlin–Göttingen–Heidelberg, vol. III/3, 1965. (*14*, *17*)

359. Tucker, M. O., Crocker, A. G., in *Mechanics of Generalized Continua* (Proc. IUTAM Symposium, Freudenstadt–Stuttgart, 1967) E. Kröner, Ed., Springer, Berlin, p. 286. (*10*)

360. Tucker, M. O., Phil. Mag., **19** (1969), 1141. (*10*)

361. Tucker, M. O., in *Fundamental Aspects of Dislocation Theory*, J. A. Simmons, R. de Wit, and R. Bullough, Eds. Nat. Bur. Stand. Spec. Publ. 317, Washington, vol. 1, p. 163, 1970. (*12*)

362. TURTELTAUB, M. J., STERNBERG, E., Arch. Rational Mech. Anal., **29** (1968), 193. (*6*)
363. VAGERA, I., Czech. J. Phys. (B), **20** (1970), 702, 1278. (*9*)
364. VAISMAN, A. M., KUNIN, I. A., Dokl. Akad. Nauk SSSR, **173** (1967), 1024. (*18*)
365. VAN BUEREN, H. C., *Imperfections in Crystals*, North-Holland, Amsterdam, 1961. (*P*)
366. VAN DER MERWE, J. H., Proc. Phys. Soc. (A), **63** (1950), 613. (*16*)
367. VAN HULL, A., WEERTMAN, J., J. Appl. Phys., **33** (1962), 1636. (*13*)
368. VITEK, V., Phil. Mag., **18** (1968), 773. (*16*)
369. VITEK, V., in *Proc. 2nd Int. Conf. on the Strength of Metals and Alloys*, ASM, Asilomar, vol. 2, p. 389, 1970. (*16*)
370. VITEK, V., PERRIN, R. C., BOWEN, D. K., Phil. Mag., **21** (1970), 1049. (*16*)
371. VOGEL, S. M., RIZZO, F. J., J. Elasticity, 3 (1973), 203. (*6*)
372. VOIGT, W., *Lehrbuch der Kristallphysik*, Teubner, Leipzig, 1910. (*4*)
373. VOLTERRA, V., Ann. École Norm. Sup. (3), **24** (1907), 401. (*2*, *6*, *7*, *9*, *16*)
374. WASHIZU, K., J. Math. Phys., **36** (1958), 306. (*2*)
375. WEEKS, R., DUNDURS, J., STIPPES, M., Int. J. Engng. Sci., **6** (1968), 365. (*8*)
376. WEERTMAN, J., in *Response of Metals to High-Velocity Deformation*, P. G. Shewmon and V. F. Zackary, Eds., Interscience, New York, p. 205, 1961. (*13*)
377. WEERTMAN, J., Phil. Mag., **7** (1962), 617. (*13*)
378. WEERTMAN, J., J. Appl. Phys., **33** (1962), 1631. (*13*)
379. WEERTMAN, J., WEERTMAN, J. R., *Elementary Dislocation Theory*, Macmillan, New York—London, 3rd ed., 1966. (*7*)
380. WEINGARTEN, G., Atti Accad. Lincei Rend. (5), **10** (1901), 57. (*2*)
381. WILLIS, J. R., Q. J. Mech. Appl. Math., **18** (1965), 419. (*6*)
382. WILLIS, J. R., Int. J. Engng. Sci., **5** (1967), 171. (*14*, *18*, *19*)
383. WILLIS, J. R., Phil. Mag., **21** (1970), 931. (*10*, *11*, *24*)
384. WIT, R. DE, Acta Met., **13** (1965), 1210. (*7*)
385. WIT, R. DE, *Continuum Theory of Stationary Dislocations*, in *Solid State Physics*, F. Seitz and D. Turnbull, Eds., Academic Press, New York, Vol. 10, p. 249, 1960. (*P*, *9*, *12*, *17*, *18*)
386. YAMAGUCHI, M., VITEK, V., J. Phys., **3** (1973), 523, 537. (*16*)
387. ZARKA, J., *Sur la viscoplasticité des métaux*, Thèse, Paris, 1968. (*7*, *17*, *20*)
388. ZARKA, J., Mémorial de l'Artillerie Française, 2-ème fasc. (1970), 223, (*7*, *17*, *20*)
389. ZARKA, J., J. Mech. Phys. Solids, **20** (1972), 179. (*7*, *17*, *20*)
390. ZEILON, N., Ark. Mat. Ast. Fiz., **6** (1911), 1.(*6*)
391. ZENER, C., Trans. Amer. Inst. Min. Met. Eng., **147** (1942), 361. (*15*)


*Addendum to the English Edition*


392. ALSHITS, V. I., KOTOWSKI, R. K., phys. stat. sol. (b), **68** (1975), K 171. (*20*)
393. ANTHONY, K.-H., KRÖNER, E., in *Deformation and Fracture of High Polymers*, H. H. Kausch, J. A. Hassel, and R. I. Jaffee, Eds., Plenum Press, New York—London, p. 429, 1974. (*7*)
394. BACON, D. J., SCATTERGOOD, R. O., J. Phys. F: Metal Phys., **4** (1974), 2126. (*11*, *12*)
395. BACON, D. J., SCATTERGOOD, R. O., J. Phys. F.: Metal Phys., **5** (1975), 193. (*11*, *12*)
396. BACON, D. J., BARNETT, D. M., SCATTERGOOD, R. O., Phil. Mag. A, **39** (1979), 231. (*11*)
397. BAHR, H.-A., SCHÖPF, H.-G., Ann. Physik, **21** (1968), 57. (*13*)
398. BARNETT, D. M., LOTHE, J., J. Phys. F: Metal Phys., **4** (1974), 1618. (*10*)
399. BELZNER, V., GRANZER, F., phys. stat. sol. (a), **39** (1977), 183. (*16*)
400. BOGDANOFF, J. L., J. Appl. Phys., **21** (1950), 1258. (*7*)
401. BOGGIO, T., Ann. Mat. pura ed appl. (3), **8** (1903), 181. (*14*)

402. Brendel, R., J. Physique, **40** (1979), Colloque C8, suppl. 11, C8-189. (*5*)

403. Bross, H., phys. stat. sol., **2** (1962), 481. (*20*)

404. Bullough, R., Sinclair, J. F., A. E. R. E. Progress Report, T. P. 29, (1974). (*16*)

405. Bullough, R., Willis, J. R., Phil. Mag., **31** (1975), 855. (*24*)

406. Chang, Z. P., Barsch, G. R., Phys. Rev. Letters, **19** (1967), 1381. (*5*)

407. Chang, Z. P., Barsch, G. R., Appl. Physics, **39** (1968), 3276. (*5*)

408. Chou, Y. T., phys. stat. sol., **15** (1966), 123. (*10*)

409. Chou, Y. T., Pande, C. S., J. Appl. Phys., **44** (1973), 3355, 5647. (*10*)

410. Chou, Y. T., Pande, C. S., Yang, H. C., J. Appl. Phys., **46** (1975) 5. (*10, 12*)

411. Duesbery, M. S., Vitek, V., Bowen, D. K., Proc. Roy. Soc. London (A), **332** (1973), 85. (*16*)

412. Dundurs, J., in *Mathematical Theory of Dislocations*, T. Mura, Ed., A. S. M. E., New York, p. 70, 1969. (*8*)

413. Dupeux, M., Bonnet, R., Acta Met., **28** (1980), 721. (*10*)

414. Eckhardt, D., Thesis, University of Stuttgart, 1975. (*20*)

415. Eckhardt, D., Wasserbäch, W., Phil. Mag. A., **37** (1978), 621. (*20*)

416. Eshelby, J. D., *Boundary Problems*, in *Dislocations in Solids*, F. R. N. Nabarro, Ed., North-Holland, Amsterdam–New York–Oxford, vol. 1, pp. 167–221, 1979. (*8*)

417. Featherstone, F. H., Neighbours, J. R., Phys. Rev., **130** (1963), 1324. (*5*)

418. Gairola, B. K. D., *Nonlinear Elastic Problems*, in *Dislocations in Solids*, F. R. N. Nabarro, Ed., North–Holland, Amsterdam–New York–Oxford, vol. 1, pp. 223–242, 1979. (*7, 14, 15*)

419. Gavazza, S. D., Barnett, D. M., J. Mech. Phys. Solids, **24** (1976), 171. (*11, 12*)

420. Goursat, É., *Cours d'Analyse Mathématique*, 5th ed., Gauthier-Villars, Paris, 1925. (*2*)

421. Graham, R. A., J. Acoust. Soc. Am., **51** (1972), 1576. (*5*)

422. Gruner, P., Bross, H., Phys. Rev., **172** (1968), 583. (*20*)

423. Gruner, P. P., in *Fundamental Aspects of Dislocation Theory*, J. A. Simmons. R. de Wit, and R. Bullough, Eds., Nat. Bur. Stand. Spec. Publ. 317, Washington, vol. 1, p. 363, 1970. (*20*)

424. Heinisch, H. L., Jr., Sines, G., Phil. Mag., **34** (1976), 945, 961. (*16*)

425. Heinisch, H. L., Jr., Sines, G., Phil. Mag., **36** (1977), 733. (*24*)

426. Heinrich, R., Schellenberger, W., Pegel, B., phys. stat. sol., **39** (1970), 493. (*16*)

427. Heinrich, R., Schellenberger, W., phys. stat. sol., **47** (1971), 81. (*16*)

428. Hirth, J. P., Wagoner, R. H., Int. J. Solids and Structures, **12** (1976), 117. (*8*)

429. Hirth, J. P., Barnett, D. M., Lothe, J., Phil. Mag. A, **40** (1979), 39. (*12*)

430. Hoagland, R. G., Ph. D. Thesis, Ohio State University, 1973. (*16*)

431. Hoagland, R. G., Hirth, J. P., Gehlen, P. C., Phil. Mag., **34** (1976), 413. (*16*)

432. Indenbom, V. L., Dubnova, G. N., Fiz. Tverdogo Tela, **9** (1967), 1171 (in Russian); Sov. Phys. Solid State (English transl.), **9** (1967), 915. (*11*)

433. Jaunzemis, W., *Continuum Mechanics*, Macmillan, 1967. (*1*)

434. Ju, F. D., J. Appl. Mech., **27** (1960), 423. (*7*)

435. Klemens, P. G., Solid St. Phys., **7** (1958), 1. (*20*)

436. Knopp, K., *Theory of Functions*, transl. from 4th German ed., Dover, New York, 1947. (*8*)

437. Korner, A., Svoboda, P., Kirchner, H. O. K., phys. stat sol. (b), **80** (1977), 441. (*12*)

438. Korner, A., Karnthaler, H. P., Kirchner, H. O. K., phys. stat. sol. (b), **81** (1977), 191. (*12*)

439. Korner, A., Prinz, F., Kirchner, H. O. K., Phil. Mag., **37** (1978), 447. (*11*)

440. Kosevich, A. M., *Crystal Dislocations and the Theory of Elasticity*, in *Dislocations in Solids*, F.R.N. Nabarro, Ed., North-Holland, Amsterdam–New York–Oxford, vol. 1, pp. 33–141, 1979. (*7, 13*)

441. Kröner, E., Appl. Mech. Rev., **15** (1962), 599. (*17*)

442. Kröner, E., J. Math. Phys., **42** (1963), 27. (*17*)

443. KRÖNER, E., *Statistical Continuum Mechanics*, Course held at the Int. Centr. Mech. Sci., Udine, 1971, Springer, New York—Wien, 1972.
444. KRÖNER, E., ANTHONY, K.-H., Annual Rev. Mat. Sci., **5** (1975), 43. (*7*)
445. KUROSAWA, T., J. Phys. Soc. Japan, **19** (1964), 2096. (*16*)
446. LIST, R. D., Proc. Camb. Phil. Soc., **6** (1969), 823. (*8*)
447. LURIE, A. I., *Theory of Elasticity* (in Russian), Nauka, Moscow, 1970. (*2*)
448. MANN, E. H., Proc. Roy. Soc. (London), Ser. A., **199** (1949), 376. (*7*)
449. MARKENSCOFF, X., J. Physique, **40** (1979), Colloque C8, suppl. 11, C8-213. (*5*)
450. MARKENSCOFF, X., J. Appl. Phys., **50** (1979), 1325. (*5*)
451. MASTROJANNIS, E. N., MURA, T., KEER, L. M., Phil. Mag., **35** (1977), 1137. (*9*)
452. MOSS, W. C., HOOVER, W. G., Phil. Mag. A, **38** (1978), 587. (*13*)
453. NAKAHARA, S., WILLIS, J. R., J. Phys. F: Metal Phys., **3** (1973), L249. (*10*)
454. NINOMIYA, T., J. Phys. Soc. Japan, **25** (1968), 830. (*20*)
455. NOWIK, A. S., BERRY, B. S., *Anelastic Relaxation in Crystalline Solids*, Academic Press, London, 1972. (*V*, *22*)
456. ORLOV, S. S., INDENBOM, V. L., Kristallografiya, **14** (1969), 780 (in Russian); Sov. Phys. Cryst. (English transl.), **14** (1970), 675. (*11*)
457. PANDE, C. S., CHOU, Y. T., J. Appl. Phys., **43** (1972), 840. (*10*)
458. PASTORI, M., Rend. Sem. Mat. Fis. Milano, **14** (1940), 170. (*2*)
459. PETRASCH, P., BELZNER, V., J. Physique, **37** (1976), Colloque C7, suppl. 12, C7-553. (*16*)
460. PETRASCH, P., Thesis, University of Frankfurt am Main, 1978. (*14*)
461. PULS, M. P., WOO, C. H., AECL — Report 5238, 1975. (*16*)
462. PULS, M. P., NORGETT, M. J., J. Appl. Phys., **47** (1976), 466. (*16*)
463. PULS, M. P., WOO, C. H., NORGETT, M. J., Phil. Mag., **36** (1977), 1457. (*24*)
464. RICE, J. R., THOMPSON, R. M., Phil. Mag., **29** (1974), 73. (*8*)
465. RIEDER, G., Z. angew. Physik, **9** (1957), 187. (*17*)
466. ROGULA, D., Bull. Acad. Polon. Sci. Ser. Sci. Techn., **13** (1965), 337. (*13*)
467. SAVINO, E. J., Phil. Mag., **36** (1977), 323. (*24*)
468. SCHOECK, G., KIRCHNER, H. O. K., J. Phys. F: Metal Physics, **8** (1973), L43. (*11*)
469. SEEGER, A., ENGELKE, H., in *Dislocation Dynamics*, A. R. Rosenfield, G. T. Hahn, A. L. Bement Jr., and R. I. Jafee, Eds., McGraw-Hill, New York—San Francisco—Toronto—London—Sydney, p. 623, 1968. (*20*)
470. SEEGER, A., in *Interatomic Potentials and Simulation of Lattice Defects*, Battelle Colloquium, P. C. Gehlen, J. R. Beeler Jr., and R. I. Jafee, Eds., Plenum Press, New York—London, p. 566, 764—765, 1972. (*16*)
471. SEKINE, H., MURA, T., Phil. Mag. A, **40** (1979), 183. (*11*)
472. SINCLAIR, J. E., HIRTH, J. P., J. Phys. F: Metal Phys., **5** (1975), 236. (*10*)
473. SINCLAIR, J. E., Phil. Mag., **31** (1975), 647. (*16*)
474. SINCLAIR, J. E., GEHLEN, P. C., HOAGLAND, R. G., HIRTH, J. P., J. Appl. Phys., **49** (1978), 3890. (*14*, *16*)
475. SOMIGLIANA, C., in *Atti del IV Congresso Internazionale dei Matematici*, Roma, vol. 3, p. 60, 1909. (*7*)
476. SOMIGLIANA, C., Rend. R. Accad. Lincei, 5$^{e}$ sér., **23** (1914), 463; **24** (1915), 655. (*7*)
477. STEEDS, J. W., WILLIS, J. R., *Dislocations in Anisotropic Media*, in *Dislocations in Solids*, F.R.N. Nabarro, Ed., vol.1, North-Holland Publ. Comp., Amsterdam—New York—Oxford, 1979. (*10*, *11*)
478. TEODOSIU, C., Int. J. Engng. Sci., **19** (1981). (*7*)

479. TEODOSIU, C., SOÓS, E., Rev. roum. sci. techn. – méc. appl., **26** (1981), 731, 785, **27** (1982). (*7*, *14*, *16*).

480. TEWARY, V. K., Adv. Phys., **22** (1973), 757. (*16*)

481. THURSTON, R. N., BRUGGER, K., Phys. Rev., **133** (1964), 1604. (*5*)

482. VITEK, V., J. Mech. Phys. Solids, **24** (1976), 67. (*8*)

483. VITEK, V., J. Mech. Phys. Solids, **24** (1976), 263. (*8*)

484. VITEK, V., SUTTON, A. P., SMITH, D. A., POND, R. C., Phil. Mag., **39** (1979), 213. (*16*)

485. VOLTERRA, V., VOLTERRA, E., *Sur les distorsions des corps élastiques (théorie et applications)*, Mém. Sci. Math., fasc. 147, Paris, Gauthier-Villars, 1960. (*7*)

486. WIT, R. DE, in *Fundamental Aspects of Dislocation Theory*, J. A. Simmons, R. de Wit, R. Bullough, Eds., Nat. Bur. Stand. Spec. Publ. 317, Washington, vol. 1, p. 651, 1970. (*2*)

487. WOO, C. H., PULS, M. P., Phil. Mag., **35** (1977), 727. (*16*)

488. YAMAGUCHI, M., VITEK, V., J. Phys. F: Metal Phys., **5** (1975), 1, 11. (*16*)

489. YOFFE, E. H. (*née* Mann), Phil. Mag., **2** (1957), 1197; **3** (1958), 8. (*7*)

# Subject Index

(lückenhaft!)